moduli of vector bundles

LECTURE NOTES IN PURE AND APPLIED MATHEMATICS

1. *N. Jacobson,* Exceptional Lie Algebras
2. *L.-Å. Lindahl and F. Poulsen,* Thin Sets in Harmonic Analysis
3. *I. Satake,* Classification Theory of Semi-Simple Algebraic Groups
4. *F. Hirzebruch, W. D. Newmann, and S. S. Koh, Differentiable Manifolds and Quadratic Forms*
5. *I. Chavel,* Riemannian Symmetric Spaces of Rank One
6. *R. B. Burckel,* Characterization of C(X) Among Its Subalgebras
7. *B. R. McDonald, A. R. Magid, and K. C. Smith,* Ring Theory: Proceedings of the Oklahoma Conference
8. *Y.-T. Siu,* Techniques of Extension on Analytic Objects
9. *S. R. Caradus, W. E. Pfaffenberger, and B. Yood,* Calkin Algebras and Algebras of Operators on Banach Spaces
10. *E. O. Roxin, P.-T. Liu, and R. L. Sternberg,* Differential Games and Control Theory
11. *M. Orzech and C. Small,* The Brauer Group of Commutative Rings
12. *S. Thomier,* Topology and Its Applications
13. *J. M. Lopez and K. A. Ross,* Sidon Sets
14. *W. W. Comfort and S. Negrepontis,* Continuous Pseudometrics
15. *K. McKennon and J. M. Robertson,* Locally Convex Spaces
16. *M. Carmeli and S. Malin,* Representations of the Rotation and Lorentz Groups: An Introduction
17. *G. B. Seligman,* Rational Methods in Lie Algebras
18. *D. G. de Figueiredo,* Functional Analysis: Proceedings of the Brazilian Mathematical Society Symposium
19. *L. Cesari, R. Kannan, and J. D. Schuur,* Nonlinear Functional Analysis and Differential Equations: Proceedings of the Michigan State University Conference
20. *J. J. Schäffer,* Geometry of Spheres in Normed Spaces
21. *K. Yano and M. Kon,* Anti-Invariant Submanifolds
22. *W. V. Vasconcelos,* The Rings of Dimension Two
23. *R. E. Chandler,* Hausdorff Compactifications
24. *S. P. Franklin and B. V. S. Thomas,* Topology: Proceedings of the Memphis State University Conference
25. *S. K. Jain,* Ring Theory: Proceedings of the Ohio University Conference
26. *B. R. McDonald and R. A. Morris,* Ring Theory II: Proceedings of the Second Oklahoma Conference
27. *R. B. Mura and A. Rhemtulla,* Orderable Groups
28. *J. R. Graef,* Stability of Dynamical Systems: Theory and Applications
29. *H.-C. Wang,* Homogeneous Branch Algebras
30. *E. O. Roxin, P.-T. Liu, and R. L. Sternberg,* Differential Games and Control Theory II
31. *R. D. Porter,* Introduction to Fibre Bundles
32. *M. Altman,* Contractors and Contractor Directions Theory and Applications
33. *J. S. Golan,* Decomposition and Dimension in Module Categories
34. *G. Fairweather,* Finite Element Galerkin Methods for Differential Equations
35. *J. D. Sally,* Numbers of Generators of Ideals in Local Rings
36. *S. S. Miller,* Complex Analysis: Proceedings of the S.U.N.Y. Brockport Conference
37. *R. Gordon,* Representation Theory of Algebras: Proceedings of the Philadelphia Conference
38. *M. Goto and F. D. Grosshans,* Semisimple Lie Algebras
39. *A. I. Arruda, N. C. A. da Costa, and R. Chuaqui,* Mathematical Logic: Proceedings of the First Brazilian Conference
40. *F. Van Oystaeyen,* Ring Theory: Proceedings of the 1977 Antwerp Conference
41. *F. Van Oystaeyen and A. Verschoren,* Reflectors and Localization: Application to Sheaf Theory
42. *M. Satyanarayana,* Positively Ordered Semigroups
43. *D. L Russell,* Mathematics of Finite-Dimensional Control Systems
44. *P.-T. Liu and E. Roxin,* Differential Games and Control Theory III: Proceedings of the Third Kingston Conference, Part A
45. *A. Geramita and J. Seberry,* Orthogonal Designs: Quadratic Forms and Hadamard Matrices
46. *J. Cigler, V. Losert, and P. Michor,* Banach Modules and Functors on Categories of Banach Spaces
47. *P.-T. Liu and J. G. Sutinen,* Control Theory in Mathematical Economics: Proceedings of the Third Kingston Conference, Part B
48. *C. Byrnes,* Partial Differential Equations and Geometry

49. *G. Klambauer,* Problems and Propositions in Analysis
50. *J. Knopfmacher,* Analytic Arithmetic of Algebraic Function Fields
51. *F. Van Oystaeyen,* Ring Theory: Proceedings of the 1978 Antwerp Conference
52. *B. Kadem,* Binary Time Series
53. *J. Barros-Neto and R. A. Artino,* Hypoelliptic Boundary-Value Problems
54. *R. L. Sternberg, A. J. Kalinowski, and J. S. Papadakis,* Nonlinear Partial Differential Equations in Engineering and Applied Science
55. *B. R. McDonald,* Ring Theory and Algebra III: Proceedngs of the Third Oklahoma Conference
56. *J. S. Golan,* Structure Sheaves Over a Noncommutative Ring
57. *T. V. Narayana, J. G. Williams, and R. M. Mathsen,* Combinatorics, Representation Theory and Statistical Methods in Groups: YOUNG DAY Proceedings
58. *T. A. Burton,* Modeling and Differential Equations in Biology
59. *K. H. Kim and F. W. Roush,* Introduction to Mathematical Consensus Theory
60. *J. Banas and K. Goebel,* Measures of Noncompactness in Banach Spaces
61. *O. A. Nielson,* Direct Integral Theory
62. *J. E. Smith, G. O. Kenny, and R. N. Ball,* Ordered Groups: Proceedings of the Boise State Conference
63. *J. Cronin,* Mathematics of Cell Electrophysiology
64. *J. W. Brewer,* Power Series Over Commutative Rings
65. *P. K. Kamthan and M. Gupta,* Sequence Spaces and Series
66. *T. G. McLaughlin,* Regressive Sets and the Theory of Isols
67. *T. L. Herdman, S. M. Rankin III, and H. W. Stech,* Integral and Functional Differential Equations
68. *R. Draper,* Commutative Algebra: Analytic Methods
69. *W. G. McKay and J. Patera,* Tables of Dimensions, Indices, and Branching Rules for Representations of Simple Lie Algebras
70. *R. L. Devaney and Z. H. Nitecki,* Classical Mechanics and Dynamical Systems
71. *J. Van Geel,* Places and Valuations in Noncommutative Ring Theory
72. *C. Faith,* Injective Modules and Injective Quotient Rings
73. *A. Fiacco,* Mathematical Programming with Data Perturbations I
74. *P. Schultz, C. Praeger, and R. Sullivan,* Algebraic Structures and Applications: Proceedings of the First Western Australian Conference on Algebra
75. *L Bican, T. Kepka, and P. Nemec,* Rings, Modules, and Preradicals
76. *D. C. Kay and M. Breen,* Convexity and Related Combinatorial Geometry: Proceedings of the Second University of Oklahoma Conference
77. *P. Fletcher and W. F. Lindgren,* Quasi-Uniform Spaces
78. *C.-C. Yang,* Factorization Theory of Meromorphic Functions
79. *O. Taussky,* Ternary Quadratic Forms and Norms
80. *S. P. Singh and J. H. Burry,* Nonlinear Analysis and Applications
81. *K. B. Hannsgen, T. L. Herdman, H. W. Stech, and R. L. Wheeler,* Volterra and Functional Differential Equations
82. *N. L. Johnson, M. J. Kallaher, and C. T. Long,* Finite Geometries: Proceedings of a Conference in Honor of T. G. Ostrom
83. *G. I. Zapata,* Functional Analysis, Holomorphy, and Approximation Theory
84. *S. Greco and G. Valla,* Commutative Algebra: Proceedings of the Trento Conference
85. *A. V. Fiacco,* Mathematical Programming with Data Perturbations II
86. *J.-B. Hiriart-Urruty, W. Oettli, and J. Stoer,* Optimization: Theory and Algorithms
87. *A. Figa Talamanca and M. A. Picardello,* Harmonic Analysis on Free Groups
88. *M. Harada,* Factor Categories with Applications to Direct Decomposition of Modules
89. *V. I. Istrățescu,* Strict Convexity and Complex Strict Convexity
90. *V. Lakshmikantham,* Trends in Theory and Practice of Nonlinear Differential Equations
91. *H. L. Manocha and J. B. Srivastava,* Algebra and Its Applications
92. *D. V. Chudnovsky and G. V. Chudnovsky,* Classical and Quantum Models and Arithmetic Problems
93. *J. W. Longley,* Least Squares Computations Using Orthogonalization Methods
94. *L. P. de Alcantara,* Mathematical Logic and Formal Systems
95. *C. E. Aull,* Rings of Continuous Functions
96. *R. Chuaqui,* Analysis, Geometry, and Probability
97. *L. Fuchs and L. Salce,* Modules Over Valuation Domains
98. *P. Fischer and W. R. Smith,* Chaos, Fractals, and Dynamics
99. *W. B. Powell and C. Tsinakis,* Ordered Algebraic Structures
100. *G. M. Rassias and T. M. Rassias,* Differential Geometry, Calculus of Variations, and Their Applications

101. *R.-E. Hoffmann and K. H. Hofmann,* Continuous Lattices and Their Applications
102. *J. H. Lightbourne III and S. M. Rankin III,* Physical Mathematics and Nonlinear Partial Differential Equations
103. *C. A. Baker and L. M. Batten,* Finite Geometrics
104. *J. W. Brewer, J. W. Bunce, and F. S. Van Vleck,* Linear Systems Over Commutative Rings
105. *C. McCrory and T. Shifrin,* Geometry and Topology: Manifolds, Varieties, and Knots
106. *D. W. Kueker, E. G. K. Lopez-Escobar, and C. H. Smith,* Mathematical Logic and Theoretical Computer Science
107. *B.-L. Lin and S. Simons,* Nonlinear and Convex Analysis: Proceedings in Honor of Ky Fan
108. *S. J. Lee,* Operator Methods for Optimal Control Problems
109. *V. Lakshmikantham,* Nonlinear Analysis and Applications
110. *S. F. McCormick,* Multigrid Methods: Theory, Applications, and Supercomputing
111. *M. C. Tangora,* Computers in Algebra
112. *D. V. Chudnovsky and G. V. Chudnovsky,* Search Theory: Some Recent Developments
113. *D. V. Chudnovsky and R. D. Jenks,* Computer Algebra
114. *M. C. Tangora,* Computers in Geometry and Topology
115. *P. Nelson, V. Faber, T. A. Manteuffel, D. L. Seth, and A. B. White, Jr.,* Transport Theory, Invariant Imbedding, and Integral Equations: Proceedings in Honor of G. M. Wing's 65th Birthday
116. *P. Clément, S. Invernizzi, E. Mitidieri, and I. I. Vrabie,* Semigroup Theory and Applications
117. *J. Vinuesa,* Orthogonal Polynomials and Their Applications: Proceedings of the International Congress
118. *C. M. Dafermos, G. Ladas, and G. Papanicolaou,* Differential Equations: Proceedings of the EQUADIFF Conference
119. *E. O. Roxin,* Modern Optimal Control: A Conference in Honor of Solomon Lefschetz and Joseph P. Lasalle
120. *J. C. Díaz,* Mathematics for Large Scale Computing
121. *P. S. Milojevič,* Nonlinear Functional Analysis
122. *C. Sadosky,* Analysis and Partial Differential Equations: A Collection of Papers Dedicated to Mischa Cotlar
123. *R. M. Shortt,* General Topology and Applications: Proceedings of the 1988 Northeast Conference
124. *R. Wong,* Asymptotic and Computational Analysis: Conference in Honor of Frank W. J. Olver's 65th Birthday
125. *D. V. Chudnovsky and R. D. Jenks,* Computers in Mathematics
126. *W. D. Wallis, H. Shen, W. Wei, and L. Zhu,* Combinatorial Designs and Applications
127. *S. Elaydi,* Differential Equations: Stability and Control
128. *G. Chen, E. B. Lee, W. Littman, and L. Markus,* Distributed Parameter Control Systems: New Trends and Applications
129. *W. N. Everitt,* Inequalities: Fifty Years On from Hardy, Littlewood and Pólya
130. *H. G. Kaper and M. Garbey,* Asymptotic Analysis and the Numerical Solution of Partial Differential Equations
131. *O. Arino, D. E. Axelrod, and M. Kimmel,* Mathematical Population Dynamics: Proceedings of the Second International Conference
132. *S. Coen,* Geometry and Complex Variables
133. *J. A. Goldstein, F. Kappel, and W. Schappacher,* Differential Equations with Applications in Biology, Physics, and Engineering
134. *S. J. Andima, R. Kopperman, P. R. Misra, J. Z. Reichman, and A. R. Todd,* General Topology and Applications
135. *P Clément, E. Mitidieri, B. de Pagter,* Semigroup Theory and Evolution Equations: The Second International Conference
136. *K. Jarosz,* Function Spaces
137. *J. M. Bayod, N. De Grande-De Kimpe, and J. Martínez-Maurica,* p-adic Functional Analysis
138. *G. A. Anastassiou,* Approximation Theory: Proceedings of the Sixth Southeastern Approximation Theorists Annual Conference
139. *R. S. Rees,* Graphs, Matrices, and Designs: Festschrift in Honor of Norman J. Pullman
140. *G. Abrams, J. Haefner, and K. M. Rangaswamy,* Methods in Module Theory
141. *G. L. Mullen and P. J.-S. Shiue,* Finite Fields, Coding Theory, and Advances in Communications and Computing
142. *M. C. Joshi and A. V. Balakrishnan,* Mathematical Theory of Control: Proceedings of the International Conference

143. *G. Komatsu and Y. Sakane*, Complex Geometry: Proceedings of the Osaka International Conference
144. *I. J. Bakelman*, Geometric Analysis and Nonlinear Partial Differential Equations
145. *T. Mabuchi and S. Mukai*, Einstein Metrics and Yang–Mills Connections: Proceedings of the 27th Taniguchi International Symposium
146. *L. Fuchs and R. Göbel*, Abelian Groups: Proceedings of the 1991 Curaçao Conference
147. *A. D. Pollington and W. Moran*, Number Theory with an Emphasis on the Markoff Spectrum
148. *G. Dore, A. Favini, E. Obrecht, and A. Venni*, Differential Equations in Banach Spaces
149. *T. West*, Continuum Theory and Dynamical Systems
150. *K. D. Bierstedt, A. Pietsch, W. Ruess, and D. Vogt*, Functional Analysis
151. *K. G. Fischer, P. Loustaunau, J. Shapiro, E. L. Green, and D. Farkas*, Computational Algebra
152. *K. D. Elworthy, W. N. Everitt, and E. B. Lee*, Differential Equations, Dynamical Systems, and Control Science
153. *P.-J. Cahen, D. L. Costa, M. Fontana, and S.-E. Kabbaj*, Commutative Ring Theory
154. *S. C. Cooper and W. J. Thron*, Continued Fractions and Orthogonal Functions: Theory and Applications
155. *P. Clément and G. Lumer*, Evolution Equations, Control Theory, and Biomathematics
156. *M. Gyllenberg and L. Persson*, Analysis, Algebra, and Computers in Mathematical Research: Proceedings of the Twenty-First Nordic Congress of Mathematicians
157. *W. O. Bray, P. S. Milojević, and Č. V. Stanojević*, Fourier Analysis: Analytic and Geometric Aspects
158. *J. Bergen and S. Montgomery*, Advances in Hopf Algebras
159. *A. R. Magid*, Rings, Extensions, and Cohomology
160. *N. H. Pavel*, Optimal Control of Differential Equations
161. *M. Ikawa*, Spectral and Scattering Theory: Proceedings of the Taniguchi International Workshop
162. *X. Liu and D. Siegel*, Comparison Methods and Stability Theory
163. *J.-P. Zolésio*, Boundary Control and Variation
164. *M. Křížek, P. Neittaanmäki, and R. Stenberg*, Finite Element Methods: Fifty Years of the Courant Element
165. *G. Da Prato and L. Tubaro*, Control of Partial Differential Equations
166. *E. Ballico*, Projective Geometry with Applications
167. *M. Costabel, M. Dauge, and S. Nicaise*, Boundary Value Problems and Integral Equations in Nonsmooth Domains
168. *G. Ferreyra, G. R. Goldstein, and F. Neubrander*, Evolution Equations
169. *S. Huggett*, Twistor Theory
170. *H. Cook, W. T. Ingram, K. T. Kuperberg, A. Lelek, and P. Minc*, Continua: With the Houston Problem Book
171. *D. F. Anderson and D. E. Dobbs*, Zero-Dimensional Commutative Rings
172. *K. Jarosz*, Function Spaces: The Second Conference
173. *V. Ancona, E. Ballico, and A. Silva*, Complex Analysis and Geometry
174. *E. Casas*, Control of Partial Differential Equations and Applications
175. *N. Kalton, E. Saab, and S. Montgomery-Smith*, Interaction Between Functional Analysis, Harmonic Analysis, and Probability
176. *Z. Deng, Z. Liang, G. Lu, and S. Ruan*, Differential Equations and Control Theory
177. *P. Marcellini, G. Talenti, and E. Vesentini*, Partial Differential Equations and Applications: Collected Papers in Honor of Carlo Pucci
178. *A. Kartsatos*, Theory and Applications of Nonlinear Operators of Accretive and Monotone Type
179. *M. Maruyama*, Moduli of Vector Bundles

Additional Volumes in Preparation

moduli of vector bundles

edited by

Masaki Maruyama
Kyoto University
Kyoto, Japan

Marcel Dekker, Inc. New York • Basel • Hong Kong

Library of Congress Cataloging-in-Publication Data

Moduli of vector bundles / edited by Masaki Maruyama.
p. cm. — (Lecture notes in pure and applied mathematics ; v. 179)
Papers from 2 conferences held in Japan, 1994.
ISBN 0-8247-9738-8 (alk. paper)
1. Vector bundles — Congresses. 2. Moduli theory — Congresses.
I. Maruyama, Masaki. II. Series.
QA612.63.M63 1996
516.3'5—dc20 96-641
CIP

The publisher offers discounts on this book when ordered in bulk quantities. For more information, write to Special Sales/Professional Marketing at the address below.

This book is printed on acid-free paper.

MARCEL DEKKER, INC.
270 Madison Avenue, New York, New York 10016

Current printing (last digit):
10 9 8 7 6 5 4 3 2 1

PRINTED IN THE UNITED STATES OF AMERICA

PREFACE

The 35th Taniguchi Symposium, Moduli of Vector Bundles, was held at Sengari Seminar House of Kanseigakiun University located in Sanda, Japan. After the symposium most of the participants moved to another conference at the Research Institute for Mathematical Science of Kyoto University. This volume is the proceedings of Taniguchi symposium together with a few articles presented at the conference in Kyoto.

Topics of discussion at both meetings ranged from differential geometry, such as magnetic monopoles (N. J. Hitchin) and hyper-Kähler manifolds (H. Ohta and K. Ono) to algebraic geometry on torus bundles (V. Brînzănescu and K. Ueno), and theory of cycles (C. Voisin), but the main interest focused on the moduli spaces of vector bundles, or instantons, and their applications.

A. Weil initiated the study of vector bundles on a compact Riemann surface C with a view to understanding the (nonabelian) extensions of the field of meromorphic functions on C. The plan received an impetus in the 1960s when D. Mumford came up with the notions of stable and semi-stable vector bundles, and the pioneering works by M. S. Narasimhan and C. S. Seshadri revealed a beautiful relationship between unitary representations of the fundamental group of a compact Riemann surface and stable vector bundles on it. Seshadri went further by developing the theory of parabolic stable bundles and extending his result with Narasimhan to a tight connection between unitary representations of a Fuchsian group and parabolic stable vector bundles. The affirmative solution of the Hitchin-Kobayashi conjecture that generalizes the above-mentioned work by Narasimhan and Seshadri afforded us a fine and wide view of a very rich field of interrelations between differential geometry and algebraic geometry. And then we happened to come to the exciting development of Seiberg-Witten theory right before our symposium.

The topology of moduli spaces of vector bundles draws the attention of quite a few mathematicians and physicists, especially in connection with Witten's S-duality conjecture. In this volume H. Nakajima, K. Yokogawa, and K. Yoshioka treat the problem from various viewpoints. The computation of Donaldson's invariant and new tools to compute it are also central tasks in this field, as found here in the articles by G. Ellingsrud, J. Le Potier and S. A. Strømme, by J. Le Potier, and by Y. Kametani. Line bundles and the space of the sections of line bundles on the moduli of vector bundles play a key role in the study of the structure of the moduli and have some beautiful applications in seemingly unrelated fields. J. Li and J. Le Potier contributed to these in this volume. S. Mukai's work is a typical example of applications of the theory of vector bundles to problems in algebraic geometry. As A. Ishii shows, the behavior of vector bundles or reflexive sheaves represents well the complexity of singularities of the base variety. New ideas and tools that are basic not only in the theory of vector bundles but also in other theories will be found in the chapters by A. Bertram, A. Fujiki, M. Maruyama, Z. Ran, F.-O. Schreyer and C. Simpson.

One month after the fruitful symposium a catastrophic earthquake hit the exotic, beautiful town of Kobe, where all the participants from abroad had enjoyed one or two nights before the symposium. It killed thousands of people and destroyed the town so much. On behalf of all the participants in the symposium I would like to express our condolences to the victims of the earthquake and hope that the town will be restored soon.

As the editor of this volume I would like to thank the referees. We wish to thank the Taniguchi Foundation for its warm hospitality and generous financial support. Special thanks are due also to Professor Shingo Murakami, to whom we owe much for the organization of the symposium.

Masaki Maruyama

CONTENTS

CONTRIBUTORS

Aaron Bertram University of Utah, Salt Lake City, Utah

Vasile Brînzănescu Institute of Mathematics of the Romanian Academy, Bucharest, Romania

Geir Ellingsrud University of Oslo, Oslo, Norway

Akira Fujiki Osaka University, Osaka, Japan

N. J. Hitchin University of Cambridge, Cambridge, England

Akira Ishii Kyoto University, Kyoto, Japan

Yukio Kametani Saga University, Saga, Japan

Joseph Le Potier University of Paris 7, Paris, France

Jun Li University of California at Los Angeles, Los Angeles, California

Masaki Maruyama Kyoto University, Kyoto, Japan

Shigeru Mukai Nagoya University, Nagoya, Japan

Hiraku Nakajima University of Tokyo, Tokyo, Japan

Hiroshi Ohta Nagoya University, Nagoya, Japan

Kaoru Ono Ochanomizu University, Tokyo, Japan

Ziv Ran University of California at Riverside, Riverside, California

Frank-Olaf Schreyer University of Bayreuth, Bayreuth, Germany

Carlos Simpson Paul Sabatier University, Toulouse, France

Stein A. Strømme University of Bergen, Bergen, Norway

Kenji Ueno Kyoto University, Kyoto, Japan

Claire Voisin University of Orsay, Orsay, France

Kôji Yokogawa Osaka University, Osaka, Japan

Kōta Yoshioka Kyoto University, Kyoto, Japan

COMPUTING SCHUBERT'S CALCULUS WITH SEVERI RESIDUES: AN INTRODUCTION TO QUANTUM COHOMOLOGY

Aaron Bertram

University of Utah, Salt Lake City, UT 84112, USA
e-mail: bertram@math.utah.edu

0. Introduction

My purpose in writing this note is to give a non-standard introduction to (part of) the theory of quantum cohomology, due to Witten ([W1]) with important contributions by Kontsevich and Manin ([KM]) and others (see e.g. [RT]). Rather than begin with the definitions and axioms for quantum cohomology, as is usually done, I want to show how a new approach to computing intersection numbers on a Grassmannian leads one naturally to contemplate a certain deformation of its cohomology ring. This deformation may be deduced from analogous intersection data on the moduli spaces of maps from a rational curve to a Grassmannian, leading to the remarkable fact, first observed by Vafa and Intriligator [I], that a single residue formula computes all such intersection numbers at once.

In fact, the theory of quantum cohomology provides **two** deformations of the cohomology ring of a Grassmannian (indeed of any "semipositive" symplectic manifold...see [RT]). To clear up potential confusion, let me clarify that in this note, we shall consider only the small quantum deformation, which will produce for us an algebra over the polynomial ring in one variable. The large quantum deformation, which contains much of the intersection data of interest to enumerative geometers, involves a further deformation to an algebra over a formal-power-series ring in several variables. The interested reader is urged to look at the already-mentioned references, as well as [DFI], [F] and [P] for accounts of the large quantum deformations.

The results in this note are not new. Rather, it was my intention in writing this to highlight the role of Severi residues in the small quantum cohomology ring of a Grassmannian. These residues appear quite naturally in the formulation of this ring, as well as providing a nice computational tool for calculating intersection numbers. Additionally, there is recent evidence that similar residues arise in the small quantum cohomology rings of other moduli spaces, for example, moduli of vector bundles on a Riemann surface [W2]. Of course, it would be very nice to know if residue formulas could be linked in a similar way to the intersection numbers arising in the large quantum deformations (even for the projective plane!).

In §2 we also present a quantum Schubert calculus, first derived in [B1], and challenge the reader to come up with a direct proof of the quantum Giambelli and quantum Pieri formulas of Corollary 2.3. Finally, some generalizations are discussed, including extensions of the intersection computations to maps from positive genus curves, as well as some recent work of Ciocan-Fontanine [C] on a conjecture of Givental-Kim [GK] for the full flag varieties.

Partially supported by NSF grant DMS-9218215

This paper arose out of my musings on quantum cohomology following a lecture at the Taniguchi symposium. I want to thank Professor Maruyama and the Taniguchi foundation for their warm hospitality.

1. Intersection Numbers on a Grassmannian

Let $G = G(k,n)$ be the Grassmann variety of k-planes in $\mathbf{C}^n$, and let $W_{\vec{a}} \subset G(k,n)$ be a Schubert cycle, depending upon a choice of a full flag of $\mathbf{C}^n$ (suppressed in the notation) and $n-k$-tuple $\vec{a} = (a_1, ..., a_{n-k})$ satisfying $k \geq a_1 \geq ... \geq a_{n-k} \geq 0$. Whenever we intersect two or more Schubert cycles, we always assume that each has been chosen with respect to a general full flag. The precise definitions are:

Definition 1.1: Let $S \subset \mathbf{C}^n \otimes \mathcal{O}_G$ be the universal subbundle on $G(k,n)$. Let $0 = V_0 \subset V_1 \subset ... \subset V_n = (\mathbf{C}^n)^*$ be a full flag on the dual space to $\mathbf{C}^n$, and let $f_i : V_i \otimes \mathcal{O}_G \to S^*$ be the induced maps. Then with respect to this flag, one defines:

$D_{i,a} :=$ (scheme-theoretic) locus in $G(k,n)$ where $\dim(\ker(f_{k+i-a})) \geq i$,

$W_{\vec{a}} := \cap_{i=1}^{n-k} D_{i,a_i} \subset G(k,n)$.

Some of the basic results on Schubert cycles which we will use are:

• Each $W_{\vec{a}} \subset G(k,n)$ has pure codimension $\deg(\vec{a}) := \sum_{i=1}^{n-k} a_i$.

• If we let $\vec{a}^c = (k - a_{n-k}, ..., k - a_1)$, then $W_{\vec{a}}$ and $W_{\vec{a}^c}$ are Poincaré dual, i.e.

$$\int_G W_{\vec{a}} \cap W_{\vec{b}} = \delta_{\vec{b},\vec{a}^c}$$

(where $\int_G V = 0$ by convention unless V is top-codimensional.)

• Via $V \mapsto \int_G V \cap W_{\vec{a}}$, the Schubert cycles determine an additive basis for the cohomology vector space $\mathrm{H}^*(G, \mathbf{C})$ as $\vec{a}$ runs over the allowable $n-k$-tuples.

The first question which I want to answer using Severi residues is:

Question: If $\vec{a}(1), ..., \vec{a}(m)$ satisfy $\sum_{j=1}^m \deg(\vec{a}(j)) = \dim(G(k,n))$, then how can we compute

$$\int_G W_{\vec{a}(1)} \cap ... \cap W_{\vec{a}(m)}$$

in terms of the $n-k$-tuples $\vec{a}(1), ..., \vec{a}(m)$?

Before I do this, though, let's recall a classical approach to answering this question using Schubert calculus.

Schubert Step 1: Reduction to special Schubert cycles.

Recall that the special Schubert cycles are those $W_{\vec{a}}$ with the property that $a_2 = ... = a_{n-k} = 0$. If $a_1 = a$, we will denote such a Schubert cycle by W_a. Notice that by definition, special Schubert cycles, since they are of the expected codimension, have the property that each W_a is a geometric representative for the chern class $c_a(S^*)$ of the dual of the universal subbundle.

The following tells us how to construct general Schubert cycles from the special ones. It is due to Giambelli (with a modern version due to Kempf and Laksov, which we will need in §2):

Giambelli's Formula: Any $W_{\vec{a}}$ determines the same element of the cohomology ring of $G(k,n)$ as does the determinant:

$$g_{\vec{a}}(W_*) := \det \begin{pmatrix} W_{a_1} & W_{a_1+1} & \cdots & W_{a_1+n-k} \\ W_{a_2-1} & W_{a_2} & \cdots & W_{a_2-1+n-k} \\ \vdots & \vdots & \vdots & \vdots \\ W_{a_{n-k}-n+k} & W_{a_{n-k}-n+k-1} & \cdots & W_{a_{n-k}} \end{pmatrix}$$

where the product is intersection, and any W_a is zero if the subscript a does not fall between 0 and $n-k$.

By using Giambelli's formula, we see that substituting $g_{\vec{a}}(W_*)$ for each occurrence of $W_{\vec{a}}$ in our question, we have reduced the question to a computation of intersections of *special* Schubert cycles:

$$\int_G W_{a(1)} \cap \ldots \cap W_{a(m)}$$

in terms of the integers $0 \leq a(1), ..., a(m) \leq k$.

Schubert Step 2: An explicit product formula:

There is an explicit classical formula for multiplying a special Schubert cycle with a general one in the cohomology ring of $G(k,n)$. It is the following:

Pieri's Formula: The intersection $W_a \cap W_{\vec{a}}$ determines the same element in $\mathrm{H}^*(G(k,n), \mathbf{C})$ as does the sum:

$$p_{a,\vec{a}}(W_*) := \sum_{\substack{\deg(\vec{b})=a+\deg(\vec{a}); \\ k \geq b_1 \geq a_1 \geq b_2 \geq \cdots \geq a_{n-k} \geq 0}} W_{\vec{b}}$$

Evidently this yields an algorithm for computing $\lambda(a(1), ..., a(m)) \in \mathbf{Z}$, the coefficient of $W_{k,k,...,k}$ when Pieri's formula is applied repeatedly to the intersection $W_{a(1)} \cap \ldots \cap W_{a(m)}$. The answer to the question for special Schubert cycles is then this $\lambda(a(1), ..., a(m))$, since $\int_G W_{k,...,k} = 1$..

Severi residues give us a surprising alternative to Step 2:

Severi Step 2: Applying the theory of residues to the cohomology ring:

Recall the following presentation for the cohomology ring of $G(k,n)$:

$$\mathrm{H}^*(G(k,n), \mathbf{C}) \cong \mathbf{C}[x_1, ..., x_k]/\langle f_1, ..., f_k \rangle$$

where the x_i are the chern classes of S^*, and the f_i are the "over-the-top" chern classes of the bundle Q^*, where $Q = (\mathbf{C}^n \otimes \mathcal{O}_G/S)$ is the universal quotient bundle of rank $n-k$ on $G(k,n)$. That is, $f_i = c_{n-k+i}(Q^*)$, thought of as a polynomial in $x_1, ..., x_k$. (See e.g. [BT]).

Alternatively, a useful way to construct the ideal of relations: $\langle f_1, ..., f_k \rangle$ is as follows. Let $q_1, ..., q_k$ be the (formal) chern roots of the chern polynomial $c_t(S^*)$, and let $P = P(x_1, ..., x_k)$ be the polynomial expressing the $n+1$st (normalized) Newton polynomial $\frac{1}{n+1}\sum_{i=1}^k q_i^{n+1}$ in terms of $x_1, ..., x_k$. Then the ideal of relations satisfies:

$$\langle f_1, ..., f_k \rangle = \langle \frac{\partial P}{\partial x_1}, \ldots \frac{\partial P}{\partial x_k} \rangle \subset \mathbf{C}[x_1, ..., x_k]$$

(This is an easy calculation. See e.g. [BDW].)

Another easy observation is necessary. Note that the common zero locus of the ideal of relations is just the origin in $\mathbf{C}^k$ (with high multiplicity). This follows from the fact that $\mathrm{H}^*(G(k,n),\mathbf{C})$ has a unique maximal ideal.

We are now ready to invoke the theory of residues. (All results we will state about Severi residues have been taken from Chapter 5 of [GH].) Given special Schubert cycles $W_{a(1)},\ldots,W_{a(m)}$ such that $\sum_{j=1}^m a(j) = \dim(G(k,n))$, consider the following meromorphic forms on $\mathbf{C}^k$:

$$\tau = \frac{\prod_{j=1}^m x_{a(j)} dx_1 \wedge \ldots \wedge dx_k}{\prod_{i=1}^k \partial P/\partial x_i} \quad \text{and} \quad \omega = \frac{x_k^{n-k} dx_1 \wedge \ldots \wedge dx_k}{\prod_{i=1}^k \partial P/\partial x_i}$$

Then the residues at the origin of τ and ω exist, and satisfy:

$$\mathrm{Res}_{(0,\ldots,0)}(\tau) = \Big(\int_G W_{a(1)} \cap \ldots \cap W_{a(m)}\Big)\, \mathrm{Res}_{(0,\ldots,0)}(\omega).$$

Thus we have expressed the answer to our question as a ratio of residues.

Next, consider the deformation of the ideal of relations given by replacing P by $P_\epsilon := P - \epsilon x_1$ for some complex number ϵ and taking derivatives (of which only the derivative with respect to x_1 changes). We claim the following:

Proposition 1.2: (a) *If $\epsilon \neq 0$, then the zero scheme $V(\frac{\partial P_\epsilon}{\partial x_1},\ldots,\frac{\partial P_\epsilon}{\partial x_k}) \subset \mathbf{C}^k$ consists of $\binom{n}{k}$ reduced points.*

(b) *Let V_ϵ be the zero scheme of (a) thought of as a set. Let τ_ϵ and ω_ϵ be defined in the same way as τ and ω, but with P replaced by P_ϵ. Then the sums:*

$$\sum_{z\in V_\epsilon} \mathrm{Res}_z(\tau_\epsilon) \quad \textit{and} \quad \sum_{z\in V_\epsilon} \mathrm{Res}_z(\omega_\epsilon)$$

are independent of ϵ.

Proof: We can explicitly write down the zeroes V_ϵ as follows. In the q-variables, we have $P_\epsilon = \sum_{i=1}^k \frac{1}{n+1} q_i^{n+1} - \epsilon q_i$, so $\partial P_\epsilon/\partial q_i = q_i^n - \epsilon$, and the q-derivatives of P_ϵ vanish simultaneously at the n^k possible k-tuples of nth roots of ϵ. By the chain rule, $\partial P_\epsilon/\partial q_i = \sum_{j=1}^k (\partial P_\epsilon/\partial x_j)(\partial x_j/\partial q_i)$. Now since the Vandermonde determinant $\det(\partial x_j/\partial q_i)$ is invertible at distinct values of the q_i, it follows that if z is the image under the symmetric polynomials $x_j(q_1,\ldots,q_k)$ of a k-tuple of *distinct* nth roots of ϵ, then $z \in V_\epsilon$. There are $\binom{n}{k}$ of these points, of course, and since the sum of the betti numbers of $G(k,n)$ is also $\binom{n}{k}$, and these points have limit $(0,\ldots,0)$ as $\epsilon \to 0$, it follows that they are reduced points of V_ϵ, and account for the entire scheme. (One can also deduce this from the degree of the polynomials.) This gives (a).

As for (b), suppose $\epsilon' = \lambda\epsilon$, suppose we fix an nth root $\lambda^{1/n}$, and use it to establish a correspondence ϕ between the points of V_ϵ and $V_{\epsilon'}$ via $\phi(x_1,\ldots,x_k) = (\lambda^{1/n}x_1,\ldots,\lambda^{k/n}x_k)$. Then it is immediate by change-of-variables that $\mathrm{Res}_{\phi(z)}(\tau_{\epsilon'}) = \mathrm{Res}_z(\tau_\epsilon)$ for all $z \in V_\epsilon$ and likewise for the residues of $\omega_{\epsilon'}$ and ω_ϵ. It follows, of course, that the sums do not depend upon ϵ.

Now, by the principle of continuity for residues, we have:

$$\mathrm{Res}_{(0,\ldots,0)}(\tau) = \lim_{\epsilon\to 0} \sum_{z\in V_\epsilon} \mathrm{Res}_z(\tau_\epsilon) = \sum_{z\in V_1} \mathrm{Res}_z(\tau_1)$$

and similarly for ω, with the last equality following from Proposition 1.2(b).

Next, there is an explicit formula for residues at reduced points, yielding:

$$\sum_{z\in V_1} \mathrm{Res}_z(\tau_1) = \sum_{z\in V_1} (\prod_{j=1}^{m} x_{a(j)})(\det(\partial^2 P_1/\partial x_i \partial x_j))^{-1}$$

and similarly for ω_1, where x_a in this context is just the ath coordinate of z. Finally, if we compute the residues in terms of the q-variables, then we find that:

$$\mathrm{Res}_{(0,\ldots,0)}(\omega) = \frac{1}{k!n^k} \sum_{\{(\rho_1,\ldots,\rho_k)|\rho_i^n=1\}} \prod_{i=1}^{k} \rho_i^{-k+1} \prod_{i\neq j}(\rho_i - \rho_j) = (-1)^{\binom{k}{2}}$$

and we get the first case of the formula of Vafa-Intriligator by rewriting the τ residues and dividing by the residue of ω:

$$(*) \quad \int_G W_{a(1)} \cap \ldots \cap W_{a(m)} = \frac{(-1)^{\binom{k}{2}}}{k!n^k} \sum_{\{(\rho_1,\ldots,\rho_k)|\rho_i^n=1\}} \prod_{j=1}^{m} x_{a(j)} \prod_{i=1}^{k} \rho_i \prod_{i\neq j}(\rho_i - \rho_j)$$

where x_a in this context is the ath symmetric function in the ρ_i.

Notice that the summands are homogeneous of weight kn, which is in particular divisible by n (otherwise the sum would be zero!)

2. Intersection Numbers on Spaces of Maps to a Grassmannian

In this section, we will show, following Witten, how to attach geometric significance to the deformations P_ϵ of §1 in terms of intersections on moduli spaces of maps from a rational curve to the Grassmannian. It will follow that the *identical* formula $(*)$ computes certain intersection numbers on these moduli spaces. (This is the genus 0 case of the formula of Vafa-Intriligator.) We begin by introducing the relevant objects.

- $\mathrm{Mor}_d(\mathbf{P}^1, G)$ = the moduli space of morphisms of degree d from $\mathbf{P}^1$ to the Grassmannian $G(k,n)$. This is a smooth, quasi-projective variety of dimension $nd + k(n-k)$. It admits various "natural" compactifications, for example by the Hilbert scheme of the graph (see [LT]) or by the Grothendieck scheme of quotient sheaves of a trivial bundle on $\mathbf{P}^1$ (see [B1] or [BDW]).

- $\mathrm{ev} : \mathbf{P}^1 \times \mathrm{Mor}_d(\mathbf{P}^1, G) \to G(k,n)$, the evaluation morphism, sending $(p,f) \mapsto f(p)$. This exists as a morphism of schemes because of the representability of the functor Mor.

- $W_{\vec{a}}(p) := \mathrm{ev}^{-1}(W_{\vec{a}}) \cap \Big(\{p\} \times \mathrm{Mor}_d(\mathbf{P}^1, G)\Big)$. Set-theoretically, $W_{\vec{a}}(p)$ is just the set of morphisms f such that $f(p) \in W_{\vec{a}}$. We think of $W_{\vec{a}}(p)$ as a subscheme (of codimension $\deg(\vec{a})$) of $\mathrm{Mor}_d(\mathbf{P}^1, G)$ via the projection.

The following is not hard to prove (see e.g.[B2]):

Proposition 2.1: *If the $W_{\vec{a}(1)}, \ldots, W_{\vec{a}(m)} \subset G(k,n)$ are Schubert cycles (in general position, as always) such that $\sum_{j=1}^{m} \deg(\vec{a}(j)) = \dim(\mathrm{Mor}_d(\mathbf{P}^1, G))$, and if $p_1, \ldots, p_m \in \mathbf{P}^1$ are distinct points, then the intersection:*

$$W_{\vec{a}(1)}(p_1) \cap \ldots \cap W_{\vec{a}(m)}(p_m)$$

is a reduced zero-dimensional scheme, whose length, which we denote by:

$$\int_{M_d} W_{\vec{a}(1)} \cap \ldots \cap W_{\vec{a}(m)}$$

only depends upon the data: $\vec{a}(1), ..., \vec{a}(m)$. *(As usual, we set the integral to zero if* $\sum_{j=1}^{m} \deg(\vec{a}(j)) \neq \dim(\mathrm{Mor}_d(\mathbf{P}^1, G))$.*)*

We use this Proposition to define the quantum intersection of Schubert cycles. Recall first that the ordinary intersection may be defined in terms of intersection numbers. In other words:

$$W_{\vec{a}(1)} \cap ... \cap W_{\vec{a}(m)} = \sum_{\vec{a}} \left(\int_G W_{\vec{a}^c} \cap W_{\vec{a}(1)} \cap ... \cap W_{\vec{a}(m)} \right) W_{\vec{a}}$$

where by equality, we mean as usual that both sides define the same class in the cohomology ring $\mathrm{H}^*(G(k,n), \mathbf{C})$. Similarly, we define the quantum intersections from intersection numbers on the moduli spaces of maps. Namely, we define:

$$W_{\vec{a}(1)} \cap' ... \cap' W_{\vec{a}(m)} := \sum_{d=0}^{\infty} q^d \sum_{\vec{a}} \left(\int_{M_d} W_{\vec{a}^c} \cap W_{\vec{a}(1)} \cap ... \cap W_{\vec{a}(m)} \right) W_{\vec{a}}$$

where q is a variable. Before we say how this induces a deformation of the cohomology ring of $G(k,n)$, a few comments are in order:

• The moduli space $\mathrm{Mor}_0(\mathbf{P}^1, G)$ is just the Grassmannian itself, from which it follows that the constant term (with respect to q) of the quantum intersection is just the usual intersection.

• For any given Schubert cycles $W_{\vec{a}(1)}, ..., W_{\vec{a}(m)}$, the sum on the right is finite. This is simply because the dimensions of $\mathrm{Mor}_d(\mathbf{P}^1, G)$ grow with d. Thus every quantum intersection is a polynomial in q with coefficients in $\mathrm{H}^*(G, \mathbf{C})$.

• On any moduli space $\mathrm{Mor}_d(\mathbf{P}^1, G)$ for $d > 1$, and for any pair of $n-k$-tuples $\vec{a}$ and $\vec{b}$, the moduli intersection number $\int_{M_d} W_{\vec{a}} \cap W_{\vec{b}} = 0$. This is because if $p \neq q \in \mathbf{P}^1$ and $f : \mathbf{P}^1 \to G(k,n)$ satisfies $f(p) \in W_{\vec{a}}$ and $f(q) \in W_{\vec{b}}$, then so does $f \circ \alpha$, where $\alpha : \mathbf{P}^1 \to \mathbf{P}^1$ is an automorphism fixing p and q. As a consequence of this, the quantum intersection applied to a single Schubert cycle is just the identity.

Theorem 2.2 (The Small Quantum Deformation):
(a) *The pairing:*

$$\mathrm{H}^*(G, \mathbf{C})[q] \times \mathrm{H}^*(G, \mathbf{C})[q] \to \mathrm{H}^*(G, \mathbf{C})[q]$$

determined by $(W_{\vec{a}}\, q^i, W_{\vec{b}}\, q^j) \mapsto (W_{\vec{a}} \cap' W_{\vec{b}})\, q^{i+j}$ *puts the structure of a* $\mathbf{C}[q]$*-algebra on* $\mathrm{H}^*(G, \mathbf{C})[q]$. *We will denote this ring by* $\mathrm{H}_q^*(G, \mathbf{C})$ *in order not to confuse it with the polynomial ring.*

(b) *The ring* $\mathrm{H}_q^*(G, \mathbf{C})$ *has the following presentation:*

$$\mathrm{H}_q^*(G, \mathbf{C}) \cong \mathbf{C}[x_1, ..., x_k, q] / \langle \partial P_q / \partial x_1, ..., \partial P_q / \partial x_k \rangle$$

where $P_q = P - (-1)^{k-1} q x_1$ *and* $x_1, ..., x_k, P$ *are as in* §1.

We refer the reader to [W1],[ST] or [B1] for proofs of the Theorem. Observe that the tricky part of (a) is the proof that the pairing is associative. Once (a) is established, the standard proofs of (b) reduce to a single computation on the space of lines in $G(k,n)$ to get the right coefficient of qx_1 (but see also [B1]).

The following "quantum" Schubert calculus is a corollary of the proof of the Theorem found in [B1], in particular using the strong form of Giambelli's formula due to Kempf and Laksov [KL]. However, it ought to be derivable directly from Theorem 2.2(b).

Corollary 2.3: (a) (Quantum Giambelli) *If the determinant $g_{\vec{a}}(W_*)$ is computed taking quantum intersections $\cap'$ instead of ordinary intersections, then the formula:*

$$W_{\vec{a}} = g_{\vec{a}}(W_*)$$

is valid in $\mathrm{H}_q^*(G, \mathbf{C})$, i.e. there are no higher order terms.

(b) (Quantum Pieri) *The following equality holds in $\mathrm{H}_q^*(G, \mathbf{C})$:*

$$W_a \cap' W_{\vec{a}} = p_{a,\vec{a}}(W_*) + q \sum_{i=1}^{k} \sum_{\{\vec{w} | \deg(\vec{w}) = k-i\}} W_{a-i} \cap' W_{\vec{a}-(1,\dots,1)-\vec{w}}$$

where of course we only sum over allowable integers and $n-k$-tuples.

Now we turn to the problem of actually computing the numbers:

$$\int_{M_d} W_{\vec{a}(1)} \cap \dots \cap W_{\vec{a}(m)}$$

By Corollary 2.3(a), we may reduce as before to the computation of

$$\int_{M_d} W_{a(1)} \cap \dots \cap W_{a(m)}.$$

Now by the definition of quantum intersection and the fact that the moduli spaces $\mathrm{Mor}_d(\mathbf{P}^1, G)$ are increasing in dimension with d, we see that if $\sum_{j=1}^{m} a(j) = \dim(\mathrm{Mor}_d(\mathbf{P}^1, G))$, then the quantum intersection satisfies:

$$W_{a(1)} \cap' \dots \cap' W_{a(m)} = \left(\int_{M_d} W_{a(1)} \cap \dots \cap W_{a(m)} \right) q^d \, W_{k,\dots,k} \mod (q^{d+1})$$

from which it follows that if we consider the two meromorphic forms on the space $\mathbf{C}^{k+1}$:

$$\tau_q(d) = \frac{\prod_{j=1}^{m} x_{a(j)} dx_1 \wedge \dots \wedge dx_k \wedge dq}{q^{d+1} \, \prod_{i=1}^{k} \partial P_q / \partial x_i} \quad \text{and} \quad \omega_q(d) = \frac{x_k^{n-k} dx_1 \wedge \dots \wedge dx_k \wedge dq}{q \, \prod_{i=1}^{k} \partial P_q / \partial x_i}$$

then considering the ring $\mathbf{C}[x_1, \dots, x_k, q] / \langle \partial P_q / \partial x_1, \dots \partial P_q / \partial x_k, q^{d+1} \rangle$, we get:

$$\mathrm{Res}_{(0,\dots,0)}(\tau_q(d)) = \left(\int_{M_d} W_{a(1)} \cap \dots \cap W_{a(m)} \right) \mathrm{Res}_{(0,\dots,0)}(\omega_q(d))$$

so our intersection is again expressed as a quotient of residues.

Notice that if we perturb our ideal to $\langle \partial P_q / \partial x_1, \dots \partial P_q / \partial x_k, q^{d+1} - \delta \rangle$, then the zero locus of the perturbed ideal is:

$$V(\langle \partial P_q / \partial x_1, \dots \partial P_q / \partial x_k, q^{d+1} - \delta \rangle) = \bigcup_{((-1)^{k-1}\epsilon)^{d+1} = \delta} \; \bigcup_{z \in V_\epsilon} \{(z, (-1)^{k-1}\epsilon)\}$$

and it is reduced by Proposition 1.2. The sign in front of ϵ arises because of the presentation of the quantum cohomology ring.

Once again, we introduce perturbed meromorphic forms:

$$\tau_{q,\delta}(d) = \frac{\prod_{j=1}^{m} x_{a(j)} dx_1 \wedge \dots \wedge dx_k \wedge dq}{(q^{d+1} - \delta) \, \prod_{i=1}^{k} \partial P_q / \partial x_i}, \omega_{q,\delta}(d) = \frac{q^d x_k^{n-k} dx_1 \wedge \dots \wedge dx_k \wedge dq}{(q^{d+1} - \delta) \prod_{i=1}^{k} \partial P_q / \partial x_i}$$

the sum of whose residues at the zeroes of the perturbed ideal approach the residues of $\tau_q(d)$ and $\omega_q(d)$ at the origin. We observe as in Proposition 1.2 that the sum of perturbed residues does not depend upon δ. In fact, each partial sum for fixed ϵ satisfying $((-1)^{k-1}\epsilon)^{d+1} = \delta$

does not depend upon ϵ, so we may as well assume that all the $\epsilon = 1$. Finally, explicit computation yields:

$$\int_{M_d} W_{a(1)} \cap \ldots \cap W_{a(m)} = \frac{(-1)^{d(k-1)+\binom{k}{2}}}{k!n^k} \sum_{\{(\rho_1,\ldots,\rho_k)|\rho_i^n=1\}} \prod_{j=1}^{m} x_{a(j)} \prod_{i=1}^{k} \rho_i \prod_{i\neq j}(\rho_i - \rho_j)$$

which is the genus zero Vafa-Intriligator formula, the only difference between it and the formula in §1 being a sign and the fact that here we choose $\sum_{j=1}^m a(j) = dn + k(n-k)$.

The point I want to stress is that the presentation of the quantum cohomology ring of the Grassmannian is, (except for the pesky sign $(-1)^{k-1}$) exactly the "smoothing" of its cohomology ring that we used to compute Schubert calculus by Severi residues in §1. It is in this sense that I see the residues as being linked to the small quantum deformation of the cohomology ring.

3. Generalizations

Clearly, there are many potential generalizations, but I want to consider two of them here, namely, how to use residues to compute intersection data on the maps from a curve of higher genus to the Grassmannian, and the form of the quantum cohomology ring for the full flag varieties. One of the features pointed out by these generalizations is the fact that the quantum Giambelli result of Corollary 2.3(a) seems to be a very rare phenomenon, perhaps only applying to maps from a rational curve to a Grassmannian.

The Vafa-Intriligator formula in full generality is the following:

Fix C a compact Riemann surface of genus g, and consider the auxiliary objects $\mathrm{Mor}_d(C,G)$, $ev : C \times \mathrm{Mor}_d(C,G) \to G$ and $W_{\vec{a}}(p) \subset \mathrm{Mor}_d(C,G)$ defined as in §2. For small values of d, these objects are rather hard to understand (even if G is projective space) basically because of Brill-Noether theory. However, if $d >> 0$, then $\mathrm{Mor}_d(C,G)$ has only local-complete-intersection singularities, and is of the expected dimension $dn - (g-1)k(n-k)$. Moreover, given Schubert cycles $W_{\vec{a}(1)}, \ldots, W_{\vec{a}(m)}$ satisfying $\sum_{j=1}^m \deg(\vec{a}(j)) = dn - (g-1)k(n-k)$, it is true that the intersection $W_{\vec{a}(1)}(p_1) \cap \ldots \cap W_{\vec{a}(m)}(p_m)$ consists of reduced points in the smooth locus of the moduli space, and that the length of this scheme does not depend upon the choice of the distinct points $p_1, \ldots, p_m \in C$. Moreover, it does not depend upon the choice of C for a given genus. (See [BDW] and [B2] for proofs of all this.) Thus, it makes sense to ask for the intersection number:

$$\int_{M_d(g)} W_{\vec{a}(1)} \cap \ldots \cap W_{\vec{a}(m)}$$

for large values of d, observing that it only depends upon g and $\vec{a}(1), \ldots, \vec{a}(m)$.

The problem here is that it is unknown whether Giambelli still holds. In fact, it is unclear whether such an intersection number will be determined from the special intersections:

$$\int_{M_d(g)} W_{a(1)} \cap \ldots \cap W_{a(m)}$$

and if so, in what way.

So we'll be satisfied instead with computing the special intersections:

The Formula of Vafa-Intriligator: $\int_{M_d(g)} W_{a(1)} \cap \ldots \cap W_{a(m)} =$

$$\frac{(-1)^{d(k-1)+(g-1)\binom{k}{2}}}{k!} \sum_{\{(\rho_1,\ldots,\rho_k)|\rho_i^n=1;\rho_i\neq\rho_j\}} \prod_{j=1}^{m} x_{a(j)} \left(\frac{\prod_{i=1}^{k}\rho_i\prod_{i\neq j}(\rho_i-\rho_j)}{n^k}\right)^{1-g}$$

(where here we need to require that the ρ_i be distinct, versus in the genus zero formulation where nondistinct choices contributed zero.)

The idea of proof of this is very intuitive. Namely, if we consider an irreducible curve X with one node and otherwise smooth, of arithmetic genus g, then the intersection numbers in the formula ought to be computable on the objects associated to X, and ought to agree with the corresponding intersection numbers for a smooth curve. On the other hand, via the normalization, the maps from X to G are just those maps from the genus $g-1$ curve $\widetilde{X}$ to G which map the two marked points to the same image point. It follows from this that a relationship is set up between the intersection data for genus g and $g-1$, and it is shown in [B2] that this relationship is precisely multiplication by the inverse of the term appearing in the formula with exponent $1-g$. It is curious that this term (which is just an expression for the top chern class of the tangent bundle to G in terms of the chern roots of S^*) is the same as the "correction" term appearing in the genus zero formula. In any case, this proof is delicate, due to the presence of singularities, and is carried out in [B2] (with a correction in [BS]).

Next, let $F(n)$ be the variety of full flags of $\mathbf{C}^n$. Then there is a theory of Schubert polynomials due to Lascoux and Schützenberger ([LS]) leading to:

• An additive basis for the cohomology ring of $F(n)$ indexed by the permutations on n letters.

• A natural Giambelli formula expressing each element of this basis in terms of "special" cycles, which in this case are just the $n-1$ nef divisors arising by pulling back $\mathcal{O}_G(1)$ from the various projections $F(n) \mapsto G(k,n)$.

In addition, the cohomology ring of $F(n)$ has a presentation, analogous to the presentation of the cohomology ring of the Grassmannian:

$$\mathrm{H}^*(F(n),\mathbf{C}) \cong \mathbf{C}[y_1,\ldots,y_n]/\langle\sigma_1(y_*),\ldots,\sigma_n(y_*)\rangle$$

where the σ_i are the elementary symmetric functions, and the y_i have weight 2 in the cohomology ring. In particular, $y_1,\ldots,y_{n-1}$ are the nef divisors listed above, and y_n is irrelevant because of the first relation.

Because $\mathrm{Pic}(F(n)) = \mathbf{Z}^{n-1}$, the components of the moduli spaces of morphisms from $\mathbf{P}^1$ to $F(n)$ are indexed by $n-1$-tuples (counting the intersections with the y_i). It follows that the small quantum cohomology ring for $F(n)$, when defined analogously to the definition in §2, is an algebra over the polynomial ring $\mathbf{C}[q_1,\ldots,q_{n-1}]$. The presentation of this algebra was first conjectured by Givental-Kim ([GK]) and proved by Ciocan-Fontanine ([C]):

Theorem: *Consider the $n\times n$ matrix:*

$$M^n_{y,q} := \begin{pmatrix} y_1 & q_1 & 0 & 0 & \cdots \\ -1 & y_2 & q_2 & 0 & \cdots \\ 0 & -1 & y_3 & q_3 & \cdots \\ \vdots & & \vdots & & \end{pmatrix}$$

Then

$$\mathrm{H}_q^*(F(n), \mathbf{C}) \cong \mathbf{C}[y_1, \ldots, y_n, q_1, \ldots, q_{n-1}]/I$$

where I is the ideal generated by the coefficients of the characteristic polynomial of $M_{y,q}^n$.

The calculation of intersection numbers by residues is mentioned in [GK] as a corollary of this presentation, but it is observed by Ciocan-Fontanine that the Giambelli formula coming from the theory of Schubert polynomials does not continue to hold when intersection is replaced by quantum intersection. However, in this case the higher-order terms ought not to be too difficult to compute.

References

[B1] A. Bertram, *Modular Schubert calculus*, to appear in Advances in Mathematics.

[B2] A. Bertram, *Towards a Schubert calculus for maps from a Riemann surface to a Grassmannian*, Int. Journal of Math. **5** (1995), pp. 811-825.

[BDW] A. Bertram, G.Daskalopoulos and R. Wentworth, *Gromov invariants for holomorphic maps from Riemann surfaces to Grassmannians*, to appear in the Journal of the AMS.

[BS] A. Bertram and H. Shahrouz, *On the Cohen-Macaulayness of certain moduli spaces*, preprint (1995).

[BT] R. Bott and L. Tu, Differential Forms in Algebraic Topology, GTM **82**, Springer-Verlag New York, Heidelberg, Berlin (1982).

[C] I. Ciocan-Fontanine, *Quantum cohomology of flag varieties*, alg-geom preprint (1995).

[DFI] P. Di Franceso and C. Itzykson, *Quantum intersection rings*, hep-th preprint (1994).

[F] W. Fulton, *Enumerative geometry via quantum cohomology*, preprint (1995).

[GH] P. Griffiths and J. Harris, Principles of Algebraic Geometry, Wiley-Interscience (1978).

[GK] A. Givental and B. Kim, *Quantum cohomology of flag manifolds and Toda lattices*, hep-th preprint (1993).

[I] K. Intriligator, *Fusion residues*, Modern Physics Letters A **6** (1991), Number 38, pp. 3543-3556.

[KL] G. Kempf and D. Laksov, *The determinantal formula of Schubert calculus*, Acta. Math. **132** (1974) pp. 153-162.

[KM] M. Kontsevich and Y. Manin, *Gromov-Witten classes, quantum cohomology, and enumerative geometry*, hep-th preprint (1994).

[LS] A. Lascoux and M-P. Schützenberter, *Polynômes de Schubert*, C.R. Acad. Sci. Paris **294** (1982), pp. 447-450.

[LT] J. Li and G. Tian, *The quantum cohomology of homogeneous varieties*, alg-geom preprint (1994).

[P] R. Pandharipande, *Notes on Kontsevich's compactification of the space of maps*, University of Chicago Course Notes (1995).

[RT] Y. Ruan and G. Tian, *A mathematical theory of quantum cohomology*, alg-geom preprint (1994).

[ST] B.Siebert and G.Tian, *On quantum cohomology rings of Fano manifolds and a formula of Vafa and Intriligator*, alg-geom preprint (1994).

[V] C. Vafa, *Topological mirrors and quantum rings*, Essays on Mirror Manifolds (S.T. Yau, ed.), International Press (1992).

[W1] E. Witten, *The Verlinde algebra and the cohomology of the Grassmannian*, hep-th preprint (1993).

[W2] E. Witten, *Two dimensional gauge theories revisited*, J. Geom. Phys. **9** (1992), pp. 303-368.

NERON-SEVERI GROUP FOR TORUS QUASI BUNDLES OVER CURVES

Vasile Brînzănescu

Institute of Mathematics of the Romanian Academy, P.O.BOX 1–764, RO-70700, Bucharest, Romania

Kenji Ueno

Department of Mathematics, Graduate School of Science, Kyoto University,
Kyoto, 606–01 Japan

0. Introduction

By the Neron-Severi group of a compact complex manifold X we mean the kernel of the natural homomorphism $H^2(X,\mathbb{Z}) \to H^2(X,\mathcal{O}_X)$. It is a subgroup of $H^2(X,\mathbb{Z})$ generated by the first Chern classes of line bundles on X. In this paper we shall study the Neron-Severi group for torus quasi bundles over curves. Firstly, we study the case of torus principal bundles $X \xrightarrow{\pi} B$ over a (complex, compact, connected, smooth) curve B, whose structure group is a compact complex torus $T = V/\Lambda$. A T-principal bundle $X \xrightarrow{\pi} B$ is defined by a cohomology class $\xi \in H^1(\mathcal{O}_B(T))$, where $\mathcal{O}_B(T)$ is the sheaf of germs of locally holomorphic maps from B to T. The cohomology class ξ determines a characteristic class $c(\xi) \in H^2(B,\Lambda)$. By a Theorem of Blanchard ([1]), the total space X of such a T-principal bundle is a non-Kähler manifold if and only if $c(\xi) \neq 0$. In the first two parts of the paper we present some basic facts on torus principal bundles (see [7]) and we compute Leray spectral sequences for the sheaves $\mathbb{Z}_X$ and $\mathcal{O}_X$. In the third part we define for any line bundle $L \in Pic(T)$ an *associated $T^\vee$-principal bundle*, described by an element $\tilde{\varphi}_L(\xi) \in H^1(\mathcal{O}_B(T^\vee))$, where $T^\vee$ is the dual torus, and we compute the Neron-Severi group for torus principal bundles. We state the main result (Theorem 5):

"For a T-principal bundle $X \xrightarrow{\pi} B$, defined by a cohomology class

$$\xi \in H^1(\mathcal{O}_B(T)),$$

we have an exact sequence of free groups

$$0 \to Hom(J_B, T^\vee) \to NS(X)/F_2 \to \tilde{N}(X) \to 0,$$

where $F_2 = \pi^* NS(B)$ and $\tilde{N}(X)$ is the subgroup of the Neron-Severi group of the torus T defined by

$$\tilde{N}(X) = \{c_1(L) \in NS(T) \mid \tilde{\varphi}_L(\xi) \text{ is the trivial torus bundle}\},$$

J_B is the Jacobian variety of the curve B and $T^\vee$ is the dual torus. If X is Kähler F_2 is isomorphic to $NS(B) \simeq \mathbb{Z}$ and if X is non-Kähler, F_2 is the torsion subgroup of $NS(X)$"

In the fourth part we reinterpret the obtained results geometrically (see Theorem 6). Then, in the fifth part, we study the case of torus quasi bundles. By a quasi T-bundle $\pi : X \to B$

over a curve B we mean that π is a T-principal bundle over $B \setminus \{b_1, b_2, ..., b_\ell\}$ and that the fibre $\pi^{-1}(b_i)$ over the point b_i is of the form $m_i T_i$ where $m_i \geq 2$ and T_i is a torus (the fibre $m_i T_i$ is called a multiple fibre of the multiplicity m_i). In the Appendix we show that all torus quasi bundles are obtained from $B \times T$ by means of generalized logarithmic transformations. We associate, canonically, a T_0- principal bundle $\pi_0 : Y \to B$ to a quasi T-bundle $\pi : X \to B$ and a holomorphic mapping $f : X \to Y$, with $T_0 = T/H$, where H is a finite subgroup of the torus T. Then we extend the computation of the Neron-Severi group for torus quasi bundles (see Theorem 17). For the case of elliptic surfaces see [3], [4].

Acknowledgments. The first named author would like to thank the Alexander von Humboldt-Stiftung for support and both authors would like to express their thanks to Professor F. Hirzebruch and Max-Planck-Institut für Mathematik in Bonn for hospitality. This paper was prepared at the time we visited this institution.

1. Basic facts on torus principal bundles

Let $T = V/\Lambda$ be an n-dimensional compact complex torus, defined by a lattice $\Lambda \subset V$ in the n-dimensional complex vector space V. Canonical notation concerning the torus T will be used:

$$T_0(T) = H^0(T, \Theta_T) = V, \quad H^i(T, \Theta_T) = H^i(T, \mathcal{O}_T) \otimes V,$$
$$H^0(T, \Omega_T^1) = H^0(T, \Theta_T)^\vee = V^\vee, \quad \Lambda = H_1(T, \mathbb{Z}), \quad H^1(T, \mathbb{Z}) = \Lambda^\vee.$$

If B is a compact complex manifold of dimension m, then $X \xrightarrow{\pi} B$ denotes a T-principal bundle over B. Let $\mathcal{O}_B(T)$ denote the sheaf of germs of locally holomorphic maps from B to T. The T-principal bundles are described by cohomology classes ξ of $H^1(B, \mathcal{O}_B(T))$ (see [6]). For a Čech 1-cocycle (ξ_{ij}) the function

$$\xi_{ij} : U_i \cap U_j \to T$$

identifies $(z, t) \in U_i \times T$ with $(z, t') = (z, \xi_{ij}(z) + t) \in U_j \times T$ for all $z \in U_i \cap U_j$.

Taking local sections of the constant sheaves

$$0 \to \Lambda \to V \to T \to 0$$

one gets an exact sequence of sheaves on the manifold B

$$0 \to \Lambda \to \mathcal{O}_B \otimes V \to \mathcal{O}_B(T) \to 0, \tag{1}$$

with the induced exact cohomology sequence

$$\begin{aligned} \cdots \to & H^0(\mathcal{O}_B(T)) \to H^1(B, \Lambda) \to H^1(B, \mathcal{O}_B) \otimes V \\ & \to H^1(\mathcal{O}_B(T)) \xrightarrow{c} H^2(B, \Lambda) \to H^2(B, \mathcal{O}_B) \otimes V \quad . \end{aligned} \tag{2}$$

The cohomology class ξ of the bundle in $H^1(\mathcal{O}_B(T))$ determines a characteristic class $c(\xi) \in H^2(B, \Lambda) = H^2(B, \mathbb{Z}) \otimes \Lambda$.

Because transition functions of the T-principal bundle $X \xrightarrow{\pi} B$ act trivially on the cohomology of fibre, we get natural identifications:

$$R^q\pi_*\mathbb{Z}_X = \mathbb{Z}_B \otimes_{\mathbb{Z}} H^q(T, \mathbb{Z}), \quad R^q\pi_*\mathcal{O}_X = \mathcal{O}_B \otimes_{\mathbb{C}} H^q(T, \mathcal{O}_T). \tag{3}$$

The transgression of the fibre bundle in integral cohomology is a map

$$\delta : H^1(T, \mathbb{Z}) \to H^2(B, \mathbb{Z}).$$

Under the identification

$$H^1(T, \mathbb{Z}) = Hom(\Lambda, \mathbb{Z}) = \Lambda^\vee,$$

the characteristic class $c(\xi) \in H^2(B,\mathbb{Z}) \otimes \Lambda$ and the mapping $\delta : H^1(T,\mathbb{Z}) \to H^2(B,\mathbb{Z})$ coincide (see [7], 6.1). The first possibly nontrivial d_2-homomorphism

$$H^0(B, R^1\pi_*\mathcal{O}_X) \to H^2(B, \pi_*\mathcal{O}_X)$$

in the Leray spectral sequence of $\mathcal{O}_X$ is denoted by

$$\varepsilon : H^1(T, \mathcal{O}_T) \to H^2(B, \mathcal{O}_B).$$

Recall for convenience the following result of Höfer (see [7], 7.1 and 7.2):

Proposition *There is an injective map*

$$\Phi : Pic(B) \otimes_{\mathbb{Z}} \Lambda = H^1(\mathcal{O}_B^*) \otimes_{\mathbb{Z}} \Lambda \to H^1(\mathcal{O}_B(T))$$

compatible with taking characteristic classes, i.e. if $\sum \mathcal{L}_k \otimes \lambda_k$ *is a combination of line bundles in* $Pic(B) \otimes_{\mathbb{Z}} \Lambda$*, then the characteristic class* $c(\xi)$ *of* $\Phi(\sum \mathcal{L}_k \otimes \lambda_k)$ *equals* $\sum c_1(\mathcal{L}_k) \otimes \lambda_k \in H^2(B,\Lambda)$.

$$\begin{array}{ccc} Pic(B) \otimes_{\mathbb{Z}} \Lambda & \xrightarrow{\Phi} & H^1(\mathcal{O}_B(T)) \\ \downarrow{\scriptstyle c_1 \otimes id} & & \downarrow{\scriptstyle c} \\ H^2(B,\mathbb{Z}) \otimes \Lambda & \xrightarrow{=} & H^2(B,\Lambda) \end{array}$$

Moreover, if $H^2(B,\mathbb{C})$ *has a Hodge decomposition, then the image of* Φ*, i.e. the set of isomorphism classes of principal bundles constructed above, equals*

$$\operatorname{im}\Phi = \{\textit{Isom. classes of } T\textit{-principal bundles with } \varepsilon = 0\}.$$

Remark. If B is a curve, then ε vanishes for dimension reasons. Thus, every T-principal bundle over B comes (in an unique way) from the above construction. The construction itself is a *generalized logarithmic transformation* applied to the trivial T-principal bundle $B \times T$ (see [9]). Indeed, we can write $\mathcal{L}_k = \mathcal{O}_B(D_k)$, with D_k a divisor on B; by choosing a sufficiently fine open covering (U_i) of B the transition functions of each $\mathcal{L}_k$ are expressed by a cocycle $(f_{ij}^{(k)})$. Now, identify $(z, t_i) \in U_i \times T$ with $(z, t_j) \in U_j \times T$ if and only if

$$t_i = t_j + \left[\sum \frac{\lambda_k}{2\pi\sqrt{-1}} \log\left(f_{ij}^{(k)}\right)\right],$$

for all $z \in U_i \cap U_j$ (this is exactly Höfer's morphism Φ).

Also we can construct a T-principal bundle over B by using logarithmic transformations similar to the case of elliptic surfaces. Express the divisor D_k as

$$D_k = \sum_{j=1}^{n_k} m_j^{(k)} b_j^{(k)}.$$

Let $U_j^{(k)}$ be a coordinate neighbourhood of $b_j^{(k)}$ with local coordinate $t_j^{(k)}$. We may assume

$$U_j^{(k)} = \{t_j^{(k)} \in \mathbb{C} \mid |t_j^{(k)}| < \varepsilon\},$$

for a sufficiently small positive number ε. Let us consider a holomorphic mapping

$$\begin{aligned} l_j^{(k)} : U_j^{(k)*} \times T &\longrightarrow U_j^{(k)*} \times T \\ (t_j^{(k)}, [\zeta]) &\to (t_j^{(k)}, \left[\zeta - \frac{m_j^{(k)}\lambda_k}{2\pi\sqrt{-1}} \log t_j^{(k)}\right]). \end{aligned}$$

Note that the mapping is an isomorphism. Hence, we can patch $U_j^{(k)} \times T$'s and $(B \setminus \{b_1^{(1)}, ..., b_j^{(k)}, ...\}) \times T$ by the isomorphisms $l_j^{(k)}$ and obtain a T-principal bundle over B. We denote the T-principal bundle obtained in this way by

$$L_{b_1^{(1)}}(m_1^{(1)}\lambda_1, 1)...L_{b_{n_l}^{(l)}}(m_{n_l}^{(l)}\lambda_l, 1)(B \times T)$$

or by

$$L_{D_1}(\lambda_1, 1)...L_{D_l}(\lambda_l, 1)(B \times T).$$

Remark. By the above proposition and Blanchard's theorem ([1]) we can easily show that a T-principal bundle

$$L_{b_1}(a_1, 1)...L_{b_l}(a_l, 1)(B \times T)$$

is Kähler if and only if $\sum_{i=1}^{l} a_i = 0$.

2. Leray spectral sequences

Let $X \xrightarrow{\pi} B$ be a T-principal bundle over the manifold B. We consider the Leray spectral sequences:

$$E_2^{pq} = H^p(B, R^q\pi_*\mathbb{Z}_X) \Longrightarrow H^{p+q}(X, \mathbb{Z}) \tag{4}$$

$$\tilde{E}_2^{pq} = H^p(B, R^q\pi_*\mathcal{O}_X) \Longrightarrow H^{p+q}(X, \mathcal{O}_X). \tag{5}$$

By the results of Höfer (see [7]) the first spectral sequence (4) degenerates at E_3-level (i.e. $d_r = 0$ for $r > 2$) and the d_2-differential is determined by the map $\delta : H^1(T, \mathbb{Z}) \to H^2(B, \mathbb{Z})$ (i.e. by $c(\xi)$).

Now, we suppose that B is a curve. By (3) we have:

$$\begin{aligned} E_\infty^{02} = E_3^{02} &= \ker(E_2^{02} \xrightarrow{d_2} E_2^{21}) \\ &= \ker(H^0(B, \mathbb{Z}) \otimes H^2(T, \mathbb{Z}) \xrightarrow{d_2} H^2(B, \mathbb{Z}) \otimes H^1(T, \mathbb{Z})). \end{aligned}$$

With the natural identifications

$$H^0(B, \mathbb{Z}) = \mathbb{Z}, \quad H^2(B, \mathbb{Z}) = \mathbb{Z}, \quad H^2(T, \mathbb{Z}) = \bigwedge^2 H^1(T, \mathbb{Z}),$$

we obtain

$$E_\infty^{02} = \ker(H^2(T, \mathbb{Z}) \xrightarrow{d_2} H^1(T, \mathbb{Z})),$$

where

$$d_2(\varphi_1 \wedge \varphi_2) = \delta(\varphi_1)\varphi_2 - \delta(\varphi_2)\varphi_1, \quad \forall \varphi_1, \varphi_2 \in H^1(T, \mathbb{Z}).$$

Obviously, we have

$$E_\infty^{11} = E_2^{11} = H^1(B, \mathbb{Z}) \otimes H^1(T, \mathbb{Z}) = H^1(B, \mathbb{Z}) \otimes \Lambda^\vee.$$

Finally, we get

$$\begin{aligned} E_\infty^{20} = E_3^{20} &= \operatorname{coker}(H^0(B, \mathbb{Z}) \otimes H^1(T, \mathbb{Z}) \xrightarrow{d_2} H^2(B, \mathbb{Z})) \\ &= \operatorname{coker}(H^1(T, \mathbb{Z}) \xrightarrow{\delta} H^2(B, \mathbb{Z})). \end{aligned}$$

The cohomology class $\xi \in H^1(\mathcal{O}_B(T))$ of the T-principal bundle $X \xrightarrow{\pi} B$ has the form $\Phi(\sum \mathcal{L}_k^0 \otimes \lambda_k^0)$ and its characteristic class has the form

$$c(\xi) = \sum c_1(\mathcal{L}_k^0) \otimes \lambda_k^0 = m\lambda^0 \in \Lambda = H^2(B, \Lambda), \tag{6}$$

where $\mathcal{L}_k^0 \in Pic(B)$, $\lambda_k^0 \in \Lambda$ is a primitive element (i.e. there exists no positive integer $l \geq 2$ with $\lambda_k^0 = l\tilde{\lambda}_k^0$, $\tilde{\lambda}_k^0 \in \Lambda), m \in \mathbb{N}, m = g.c.d.(c_1(\mathcal{L}_k^0))$ and $\lambda^0 \in \Lambda$. It follows that for any $\varphi \in H^1(T, \mathbb{Z})$ we have the equality $\delta(\varphi) = m\varphi(\lambda^0)$, under the identification $H^1(T, \mathbb{Z}) = \Lambda^\vee = Hom(\Lambda, \mathbb{Z})$. We get

$$E_\infty^{20} = \begin{cases} \mathbb{Z}_m & \text{for } c(\xi) \neq 0 \\ \mathbb{Z} & \text{for } c(\xi) = 0 \end{cases}.$$

The second spectral sequence (5) degenerates at E_2-level for torus principal bundles with $\varepsilon = 0$, since the d_2-differential is determined by ε (see [7], 4. and [2]). With natural identifications, by (3) we get:

$$\tilde{E}_\infty^{20} = \tilde{E}_2^{20} = H^0(B, \mathcal{O}_B) \otimes H^2(T, \mathcal{O}_T) = H^2(T, \mathcal{O}_T).$$
$$\tilde{E}_\infty^{11} = \tilde{E}_2^{11} = H^1(B, \mathcal{O}_B) \otimes H^1(T, \mathcal{O}_T).$$
$$\tilde{E}_\infty^{20} = \tilde{E}_2^{20} = 0.$$

3. Neron-Severi group for torus principal bundles

Let $X \xrightarrow{\pi} B$ be a T-principal bundle over the curve B, defined by $\xi \in H^1(\mathcal{O}_B(T))$ with $c(\xi) \neq 0$ (i.e. X is non-Kähler). Let

$$0 \subset F_2 \subset F_1 \subset F_0 = H^2(X, \mathbb{Z})$$

be the filtration induced by the first spectral sequence (4). Then $F_2 = E_\infty^{20} \cong \mathbb{Z}_m$ is a torsion subgroup of $H^2(X, \mathbb{Z})$. Since both $F_1/F_2 = E_\infty^{11}$ and $F_0/F_1 = E_\infty^{02}$ are free, it follows $Tors\ H^2(X, \mathbb{Z}) = F_2 \cong \mathbb{Z}_m$. We get the exact sequence:

(7)
$$\begin{aligned} 0 \to & H^1(B, \mathbb{Z}) \otimes H^1(T, \mathbb{Z}) \to H^2(X, \mathbb{Z})/Tors\ H^2(X, \mathbb{Z}) \\ & \to \ker(H^2(T, \mathbb{Z}) \xrightarrow{d_2} H^1(T, \mathbb{Z})) \to 0. \end{aligned}$$

Let

$$0 \subset \tilde{F}_2 \subset \tilde{F}_1 \subset \tilde{F}_0 = H^2(X, \mathcal{O}_X)$$

be the filtration induced by the second spectral sequence (5). Then, we get the exact sequence:

(8)
$$0 \to H^1(B, \mathcal{O}_B) \otimes H^1(T, \mathcal{O}_T) \to H^2(X, \mathcal{O}_X) \to H^2(T, \mathcal{O}_T) \to 0.$$

The Neron-Severi group, denoted by $NS(X)$, is the kernel of the map in cohomology $H^2(X, \mathbb{Z}) \xrightarrow{i} H^2(X, \mathcal{O}_X)$, induced by the natural map $\mathbb{Z}_X \xrightarrow{i} \mathcal{O}_X$. Since $F_2 \xrightarrow{i} \tilde{F}_2 = 0$, we have $F_2 \subset NS(X)$ and

(9)
$$TorsNS(X) = F_2 = TorsH^2(X, \mathbb{Z}) \cong \mathbb{Z}_m.$$

Using the exact sequence of small terms of the first spectral sequence (4) we get

$$TorsNS(X) = \operatorname{im}(H^2(B, \mathbb{Z}) \xrightarrow{\pi^*} H^2(X, \mathbb{Z})).$$

By functoriality of the spectral sequences we get the following commutative diagram

$$\begin{array}{ccccccccc} 0 & \longrightarrow & F_1/F_2 & \longrightarrow & F_0/F_2 & \longrightarrow & F_0/F_1 & \longrightarrow & 0 \\ & & \downarrow{\scriptstyle i'} & & \downarrow{\scriptstyle i} & & \downarrow{\scriptstyle i''} & & \\ 0 & \longrightarrow & \tilde{F}_1 & \longrightarrow & \tilde{F}_0 & \longrightarrow & \tilde{F}_0/\tilde{F}_1 & \longrightarrow & 0 \end{array}$$

where the first line is the exact sequence (7) and the second line is the exact sequence (8). Since $NS(X)/TorsNS(X) \cong \ker(i)$, we obtain the exact sequence:

$$0 \longrightarrow \ker(i') \longrightarrow NS(X)/TorsNS(X) \longrightarrow \ker(i'') \xrightarrow{\beta} \operatorname{coker}(i'). \tag{10}$$

Lemma 1. *We have* $\ker(i') \cong Hom(J_B, T^\vee)$, *where* J_B *is the Jacobian variety of the curve* B, $T^\vee$ *is the dual torus of the torus* T *and* $Hom(J_B, T^\vee)$ *is the group of homomorphisms of group varieties.*

Proof: By [8], Chap.I, 2, we have the exact sequence

$$0 \to \Lambda^\vee \to \overline{V}^\vee \to T^\vee \to 0,$$

where

$$\Lambda^\vee = H^1(T, \mathbb{Z}), \quad \overline{V}^\vee = H^1(T, \mathcal{O}_T), \quad T^\vee = Pic^0(T).$$

Taking local sections of these constant sheaves one gets an exact sequence of sheaves on B

$$0 \to \Lambda^\vee \to \mathcal{O}_B \otimes \overline{V}^\vee \to \mathcal{O}_B(T^\vee) \to 0, \tag{11}$$

with the induced exact cohomology sequence:

$$\begin{aligned} &0 \longrightarrow H^0(B, \Lambda^\vee) \longrightarrow H^0(B, \mathcal{O}_B) \otimes \overline{V}^\vee \longrightarrow H^0(\mathcal{O}_B(T^\vee)) \longrightarrow H^1(B, \Lambda^\vee) \\ &\xrightarrow{j} H^1(B, \mathcal{O}_B) \otimes \overline{V}^\vee \longrightarrow H^1(\mathcal{O}_B(T^\vee)) \xrightarrow{c^\vee} H^2(B, \Lambda^\vee) \longrightarrow 0. \end{aligned} \tag{12}$$

But

$$\begin{aligned} H^1(B, \Lambda^\vee) &= H^1(B, \mathbb{Z}) \otimes H^1(T, \mathbb{Z}), \\ H^1(B, \mathcal{O}_B) \otimes \overline{V}^\vee &= H^1(B, \mathcal{O}_B) \otimes H^1(T, \mathcal{O}_T) \end{aligned}$$

and $j = i'$ by naturality. It follows

$$\begin{aligned} \ker(i') &= \ker(H^1(B, \Lambda^\vee) \xrightarrow{j} H^1(B, \mathcal{O}_B) \otimes \overline{V}^\vee) \\ &\cong \operatorname{im}(H^0(\mathcal{O}_B(T^\vee)) \to H^1(B, \Lambda^\vee)) \\ &\cong \operatorname{coker}(H^0(B, \mathcal{O}_B) \otimes \overline{V}^\vee \longrightarrow H^0(\mathcal{O}_B(T^\vee))). \end{aligned}$$

But $H^0(\mathcal{O}_B(T^\vee))$ is the group of global holomorphic maps $B \to T^\vee$ and

$$\operatorname{im}(H^0(B, \mathcal{O}_B) \otimes \overline{V}^\vee \to H^0(\mathcal{O}_B(T^\vee))) \cong \overline{V}^\vee/\Lambda^\vee = T^\vee$$

is the subgroup of constant maps $B \to T^\vee$, which can be identified with the points of $T^\vee$ (or, with the translations of $T^\vee$). Let $B \to J_B$ be the canonical holomorphic map (determined up to a translation of J_B). Given any holomorphic map $B \to T^\vee$ then, if we choose the proper origin on $T^\vee$, the holomorphic map $B \to T^\vee$ is the composition of the canonical map $B \to J_B$ and a homomorphism from J_B to $T^\vee$ (the universal property of the Jacobian). It follows the isomorphism

$$\ker(i') \cong Hom(J_B, T^\vee).$$

□

Lemma 2. *We have*

$$\ker(i'') = \{c_1(L) \in NS(T) \mid c_1(L)(\lambda^0) = 0\},$$

where $c(\xi) = m\lambda^0 \in \Lambda$.

Proof: From the previous diagram we get

$$\ker(i'') = \{c_1(L) \in NS(T) \mid d_2(c_1(L)) = 0\}.$$

Let $\{e_1, ..., e_{2n}\}$ be a basis of the lattice Λ and let $\{e^1, ..., e^{2n}\}$ be the dual basis in the lattice $\Lambda^\vee$. Any element $E = c_1(L) \in NS(T)$ can be written in the form

$$E = \sum_{1 \leq i < j \leq 2n} a_{ij} e^i \wedge e^j, \quad a_{ij} \in \mathbb{Z}$$

(see [8], Chap. I, 2). By direct computation we obtain

$$\begin{aligned} d_2(c_1(L)) &= \sum_{i<j} a_{ij} d_2(e^i \wedge e^j) = \sum_{i<j} a_{ij}(\delta(e^i)e^j - \delta(e^j)e^i) \\ &= m \sum_{i<j} a_{ij}(e^i(\lambda^0)e^j - e^j(\lambda^0)e^i) = mc_1(L)(\lambda^0), \end{aligned}$$

where we made the natural identifications

$$Bil(\Lambda \times \Lambda, \mathbb{Z}) = Hom_{\mathbb{Z}}(\Lambda \otimes \Lambda, \mathbb{Z}) = Hom_{\mathbb{Z}}(\Lambda, \Lambda^\vee).$$

The assertion follows. □

For any line bundle $L \in Pic(T)$ we have the homomorphism

(13) $$\varphi_L : T \to Pic^0(T) = T^\vee, \ \varphi_L(x) = \text{isom. class of } T_x^* L \otimes L^{-1},$$

where $T_x : T \to T$ is the translation with $x \in T$ (see [8]). The T-principal bundle $X \xrightarrow{\pi} B$ being fixed, we can associate to any line bundle $L \in Pic(T)$ an element in $H^1(\mathcal{O}_B(T^\vee))$ in the following way: For the Čech 1-cocycle (ξ_{ij}) defining our T-principal bundle, $\xi_{ij} : U_i \cap U_j \to T$, we put

$$\eta_{ij}^L := \varphi_L \circ \xi_{ij} : U_i \cap U_j \to T^\vee.$$

Then (η_{ij}^L) is a Čech 1-cocycle (φ_L is a homomorphism) and defines a cohomology class in $H^1(\mathcal{O}_B(T^\vee))$, denoted by $\tilde{\varphi}_L(\xi)$.

Definition Let $\xi \in H^1(\mathcal{O}_B(T))$ be fixed. For any $L \in Pic(T)$ the $T^\vee$-principal bundle described by $\tilde{\varphi}_L(\xi)$ will be called the *associated* $T^\vee$*-bundle to* L.

Lemma 3. *Let $L \in Pic(T)$ be a line bundle. Then, the obstruction to extend L to a line bundle on the total space of the fixed T-principal bundle $X \xrightarrow{\pi} B$ is the associated $T^\vee$-bundle to L, $\tilde{\varphi}_L(\xi)$.*

Proof: Let $\mathcal{L}_i$ be a line bundle on $U_i \times T$ such that for each point $x \in U_i$, we have

(14) $$c_1(\mathcal{L}_i|_{x \times T}) = c_1(L).$$

Then, for each point $x \in U_i$,

$$\mathcal{M}_x = (\mathcal{L}_i|_{x \times T}) \otimes L^{-1}$$

is a line bundle of degree zero on T, hence determines a point of $Pic^0(T) = T^\vee$. In this way, the line bundle $\mathcal{L}_i$ defines a holomorphic mapping

$$\varphi_i : U_i \to T^\vee,$$

such that the line bundle

(15) $$p_i^*(L) \otimes (\varphi_i \times id_T)^*(\mathcal{P})$$

is isomorphic to $\mathcal{L}_i$, where $p_i : U_i \times T \to T$ is the natural projection to the second factor and $\mathcal{P}$ is the Poincaré bundle of $T^\vee$ (which is a line bundle on $T^\vee \times T$). Conversely, if a holomorphic mapping $\varphi_i : U_i \to T^\vee$ is given, then (15) defines a line bundle $\mathcal{L}_i$ on $U_i \times T$

with the property (14). Patching together $\mathcal{L}_i$'s to obtain a line bundle on X, we need to have isomorphisms

$$T^*_{\xi_{ij}}\mathcal{L}_j|_{U_{ij}\times T} \cong \mathcal{L}_i|_{U_{ij}\times T} \tag{16}$$

for all $U_{ij} = U_i \cap U_j \neq \emptyset$, where $T_{\xi_{ij}}$ is an automorphism of $U_{ij} \times T$ induced by the translation of T by $\xi_{ij}(x)$ for each $x \in U_{ij}$.

Since we may assume that $\mathcal{L}_i$ has the form (15), the isomorphism (16) can be rewritten as

$$T^*_{\xi_{ij}}(p_j^*L) \otimes (\varphi_j \times id_T)^*(\mathcal{P})|_{U_{ij}\times T} \cong (p_i^*L) \otimes (\varphi_i \times id_T)^*(\mathcal{P})|_{U_{ij}\times T}. \tag{17}$$

Note that for any line bundle M of degree zero on T, we have an isomorphism $T_a^*M \cong M$ for any translation T_a of the torus T.

On the other hand, for each $x \in U_{ij}$, the line bundle

$$T^*_{\xi_{ij}(x)}(L) \otimes L^{-1}$$

defines an element of $T^\vee$ and we have a holomorphic mapping of U_{ij} to $T^\vee$. This holomorphic mapping is nothing but

$$\eta^L_{ij} = \varphi_L \circ \xi_{ij} : U_{ij} \longrightarrow T^\vee.$$

Then, the existence of an isomorphism (17) is equivalent to the equality

$$\eta^L_{ij} + \varphi_j = \varphi_i \tag{18}$$

as the equality in $H^0(U_{ij}, \mathcal{O}_{U_{ij}}(T^\vee))$.

If there exists a line bundle $\mathcal{L}$ on X such that for a point $y \in B$, $\mathcal{L}|_{\pi^{-1}(y)}$ is isomorphic to L, then

$$\mathcal{L}_i := \mathcal{L}|_{U_i\times T}, \quad i \in I$$

satisfy (14) and (16). Therefore, the equality holds for (i, j) with $U_{ij} \neq \emptyset$. Hence, the cocycle $\tilde{\varphi}_L(\xi)$ is zero in $H^1(B, \mathcal{O}_B(T^\vee))$.

Conversely, if $\tilde{\varphi}_L(\xi)$ is zero in $H^1(B, \mathcal{O}_B(T^\vee))$, by choosing a suitable open covering $\{U_i\}$ of B, we may assume that the equality (18) holds. Define a line bundle $\mathcal{L}_i$ on $U_i \times T$ by

$$\mathcal{L}_i = p_i^*L \otimes (\varphi_i \times id_T)^*(\mathcal{P}).$$

By (18) we have an isomorphism

$$g_{ij} : \mathcal{L}_j|_{U_{ij}\times T} \longrightarrow \mathcal{L}_i|_{U_{ij}\times T}.$$

Note that g_{ij} is uniquely determined up to the multiplication by an element of $H^0(U_{ij}, \mathcal{O}^*_{U_{ij}})$. For $i < j$ choose an isomorphism g_{ij} and fix it. Put

$$g_{ji} = g_{ij}^{-1}, \quad i < j$$
$$g_{ii} = id.$$

For $U_{ijk} = U_i \cap U_j \cap U_k \neq \emptyset$, put

$$g_{ijk} = g_{ki} \circ g_{ij} \circ g_{jk}.$$

Since there is a canonical isomorphism of $\mathrm{Aut}(\mathcal{L}|_{\pi^{-1}(U)})$ to $H^0(\pi^{-1}(U), \mathcal{O}^*_{\pi^{-1}(U)}) = H^0(U, \mathcal{O}^*_U)$, the automorphism g_{ijk} of $\mathcal{L}_k|_{U_{ijk}\times T}$ determines and element $\sigma(g_{ijk}) \in H^0(U_{ijk}, \mathcal{O}^*_{U_{ijk}})$. Note that we have equalities:

$$\sigma(g_{\ell k} \circ g_{ijk} \circ g_{k\ell}) = \sigma(g_{ijk}) \quad \text{on } U_{ijk\ell}$$
$$\sigma(g_{ijk} \circ g_{\ell mk}) = \sigma(g_{ijk})\sigma(g_{\ell mk}) \quad \text{on } U_{ijk\ell m}.$$

By using these equalities, it is easy to show that $\{\sigma(g_{ijk})\}$ is a two-cocycle with values in $\mathcal{O}^*_B$. Since we have $H^2(B, \mathcal{O}^*_B) = 0$, if necessarily, by choosing a finer open covering of B

and changing the isomorphism g_{ij} by the multiplication by a nowhere vanishing function, we may assume that

$$\sigma(g_{ijk}) = 1.$$

This means that $g_{ijk} = id$ and we can patch together the line bundles $\mathcal{L}_i$ by the isomorphism g_{ij} to obtain a line bundle $\mathcal{L}$ on X. We may also assume that for a point $x \in U_i$ we have $\varphi_i(x) = 0$. Then, we have an isomorphism $\mathcal{L}|_{\pi^{-1}(x)} \cong L$. This proves the lemma. □

Lemma 4. *The homomorphism* β : $\ker(i'') \to \operatorname{coker}(i')$ *is given by the correspondence* $c_1(L) \longmapsto \tilde{\varphi}_L(\xi)$.

Proof: Let $L \in Pic(T)$ be a line bundle. By Appel-Humbert Theorem (see [8], Chap. I, 2) one has $L = L(H, \alpha)$, where H is a hermitian form on V with $E(\Lambda \times \Lambda) \subset \mathbb{Z}$ ($E = \operatorname{Im} H$) and $\alpha : \Lambda \to U(1)$ is a map with

$$\alpha(\lambda_1 + \lambda_2) = e^{i\pi E(\lambda_1, \lambda_2)} \alpha(\lambda_1)\alpha(\lambda_2), \quad \lambda_i \in \Lambda.$$

Let us denote by p the canonical projection $V \to T$. By [8], Chap. II, 9, if $a \in V$ with $p(a) = x \in T$, we have

$$\varphi_{L(H,\alpha)}(x) = \text{isom. class of } L(0, \gamma_a),$$

where $\gamma_a : \Lambda \to U(1)$ is the map

(19) $$\gamma_a(\lambda) = e^{2\pi i E(a,\lambda)}, \quad \lambda \in \Lambda.$$

From the exact sequence (12) we get

$$\operatorname{coker}(i') \cong \ker(H^1(\mathcal{O}_B(T^\vee)) \xrightarrow{c^\vee} H^2(B, \Lambda^\vee)).$$

By the previous lemmas it remains to show that the condition $c_1(L)(\lambda^0) = 0$ implies the condition $c^\vee(\eta) = 0$, where $\eta = \tilde{\varphi}_L(\xi)$. For any $z \in U_i \cap U_j$ we choose $a_{ij}(z) \in V$ such that $p(a_{ij}(z)) = \xi_{ij}(z) \in T$. Then

$$\eta^L_{ij}(z) = \varphi(\xi_{ij}(z)) = L(0, \gamma_{a_{ij}(z)}),$$

where $\gamma_{a_{ij}(z)}$ is given by the formula (19) for $c_1(L) = E$. Since (ξ_{ij}) is a cocycle we have $a_{jk}(z) - a_{ik}(z) + a_{ij}(z) \in \Lambda$. More precisely, we have

$$cls(a_{jk}(z) - a_{ik}(z) + a_{ij}(z)) = m\lambda^0 = c(\xi) \in \Lambda = H^2(B, \Lambda).$$

Let us denote by $p^\vee$ the canonical projection $\overline{V}^\vee \to T^\vee$ and recall that

$$\overline{V}^\vee = Hom_{\mathbb{C}-antilin.}(V, \mathbb{C}).$$

If $l \in \overline{V}^\vee$ then $p^\vee(l) = L(0, \alpha_l)$, where $\alpha_l : \Lambda \to U(1)$ is the map

$$\alpha_l(\lambda) = e^{2\pi i \operatorname{Im} l(\lambda)}, \quad \lambda \in \Lambda,$$

(see [8], Chap. II, 9). In order to define $c^\vee(\eta)$ in Čech cohomology we can choose $l_{ij;z} \in \overline{V}^\vee$ such that

$$\operatorname{Im} l_{ij;z} = E(a_{ij}(z), \cdot).$$

Then, the characteristic class $c^\vee(\eta)$ is given by the 2-cocycle $(\rho_{ijk;z})$,where

$$\rho_{ijk;z} = l_{jk;z} - l_{ik;z} + l_{ij;z} \in \Lambda^\vee = H^2(B, \Lambda^\vee).$$

But, for all $\lambda \in \Lambda$, we have

$$\operatorname{Im} \rho_{ijk;z}(\lambda) = E(a_{jk}(z) - a_{ik}(z) + a_{ij}(z), \lambda) = E(m\lambda^0, \lambda) = 0.$$

Since a linear form $l \in \overline{V}^\vee$ is uniquely determined by its imaginary part, we get $c^\vee(\eta) = 0$ in $H^2(B, \Lambda^\vee)$. □

We have proved the following result:

Theorem 5. *Let $X \xrightarrow{\pi} B$ be a T-principal bundle over the curve B, defined by a cohomology class $\xi \in H^1(\mathcal{O}_B(T))$ with $c(\xi) \neq 0$ (i.e. X is non-Kähler). Then we have an exact sequence of free abelian groups*

$$0 \to Hom(J_B, T^\vee) \to NS(X)/TorsNS(X) \to \tilde{N}(X) \to 0,$$

where $\tilde{N}(X)$ is the subgroup of the Neron-Severi group of the torus T defined by

$$\tilde{N}(X) = \{c_1(L) \in NS(T) \mid \tilde{\varphi}_L(\xi) \text{ is the trivial torus bundle}\}.$$

Remark. In the case T is an elliptic curve we have $\tilde{N}(X) = 0$ (see [3]).

Remark. Clearly, a similar result holds in the case of a Kähler torus principal bundle for the group $NS(X)/\pi^* NS(B)$ (see also the last section).

Example. Let T be a two-dimensional complex torus with period matrix Ω, where

$$\Omega^t = \begin{pmatrix} 1 & 0 & \tau_1 & \alpha \\ 0 & 1 & 0 & \tau_2 \end{pmatrix}$$

with $\operatorname{Im} \tau_j > 0, \quad j = 1, 2$. If the complex numbers τ_1, τ_2, α are algebraically independent over the rational numbers $\mathbb{Q}$ then, it is well-known that T is not algebraic, that is, T is not an abelian variety. Let E_j be an elliptic curve with period matrix $(1, \tau_j), \quad j = 1, 2$. Then, there exists a holomorphic mapping

$$\pi : T \to E_2$$

such that π is an E_1-principal bundle over E_2.

The lattice Λ of T is generated by vectors $(1,0), (0,1), (\tau_1, 0), (\alpha, \tau_2)$. Put $\lambda^0 = (\tau_1, 0)$. Choose a point b of a curve B and make a logarithmic transformation to obtain a T-principal bundle

$$X = L_b(m\lambda^0, 1)(B \times T),$$

where m is an arbitrary positive integer. Then, we have $c(X) = m\lambda^0$ and X is non-Kähler.

Since the second coordinate of λ^0 is zero, there exists a holomorphic mapping

$$\mu : X \to B \times E_2.$$

Then, any line bundle L on T, which is the pull-back of a line bundle L_2 on E_2 by π, can be extended holomorphically to the one on X, since L_2 can be extended to a line bundle on $B \times E_2$. Hence, for our T-principal bundle X, we have $\tilde{N}(X) \neq 0$.

Similarly, we can also construct a T-principal bundle over B with $\tilde{N}(X) \neq 0$ from a period matrix Ω

$$\Omega^t = \begin{pmatrix} I_m & 0 & \tau_m & \alpha \\ 0 & I_n & 0 & \tau_n \end{pmatrix},$$

where $(I_m, \tau_m)^t$ and $(I_n, \tau_n)^t$ are period matrix of tori and α is an $m \times n$ matrix.

4. A filtration on $Pic(X)$

In this section we reinterpret the results in the previous section geometrically. We use freely the notation in the previous section. Let $\pi : X \to B$ be a T-principal bundle as in the previous section. Choose a general point $b \in B$ and fix it. In the following we identify the torus T with the fiber $\pi^{-1}(b)$. Restricting a line bundle $\mathcal{L}$ on X to the fiber $\pi^{-1}(b)$, we have a natural group homomorphism

$$Pic(X) \xrightarrow{r} Pic(\pi^{-1}(b)) = Pic(T). \tag{20}$$

Then $\ker(r)$ consists of isomorphism classes of line bundles whose restriction to the fibre $\pi^{-1}(b)$ is trivial, hence the restriction to each fiber of π is a line bundle of degree 0 on the torus under identification of the torus with each fiber.

Let $\{U_j\}$ be an open covering of B with trivialization

$$\pi^{-1}(U_j) \simeq U_j \times T. \tag{21}$$

For each line bundle $\mathcal{L}$ belonging to $\ker(r)$ there exists a holomorphic mapping

$$\varphi_j : U_j \to Pic^0(T) = T^\vee$$

with

$$\mathcal{L}|_{\pi^{-1}(U_j)} \simeq (\varphi_j \times id_T)^*(\mathcal{P}),$$

where $\mathcal{P}$ is the Poincaré bundle on $Pic^0(T) \times T$. Since any line bundle of degree 0 on the torus is invariant by the translations, on $U_j \cap U_k \neq \emptyset$ we have

$$\varphi_j = \varphi_k.$$

Hence, the line bundle $\mathcal{L}$ defines a holomorphic mapping

$$\varphi : B \to T^\vee. \tag{22}$$

Since the restriction $\mathcal{L}|_{\pi^{-1}(b)}$ is trivial, the holomorphic mapping (22) satisfies

$$\varphi(b) = [0]. \tag{23}$$

The line bundle $\mathcal{L}$ and the holomorphic mapping φ are related by

$$\mathcal{L} \simeq \pi^*(M) \otimes \varphi^*(\mathcal{P}),$$

where M is a line bundle on the curve B and $\varphi^*(\mathcal{P})$ is the line bundle on X whose restriction to $\pi^{-1}(U_j)$ is $(\varphi_j \times id_T)^*(\mathcal{P})$. Note that by the argument of the proof of Lemma 3 we can patch together $(\varphi_j \times id_T)^*(\mathcal{P})$'s to get $\varphi^*(\mathcal{P})$, since the line bundle of degree 0 on a torus is invariant under the translations. Also note that there is a one to one correspondence between the set of holomorphic mappings (22) with property (23) and $Hom(J_B, T^\vee)$.

Let us consider a group homomorphism

$$R : Pic(X) \xrightarrow{r} Pic(T) \xrightarrow{c} H^2(T, \mathbb{Z}). \tag{24}$$

The homomorphism R is essentially equivalent to a natural homomorphism

$$Pic(X) \to Pic(T)/Pic^0(T)$$

induced by the homomorphism r. A line bundle $\mathcal{L}$ belonging to $\ker(R)$ is the one whose restriction to each fiber of π is of degree 0. Note that by the proof of Lemma 3 each line bundle $L \in Pic^0(T)$ can be extended to a line bundle $\mathcal{L}$ on X in such a way that its restriction to each fiber is isomorphic to L. Hence, there is an isomorphism

$$\ker(R)/\ker(r) \simeq Pic^0(T). \tag{25}$$

Define subgroups P_j of $Pic(X)$ by

$$P_2 = \pi^* Pic(B), \quad P_1 = \ker(r), \quad P_0 = Pic(X). \tag{26}$$

Then, $\{P_\bullet\}$ defines an decreasing filtration of $Pic(X)$. By the above consideration and the arguments of the previous section we have the following theorem.

Theorem 6. *We have the following isomorphisms.*

$$P_1/P_2 \simeq Hom(J_B, Pic^0(T)) \tag{27}$$

$$P_0/P_1 \simeq \{L \in Pic(T) \mid \tilde{\varphi}_L(\xi) = 0\} \tag{28}$$

where $\xi \in H^1(B, \mathcal{O}_B(T))$ *is the cohomology class corresponding to the* T*-principal bundle* $\pi : X \to B$ *and* $\tilde{\varphi}_L(\xi) = 0$ *is defined in §3.*

Remark. Taking the Chern classes, we have

$$c_1(P_2) = F_2, \quad c_1(P_1) = F_1. \tag{29}$$

5. Neron-Severi group for torus quasi bundles

Let $T = V/\Lambda$ be an n-dimensional torus. By a quasi T-bundle $\pi : X \to B$ over a curve B we mean that π is a T-principal bundle over $B \setminus \{b_1, b_2, \dots, b_\ell\}$ and that the fiber $\pi^{-1}(b_j)$ over the point b_j is of the form $m_j T_j$ where $m_j \geq 2$ and T_j is a torus. The fiber $m_j T_j$ is called a multiple fiber of the multiplicity m_j. To construct such a quasi T-bundle we first generalize the notion of logarithmic transformation.

Choose points $b_1, b_2, \dots, b_k$ on B and put $B' = B \setminus \{b_1, b_2, \dots, b_k\}$. For each point b_i fix a positive integer m_i. We let a_i be an element of $\frac{1}{m_i}\Lambda$ such that the order of the point $[a_i]$ of the torus T corresponding to a_i is precisely m_i. Let

$$D_i = \{t_i \in \mathbb{C} \mid |t_i| < \epsilon\}$$

be a coordinate neighbourhood of the point b_i and put

$$\widehat{D_i} = \{s_i \in \mathbb{C} \mid |s_i| < \epsilon^{1/m_i}\}.$$

By the mapping

$$\begin{aligned} \lambda_i : \widehat{D_i} &\to D_i \\ s_i &\mapsto s_i^{m_i}, \end{aligned} \tag{30}$$

$\widehat{D_i}$ is an m_i-sheeted ramified covering of D_i. A holomorphic mapping $g_i : \widehat{D_i} \times T \to \widehat{D_i} \times T$ defined by

$$g_i : (s_i, [\zeta]) \mapsto (e_{m_i} s_i, \ [\zeta + a_i]) \tag{31}$$

is an analytic automorphism of order m_i and generates the cyclic group $G_i = \langle g_i \rangle$ of order m_i where

$$e_{m_i} = \exp(2\pi\sqrt{-1}/m_i).$$

Since the automorphism g_i has no fixed points, the quotient $\widehat{D_i} \times T/G_i$ is a complex manifold. Let

$$\mu_i : \widehat{D_i} \times T \to \widehat{D_i} \times T/G_i \tag{32}$$

be the canonical quotient mapping. By $[s_i, [\zeta]]$ we denote the point of the quotient manifold $\widehat{D_i} \times T/G_i$ corresponding to a point $(s_i, [\zeta])$ of $\widehat{D_i} \times T$. We have a holomorphic mapping

$$\begin{aligned} \pi_i : \widehat{D_i} \times T/G_i &\to D_i \\ [s_i, [\zeta]] &\mapsto s_i^{m_i}. \end{aligned}$$

Over the punctured disk D_i^* the holomorphic mapping π_i gives a T-principal bundle, and over the origin 0 the equation

$$\pi_i = 0$$

defines a divisor of a form m_iT_i where $T_i = T/\langle[a_i]\rangle$ is a torus obtained as the quotient by a finite subgroup generated by the point $[a_i]$.

The mapping

$$\ell_{a_i} : \widehat{D}_i^* \times T/G \to D_i^* \times T$$

(33) $$[s_i, [\zeta]] \mapsto (s_i^m, [\zeta - \frac{m_i a_i}{2\pi\sqrt{-1}} \log s_i])$$

is a well-defined holomorphic mapping and isomorphic. Therefore, we can patch together $\widehat{D}_i \times T/G_i$, $i = 1, 2, \ldots, k$ and $B' \times T$ by the isomorphisms ℓ_{a_i} to obtain a compact complex manifold X which is denoted by

(34) $$L_{b_1}(a_1, m_1)L_{b_2}(a_2, m_2) \cdots L_{b_k}(a_k, m_k)(B \times T)$$

and is called the manifold obtained from $B \times T$ by means of logarithmic transformations. There is a natural holomorphic mapping $\pi : X \to B$ given by π_i on $\widehat{D}_i \times T/G_i$ and the projection to the first factor on $B' \times T$. The fiber space $\pi : X \to B$ is a T-principal bundle over B' and has multiple fibres with multiplicity m_i, if $m_i \geq 2$. In the Appendix we shall show that all quasi T-bundles are obtained in this manner.

In the following let us consider a quasi T-bundle $\pi : X \to B$ of the form (34) and we assume that

$$m_i \geq 2, \quad i = 1, 2, \ldots, \ell, \quad m_{\ell+1} = \cdots = m_k = 1.$$

Let us consider geometrically line bundles on X. Choose a general point b and consider a natural restriction homomorphism

(35) $$r : Pic(X) \to Pic(\pi^{-1}(b)) = Pic(T)$$

Let us first consider the structure of $\ker(r)$. Note that for the multiple fiber m_iT_i the line bundle $[T_i]$ associated with the divisor T_i of X is an element of $\ker(r)$ and $[T_i]^{\otimes m_i} = [m_iT_i]$ is the pull-back of the line bundle $[b_i]$ on the curve B.

Let P_2 be a subgroup of $Pic(X)$ generated by $\pi^*Pic(B)$ and $[T_i]$, $i = 1, 2, \ldots, \ell$. A line bundle $\mathcal{L}$ belonging to P_2 is characterized by the fact that the restriction of $\mathcal{L}$ to each fiber $\pi^{-1}(c)$, $c \in B'$ is the trivial line bundle.

To a line bundle $\mathcal{L} \in \ker(r)$, by the same argument as in §4, we can associate a holomorphic mapping

$$\varphi' : B' \to Pic^0(T) = T^\vee.$$

The pull-back $\mu_i^*(\mathcal{L}|_{\pi^{-1}(D_i)})$ defines also a holomorphic mapping

$$\widehat{\varphi}_i : \widehat{D}_i \to Pic^0(T),$$

where $\mu_i : \widehat{D} \times T \to \pi^{-1}(D_i) = \widehat{D}_i \times T/G_i$ is a natural quotient mapping (32). Then, on $\widehat{D}_i^*$ we have

$$\widehat{\varphi}_i = \varphi' \circ \lambda_i,$$

where $\lambda_i : \widehat{D}_i \to D_i$ is defined in (30). This implies that the holomorphic mapping φ' can be extended to a holomorphic mapping

(36) $$\varphi : B \to Pic^0(T) = T^\vee.$$

As $\mathcal{L}|_{\pi^{-1}(b)}$ is a trivial bundle, we have

(37) $$\varphi(b) = [0].$$

Note that the set of holomorphic mappings (36) with property (37) are canonically isomorphic to $Hom(J_B, Pic^0(X))$. If $\mathcal{L}$ and $\mathcal{M}$ in $\ker(r)$ give the same holomorphic mapping

(36), then the restriction of the line bundle $\mathcal{L} \otimes \mathcal{M}^{-1}$ to each fiber $\pi^{-1}(c)$, $c \in B'$ is the trivial bundle, hence is an element of P_2.

Lemma 7. *There exists a natural group isomorphism*

$$j : \ker(r)/P_2 \simeq Hom(J_B, Pic^0(T)). \tag{38}$$

Proof: To each line bundle $\mathcal{L} \in \ker(r)$ we can associate a holomorphic mapping (36) with property (37). This defines an element of $Hom(J_B, Pic^0(T))$. If the mapping φ gives the zero element of $Hom(J_B, Pic^0(T))$, φ is the zero map. Hence, the restriction of $\mathcal{L}$ to each fiber $\pi^{-1}(c)$, $c \in B'$ is the trivial bundle. Hence, $\mathcal{L}$ belongs to P_2. This shows the injectivity.

Conversely, let $\varphi : B \to T^{\vee}$ be a non-constant holomorphic mapping with $\varphi(b) = [0]$. Then, on $X' = \pi^{-1}(B')$ we can construct a line bundle $\mathcal{L}'$ such that $\mathcal{L}'|_{\pi^{-1}(c)}$ is a line bundle of degree zero corresponding to the point $\varphi(c)$ for each $c \in B'$. For $\widehat{D}_i$, $i = 1, 2 \ldots, k$, put

$$\widehat{\varphi}_i = \varphi \circ \lambda_i.$$

Then, $\widehat{\varphi}_i$ defines a line bundle $\widehat{\mathcal{L}}_i$ such that $\widehat{\mathcal{L}}_i|_{s_i \times T}$ corresponds to $\widehat{\varphi}_i(s_i)$. As the line bundle $\widehat{\mathcal{L}}_i$ is invariant under the group G_i, it defines a line bundle $\mathcal{L}_i$ on $\widehat{D}_i \times T/G_i$. By our construction, $\mathcal{L}_i|_{\pi^{-1}(D_i^*)}$ and $\mathcal{L}'|_{\pi^{-1}(D_i^*)}$ are isomorphic. Hence, $\mathcal{L}_i$'s and $\mathcal{L}'$ define a line bundle $\mathcal{L}$ on X which corresponds to the mapping φ. This shows the surjectivity of the mapping j. □

Next let us consider the image of the homomorphism r.

Lemma 8. *If a line bundle L of T can be extended to a line bundle $\mathcal{L}$ on X, then L is invariant by the translations $T_{[a_i]}$, $i = 1, 2, \ldots, \ell$.*

Proof: The pull-back $\widetilde{\mathcal{L}}_i := \mu_i^*(\mathcal{L}|_{\pi_i^{-1}(D_i)})$ is invariant by the action of the group G_i, where $\mu_i : \widehat{D}_i \times T \to \widehat{D}_i \times T/G_i = \pi^{-1}(D_i)$ is the natural quotient mapping (32). In particular, the restriction $\widetilde{\mathcal{L}}_i|_{0 \times T}$ is invariant by the group generated by the translation $T_{[a_i]}$. Since $\widetilde{\mathcal{L}}_i|_{0 \times T}$ has a form $L \otimes M$ with degree zero line bundle M on T and M is invariant by all the translations, the line bundle L is invariant by the translation $T_{[a_i]}$. □

Let H be a subgroup of the torus T generated by $[a_1], [a_2], \ldots, [a_\ell]$. The group H is isomorphic to Λ_0/Λ where Λ_0 is the lattice generated by Λ and a_i's. To any H-invariant line bundle L on the torus T, we associate a cohomology class $\{\eta_{ij}^L\}$ in $H^1(B, \mathcal{O}_B(T^{\vee}))$ as follows.

Let $\{U_j\}$ be an open covering of the curve B such that $U_i = D_i$ for $i = 1, 2, \ldots, \ell$ and that $b_i \notin U_i \cap U_j$ for $i \neq j$. Since the line bundle L is invariant by the translation $T_{[a_i]}$, though $[\frac{a_i}{2\pi\sqrt{-1}} \log t_i]$ is multivalued

$$T^*_{[\frac{a_i}{2\pi\sqrt{-1}} \log t_i]} L \otimes L^{-1} \tag{39}$$

is a well-defined line bundle on $\pi^{-1}(U_i \cap U_j)$ for $i = 1, 2, \ldots, \ell$ and $j \neq i$. Then there exists a holomorphic mapping φ_{ij} from $U_{ij} = U_i \cap U_j$ to $T^{\vee}$ such that the line bundle (39) is the pull-back $(\varphi_{ij} \times id_T)^*(\mathcal{P})$ of the Poincaré bundle. Put

$$\eta_{ij}^L := \begin{cases} \varphi_{ij} & \text{if } 1 \le i \le \ell,\ \ell < j \\ 0 & \text{if } \ell < i, j \end{cases}. \tag{40}$$

Then, it is easy to show that $\{\eta_{ij}^L\}$ is a one cocycle and defines a cohomology class $[\{\eta_{ij}^L\}] \in H^1(B, \mathcal{O}_B(T^{\vee}))$.

Lemma 9. *An H-invariant line bundle L on the torus $T = \pi^{-1}(b)$ can be extended to the one on X if and only if the cohomology class $[\{\eta_{ij}^L\}]$ is zero.*

Proof: Assume that there exists a line bundle $\mathcal{L}$ on X which is an extension of L. Then, the pull-back $\mu_i^*(\mathcal{L}|_{\pi^{-1}(U_i)})$ of the restriction of $\mathcal{L}$ on $\pi^{-1}(U_i)$, $i = 1, 2, \ldots, \ell$, to $\widehat{D}_i \times T$ can be expressed as

$$L \otimes (\widehat{\varphi}_i \times id_T)^*(\mathcal{P}), \tag{41}$$

where $\widehat{\varphi}_i : \widehat{D}_i \longrightarrow T^\vee$ is a holomorphic mapping. Since the line bundle $\mu_i^*(\mathcal{L}|_{\pi^{-1}(U_i)})$ is invariant under the group G_i, we have

$$\widehat{\varphi}_i(s_i) = \widehat{\varphi}_i(e_{m_i} s_i).$$

Hence, there exists a holomorphic mapping $\varphi_i : U_i \longrightarrow T^\vee$ with

$$\widehat{\varphi}_i(s_i) = \varphi_i(s_i^{m_i}). \tag{42}$$

Since $\mathcal{L}$ is a global line bundle, on $U_{ij} \neq \emptyset$ we have

$$T^*_{[\frac{a_i}{2\pi\sqrt{-1}} \log t_i]} L \otimes (\varphi_i \times id_T)^*(\mathcal{P}) = L \otimes (\varphi_j \times id_T)^*(\mathcal{P}). \tag{43}$$

This implies that we have

$$\eta_{ij}^L = \varphi_j - \varphi_i. \tag{44}$$

Hence, the cohomology class is zero.

Conversely assume that the cohomology class is zero, hence we have holomorphic mappings $\varphi_j : U_j \longrightarrow T^\vee$ which satisfy (44). For $i = 1, 2, \ldots, \ell$ define $\widehat{\varphi}_i$ by (42). Then the line bundle $\widehat{\mathcal{L}}_i = L \otimes (\widehat{\varphi}_i \times id_T)^*(\mathcal{P})$ is invariant by the action of the group G_i, hence defines a line bundle $\mathcal{L}_i$ on $\pi^{-1}(U_i)$. For $j > \ell$ put $\mathcal{L}_j = L \otimes (\varphi_j \times id_T)^*(\mathcal{P})$. Since we have the equality (43), we can patch together these line bundles and obtain a line bundle $\mathcal{L}$ which is an extension of L. □

Now as in §4 we introduce a decreasing filtration $\{P_\bullet\}$ of $Pic(X)$ by

$$P_2 = \text{the subgroup generated by } \pi^* Pic(B) \text{ and } [T_i]\text{'s}, \tag{45}$$

$$P_1 = \ker(r), \quad P_0 = Pic(X), \tag{46}$$

where $m_i T_i$, $i = 1, 2, \ldots, \ell$ are all the multiple fibers of the quasi T-bundle $\pi : X \longrightarrow B$. By the above arguments we have the following theorem.

Theorem 10. *We have the following isomorphisms.*

$$P_1/P_2 \simeq Hom(J_B, Pic^0(T)) \tag{47}$$

$$P_0/P_1 \simeq \{L \in Pic(T)^H \mid [\{\eta_{ij}^L\}] = 0\}. \tag{48}$$

Let us reinterpret the group $\{L \in Pic(T)^H \mid [\{\eta_{ij}^L\}] = 0\}$ by means of a torus principal bundle associated with the quasi T-bundle $\pi : X \rightarrow B$.

Let Λ_0 be a lattice in the vector space V generated by Λ and a_i, $i = 1, 2, \ldots, \ell$ and put

$$T_0 = V/\Lambda_0. \tag{49}$$

Then, we have

$$T_0 = T/H,$$

where H is a subgroup of the torus T generated by $[a_1], [a_2], \ldots, [a_\ell]$. There is a canonical surjective homomorphism

$$h : T \longrightarrow T_0 \tag{50}$$

of complex tori. The following lemma is well-known and easy to prove.

Lemma 11. *A line bundle L on the torus T is invariant by the translations $T_{[a_i]}$, $i = 1, 2, \ldots, \ell$, if and only if there exists a line bundle L_0 on T_0 with*

$$L = h^* L_0.$$

Put

$$Y = L_{b_1}(a_1, 1) L_{b_2}(a_2, 1) \cdots L_{b_\ell}(a_\ell, 1)(B \times T_0) \tag{51}$$

with structure morphism $\pi_0 : Y \to B$, which is a T_0-principal bundle.

Lemma 12. *There exists a holomorphic mapping*

$$f : X \to Y$$

such that the following diagram is commutative.

$$\begin{array}{ccc} X & \xrightarrow{f} & Y \\ {\scriptstyle \pi}\downarrow & & \downarrow{\scriptstyle \pi_0} \\ B & = & B \end{array}$$

Moreover, f is unramified outside the multiple fibers.

Proof: There is a natural unramified holomorphic mapping

$$f' : B' \times T \to B' \times T_0.$$

We need to show that f' can be extended to a holomorphic mapping f of X to Y. On $\widehat{D}_i \times T/G_i$ let us define a holomorphic mapping f_i by

$$\begin{aligned} f_i : \widehat{D}_i \times T/G_i &\to D_i \times T_0 \\ [s_i, [\zeta]] &\mapsto (s_i^{m_i}, h([\zeta])). \end{aligned}$$

We need to show that these holomorphic mappings are compatible to f'. By our definition of the logarithmic transformation we have the following commutative diagram.

$$\begin{array}{ccccc} \ell_i : \widehat{D}_i^* \times T/G_i & & \longrightarrow & & D_i^* \times T \\ \Big| & [s_i, [\zeta]] & \mapsto & (s_i^{m_i}, [\zeta - \frac{m_i a_i}{2\pi\sqrt{-1}} \log s_i]) & \Big| \\ f' \Big| & \downarrow & & \downarrow & \Big| f_i \\ \downarrow & (s_i^{m_i}, [\zeta]_0) & \mapsto & (s_i^{m_i}, [\zeta - \frac{a_i}{2\pi\sqrt{-1}} \log(s_i^{m_i})]_0) & \downarrow \\ \ell_i^{(0)} : D_i^* \times T_0 & & \longrightarrow & & D_i^* \times T_0 \end{array}$$

Here, $[\zeta]_0$ means the point of the torus T_0 corresponding to ζ. The commutativity of the above diagram shows that the mappings f' and f_i's are compatible and define a holomorphic mapping $f : X \to Y$ over B. □

Lemma 13. *The quasi T-bundle X is Kähler if and only if Y is Kähler. The condition is equivalent to the equality*

$$\sum_{i=1}^{k} a_i = 0 \tag{52}$$

Proof: Assume that the equality (52) holds, hence, Y is Kähler. Let ω be a Kähler form of Y. Note that $f : X \to Y$ is an abelian covering ramified along the support of T_i of the multiple fibers. Hence, the pull-back $f^*\omega$ is positive definite on $X \setminus \cup_{i=1}^{\ell} T_i$ and at each point of T_i it is positive semi-definite. Near the multiple fiber $m_i T_i$, X is isomorphic to $\widehat{D_i} \times T/G_i$. As a (1,1)-form

$$\frac{\sqrt{-1}}{2\pi}\partial\bar{\partial}(\sum_{\nu=1}^{n} |\zeta_\nu|^2 + |s_i|^2)$$

is G_i-invariant, it defines a Kähler form on $\widehat{D_i} \times T/G_i$. Let ρ_i be a non-negative C^∞-function in $|s_i|^2$ satisfying

$$\rho_i(t) = \begin{cases} 1 & |t| < \epsilon^{2/m_i}/3 \\ 0 & |t| \geq 2\epsilon^{2/m_i}/3. \end{cases}$$

Then, a form

$$\omega_i = \frac{\sqrt{-1}}{2\pi}\partial\bar{\partial}\{\rho_i(|s_i|^2)(\sum_{\nu=1}^{n} |\zeta_\nu|^2 + |s_i|^2)\}$$

is positive definite on $\pi^{-1}(D_i(\epsilon^{2/m_i}/3)$ and $\omega_i \equiv 0$ on $\pi^{-1}(D_i(2\epsilon^{2/m_i}/3)$, where we put $D_i(r) = \{s_i \mid |s_i| < r\}$. Hence, we may regard ω_i as a global (1,1)-form on X. Since $f^*\omega$ is positive definite on $X \setminus \cup_{i=1}^{\ell} T_i$, and ω_i is positive definite in a neighbourhood of T_i and zero outside a certain neighbourhood of T_i, the form

$$\alpha f^*\omega + \sum_{i=1}^{\ell} \omega_i$$

is positive definite on X, if we choose α sufficiently large. Hence, X is Kähler.

Conversely, assume that X is Kähler. Put

$$d = m_1 \cdot m_2 \cdots m_\ell, \quad m_0 = LCM\{m_1, m_2, \dots, m_\ell\}.$$

We can always find a d-fold abelian covering $\sigma : \widetilde{B} \to B$ of the curve B branched at b_1, $b_2, \dots, b_\ell$ and a point $b_0 \in B \setminus \{b_1, b_2, \dots, b_\ell\}$ such that σ has d/m_i ramification points $\{b_i^{(m)}\}$, $m = 1, 2, \dots, d/m_i$, $i = 0, 1, 2, \dots, \ell$. Over the points b_j, $\ell < j \leq k$, σ is unramified. Put $\sigma^{-1}(b_j) = \{b_j^{(1)}, b_j^{(2)}, \dots, b_j^{(d)}\}$. Then, the normalization $\widetilde{X}$ of $X \times_B \widetilde{B}$ has a natural structure of a principal T-bundle over $\widetilde{B}$ and it is isomorphic to

$$\prod_{i=1}^{k}\prod_{m=1}^{d/m_i} L_{b_i^{(m)}}(m_i a_i, 1)(\widetilde{B} \times T) \tag{53}$$

The natural holomorphic mapping $\widetilde{\sigma} : \widetilde{X} \to X$ is only branched over $\pi^{-1}(b_0)$. By the similar argument as above we can show that $\widetilde{X}$ is Kähler if X is Kähler. Then, by (52), $\widetilde{X}$ is Kähler if and only if

$$\sum_{i=1}^{k}\sum_{m=1}^{d/m_i} m_i a_i = 0.$$

The equality can be rewritten as

$$\sum_{i=1}^{k} \frac{d}{m_i} m_i a_i = d \sum_{i=1}^{k} a_i = 0.$$

Hence, the equality (52) holds and Y is also Kähler. This proves the lemma. □

Lemma 14. *The subgroup $\pi^* H^2(B,\mathbb{Z})$ of $H^2(X,\mathbb{Z})$ is a finite group if and only if*

$$\sum_{i=1}^{k} a_i \neq 0.$$

Proof: Since the holomorphic mapping $f : X \to Y$ is finite, $\pi^* H^2(B,\mathbb{Z})$ is finite if and only if the subgroup $\pi_0^* H^2(B,\mathbb{Z})$ in $H^2(Y,\mathbb{Z})$ is finite. The latter group is finite if and only if Y is non-Kähler. On the other hand, Y is non-Kähler if and only if

$$\sum_{i=1}^{k} a_i \neq 0.$$

This proves the lemma. □

Put

$$N(X) = \{L \in Pic(T)^H \mid [\{\eta_{ij}^L\}] = 0\} \tag{54}$$

$$N(Y) = \{L_0 \in Pic(T_0) \mid \tilde{\varphi}_{L_0}(\xi_0) = 0\} \tag{55}$$

where $\xi_0 \in H^1(B, \mathcal{O}_B(T_0))$ is the cohomology class corresponding to the T_0-principal bundle $\pi_0 : Y \to B$. Taking the dual of the homomorphism $h : T \to T_0$ (50) we have an exact sequence

$$0 \to T_0^\vee \xrightarrow{h^\vee} T^\vee \to H^\vee \to 0, \tag{56}$$

where $H^\vee$ is a finite abelian group. Sheafifying the exact sequence (56) and taking the cohomology, we obtain the following exact sequence.

$$0 \to H^1(B, \mathcal{O}_B(T_0^\vee)) \xrightarrow{h^\vee} H^1(B, \mathcal{O}_B(T^\vee)) \to H^1(B, H^\vee) \to . \tag{57}$$

Lemma 15. *For a line bundle L_0 on the torus T_0 put $L = h^* L_0$. Then we have*

$$h^\vee(\tilde{\varphi}_{L_0}(\xi_0)) = [\{\eta_{ij}^L\}].$$

Proof: We use the same open covering $\{U_j\}$ of the curve B defined above. Then, the cohomology class ξ_0 is given by a cocycle

$$\zeta_{ij} := \begin{cases} \frac{a_i}{2\pi\sqrt{-1}} \log t_i & \text{if } 1 \le i \le \ell,\ \ell < j \\ 0 & \text{if } \ell < i, j. \end{cases} \tag{58}$$

Hence $\tilde{\varphi}_{L_0}(\xi_0)$ is given by a cocycle

$$\zeta_{ij}^L := \begin{cases} \phi_{ij} & \text{if } 1 \le i \le \ell,\ \ell < j \\ 0 & \text{if } \ell < i, j \end{cases}$$

where ϕ_{ij} is given by

$$T^*_{[\frac{a_i}{2\pi\sqrt{-1}} \log t_i]} L_0 \otimes L_0^{-1} = (\phi_{ij} \times id_T)^*(\mathcal{P}_0).$$

Here $\mathcal{P}_0$ is the Poincaré bundle on $Pic^0(T_0) \times T_0$. Then it is easy to show that we have

$$h^\vee(\phi_{ij}) = \varphi_{ij}.$$

This is the desired result. □

Lemma 16.

$$h^*(N(Y)) = N(X).$$

Proof: For a line bundle $L_0 \in N(Y)$ we let $\mathcal{L}_0$ be a line bundle on Y which is an extension of L_0. Then, $f^*\mathcal{L}_0$ is a line bundle on X which is an extension of the line bundle h^*L_0, where $f : X \to Y$ is the holomorphic mapping in Lemma 12. Hence, we have $h^*(N(Y)) \subset N(X)$.

Conversely, take a line bundle $L \in N(X)$ and choose a line bundle L_0 on T_0 with $h^*L_0 = L$. By the above Lemma 15 and the exact sequence (57), $\tilde{\varphi}_{L_0}(\xi_0) = 0$. Hence, $L_0 \in N(Y)$. This shows $N(X) \subset h^*(N(Y))$. □

By the above argument and the arguments in the previous sections we have the following exact sequences.

(59) $$0 \to Hom(J_B, T^\vee) \to Pic(X)/P_2 \to N(X) \to 0$$

(60) $$0 \to Hom(J_B, T_0^\vee) \to Pic(Y)/\pi_0^* Pic(B) \to N(Y) \to 0.$$

Taking the Chern classes of the line bundles, finally we obtain the following theorem.

Theorem 17. *There exists an exact sequence*

(61) $$0 \to Hom(J_B, T^\vee) \to NS(X)/\widetilde{F}_2 \to \widetilde{N}(X) \to 0,$$

where $\widetilde{F}_2$ is a subgroup of $H^2(X, \mathbb{Z})$ generated by $c_1([T_i])$, $i = 1, 2, \dots, \ell$, and

(62) $$\widetilde{N}(X) = \{c_1(L) \mid L \in Pic(X)^H, \quad [\{\eta_{ij}^L\}] = 0\}.$$

The subgroup $\widetilde{F}_2$ is finite if and only if X is non-Kähler. Moreover, we have

$$\widetilde{N}(X) = h^*\widetilde{N}(Y)$$

where

$$\widetilde{N}(Y) = \{c_1(L_0) \mid L_0 \in Pic(Y), \quad \tilde{\varphi}_{L_0}(\xi_0) = 0\}.$$

Proof: To each homomorphism

$$\varphi \in Hom(J_B, T^\vee)$$

we can associate a line bundle $\mathcal{L}$ on X such that for each point $c \in B'$ the restriction $\mathcal{L}|_{\pi_{-1}}(c)$ corresponds to $\varphi(c)$. Let us consider the first Chern class $c_1(\mathcal{L})$ of $\mathcal{L}$. Note that we have an exact sequence

$$0 \to R^1\pi_*\mathbb{Z} \to R^1\pi_*\mathcal{O}_X \to \mathcal{O}_B(T^\vee) \to 0$$

and from this exact sequence we have the exact sequence

(63) $$\to H^0(B, R^1\pi_*\mathcal{O}_X) \to H^0(B, \mathcal{O}_B(T^\vee)) \xrightarrow{c} H^1(B, R^1\pi_*\mathbb{Z}) \to .$$

The element $\varphi \in Hom(J_B, \mathcal{O}_B(T^\vee))$ gives an element $\tilde{\varphi} \in H^0(B, \mathcal{O}_B(T^\vee))$ with $\tilde{\varphi}(b) = [0]$. Then the image of $c(\tilde{\varphi}) \in H^1(B, R^1\pi_*\mathbb{Z})$ to $H^2(X, \mathbb{Z})/\pi^*H^2(B, \mathbb{Z})$ is $c_1(\mathcal{L}) \mod \pi^*H^2(B, \mathbb{Z})$. Since we have an isomorphism

$$H^0(B, \mathcal{O}_B(T^\vee))/\operatorname{Im} H^0(B, R^1\pi_*\mathcal{O}_X) \simeq Hom(J_B, T^\vee),$$

by the exact sequence (63) we have an inclusion

$$Hom(J_B, \mathcal{O}_B(T^\vee)) \hookrightarrow H^1(B, R^1\pi_*\mathbb{Z}).$$

To show that the natural mapping

$$H^1(B, R^1\pi_*\mathbb{Z}) \to H^2(X, \mathbb{Z})/\pi^*H^2(B, \mathbb{Z})$$

is injective, we need to consider the spectral sequence

$$E_2^{p,q} = H^p(B, R^q\pi_*\mathbb{Z}) \Longrightarrow H^{p+q}(X, \mathbb{Z}).$$

By the dimension reason, we have

$$E_\infty^{0,2} = E_3^{0,2} = \ker(H^0(B, R^2\pi_*\mathbb{Z}) \to H^2(B, R^1\pi_*\mathbb{Z}))$$
$$E_\infty^{1,1} = E_2^{1,1} = H^1(B, R^1\pi_*\mathbb{Z})$$
$$E_\infty^{2,0} = E_3^{2,0} = \mathrm{coker}(H^0(B, R^1\pi_*\mathbb{Z}) \to H^2(B, \mathbb{Z})).$$

The spectral sequence defines the filtration $\{F_\bullet\}$ on the cohomology group $H^2(X, \mathbb{Z})$ such that there are canonical isomorphisms

(64) $$E_\infty^{2,0} \simeq F_2,$$

(65) $$E_\infty^{1,1} \simeq F_1/F_2,$$

(66) $$E_\infty^{0,2} \simeq F_0/F_1.$$

It is easy to see that $F_2 = \pi^* H^2(B, \mathbb{Z})$, hence by the above isomorphism (65) the natural mapping

$$H^1(B, R^1\pi_*\mathbb{Z}) \to H^2(X, \mathbb{Z})/\pi^* H^2(B, \mathbb{Z})$$

is injective. Therefore, the natural mapping

$$Hom(J_B, \mathcal{O}_B(T^\vee)) \to H^1(B, R^1\pi_*\mathbb{Z}) \to H^2(X, \mathbb{Z})/\pi^* H^2(B, \mathbb{Z})$$

is also injective. The rest of the statements follows from the above arguments. This proves the theorem. □

Remark. By the similar arguments as in [5, Chap. II, Lemma 1.6 and Lemma 7.3], the structure of the first homology group $H_1(X, \mathbb{Z})$ is given by

$$H_1(X, \mathbb{Z}) \simeq \mathbb{Z}\alpha_1 \oplus \cdots \oplus \mathbb{Z}\alpha_g \oplus \mathbb{Z}\beta_1 \oplus \cdots \oplus \mathbb{Z}\beta_g \oplus (\Lambda_0/(\sum_{i=1}^{k} a_i)),$$

where Λ_0 is the lattice in the vector space V generated by Λ and a_i's and

$$H_1(B, \mathbb{Z}) \simeq \mathbb{Z}\alpha_1 \oplus \cdots \oplus \mathbb{Z}\alpha_g \oplus \mathbb{Z}\beta_1 \oplus \cdots \oplus \mathbb{Z}\beta_g.$$

By virtue of Lemma 13, $H_1(X, \mathbb{Z})$ has torsion if and only if X is non-Kähler. Moreover, if X is non-Kähler, there is a non-canonical isomorphism

$$Tor\ H^2(X, \mathbb{Z}) \simeq Tor\ \Lambda_0/(\sum_{i=1}^{k} a_i).$$

Thus, in this case, since $R^1\pi_*\mathbb{Z}$ and $R^2\pi_*\mathbb{Z}$ are constant sheaves of finite free $\mathbb{Z}$-modules, by the isomorphisms (64), (65) and (66), we conclude that

$$Tor\ H^2(X, \mathbb{Z}) = \pi^* H^2(B, \mathbb{Z}).$$

Appendix

In this appendix we shall show that all quasi T-bundle over a curve B are obtained from the product $B \times T$ by means of logarithmic transformations. Let $\pi : X \to B$ be a quasi T-bundle over the curve B. We let $m_1T_1, m_2T_2, \ldots, m_\ell T_\ell$ be all the multiple fibers of π. Put

$$b_i = \pi(T_i), \quad i = 1, 2, \ldots, \ell.$$

Choose a coordinate neighbourhood D_i of b_i and a local coordinate t_i with center b_i. We may assume

$$D_i = \{t_i \in \mathbb{C} \mid |t_i| < \epsilon\}.$$

Put

$$\widehat{D_i} = \{s_i \in \mathbb{C} \mid |s_i| < \epsilon^{1/m_i}\}.$$

Then a homomorphism

$$\begin{aligned} \widehat{D_i} &\to D_i \\ s_i &\mapsto s_i^{m_i} \end{aligned}$$

is an m_i-sheeted cyclic covering. We let $\widehat{X_i}$ be the normalization of the fiber product $X|_{D_i} \times_{D_i} \widehat{D_i}$ with a natural holomorphic mapping

$$\mu_i : \widehat{X_i} \to X_i = \pi^{-1}(D_i).$$

At a point $p \in \pi^{-1}(b_i)$ we can choose local coordinates $(x, y_1, \ldots, y_n)$ where the holomorphic mapping π is expressed as

$$t_i = \pi((x, y_1, \ldots, y_n)) = x^{m_i}.$$

Then, $\widehat{X_i}$ is locally given by the normalization of

$$s_i^{m_i} - x^{m_i} = 0.$$

Hence, μ_i is a unramified covering. Also the complex manifold $\widehat{X_i}$ has a structure of a fiber space

$$\widehat{\pi}_i : \widehat{X_i} \to \widehat{D_i}$$

over $\widehat{D_i}$ which is smooth over $\widehat{D_i}$. Since $X_i \to D_i$ is a T-principal bundle over the punctured disk D_i^*, it is easy to show that $\widehat{\pi}_i$ is a T-principal bundle, hence $\widehat{\pi}_i$ is isomorphic to the product $D_i \times T$ with the projection to the first factor.

By our construction $\mu_i : \widehat{X_i} \to X_i$ is an m_i-sheeted cyclic unramified covering and the cyclic G_i of order m_i operates on $\widehat{X_i}$. A generator g_i of the group G_i has a form

$$\begin{aligned} g_i : \widehat{D_i} \times T &\to \widehat{D_i} \times T \\ (s_i, [\zeta]) &\mapsto (e_{m_i} s_i, [\zeta + a_i]) \end{aligned} \tag{67}$$

where $[a_i]$ is a point of the torus T of order m_i. Then, the quotient manifold $\widehat{D_i} \times T/G_i$ is isomorphic to $X_i = \pi^{-1}(D_i)$. There is an analytic isomorphism

$$\begin{aligned} \ell_{a_i} : \widehat{D_i^*} \times T/G_i &\to D_i^* \times T \\ [s_i, [\zeta]] &\to (s_i^{m_i}, [\zeta - \frac{m_i a_i}{2\pi\sqrt{-1}} \log s_i]) \end{aligned} \tag{68}$$

We let $\widetilde{X}$ be a complex manifold obtained by patching together $X \setminus \cup_{i=1}^{\ell} \pi^{-1}(b_i)$ and $D_i \times T$'s by the isomorphisms $\ell_{a_i}^{-1}$:

$$\widetilde{X} = (X \setminus \cup_{i=1}^{\ell} \pi^{-1}(b_i)) \bigcup_{i=1}^{\ell} D_i \times T. \tag{69}$$

Then, the complex manifold $\widetilde{X}$ has a natural structure $\widetilde{\pi} : \widetilde{X} \to B$ of a T- principal bundle over the curve B.

Conversely, the quasi T-bundle $\pi : X \to B$ is obtained from the T-principal bundle $\widetilde{\pi} : \widetilde{X} \to B$ by means of the logarithmic transformations:

$$X = L_{b_1}(a_1, m_1) L_{b_2}(a_2, m_2) \cdots L_{b_\ell}(a_\ell, m_\ell)(\widetilde{X}), \tag{70}$$

by patching together $(\widetilde{X} \setminus \cup_{i=1}^{\ell} \widetilde{\pi}^{-1}(b_i))$ and $\widehat{D}_i^* \times T/G_i$'s by the isomorphisms ℓ_{a_i}.

By the remark in §1, the T-principal bundle $\widetilde{\pi} : \widetilde{X} \to B$ is obtained from $B \times T$ by means of logarithmic transformations

$$\widetilde{X} = L_{b_{\ell+1}}(a_{\ell+1}, 1)L_{b_{\ell+2}}(a_{\ell+2}, 1) \cdots L_{b_k}(a_k, 1)(B \times T). \tag{71}$$

Hence, by (70) and (71) the quasi T-bundle $\pi : X \to B$ is obtained from $B \times T$ by means of logarithmic transformations

$$X = L_{b_1}(a_1, m_1) \cdots L_{b_\ell}(a_\ell, m_\ell)L_{b_{\ell+1}}(a_{\ell+1}, 1) \cdots L_{b_k}(a_k, 1)(B \times T).$$

Thus, any quasi T-bundle over the curve B is obtained from $B \times T$ by means of logarithmic transformations.

REFERENCES

[1] Blanchard, A. : Espaces fibrés kählériens compacts, C.R. Acad. Sci., **238** (1954), 2281–2283.
[2] Borel, A. : A spectral sequence for complex analytic bundles, Appendix II to [5].
[3] Brînzănescu, V. : Neron-Severi group for nonalgebraic elliptic surfaces I: elliptic bundle case, Manuscripta Math., **79** (1993), 187–195.
[4] Brînzănescu, V. : Neron-Severi group for nonalgebraic elliptic surfaces II: non-kählerian case, to appear in Manuscripta Math. (1994).
[5] Friedman, F. & Morgan, J.W. : Smooth Four-Manifolds and Complex Surfaces, Springer-Verlag, 1994.
[6] Hirzebruch, F. : Topological Methods in Algebraic Geometry. 3 rd Edition, Springer-Verlag: Berlin-Heidelberg-New York, 1966.
[7] Höfer, T. : Remarks on torus principal bundles, J. Math. Kyoto Univ. (JMKYAZ), **33**-1 (1993), 227–259.
[8] Mumford, D. : Abelian Varieties, Oxford University Press, 1974.
[9] Ueno, K. : Degeneration of elliptic surfaces and certain non-Kähler manifolds, Progr. Math. Vol. **39**, 545–566, Birkhäuser: Boston-Basel-Stuttgart, 1983.

SOME DONALDSON INVARIANTS OF $\mathbf{CP}^2$

Geir Ellingsrud

Mathematical Institute, University of Oslo, P. O. Box 1053 N–0316 Oslo, Norway
e-mail: ellingsr@math.uio.no

Joseph Le Potier

Université Paris 7, UFR de Mathématiques et Institut de Mathématiques de Jussieu, Case Postale 7012, 2, Place Jussieu, F–75251 Paris Cedex 05
e-mail: jlp@mathp7.jussieu.fr

Stein A. Strømme

Mathematical Institute, University of Bergen, Alleg 55, N–5007 Bergen, Norway
e-mail: stromme@mi.uib.no

In memory of the victims of the Kobe earthquake

INTRODUCTION

For an integer $n \geq 2$, let q_{4n-3} be the coefficient of the Donaldson polynomial of degree $4n-3$ of $P = \mathbf{CP}^2$. An interpretation of q_{4n-3} in an algebro-geometric context is the following. Let M_n denote the Gieseker-Maruyama moduli space of semistable coherent sheaves on P with rank 2 and Chern classes $c_1 = 0$ and $c_2 = n$. For such a sheaf F, the Grauert-Mülich theorem implies that the restriction of F to a general line $L \subseteq P$ splits as $F_L \simeq \mathcal{O}_L \oplus \mathcal{O}_L$, and that the exceptional lines form a curve $J(F)$ of degree n in the dual projective plane $P^\vee$. The association $F \mapsto J(F)$ is induced from a morphism of algebraic varieties, called the Barth map, $f_n \colon M_n \to P_n$. Here $P_n = \mathbf{P}^{n(n+3)/2}$ is the linear system parameterizing all curves of degree n in $P^\vee$. Let $H \in \operatorname{Pic}(P_n)$ be the hyperplane class and let $\alpha = f_n^* H$. The interpretation of the Donaldson invariant is:

$$q_{4n-3} = \int_{M_n} \alpha^{4n-3}.$$

Thus q_{4n-3} is the degree of f_n times the degree of its image. From [1] it follows that f_n is generically finite for all $n \geq 2$, that f_2 is an isomorphism and $q_5 = 1$, and that f_3 is of degree 3 and $q_9 = 3$. Le Potier [9] proved that f_4 is birational onto its image and that $q_{13} = 54$. The value of q_{13} has also been computed independently by Tikhomirov and Tyurin [13, prop. 4.1] and by Li and Qin [10, thm. 6.29].

The main result in the present note is the following

Theorem 0.1. *$q_{17} = 2540$ and $q_{21} = 233208$.*

1991 *Mathematics Subject Classification.* 14.
Key words and phrases. Donaldson polynomial, Hilbert scheme.

The proof consists of two parts. The first part, treated in this note, is to express q_{4n-3} in terms of certain classes on the Hilbert scheme of length-$(n+1)$ subschemes of P. This is theorems 0.2 and 0.3 below. The second part is to evaluate these classes numerically. This has been carried out in [4, prop. 4.2].

Let $H_{n+1} = \mathrm{Hilb}_P^{n+1}$ denote the Hilbert scheme parameterizing closed subschemes of P of length $n+1$. There is a universal closed subscheme $\mathcal{Z} \subseteq H_{n+1} \times P$. Consider the vector bundles

$$\mathcal{E} = R^1 p_{1*}(\mathcal{I}_\mathcal{Z} \otimes p_2{}^* \mathcal{O}_P(-1)) \text{ and } \mathcal{G} = R^1 p_{1*} \mathcal{I}_\mathcal{Z}$$

on H_{n+1} of ranks $n+1$ and n, respectively, and the linebundle

$$\mathcal{L} = \det(\mathcal{G}) \otimes \det(\mathcal{E})^{-1}.$$

Theorem 0.2. *Let the notation be as above. Then*

$$q_{17} = \int_{H_6} s_{12}(\mathcal{E} \otimes \mathcal{L}) \quad \textit{and} \quad q_{21} = \frac{2}{5} \int_{H_7} s_{14}(\mathcal{E} \otimes \mathcal{L}).$$

This result was obtained both by Tikhomirov and Tyurin [12], using the method of "geometric approximation procedure" and by Le Potier [7], using "coherent systems". We present in this note what we believe is a considerably simplified proof, which is strongly hinted at on the last few pages of [12].

The formula for q_{17} is a special case of the following formula:

Theorem 0.3. *For $2 \leq n \leq 5$, we have*

$$q_{4n-3} = \frac{1}{2^{5-n}} \int_{H_{n+1}} c_1(\mathcal{L})^{5-n} s_{3n-3}(\mathcal{E} \otimes \mathcal{L}).$$

With this it is also easy to recompute q_5, q_9, and q_{13} using similar techniques as in [4].

Notation. We let h, $h^\vee$, and H be the hyperplane classes in P, $P^\vee$, and P_n, respectively. In general, if ω is a divisor class, we denote by $\mathcal{O}(\omega)$ the corresponding linebundle and its natural pullbacks.

Acknowledgements. This work is heavily inspired by conversations with A. Tyurin, and we thank him for generously sharing his ideas. We would also like to express our gratitude towards the Taniguchi Foundation.

1. Hulsbergen sheaves

Barth [1] used the term Hulsbergen bundle to denote a stable rank-2 vector bundle F on P with $c_1(F) = 0$ and $H^0(P, F(1)) \neq 0$. We modify this definition a little as follows:

Definition 1.1. A *Hulsbergen sheaf* is a coherent sheaf F on P which admits a non-split short exact sequence (*Hulsbergen sequence*)

$$0 \to \mathcal{O}_P \to F(1) \to \mathcal{I}_Z(2) \to 0, \tag{1.1}$$

where $Z \subseteq P$ is a closed subscheme of finite length (equal to $c_2(F) + 1$).

Note that a Hulsbergen sheaf is not necessarily semistable or locally free. However:

Lemma 1.2. *Let F be a Hulsbergen sheaf with $c_2(F) = n > 0$. Then the set $J(F) \subseteq P^\vee$ of exceptional lines for F is a curve of degree n, defined by the determinant of the bundle map*

$$m\colon H^1(P, F(-2)) \otimes \mathcal{O}_{P^\vee}(-1) \to H^1(P, F(-1)) \otimes \mathcal{O}_{P^\vee}$$

induced by multiplication with a variable linear form.

Proof. First note from the Hulsbergen sequence that the two cohomology groups have dimension n. It is easy to see that any Hulsbergen sheaf is slope semistable, in the sense that it does not contain any rank-1 subsheaf with positive first Chern class. Thus by [2, thm. 1], $F_L \simeq \mathcal{O}_L \oplus \mathcal{O}_L$ for a general line L. On the other hand, it is clear that a line L is exceptional if and only if m is not an isomorphism at the point $[L] \in P^\vee$. □

It is straightforward to construct a moduli space for Hulsbergen sequences. For any length-$(n+1)$ subscheme $Z \subseteq P$, the isomorphism classes of extensions (1.1) are parameterized by $\mathbf{P}(\mathrm{Ext}^1_P(\mathcal{I}_Z(2), \mathcal{O}_P)^\vee)$. By Serre duality,

$$\mathrm{Ext}^1_P(\mathcal{I}_Z(2), \mathcal{O}_P)^\vee \simeq H^1(P, \mathcal{I}_Z(-1)).$$

For varying Z, these vector spaces glue together to form the vector bundle $\mathcal{E}$ over H_{n+1}, hence $D_n = \mathbf{P}(\mathcal{E})$ is the natural parameter space for Hulsbergen sequences. Let $\mathcal{O}(\tau)$ be the associated tautological quotient linebundle. For later use, note that for any divisor class ω on H_{n+1}, we have $\pi_*(\tau + \pi^*\omega)^{k+n} = s_k(\mathcal{E}(\omega))$, where $\pi\colon D_n \to H_{n+1}$ is the natural map [5].

The tautological quotient $\pi^*\mathcal{E} \to \mathcal{O}(\tau)$ gives rise to a short exact sequence on $D_n \times P$:

$$0 \to \mathcal{O}(\tau) \to \mathcal{F}(h) \to (\pi \times 1)^*\mathcal{I}_Z(2h) \to 0$$

which defines a complete family $\mathcal{F}$ of Hulsbergen sheaves.

As we noted earlier, a Hulsbergen sheaf is not necessarily semistable. On the other hand, the *generic* Hulsbergen sheaf is stable if $n \geq 2$. It follows that the family $\mathcal{F}$ induces a *rational* map $g_n\colon D_n \to M_n$. By lemma 1.2 above, there is also a Barth map $b_n\colon D_n \to P_n$, defined everywhere, and by construction, the following diagram commutes:

$$\begin{array}{ccc} D_n & \xrightarrow{b_n} & P_n \\ {\scriptstyle g_n}\downarrow & & \downarrow \| \\ M_n & \xrightarrow[f_n]{} & P_n \end{array} \tag{1.2}$$

Proposition 1.3. *Put $\lambda = c_1(\pi^*\mathcal{L})$. Then $b_n^*H = \tau + \lambda$.*

Proof. Let $L \subseteq P$ be a line. Twist the universal Hulsbergen sequence by $-2h$ and $-3h$ respectively. Multiplication by an equation for L gives rise to the vertical arrows in a commutative diagram with exact rows on $D_n \times P$:

$$\begin{array}{ccccccccc} 0 & \longrightarrow & \mathcal{O}(\tau - 3h) & \longrightarrow & \mathcal{F}(-2h) & \longrightarrow & (\pi \times 1)^*\mathcal{I}_Z(-h) & \longrightarrow & 0 \\ & & \downarrow & & \downarrow & & \downarrow & & \\ 0 & \longrightarrow & \mathcal{O}(\tau - 2h) & \longrightarrow & \mathcal{F}(-h) & \longrightarrow & (\pi \times 1)^*\mathcal{I}_Z & \longrightarrow & 0 \end{array}$$

Pushing this down via the first projection, we get the following exact diagram on D_n:

$$\begin{array}{ccccccccc} 0 & \longrightarrow & R^1p_{1*}\mathcal{F}(-2h) & \longrightarrow & \pi^*\mathcal{E} & \longrightarrow & \mathcal{O}(\tau) & \longrightarrow & 0 \\ & & {\scriptstyle m_L}\downarrow & & \downarrow & & \downarrow & & \\ 0 & \longrightarrow & R^1p_{1*}\mathcal{F}(-h) & \xrightarrow{\simeq} & \pi^*\mathcal{G} & \longrightarrow & 0 & & \end{array}$$

Here the last map of the top row is nothing but the tautological quotient map on $\mathbf{P}(\mathcal{E})$. Let $A(L) \subseteq D_n$ be the set of Hulsbergen sequences where L is an exceptional line for the middle term. Clearly, $A(L)$ is the degeneration locus of the left vertical map m_L above. Hence the divisor class of $A(L)$ is

$$\begin{aligned}[A(L)] &= c_1(R^1p_{1*}\mathcal{F}(-h)) - c_1(R^1p_{1*}\mathcal{F}(-2h)) \\ &= \pi^*c_1(\mathcal{G}) - \pi^*c_1(\mathcal{E}) + \tau \\ &= \tau + \lambda.\end{aligned}$$

On the other hand, $A(L)$ is the inverse image of a hyperplane in P_n under b_n, so its divisor class is b_n^*H. □

2. The case $n \leq 5$

Proposition 2.1. *For $2 \leq n \leq 5$, the rational map g_n is dominating, and the general fiber is isomorphic to $\mathbf{P}^{n-5}$. For $n \geq 5$, the map g_n is generically injective with image of codimension $n-5$. In particular, g_5 is birational.*

Proof. Everything follows from the observation that the fiber over a point $[F] \in M_n$ in the image of g_n is the projectivization of $H^0(P, F(1))$, and that for general such F, this vector space has dimension $h^0(F(1)) = \max(1, 6-n)$, which is easily seen from (1.1). The assertion about the codimension follows from a dimension count: $\dim(M_n) = 4n-3$ and $\dim(D_n) = 3n+2$. □

The first half of theorem 0.2 now follows: First of all, since g_5 is birational, the two morphisms f_5 and b_5 have the same image and the same degree. Therefore q_{17} can be computed as

$$q_{17} = \int_{D_5} H^{17} = \int_{D_5} (\tau+\lambda)^{17} = \int_{H_6} s_{12}(\mathcal{E} \otimes \mathcal{L}).$$

For theorem 0.3, let $L_1, \ldots, L_{5-n}$ be general lines in P, and let $B_n \subseteq D_n$ be the locus of Hulsbergen sequences where the closed subscheme Z meets all these $5-n$ lines. The cohomology class of B_n in $H^*(D_n)$ is λ^{5-n}.

Lemma 2.2. *Let $2 \leq n \leq 5$. The general nonempty fiber of g_n meets B_n in 2^{5-n} points, hence the rational map $g_n|_{B_n} \colon B_n \to M_n$ is dominating and generically finite, of degree 2^{5-n}.*

Proof. The general nonempty fiber is of the form $\mathbf{P}(H^0(P, F(1))^\vee)$. It suffices to show that the restriction of $\mathcal{L}$ to this fiber has degree 2 (if $n < 5$). For this, it suffices to consider a linear pencil in the fiber. So let σ_0 and σ_1 be two independent global sections of $F(1)$, and consider the pencil they span. Now $\sigma_0 \wedge \sigma_1 \in H^0(P, \wedge^2 F) = H^0(P, \mathcal{O}_P(2))$ is the equation of a conic $C \subseteq P$ which contains the zero scheme $V(t_0\sigma_0 + t_1\sigma_1)$ of each section in the pencil, $(t_0, t_1) \in \mathbf{P}^1$. Since C meets a general line in two points, it follows that there are exactly two members of the pencil whose zero set meets a general line. □

To complete the proof of theorem 0.3, by lemma 2.2 we now have for $2 \leq n \leq 5$:

$$\begin{aligned}2^{5-n}\, q_{4n-3} &= 2^{5-n} \int_{M_n} H^{4n-3} \\ &= \int_{B_n} (\tau+\lambda)^{4n-3} \\ &= \int_{D_n} \lambda^{5-n} (\tau+\lambda)^{4n-3} \\ &= \int_{H_{n+1}} c_1(\mathcal{L})^{5-n}\, s_{3n-3}(\mathcal{E} \otimes \mathcal{L}).\end{aligned}$$

This completes the proof of the theorems for $n \leq 5$.

3. The case $n = 6$

For $n \geq 6$ the techniques above will say something about the restriction of the Barth map to the Brill-Noether locus $B \subseteq M_n$ of semistable sheaves whose first twist admit a global section. For general n this locus is too small to carry enough information about M_n, but in the special case $n = 6$, it is actually a divisor, whose divisor class $\beta = [B]$ we can determine. Now $\mathrm{Pic}(M_n) \otimes \mathbf{Q}$ has rank 2, generated by α and $\delta = [\Delta]$, the class of the locus $\Delta \subseteq M_n$ corresponding to non-locally free sheaves [8].

Proposition 3.1. *In* $\mathrm{Pic}(M_6) \otimes \mathbf{Q}$, *the following relation holds:*

$$\beta = \frac{5}{2}\,\alpha - \frac{1}{2}\,\delta.$$

Proof. Let $\xi \colon X \to M_6$ be a morphism induced by a flat family $\mathcal{F}$ of semistable sheaves on P, parameterized by some variety X. For certain divisor classes a and d on X, the second and third Chern classes of $\mathcal{F}$ can be written in the form

$$c_2(\mathcal{F}) = a\,h + 6\,h^2, \quad c_3(\mathcal{F}) = d\,h^2$$

modulo higher codimension classes on X. The Grothendieck Riemann-Roch theorem for the projection $p \colon X \times P \to X$ easily gives (for example using [6]) that

$$-c_1(p_!\mathcal{F}(h)) = \frac{5}{2}\,a - \frac{1}{2}\,d.$$

The locus $\xi^{-1}B \subseteq X$ is set-theoretically the support of $R^1p_*\mathcal{F}(h)$. It is not hard to see that one can take the family X in such a way that the 0-th Fitting ideal of $R^1p_*\mathcal{F}(h)$ is actually reduced. Therefore the left hand side of the equation above is $\xi^*\beta$. On the other hand, $a = \xi^*\alpha$ by the usual definition of the μ map of Donaldson [3], and $d = \xi^*\delta$. Since the family $\mathcal{F}/X$ was arbitrary, the required relation is actually universal, and so holds also in $\mathrm{Pic}(M_6) \otimes \mathbf{Q}$. (It suffices to take a family with the properties that (i) $\xi^* \colon \mathrm{Pic}_{\mathbf{Q}}(M_6) \to \mathrm{Pic}_{\mathbf{Q}}(X)$ is injective, (ii) the Fitting ideal above is reduced, and (iii) the general non-locally free sheaf in the family has colength 1 in its double dual.) □

With this, we complete the proof of the second part of theorem 0.2 in the following way. The general fiber of f_6 restricted to Δ has dimension 1, so $f_6(\Delta)$ has dimension 19, see e.g. [11]. Therefore we get

$$\begin{aligned}
\int_{H_7} s_{14}(\mathcal{E} \otimes \mathcal{L}) &= \int_{D_6} (\lambda + \tau)^{20} \\
&= \int_{M_6} \beta\,\alpha^{20} \\
&= \int_{M_6} (\frac{5}{2}\,\alpha - \frac{1}{2}\,\delta)\,\alpha^{20} \\
&= \frac{5}{2}\int_{M_6} \alpha^{21} - \frac{1}{2}\int_{\Delta} \alpha^{20} = \frac{5}{2}\,q_{21}.
\end{aligned}$$

4. A geometric interpretation

Definition 4.1. A *Darboux configuration* in $P^\vee$ consists of a pair (Π, C) where $\Pi \subseteq P^\vee$ is the union of $n+1$ distinct lines, no three concurrent, and $C \subseteq P^\vee$ is a curve of degree n passing through all the nodes of Π.

If we let $Z \subseteq P$ consist of the $n+1$ points dual to the components of Π, we have by Hulsbergen's theorem [1, thm. 4] a natural 1-1 correspondence between Hulsbergen sequences (1.1) and Darboux configurations (Π, C), by letting $C = J(F)$. Therefore D_n can be used as a compactification of the set of Darboux configurations, and the intersection number

$$\int_{D_n} \lambda^i(\tau+\lambda)^{3n+2-i} = \int_{H_{n+1}} c_1(\mathcal{L})^i s_{2n+2-i}(\mathcal{E}\otimes\mathcal{L})$$

can be interpreted as the number of Darboux configurations (Π, C) where Π passes through i given points and C passes through $3n+2-i$ given points.

It is not known whether the Barth map has degree 1 for $n \geq 5$. A related question is the following: Let (Π, C) be a general Darboux configuration $(n \geq 5)$. Is the inscribed polygon Π uniquely determined by C?

References

[1] W. Barth. Moduli of vector bundles on the projective plane. *Invent. Math.*, 42:63–91, 1977.

[2] W. Barth. Some properties of stable rank-2 vector bundles on $\mathbf{P}_n$. *Math. Ann.*, 226:125–150, 1977.

[3] S. K. Donaldson. Polynomial invariants for smooth 4-manifolds. *Topology*, 29:257–315, 1990.

[4] G. Ellingsrud and S. A. Strømme. Bott's formula and enumerative geometry. To appear in Journal of the AMS.

[5] W. Fulton. *Intersection Theory*. Number 2 in Ergebnisse der Mathematik und ihrer Grenz-Gebiete. Springer-Verlag, Berlin-Heidelberg-New York, 1984.

[6] S. Katz and S. A. Strømme. "Schubert", a Maple package for intersection theory and enumerative geometry. Software and documentation available from the authors or by anonymous ftp from ftp.math.okstate.edu or linus.mi.uib.no, 1992.

[7] J. Le Potier. Systèmes cohérent et polynômes de Donaldson. In these proceedings.

[8] J. Le Potier. Sur le groupe Picard de l' espace de modules des fibrés stables sur $\mathbf{P}_2$. *Ann. scient. Éc. Norm. Sup., 4. série*, 13:141–155, 1981.

[9] J. Le Potier. Fibrés stables sur le plan projectif et quartiques de Lüroth. Preprint, Oct 1989.

[10] W.-P. Li and Z. Qin. Lower-degree Donaldson polynomial invariants of rational surfaces. *J. Alg. Geom.*, 2:413–442, 1993.

[11] S. A. Strømme. Ample divisors on fine moduli spaces on the projective plane. *Math. Z.*, 187:405–523, 1984.

[12] A. Tikhomirov and A. N. Tyurin. Application of geometric approximation procedure to computing the Donaldson's polynomials for $\mathbf{CP}^2$. *Mathematica Gottingensis Schriftenreihe des Sonderforschungsbereichs Geometrie und Analysis*, 12:1–71, 1994.

[13] A. N. Tyurin. The moduli spaces of vector bundles on threefolds, surfaces and curves I. Erlangen preprint, 1990.

KÄHLER QUOTIENT AND EQUIVARIANT COHOMOLOGY

Akira Fujiki

Department of Mathematics, Osaka University, Toyonaka, Osaka 560, JAPAN

Introduction

Let X be a compact Kähler manifold and G a connected complex reductive algebraic group acting biholomorphically and meromorphically on X. Fix a maximal compact subgroup K of G and a K-invariant Kähler form ω on X. Then the action of K on the Kähler manifold (X, ω) admits a moment map $\mu : X \to \underline{k}^*$. In this situation we have the notions of stability and semistability of a point of X depending on the data X, K, ω, and μ as above. Then we can form the associated categorical projective quotient $X^{ss}//G$ which contains the geometric quotient X^s/G as a Zariski open subset, where X^{ss} and X^s are the set of semistable and stable points of X ([8]) (cf. Sect. 2). They specialize to the geometric invariant theory quotients of Mumford when X is projective, the de Rham class $[\omega]$ of ω is an integral class and when μ is associated to a connection form of an ample line bundle L with $c_1(L) = \omega$ (cf. Sect. 1).

In this note, with X and K fixed and letting ω and μ vary we study the resulting variations of the quotients $X^s/G \subseteq X^{ss}//G$. It turns out that such quotients depends only on the cohomology class of the pair (ω, μ) in the real equivariant cohomology group $H^2_K(X)$ associated to the action of K on X (Sect. 3 and Sect. 4). Moreover, we can define a certain finite collection of linear subspaces of $H^2_K(X)$ such that the quotients remain invariant as long as the class (ω, μ) stays inside one of the chambers arising from the collection of these subspaces. When specialized to the projective and rational case as mentioned above, this result is equivalent to that of Dolgacev and Hu [4], who treated the problem algebraically in terms of the G-Neron-Severi group of X.

In a good case where $X^s = X^{ss}$ (which is the generic case when G is an algebraic torus), we then proceed to describe the second cohomology group and the Kähler cone of the quotient X^{ss}/G in terms of the equivariant cohomology group of X (Sect. 5). This also leads us to a certain generalization of a result of Duistermaat and Heckman [5]. Finally in Sect. 6, we also compare the Kähler cones of the quotient associated to adjacent chambers in a certain case.

1. Review of the Projective Case

1. Let X be a complex projective manifold and G a connected complex reductive algebraic group acting regularly on X. Let L be an ample line bundle on X. Assume that the action of G on X lifts to a bundle action of G on L. Then we have the usual notion in geometric invariant theory of stable and semistable points of X with respect to this action of G on the pair (X, L) (cf. [14][10]). Let X^s (resp. X^{ss}) be the set of stable (resp. semistable) points of X which is a Zariski open subset of X. Moreover, we may take the geometric (resp. categorical) quotient X^s/G (resp. $X^{ss}//G$) of X^s (resp. X^{ss}) by G which fits into the diagram of algebraic varieties

$$\begin{array}{ccccc} X^s & \subseteq & X^{ss} & \subseteq & X \\ \downarrow & & \downarrow \pi & & \\ X^s/G & \subseteq & X^{ss}//G & & \end{array} \tag{1}$$

where $X^{ss}//G$ is projective and X^s/G is a Zariski open subset of it. It is also known that if m is a sufficiently large integer, then the restriction $L^m|X^{ss}$ of L^m descends to an ample line bundle $\bar{L}_m$ on $X^{ss}//G$.

2. Dolgacev and Hu [4] has studied how the above quotient construction depends on the given data, i.e., the G-linearized line bundle L on X as above. Let $\mathrm{Pic}^G(X)$ be the G-equivariant Picard group of X with the given G action, that is, $\mathrm{Pic}^G(X)$ is the set of isomorphism classes of the pairs (F, σ) consisting of a holomorphic line bundle F and a bundle action of G on F lifting the given action on X. Similarly, let $NS^G(X)$ be the G-equivariant Neron-Severi group of X, which is by definition of the quotient of $\mathrm{Pic}^G(X)$ with respect to a natural numerical equivalence relation. We may then speak of the (G-equivariant) ample cone A^G in $NS^G(X)$ consisting of those elements which can be represented by a pair (F, σ) with F ample. Then one of their results is that the quotient data (1) actually depends only on the image (L, τ) in A^G, where τ denotes the given action of G on L. So we may speak of the quotient associated to any point of A^G. More precisely, they define a certain system of hyperplanes $\{H_\lambda\}_{\lambda \in \Lambda}$ in $NS^G(X)$ leading to a certain polyhedral decomposition of A^G such that in each of the resulting chambers inside A^G the associated quotient diagram (1) is unchanged, although the ample line bundle $\bar{L}_m$ to which $L^m|X^{ss}$ descends of course varies.

3. In this note we treat a Kählerian analogue of the above result of Dolgacev and Hu. Namely, we start with a compact Kähler manifold X and a connected complex reductive algebraic group G acting biholomorphically and meromorphically on X. Here, the action is said to be *meromorphic* if the given action $G \times X \to X$ extends to a meromorphic map $G^* \times X \to X$, where G^* is any algebraic compactification of G into a projective variety. In this case we have also the notions which turn out to be equivalent to those of the stability and semistability when specialized to the projective case as in **1** (cf. [10]).

4. To explain this, we go back to the projective situation of 1.1. We fix a maximal compact subgroup K of G. Then we may find by using the standard averaging process a K-invariant Kähler form ω representing the real first chern class $c_1(L) \in H^2(X, \mathbf{R})$ of L. Then there exists a unique K-invariant hermitian metric h on L such that the associated chern form coincides with ω; $\sqrt{-1}/2\pi\, \partial\bar{\partial} \log h = \omega$. We consider the associated connection form θ as a $\mathbf{C}^*$-invariant complex valued 1-form on $L^* - 0$, where L^* is the dual of L and 0 denotes the zero section. Let $\underline{k}$ be the Lie algebra of K, and for any element a of $\underline{k}$ we shall denote by $\underline{a}$ (resp. $\tilde{\underline{a}}$) the holomorphic vector field on X (resp. L^*) induced by the action. Since $\tilde{\underline{a}}$ is $\mathbf{C}^*$-invariant, the C^∞ function $\theta(\tilde{\underline{a}})$ descends to a unique C^∞ function on X, which we shall denote by $\mu_a; \theta(\tilde{\underline{a}}) = p^*\mu_a$, where $p : L^* - 0 \to X$ is the natural projection. These μ_a for various a in turn determines a $\underline{k}^*$-valued C^∞ function

$$\mu : X \to \underline{k}^*$$

having the properties

(1) μ is K-equivariant and

(2) $d\mu_a = \iota_{\underline{a}}\omega$ for every $a \in \underline{k}$,

where $\underline{k}^*$ is the dual of $\underline{k}$ as a vector space with coadjoint action of K, and $\iota_{\underline{a}}$ is the interior product with $\underline{a}$. Thus, μ gives a canonical *moment map* for the action of the Lie group K on the Kähler manifold (X, ω). This description of a moment map is in fact a special case of that of Berligne-Vergne valid for general principal bundles ((cf. [3, Chap.7]).

5. The result established by Kirwan [10, §8] is the following:

A. A point $x \in X$ is stable if and only if $Gx \bigcap \mu^{-1}(0)$ is nonempty and the stabilizer G_x of x in G is finite. Similarly, a point $x \in X$ is semistable if and only if $\overline{Gx} \bigcap \mu^{-1}(0)$ is nonempty, where $\overline{Gx}$ is the topological closure of the orbit Gx in X.

B. Let $\mu^{-1}(0)_f$ be the set of points of X with finite stabilizers so that *via* the Marsden-Weinstein method the quotient $\mu^{-1}(0)_f/K$ has a natural structure of a Kähler (orbi)fold with a naturally induced Kähler (orbi)form ω_{MW} on it. Then we have the following commutative diagram

$$\begin{array}{ccc} \mu^{-1}(0)_f/K & \subseteq & \mu^{-1}(0)/K \\ \varphi_f \downarrow & & \downarrow \varphi \\ X^s/G & \subseteq & X^{ss}//G \end{array} \qquad (2)$$

where the naturally induced map φ is homeomorphic and φ_f is biholomorphic if $\mu^{-1}(0)_f/K$ is endowed with the Marsden-Weinstein complex structure above.

2. Definition of Stability and Existence of Quotients in the Kähler case

6. We return now to a general Kähler case with X and G as in **3**. Moreover, we fix a maximal compact subgroup K of G and a K-invariant Kähler form ω on X. First we note the following result which guarantees the existence of a moment map under our assumption of the meromorphicity of the action. We denote by $\underline{c}$ the center of $\underline{k}$ considered naturally as a direct summand of $\underline{k}$.

Lemma 2.1. *The notations being as above, there always exists a moment map $\mu : X \to \underline{k}^*$ associated to the given action of K on the Kähler manifold (X, ω), which is defined up to additions of (constant) elements of $\underline{c}^*$, naturally identified with a subspace of $\underline{k}^*$.*

Proof. G is finitely covered by a product of a semisimple algebraic group and an algebraic torus. Since for the semisimple group the existence (and the uniqueness) of the moment map is well-known, the proof is reduced to the case where G is an algebraic torus $\mathbf{C}^{*k}$; and this case is obviously reduced to the case where $k = 1$. Now it is also a known fact that $G = \mathbf{C}^*$ acts meromorphically in the sense defined above if and only if $\underline{a}$ admits a zero on X (cf. e.g. [6, Prop.6.8]) and that the latter is equivalent to the fact that the interior product $\iota_{\underline{a}}\omega$ is d-exact, i.e., $\underline{a}$ is Hamiltonian, where $\underline{a}$ is any element of $\underline{k}$ (cf. [11, Ch.3, §4]). The lemma follows. □

Remark. As the above proof shows, the converse is also true, namely, if a moment map exists, then the action of G on X is meromorphic in the sense defined in **3**.

7. Now we take and fix a moment map $\mu : X \to \underline{k}^*$ associated to the action of K on (X, ω). Then we may speak of the notions of stability and semistablity of a point according to A in **5**. More precisely,

Definition. Let G, X, K, and ω be as in **6** and suppose that a moment map $\mu : X \to \underline{k}^*$ is given as above. Then:

(1) A point $x \in X$ is called *quasi-stable* if $Gx \bigcap \mu^{-1}(0)$ is nonempty.
(2) A point $x \in X$ is called *stable* if it is quasi-stable and its stabilizer G_x in G is finite.
(3) A point $x \in X$ is called *semistable* if $\overline{Gx} \bigcap \mu^{-1}(0)$ is nonempty.

8. We denote by X^s (resp. X^{ss}) the set of stable (resp. semistable) points of X. Then analogously to the projective case we have the following existence and comparison results of the quotients :

Theorem 2.2. *1) There exists a geometric (resp. categorical) quotient X^s/G (resp. $X^{ss}//G$) of X^s (resp. X^{ss}) by G which fits into the following commutative diagram of normal complex spaces*

$$\begin{array}{ccccc} X^s & \subseteq & X^{ss} & \subseteq & X \\ \pi\downarrow & & \downarrow\pi & & \\ X^s/G & \subseteq & X^{ss}//G & & \end{array} \tag{3}$$

and

$$\begin{array}{ccc} \mu^{-1}(0)_f/K & \subseteq & \mu^{-1}(0)/K \\ \varphi_f\downarrow & & \downarrow\varphi \\ X^s/G & \subseteq & X^{ss}//G \end{array} \tag{4}$$

where $X^{ss}//G$ is compact and when the naturally induced map φ is homeomorphic and φ_f is moreover biholomorphic with respect to the natural complex structures on both spaces.

2) There exists a unique continuous Kähler form $\bar{\omega}^{ss}$ on $X^{ss}//G$ such that its restriction to the Zariski open subset X^s/G coincides with $\varphi_f^*\omega_{\mathrm{MW}}$, where ω_{MW} is the Marsden-Weinstein Kähler (orbi)form on $\mu^{-1}(0)_f/K$.

3) $\pi : X^{ss} \to X^{ss}//G$ extends to a meromorphic map $X \to X^{ss}//G$.

Here, in general a continuous Kähler form on a complex space Y is given by a data $(\{U_\alpha\}, \{\psi_\alpha\})$, where $\{U_\alpha\}$ is an open covering of Y and each ψ_α is a continuous strictly plurisubharmonic function defined on U_α such that the difference $\psi_\alpha - \psi_\beta$ is a (smooth) pluriharmonic function on $U_\alpha \bigcap U_\beta$. By Varouchas [16], in this case by perturbing this Kähler form within the same cohomology class in $H^1(Y, \mathcal{P}_Y)$ we can get a C^∞ Kähler form (defined in the same way), where $\mathcal{P}_Y$ is the sheaf of pluriharmonic functions on Y.

1) and 2) have been shown recently by Heinzner and Loose [8] [1].

9. Brief comments on the proof of the theorem are in order. Take any point x of $\mu^{-1}(0)$. The first main point of the proof of 1) is to find a G-invariant Stein neighborhood U of x in X. Then by Snow [15] there would exist a categorical quotient $U//G$ of U by G with all the favourable properties; then by covering X by such G-invariant open subsets the quotient $X//G$ is obtained by patching such Stein quotients $U//G$ together. Here, note that the stabilizer G_x is a complexification of the stabilizer K_x ([10, §7]) so that the orbit $Gx \cong G/G_x$ is Stein.

On 2): We prove the existence of a Kähler potential p of the restricted Kähler form $\omega|U$ which is compatible with the moment map μ in the sense that $\mu_a = (d^c p)(\underline{a})$ for any $a \in \underline{k}$, where $d^c = i/2\pi(\partial - \bar{\partial})$. Then $p_U|U \bigcap \mu^{-1}(0)$ descends to a strictly plurisubharmonic

[1] In the symposium I have stated the results as new. Later, A. Huckleberry kindly communicated to me that they have already been shown by the above authors.

function on $U//G \sim U \cap \mu^{-1}(0)/K$ (homeo.), and p_U for varying U altogether give a desired continuous Kähler form as defined above. 3) is obtained by comparing the quotient map π with the so-called "Chow quotient map" though we omit the detail here.

10. Note that the theorem is of interest even when X is projective, if the Kähler class $[\omega] \in H^2(X, \mathbf{R})$ is not a rational class. Related with such a situation we might ask the following question. In general when $X^s = X^{ss}$, or equivalently, when the stabilizer at each point of X^{ss} is finite, we shall say that we are dealing with the *good case*. In the good case, the geometric quotient $Y := X^s/G$ is already compact. If, further, the action of G on X^s is free, then in view of 3) of the above theorem Y is a (compact) Kählerian Moishezon manifold. Hence, by a theorem of Moishezon Y is in fact projective. We may then ask if this holds true even when the action is not free, or how the ample class on Y arises from the given data even when the action is free[2]. This in general would involve a general study of the ample cone of the quotient manifold Y in terms of that of the original manifold X. But in the latter case in stead of the usual Kähler cone in the ordinary cohomology group $H^2(X, \mathbf{R})$, it turns out to be more appropriate to use the equivariant cohomology group $H^2_K(X, \mathbf{R})$. This is also in accordance with the consideration of G-equivariant Neron-Severi group $NS^G(X)$ by Dolgacev and Hu as in **2** as we now explain.

3. Equivariant Cohomology and Moment Maps

11. Let X be a compact Kähler manifold and G a connected complex reductive algebraic group acting biholomorphically and meromorphically on X as in **3**. We fix a maximal compact subgroup K of G as in **6**. The (real) *equivariant cohomology group* $H^*_K(X) = H^*_K(X, \mathbf{R})$ associated to the action of K on X is defined as follows. Let $E_K \to B_K$ be a universal K-bundle, where B_K is the classifying space of K. Consider the associated fiber bundle $X_K := E_K \times_K X \to B_K$ with typical fiber X. Then we define $H^*_K(X)$ to be the ordinary cohomology group of X_K:

$$H^*_K(X) := H^*(X_K, \mathbf{R}).$$

Similar definitions are obtained also for G instead of K, or for $\mathbf{Q}$ instead of $\mathbf{R}$. However, since G/K is a cell, we have a natural isomorphism $H^*_K(X) \cong H^*_G(X)$. In particular, $H^*_K(X)$ is actually an invariant of the G-action on X. For our purpose here the de Rham description of the equivariant cohomology group is more relevant [2, §§4, 6][3, Ch.7]. For the second cohomology group $H^2_K(X)$ this takes the following form: Let $A^i_K(X)$ be the space of K-invariant C^∞ i-forms and $C^\infty(X, \underline{k}^*)_K$ that of K-equivariant C^∞ maps of X into $\underline{k}^*$. Then $H^2_K(X)$ is the quotient $H^2_K(X) = Z^2_K(X)/B^2_K(X)$, where $Z^2_K(X)$ and $B^2_K(X)$ are vector subspaces of the direct sum $A^2_K(X) \bigoplus C^\infty(X, \underline{k}^*)_K$ defined by

$$Z^2_K = \{(\eta, \nu); d\eta = 0, d\nu_a = \iota_{\underline{a}}\eta, \forall a \in \underline{k}\} \tag{5}$$

where ν_a is the function on X defined by ν and a (cf. **4**), and

$$B^2_K = \{(d\beta, \mu^\beta); \exists \beta \in A^1_K(X)\}$$

where by definition $\mu^\beta_a = \beta(\underline{a})$.

In our case where X is a compact Kähler manifold, $H^*_K(X)$ naturally admits a Hodge structure (cf. [10, §14]). In particular, $H^2_K(X)$ admits a pure Hodge structure of weight 2;

[2]Answers for these questions will be given elsewhere

in fact we have the Hodge decomposition into types

$$H_K^2(X,\mathbf{C}) = H_K^{2,0}(X)\bigoplus H_K^{1,1}(X)\bigoplus H_K^{0,2}(X).$$

It is easy to check that in the de Rham description, $H_K^{1,1}(X)$ is the subspace of $H_K^2(X,\mathbf{C})$ consisting of those elements which are represented by the "(equivariant) cocycles" (η,ν) in Z_K^2 with η of type $(1,1)$.

12. For any holomorphic line bundle L with a holomorphic lift of the action of G on X, by taking the equivariant chern class $c_1^K(L)\in H_K^2(X)$ we obtain a homomorphism $\mathrm{Pic}^G(X)\to H_K^2(X)$ which factors through an inclusion $NS^G(X)\hookrightarrow H_K^2(X)$ inducing an isomorphism

$$NS^G(X)_{\mathbf{Q}}\cong H_K^{1,1}(X,\mathbf{Q}) := H^2(X,\mathbf{Q})\bigcap H^{1,1}(X)$$

as in the absolute case. Indeed, in the de Rham description $c_1^K(L)$ is the class represented by the cocycle (ω,μ) in Z_K^2, where ω and μ are defined as in **4** (cf. [2, §8][3, Ch.7]).

13. A basic structure of $H_K^2(X)$ is given by the following :

Lemma 3.1. *There exists a natural commutative diagram of exact sequences of real vector spaces*

$$\begin{array}{ccccccccc} 0 & \to & \underline{c}^* & \to & H_K^{1,1}(X) & \to & H^{1,1}(X) & \to & 0 \\ & & \| & & \cap & & \cap & & \\ 0 & \to & \underline{c}^* & \xrightarrow{j} & H_K^2(X) & \xrightarrow{r} & H^2(X) & \to & 0. \end{array} \tag{6}$$

Moreover, the bottom sequence is defined over $\mathbf{Q}$, *where* c^* *is endowed with a natural rational structure as the dual of a Lie algebra of a real torus, i.e., the center* C *of* K.

Proof. Consider $\underline{c}^*$ naturally as a direct summand of k^*. In the de Rham description the maps j and r are given by $j(\nu) = [0,\nu]\in H_K^2(X)$ for $\nu\in\underline{c}^*$, and by $r([\eta,\nu]) = [\eta]$ for $[\eta,\nu]\in H_K^2(X)$ respectively, where $[*]$ denotes the (equivariant) de Rham class of $*$. The exactness of the sequences is easily checked. For the rationality of the sequence it is more appropriate to use the original topological description of $H_K^2(X)$, but here we omit the standard details. □

14. Let C_K be the (equivariant) *Kähler cone*, namely, the set of (equivariant) *Kähler classes* of $H_K^{1,1}(X)$, i.e., those elements which can be represented by a *Kähler cocycle* (η,ν) in the sense that η is a Kähler form. In other words, C_K is the inverse image by r of the ordinary Kähler cone $C = C(X)$ in $H^{1,1}(X)$, and the (rational) ample cone $A_{\mathbf{Q}}^G$ is given by $A_{\mathbf{Q}}^G = C_K\bigcap H^{1,1}(X)_{\mathbf{Q}}$ in the projective case. Thus C_K is an open convex cone in $H_K^{1,1}(V)$ which is a principal bundle over C with fiber the vector space $\underline{c}^*$. By the definition of the cocycle condition (5), if (ω,μ) is a Kähler cocycle, then μ is nothing but a moment map associated to the induced action of K on the Kähler manifold (X,ω). Thus we get the associated quotient diagrams (3) and (4) and we are interested in how the diagrams depend on the given Kähler cocycle (ω,μ).

15. We observe first that the underlying symplectic quotient $(\mu^{-1}(0)_f/K,\omega_{\mathrm{MW}})$ depends only on the associated (equivariant) Kähler class $[\omega,\mu]$ in $H_K^2(X)$.

Proposition 3.2. *Let* (ω,μ) *and* (ω',μ') *be two Kähler cocycles in the same Kähler class in* $H_K^2(X)$. *Then there exists a diffeomorphism* φ *of* X *such that* φ *is compatible with the* K*-action and that* $\varphi^*\omega=\omega'$ *and* $\varphi^*\mu=\mu'$. *In particular, two symplectic quotients*

$\mu^{-1}(0)_f/K, \omega_{\rm MW})$ and $({\mu'}^{-1}(0)_f/K, \omega'_{\rm MW})$ are isomorphic as symplectic manifolds, and $\mu^{-1}(0)/K$ and ${\mu'}^{-1}(0)/K$ are homeomorphic.

In view of the connectedness of the space of the equivariant Kähler cocycles in the same cohomology class this follows immediately from the following equivariant version of Moser's theorem [12, Th.2] which is obtained by a straightforward modification of the proof of the original result.

Proposition 3.3. *Let X be a compact connected smooth manifold on which a compact Lie group K acts smoothly. Let $\{\omega_t\}_{0\le t\le 1}$ be a smooth family of K-invariant symplectic forms ω_t on X, which are mutually cohomologous. Then there exists a family $\{\varphi_t\}_{0\le t\le 1}$ of diffeomorphisms φ_t of X such that $\varphi_t^*\omega_t = \omega_0$ and that each φ_t commutes with the given K-action.*

4. Equivariant Cohomology and Quotients

16. The map φ in Theorem 3.2 does not preserve the complex structures in general. To understand the (in)variance of complex structures we have to deal directly with how the stability condition of a point changes when the Kähler cocycles (ω, μ) vary (in a fixed cohomology class). The result is:

Theorem 4.1. *Let β be any equivariant Kähler class in $H^2_K(X)$. Then the quotient diagram (3) depends only on the Kähler class β and is independent of the choice of a Kähler cocycle (ω, μ) which represents β.*

Recall that set-theoretically the quotient map $\pi : X^{ss} \to X^{ss}//G$ is characterized by the following relation:

$\pi(x) = \pi(y)$ if and only if $\overline{Gx} \cap \overline{Gy} \neq \emptyset$.

Furthermore, the topology of $X^{ss}//G$ is precisely the quotient topology with respect to the map π (cf. [9]), and the structure sheaf of the quotient as an analytic space is defined to be the G-invariant direct image sheaf by π of the structure sheaf of X^{ss}. Moreover, each fiber contains a unique closed orbit, and a point x is quasi-stable if and only if x belongs to the unique closed orbit in $\pi^{-1}\pi(x)$. X^s are thus the union of the fibers of π consisting only of a closed orbit. From all these we deduce that the above theorem results from the following:

Proposition 4.2. *Let (ω, μ) and (ω', μ') be two Kähler cocycles representing the same (equivariant) cohomology class. Then for any point x of X, x is semistable with respect to (ω, μ) if and only if so is x with respect to (ω', μ').*

Proof. Suppose first that G is an algebraic torus. In this case the result follows immediately from the following result of Atiyah [1, Th.2]:

> Let $F_1, \dots, F_k$ be the set of connected components of the fixed point set F of G. Then each $\mu(F_i)$ is a point, say p_i, in $\underline{k}^*$ and the image $\mu(\overline{Gx})$ in $\underline{k}^*$ of the closure $\overline{Gx}$ of the orbit Gx is a convex hull of the points $p_i, 1 \le i \le k$.

Indeed, since (ω, μ) and (ω', μ') are cohomologous to each other, we may write $\mu' = \mu + \mu^\theta$ for some K-invariant 1-form θ on X, where $\mu^\theta_{\underline{a}} = \theta(\underline{a})$. Then since $\underline{a}$ vanishes along F, so does the map μ^θ also. Thus $\mu'(F_i) = \mu(F_i)$ for all i. In view of the above result of Atiyah (applied to μ') this implies that the images of $\overline{Gx}$ under μ and μ' are the same in $\underline{k}^*$. Hence, by the definition of semistability the assertion follows in the torus case.

In the general case we shall use the Hilbert-Mumford-Kirwan criterion [10, §8]: Let (η, ν) be a Kähler cocycle. For a real 1-parameter subgroup $\lambda : S^1 \to K$ of K denote by (η, ν_λ)

the induced Kähler cocycle with respect to the induced action of S^1 on (X, ω). Obviously, λ complexifies to give a complex 1-parameter subgroup $\lambda_C : \mathbf{C}^* \to G$ with the induced $\mathbf{C}^*$-action on X so that we may speak of the semistability of x with respect to (η, ν_λ) in the obvious sense. Then the criterion is:

> x is a semi-stable point with respect to the Kähler cocycle (η, ν) if and only if so is x with respect to the induced Kähler cocycles (η, ν_λ) for any 1-parameter subgroup λ as above.

Now suppose that x is semistable with respect to (ω, μ). Then so is x with respect to (ω, μ_λ) for every λ by the above criterion. Since (ω, μ_λ) and (ω', μ'_λ) are cohomologous, by the first part of the proof x is semistable also with respect to (ω', μ'_λ). Since this is true for any λ, again by the above criterion x is also semistable with respect to (ω', μ'_λ). □

17. Next we let the cohomology classes vary and study how the diagrams (3) vary accordingly. Henceforce, we assume for simplicity that the action of G or K is generically free so that a general point of X admits only a trivial stabilizer. For this purpose we define a system of linear subspaces of $H^2_K(X)$. We consider the set $\mathcal{S}$ of conjugacy classes of connected reductive subgroups of G which are connected components of the stabilizers in G of some point of X. By Mostow [13] this is a finite set; write $\mathcal{S} = \{\Gamma_0, \Gamma_1, \ldots, \Gamma_k\}$, where Γ_0 is the class of the trivial group $\{e\}$.

For a point x of X denote by G^0_x (resp. K^0_x) the identity component of the stabilizer of x in G (resp. K). For each i, W_i will denote the set of points of x such that the complexfication $K^{0\mathbf{C}}_x$ of K^0_x belongs to Γ_i. The disjoint union $W := \bigcup W_i$ is the set of points of X whose stabilizer G_x in G is a complexification of its intersection with K. Let $W_{i\alpha}, 1 \le \alpha \le s_i$, be the connected components of W_i. It is not difficult to see that each $W_{i\alpha}$ is a locally closed C^∞ submanifold of X. The significance of the set W lies in the following:

Lemma 4.3. *Suppose that $\mu : X \to \underline{k}^*$ is a moment map associated to a K-action on a Kähler manifold (X, ω). Then $\mu^{-1}(0) \subseteq W$.*

Proof. By [10, Lemma 7.2] for any point x of $\mu^{-1}(0)$ its stabilizer G_x in G is a complexfication of the stabilizer K_x in K. □

For any i and α, take a point $x \in W_{i\alpha}$. We consider the linear map

$$q_{i\alpha} : H^2_K(X) \to H^2_{K^0_x}(\{x\}) \cong \underline{c}^*_i,$$

where $\underline{c}_i$ is the Lie algebra of the common center of the subgroups K_i whose complexification belongs to Γ_i (cf. Lemma 3.1). In the de Rham description we may write

$$q_{i\alpha}([\omega, \mu])(a) = \mu_a(x) \quad \text{for any} \quad a \in \underline{c}_i.$$

Lemma 4.4. *The map $q_{i\alpha} = q_{i\alpha x}$ is actually independent of the choice of the point x of $W_{i\alpha}$ and depends only on i and α.*

Proof. In the de Rham description this follows immediately from the following two facts:

(1) For any $a \in \underline{c}_i$ and any $k \in K$ we have

$$q_{i\alpha kx}([\omega, \mu])(a) = \mu_a(kx) = \mu_{ad(k)a}(x) = \mu_a(x) = q_{i\alpha x}([\omega, \mu]).$$

(2) If B is a connected submanifold of X such that K^0_x is independent of $x \in B$. Then μ_a is constant for any $a \in \underline{c}_i$ since $d\mu_a = \iota_{\underline{a}}\omega = 0$ on B. □

Now denote by $H_{i\alpha}$ the kernel of $q_{i\alpha}$. Then $\{H_{i\alpha}\}_{0\le i\le k, 1\le\alpha\le s_i}$ forms a finite collection of vector subspaces of $H^2_K(X)$.

This in turn gives rise to a system $\{H_s\}_{s\in S}$ of vector subspaces of $H^2_K(X)$, where H_s run over all the subspaces which are the intersections of the subspaces of $\{H_{i\alpha}\}$. (The total space $H^2_K(X)$ is included and write $H_0 = H^2_K(X)$.) Let C_K be the Kähler cone in $H^2_K(X)$ and set $C_s := H_s \bigcap C_K - \bigcup_s H_{i\alpha}$ for each $s \in S$, where the union $\bigcup_s$ is taken over all $H_{i\alpha}$ with $H_s \not\subseteq H_{i\alpha}$. C_s is an open subset of H_s and we consider the connected components $C_{sp}, 1 \le p \le \ell_s$ of C_s. Especially, $C_0 := C_K - \bigcup_0 H_{i\alpha}$ and C_{0p} will be called the *chambers* of C_K.

18. In the case where $G = \mathbf{C}^{*k}$ is an algebraic torus, we may identify the conjugacy class Γ_i with its representative subtorus S_i of $K = (S^1)^k$. Set $r_i = \dim S_i$. Renumbering the indices we may assume that $r_i = 1$ for $1 \le i \le c$ and $r_i > 1$ for $c > i$ for some c with $1 \le c \le k$. A simplifying fact in the torus case is:

Lemma 4.5. *For any i we have* $\mathrm{codim} H_{i\alpha} = r_i$ *and $H_{i\alpha}$ is an intersection of r_i hyperplanes $H_{j\beta}$ for some $1 \le j \le c$ and β.*

Proof. Pick any point $x_{i\alpha}$ from $W_{i\alpha}$ (which in this case coincides with a connected component of the fixed point set of S_i.) Then the first assertion follows easily from the commutative diagram

$$\begin{array}{ccccccccc} 0 & \longrightarrow & \underline{k}^* & \longrightarrow & H^2_K(X) & \longrightarrow & H^2(X) & \longrightarrow & 0 \\ & & \downarrow{\scriptstyle p_i} & & \downarrow{\scriptstyle q_{i\alpha}} & & \downarrow & & \\ 0 & \longrightarrow & \underline{s}_i^* & \xrightarrow{j} & H^2_{S_i}(\{x_{i\alpha}\}) & \xrightarrow{r} & H^2(\{x_i\}) & \longrightarrow & 0 \end{array} \tag{7}$$

of exact sequences of vector spaces (cf. Lemma 3.1), where $H^2(\{x_i\})$ vanishes and p_i is a natural surjection. In a neighborhood of $x_{i\alpha}$, the action of S_i is linearized and the following is true: $W_{i\alpha}$ is locally a complete intersection of smooth divisors $D_1, \dots, D_{r_i}$ on X, pointwise fixed by some 1-dimensional subtorus of S_i. It follows that for each ℓ $D_\ell = W_{j\alpha_j}$ for some $j = j(\ell)$ and α_j with $1 \le j \le c$. Then, by using the commutative diagrams similar to (7) we see readily that $H_{i\alpha} = \bigcap_{1\le\ell\le r_i} H_{j\alpha_j}$. □

In particular, if we denote by $\{L_i\}_{1\le i\le d}$ the set of $H_{i\alpha}$'s with $1 \le i \le c$, then $\{H_s\}$ may be written as $\{L_{i_1\dots i_u} := L_{i_1} \bigcap \cdots \bigcap L_{i_u}\}_{1\le i_1<\dots<i_u\le d}$ and C_s as $C_s = L_{i_1\dots i_u} \bigcap C_K - \bigcup_{i\ne i_s} L_i$. C_s are thus open cones in H_s.

Moreover, when $G = \mathbf{C}^*$, the set $\{L_i\}$ is simply described as follows. Let $F_1, \dots, F_b$ be the connected components of the set F of fixed points of $K = S^1$. Then the set $\{q_{i\alpha}\}$ is naturally identified with the set of maps

$$q_i : C_K \to \mathbf{R} = \underline{k}; q_i([\omega, \mu]) = \mu(F_i)$$

and the space L_i with $\mathrm{Ker} q_i$.

19. Now we show the following :

Theorem 4.6. *If two Kähler classes $[\omega, \mu]$ and $[\omega', \mu']$ in C_K are in one and the same chamber, then for any point x in X, x is semistable with respect to $[\omega, \mu]$ if and only if so is x with respect to $[\omega', \mu']$. Moreover, when G is an algebraic torus, if $[\omega, \mu]$ and $[\omega', \mu']$ are in one and the same connected component C_{sp} of C_s for some $s \in S$ and p, then the same conclusion holds.*

Proof. First consider the case where $G = \mathbf{C}^*$. Suppose that C_s is of the form $C_s := L_{i_1 \dots i_u} \bigcap C_K - \bigcup_{i \neq i_s} L_i$. Then for any $[\omega, \mu]$ in C_s, $\mu(x) = 0$ if $x \in F_{i_p}$ for some $1 \leq p \leq u$, and otherwise $\mu(x) \neq 0$ and so by continuity is either always positive, or always negative (independently of the choice of $[\omega, \mu]$). In particular, if $x \in F$, then x is either always unstable, (i.e., not semistable), or always quasi-stable, for all $[\omega, \mu]$ in a fixed C_s. On the other hand, if $x \notin F$, then we may write $\overline{Gx} = Gx \bigcup 0(x) \bigcup \infty(x)$, where $0(x)$ and $\infty(x)$ are points with $\mu(0(x)) < \mu(\infty(x))$, and then its image $\mu(\overline{Gx})$ is the closed interval $[\mu(0(x)), \mu(\infty(x))]$. Thus by what is mentioned above, the semistability of a point x is independent of the choice of the points $[\omega, \mu]$ inside C_s (not necessarily of C_{sp}).

In the case where $G = \mathbf{C}^{*k}$ is a general algebraic torus, we prove the theorem by continuity argument.

We put a natural topology on C_{sp} as a locally closed submanifold of the Euclidian space $H_K^2(X)$. Let $A = A(x)$ (resp. $B = B(x)$, resp. $C = C(x)$) be the set of those points β of C_{sp} with respect to which x is semistable (resp. unstable, resp. quasi-stabile). We show that both A and B are open in C_{sp}; then $C_{sp} = A$ or $C_{sp} = B$ as desired. First of all by the definition of semistability it is clear that B is open in C_{sp}.

For the openness of A it suffices to show the openness of $C = C(y)$ (for all $y \in X$); indeed, if x is semistable for some $\beta \in C_{sp}$, then the closure $\overline{Gx}$ contains a quasi-stable orbit, say Gy. Then by the openness of $C(y)$ there exists a neighborhood N of β in C_{sp} such that Gy is still a quasi-stable orbit with respect to any β' in N. Then x is semistable with respect to β', too. Thus $A = A(x)$ is open, too.

Now we show that $C = C(x)$ is open. We may write $K_x^0 = S_i$ for some i in the notation of **18**. Then S_i fixes every point of Gx and μ_a is constant on Gx for any $a \in \underline{s}_i$ and for any Kähler class $[\omega, \mu]$ in C_s. Thus $\mu|Gx : Gx \to \underline{k}^*$ factors through the subspace $\underline{s}_i^{\perp}$ in $\underline{k}^*$ of annihilators of s_i. The defining equation of the moment map then shows that as a map from Gx to $\underline{s}_i^{\perp}$, $\mu|Gx$ is a submersion. Suppose first that $S_i \neq K$ so that dimension $Gx > 0$. Then if $[\omega, \mu] \in C(x)$ and μ admits a zero on Gx then so does μ' on Gx for all $[\omega', \mu'] \in C_s$ which are sufficiently close to $[\omega, \mu]$. If, on the other hand, $S_i = K$ then either x is quasi-stable for all $[\omega, \mu]$ or x is unstable for all $[\omega, \mu]$ in C_s as in the argument in the case $k = 1$. Thus, the openness of C is proved.

Finally we treat the general case: if $\dim K_x > 0$, then by the definition of a chamber either $\mu(x)$ never vanishes for any Kähler classes $[\omega, \mu]$ in C_{sp}, or $\mu(x) = 0$ for all the Kähler classes $[\omega, \mu]$. Thus, we may assume that K_x is finite. Then $\mu|Gx : X \to \underline{k}^*$ is a submersion for all $[\omega, \mu]$ in some chamber. Thus, the continuity argument as in the torus case applies also in this case. □

5. Variation of the Kähler Classes on the Quotient

20. Fix a Kähler cocycle (ω, μ). We consider a good case $X^s = X^{ss}$. We denote this set by U. Then $V := U/G = \mu^{-1}(0)/K$ is a compact Kähler orbifold. We want to give a cohomological description of $H^2(V)$ in terms of the equivariant cohomology group of the action.

Let $D_i, 1 \leq i \leq d$, be the irreducible components of $X - U$ of codimension one. Then the given K-action canonically lifts to a holomorphic action on the line bundle $[D_i]$ such that the action preserves the canonical sections of $[D_i]$. Let $c_1^K(D_i)$ be the equivariant chern class of the line bundle $[D_i]$. Then we prove the following:

Theorem 5.1. *There exists a natural exact sequence of real vector spaces*

$$0 \to \bigoplus_{1\le i\le d} \mathbf{R}c_1^K(D_i) \xrightarrow{c} H_K^2(X) \xrightarrow{r} H^2(V) \to 0$$

which is defined over $\mathbf{Q}$ *and is compatible with the Hodge structures, where* $c_1^K(D_i)$ *are purely of type* $(1,1)$.

21. Since the G-action is locally free on U in the sense that the stabilizer at each point is finite, for the real cohomology we have natural isomorphisms $H_K^2(U) \cong H_G^2(U) \cong H^2(V)$. Hence the theorem results from the following:

Lemma 5.2. *We have a natural exact sequence of real vector spaces*

$$0 \to \bigoplus_{1\le i\le d} \mathbf{R}c_1^K(D_i) \xrightarrow{c} H_K^2(X) \xrightarrow{r} H_K^2(U) \to 0$$

which is defined over $\mathbf{Q}$ *and is compatible with the Hodge structures, where* $c_1^K(D_i)$ *is purely of type* $(1,1)$ *and* r *is the restriction map.*

Proof. This follows from the equivariant perfectness of a certain G-invariant stratification of X proved by Kirwan [10]. By [10, §6] there exists a stratification $\{S_\alpha\}_{1\le\alpha\le\ell}$ of X by connected Zariski locally closed submanifold S_α of X with the following properties:

(1) $S^\alpha := \bigcup_{1\le\beta\le\alpha} S_\beta$ is an analytic subset of X, and hence, $U_\alpha = X - S^\alpha$ is Zariski open subset of X.

(2) The Thom-Gysin long exact sequence associated to the pair $(U_\alpha, S^\alpha), 0\le\alpha<\ell$, splits into short exact sequences as follows;

$$0 \to H_K^{i-2d(\alpha+1)}(S_{\alpha+1}) \to H_K^i(U_\alpha) \xrightarrow{r_i} H_K^i(U_{\alpha+1}) \to 0, \tag{8}$$

where r_i is the restriction map and $d(\alpha)$ is the complex codimension of S_α in X.

(3) S_ℓ is the unique open stratum and coincides with X^{ss}. We have $S_\ell = U_{\ell-1} = U$.

Moreover, we may rearrange the indices α so that $d(\alpha)\ge 2$ if and only if $1\le\alpha\le k$ for some k. In particular, the set $\{D_i\}$ coincides with the set $\{\bar{S}_\alpha\}_{k<\alpha\le\ell-1}$, where $\bar{S}_\alpha$ is the closure of S_α. We now consider the case $i=2$. By (8) for $1\le\alpha<k$ we have $H_K^2(U_\alpha) \overset{r}{\cong} H_K^2(U_{\alpha+1})$ and therefore $H_K^2(X) \overset{r}{\cong} H_K^2(U_k)$, while for $k<\alpha\le\ell-1$ we have an exact sequence

$$0 \to \mathbf{R} \cong H_K^0(S_\alpha) \to H_K^2(U_\alpha) \xrightarrow{r} H_K^2(U_{\alpha+1}) \to 0.$$

Combining all these we get the bottom exact sequence in the following commutative diagram of vector spaces

$$\begin{array}{ccccccccc} 0 & \longrightarrow & \bigoplus_{k<\alpha\le\ell-1} H_K^0(\overline{S}_\alpha) & \longrightarrow & H_K^2(X) & \xrightarrow{r} & H_K^2(X - \bigcup_{k<\alpha\le\ell-1} S_\alpha) & \longrightarrow & 0 \\ & & \downarrow & & \downarrow & & \downarrow & & \\ 0 & \longrightarrow & \bigoplus_{k<\alpha\le\ell-1} H_K^0(S_\alpha) & \longrightarrow & H_K^2(U_k) & \xrightarrow{r} & H_K^2(U) & \longrightarrow & 0. \end{array}$$

where the vertical arrows are restriction maps, and all isomorphic by the dimensional reason as above. The lemma is proved. □

22. By the above theorem we have

$$H^{1,1}(V) \cong H_K^{1,1}(X)/\bigoplus \mathbf{R}c_1^K(D_i).$$

Let $\bar{C}_K = r(C_K)$ be the image of the Kähler cone C_K in $H_K^{1,1}(X)$. This is an open convex cone in $H^{1,1}(V)$. Now we consider the class $[\omega,\mu]$ of the Kähler cocycle (ω,μ) which we have started with. Since we have assumed that we are in a good case, $[\omega, \mu]$ is in a chamber, say B, of the Kähler cone C_K.

Proposition 5.3. *Let $\bar{B}$ be the image of B in $\bar{C}_K$. Then $\bar{B}$ is contained in the Kähler cone $C(V)$ of V. More precisely, the induced map $r|B : B \to \bar{B}$ is obtained by $r([\omega, \mu]) = [\omega_{\mathrm{MW}}]$, where ω_{MW} is the natural Kähler form on U/G identified naturally with $\mu^{-1}(0)/K$.*

Proof. Let (η, ν) be an arbitrary element of $\bar{B}$. Since $U = X^{ss}$ also for the cocycle (η, ν) by Theorem 4.6, we may assume that $(\eta, \nu) = (\omega, \mu)$. The map r is factored as $H_K^2(X) \to H_K^2(U) \cong H_K^2(\mu^{-1}(0)) \cong H^2(V)$ (cf. [10, §10]). Then the image of $[\omega, \mu]$ in $H_K^2(\mu^{-1}(0))$ is represented by a pair $(\omega|\mu^{-1}(0), 0)$, whose image in $H^2(V)$ is in turn represented by the Marsden-Weinstein Kähler form ω_{MW} with $\pi^*\omega_{\mathrm{MW}} = \omega|\mu^{-1}(0)$, where $\pi : \mu^{-1}(0) \to V$ is the natural projection. This proves the proposition. □

23. In the case of an algebraic torus K acting quasi-freely on a compact Kähler manifold (X, ω) with a moment map μ Duistermaat-Heckman [5] studied the variation of Kähler classes $[\omega_{\mathrm{MW}}]$ on V when the original moment map μ is replaced by $\mu + s$ for a constant element $s \in \underline{k}^*$. Their result is that then $[\omega_{\mathrm{MW}}]$ is changed into

$$[\omega_{\mathrm{MW}}] + \langle c_1(\mu^{-1}(0)/V), s\rangle \tag{9}$$

on V, where $c_1(U/V) \in H^2(V, \underline{k})$ is the real chern class of the principal bundle with structure group the torus $K = (S^1)^k$ and $\langle\ ,\ \rangle$ denotes the natural pairing $\underline{k} \times \underline{k}^* \to \mathbf{R}$. In particular, the variation is a linear function with respect to s.

In our formulation the result is interpreted in terms of equivariant cohomology as follows. This enable us to generalize partly the above result of [5, Th. 1.1] for a general variation of the original Kähler classes, which is not necessarily of the form $[\omega, \mu] \to [\omega, \mu + s]$. This is essentially contained in the following:

Proposition 5.4. *There exists a commutative diagram of exact sequences of real vector spaces*

$$\begin{array}{ccccc} \longrightarrow c^* & \xrightarrow{q} & H^2(V) & \xrightarrow{\pi^*} & H^2(U) \\ \| & & b\downarrow\| & & \downarrow\| \\ \longrightarrow c^* & \xrightarrow{a} & H_K^2(\mu^{-1}(0)) & \longrightarrow & H^2(\mu^{-1}(0)). \end{array} \tag{10}$$

Proof. The top line is a part of the five term exact sequence associated to the Leray spectral sequence for the map $\pi : U \to V$ if we take into account the natural isomorphisms

$$H^1(G) \cong H^1(K) \cong H^1(C) \cong \underline{c}^*.$$

The bottom sequence arises from the Leray spectral sequence for the map $\mu^{-1}(0)_K \to B_K$ (cf. the proof of Lemma 3.1 for the de Rham description of the maps.) Furthermore, b is the isomorphism coming from the relation $V = \mu^{-1}(0)/K$.

We check the commutativity of the diagram. First, in the de Rham description b is induced by the pull-back of differential forms $\phi \to (\pi^*\phi, 0)$. Hence the commutativity of the right square of (10) is clear.

We shall show the commutativity of the left square by giving the de Rham description of the map q. Fix a connection θ on $\mu^{-1}(0)$ with curvature form F. (All the arguments here can be carried out in the orbifold category under a suitable definition of a connection, a curvature form etc., although one may restrict oneself to the case of quasi-free action as in [5].) Identify $\underline{c}^*$ with the space $\hat{k}$ of characters of the Lie algebra $\underline{k}$. Then for any element

$\xi \in \underline{c}^*$, $\xi(F)$ is a K-invariant 2-form on $\mu^{-1}(0)$ which annihilate all the vertical vector fields. Moreover, since $F = d\theta + \frac{1}{2}[\theta, \theta]$, we have

$$\begin{aligned}\xi(F) &= \xi(d\theta) + \frac{1}{2}\xi([\theta,\theta]) \\ &= \xi(d\theta) \text{ because } \xi \text{ is a character} \\ &= d(\xi(\theta))\end{aligned}$$

Thus $\xi(F)$ descends to a d-closed 2-form, say ϕ, on V. Then the assignment $\xi \to \phi$ induces a map $q : \underline{c}^* \to H^2(V)$, which, composed with π^*, is mapped to zero in $H^2(\mu^{-1}(0))$ since $\pi^*\phi = d(\xi(\theta))$

We shall check that the definition of q is independent of the choice of the connection θ. Let θ' be another connection and F' its curvature form. Then we may write $\theta' = \theta + \pi^*\alpha$ for some $\underline{k}$-valued 1-form α on V. Then F' is given by

$$F' = F + \pi^* d\alpha + \frac{1}{2}\pi^*[\alpha, \alpha]$$

and hence

$$\xi(F') = \xi(F) + \pi^*\xi(d\alpha) = \xi(F) + \pi^* d(\xi(\alpha)).$$

Therefore, if ϕ' corresponds to ξ', we have $\phi' = \phi + d\xi(\alpha)$ and q is well-defined.

We now show the commutativity of the left square of (10). First of all, in the de Rham description bq is given by the composition

$$\xi \to \phi \to (\pi^*\phi, 0),$$

while that of a is given by $\xi \to (0, \xi)$. Thus it suffices to show that $(\pi^*\phi, -\xi)$ is a coboundary of the K-invariant 1-form $\xi(\theta)$. In fact:

$$d_K(\xi(\theta)) = (d(\xi(\theta)), -\iota_{\underline{a}}(\xi(\theta))) = (d\xi(\theta), -\xi(\iota_{\underline{a}}(\theta))) = (d(\xi(\theta)), -\xi(a)) = (\pi^*\phi, -\xi),$$

where a is considered as a variable in $\underline{k}$ by some abuse of notation. Thus the proposition is proved. □

24. We explain the relation of the above proposition with the above cited result of [5]. The de Rham description shows that the change of the moment map $\mu \to \mu + s$ corresponds to the translation of the Kähler class $[\omega, \mu]$ by the image by a of an element $s \in \underline{c}^*$. Furthermore, the commutativity of the left square is precisely the formula (9) of [5]:

$$b^{-1}([\omega, \mu + s]) = [\omega_{\mathrm{MW}}(\mu + s)] = [\omega_{\mathrm{MW}}(\mu) + \langle c_1(U/V), s\rangle]$$

in the case of a torus. Thus the isomorphism b is considered to be giving a certain generalization of the formula (9) to a general variation of Kähler classes. In particular, we have

Theorem 5.5. *Any linear change of (equivariant) Kähler classes in a fixed chamber B on V results in a linear change in the corresponding Marsden-Weinstein Kähler classes. More precisely, the MW correspondence $[\omega, \mu] \to [\omega_{\mathrm{MW}}(\mu)]$ for $[\omega, \mu] \in B$, extends to a linear map $H_K^{1,1}(X) \to H^{1,1}(V)$.*

6. Comparison of the Kähler Cones of Adjacent Quotients

25. Given two adjacent chambers $B_i, i = 1, 2$, of the equivariant Kähler cone C_K in $H_K^{1,1}(X)$ we are interested in the relations between the corresponding quotients $\pi_i : U_i \to V_i$. In this section we give some observations in this direction. We assume that G is an algebraic torus and the actions are good so that V_i is compact. We shall study the relation of the images $\bar{B}_i$ of B_i in the Kähler cone $C_i := C(V_i)$ of $H^{1,1}(V_i)$.

Let W be the wall separating B_i. Then there exist a 1-dimensional subtorus S of K and a connected component F of the fixed point set of S on X such that there exists a Zariski open subset F^0 of F and that W is the hyperplane defined by the pair (S, F^0) in the sense of **17.**. Let N be the normal bundle of F in X with the natural action of $S^{\mathbf{C}} \cong S^1 \times \mathbf{R}^+$. N is naturally decomposed into a direct sum $N = N^+ \bigoplus N^-$ of the subbundles $N^{\pm}$, where the subgroup $\mathbf{R}^+$ acts via a positive (resp. negative) character on N^+ (resp. N^-). Let p (resp. q) be the rank of N^+ (resp. N^-). We assume for simplicity that $p \geq q$. Then $p+q$ is the codimension of F in X. By our assumption that W intersect with the Kähler cone it follows easily that $q > 0$. V_1 and V_2 are naturally bimeromorphic to each other (cf. Theorem 2.2, 3)). From the properties of the quotients π_i as described before Proposition 4.2 the following are immediately seen.

Lemma 6.1. *Let E^+ (resp. E^-) be the set of points x of X such that $0(x) := \lim_{t \to 0} tx$ (resp. $\infty(x) := \lim_{t\to\infty} tx$) belongs to F, where $t \in \mathbf{R}^+$. Then $V_1 - \bar{E}^+$ and $V_2 - \bar{E}^-$ are naturally isomorphic to each other (if we interchange V_1 and V_1 if necessary), where $\bar{E}^+$ is the closure of the image of E^+ in V_1, and is an irreducible analytic subset of V_1 and similarly for $\bar{E}^-$. Moreover, $\dim \bar{E}^+ = \dim V_1 - q$ and $\dim \bar{E}^- = \dim V_2 - p$.*

26. It is convenient to distinguish the three cases:
Case A. $p, q \geq 2$
Case B. $p > q = 1$
Case C. $p = q = 1$.
Case A is the most interesting case and here we shall just check the following:

Theorem 6.2. *In Case A there exists a natural identification of $H^{1,1}(V_1)$ and $H^{1,1}(V_2)$. (Denote the identified space by H and consider $\bar{B}_i$ as a subset of $H^{1,1}$.) Moreover, C_1 and C_2 have no common point. Let $\bar{W}$ be the image of W in H. Then the part of $\bar{W}$ which is the common boundary of $\bar{B}_i$ forms a part of the boundary of the Kähler cone of $H^{1,1}(V_i)$.*

Proof. The first assertion is clear since $\bar{E}^+$ and $\bar{E}^-$ is of codimension ≥ 2 in V_1 and V_2 respectively. Suppose that C_1 and C_2 have a common element. Then there exist Kähler classes $[\omega_i, \mu_i]$ in $\bar{B}_i$ such that $[\omega_1, \mu_1] = [\omega_2, \mu_2] + \sum a_j c_1^K(D_j)$ for some real numbers a_j, where D_j are irreducible divisors in the boundary of U_i in X, which are common for $i = 1$ and 2. Then for suitable K-invariant hermitian connections θ_j of the bundle $[D_j]$ we may write $(\omega_1, \mu_1) = (\omega_2, \mu_2) + \sum a_j (i/2\pi d\theta_j, \mu^{\theta_j})$ in the notation of **4**. We show that such a relation is impossible.

Since F^0 is contained in X^{ss} for both of $\bar{B}_i$, F has the empty intersection with D_j. Hence the canonical section trivializes the restricted bundle $[D_j]|F$. Since the canonical section is left fixed by the natural K-action on $[D_j]$, the vertical component of the lifted vector fields $\tilde{\underline{a}}$ on $[D_j]|F$ with respect to this trivialization must vanishes identically for any $a \in \underline{k}$. Thus $\tilde{\underline{a}}$ must vanish identically on $[D_j]|F$ and hence so does $\pi_j^* \mu_a^{\theta_j} = \theta_j(\tilde{\underline{a}})$, where $\pi_j : [D_j] \to X$ is the natural projection. This implies that μ_{1a} and μ_{2a} have the same (constant) value on F. This contradicts the fact that $[\omega_1, \mu_1]$ and $[\omega_2, \mu_2]$ belong to the two chambers separated by the wall W defined by the value of μ on F.

To show the last assertion we assume, say, that the Kähler cone of $H = H^{1,1}(V_1)$ extends across $\bar{W}$ from C_1 toward C_2. It follows then that there exists an element $[\bar{\omega}]$ of $H^{1,1}$, which represents a Kähler class on each of the (V-)manifolds V_i. But since V_i are mutually bimeromorphic and isomorphic in codimension one, this is impossible (cf. [7, Th. 3.2]). □

Case B. From Lemma 6.1 we deduce that in this case $\bar{E}^-$ is an irreducible divisor in V_2 and there exists a natural direct sum decomposition $H^{1,1}(V_2) = H^{1,1}(V_1) \bigoplus \mathbf{R}c_1^K(\bar{E}-)$ and C_1 is a part of the boundary of the Kähler cone of $H^{1,1}(V_2)$.

Case C. In this case we can prove the following:

Proposition 6.3. *Suppose that* $\dim \underline{k} = 1$ *so that* $K = S$. *Then* V_1 *and* V_2 *are canonically isomorphic. There exists a natural isomorphism* $H_K^2(U) \cong H_K^2(U)$ *inducing a natural isomorphism* $H^2(V_1) \cong H^2(V_2)$ *by passing to the quotients.*

References

[1] Atiyah, M.F.: Convexity and commuting Hamiltonians, *Bull. London Math. Soc.* **14**, 1–15.

[2] Atiyah, M.F., and Bott, R.: The Moment Map and Equivariant Cohomology, *Topology* **23** (1984), 1–28.

[3] Berligne, N., Getzler, E., and Vergne, M.: *Heat kernels and Dirac operators*, Grundlehren der math. Wiss. 298, 1992.

[4] Dolgacev, I. and Hu, Y.: Variation of quotients in geometric invariant theory and applications to moduli problems, Michigan preprint 1992.

[5] Duistermaat, J.J., and Heckman, On the variation in the cohomology in symplectic form on the reduced phase space, *Invent. math.* **69** (1982), 259–268.

[6] Fujiki, A.: On automorphism groups of compact Kähler manifolds, *Inventiones math.* **44** (1978), 225–258.

[7] Fujiki, A.: A theorem of bimeromorphic maps of Kähler manifolds and its applications, *Publ. RIMS, Kyoto Univ.* **17** (1981), 735–754.

[8] Heinzner, P., and Loose, F.: Reduction of complex Hamiltonian G-spaces, *Geometric and Functional Analysis* **4** (1994), 288–297.

[9] Heinzner, P., and Huckleberry, A.T.: Invariant plurisubharmonic exhaustion and retractions, *Manuscripta math.* **83** (1994), 19–30.

[10] Kirwan, F.C.: *Cohomology of quotients in symplectic and algebraic geometry*, Math. Notes 31, Princeton Univ. Press, 1984.

[11] S. Kobayashi: *Transformation groups in differential geometry*, Berlin Heidelberg New York, Springer Verlag, 1972.

[12] Moser, J.: On the volume elements on a manifold, *Trans. Amer. Math. Soc.* **120** (1965), 286–294.

[13] Mostow, G.: On a conjecture of Montgomery, *Ann. of Math.* **65** (1957), 513–516.

[14] Mumford, D.: *Geometric invariant theory*, Berlin Heidelberg New York, Springer-Verlag, 1965.

[15] Snow, D.: Reductive group actions on Stein spaces, *Math. Ann.* **259** (1982), 79–97.

[16] Varouchas, J.: Kähler space and proper open morphisms, *Math. Ann.* **283** (1989), 13–52.

MAGNETIC MONOPOLES WITH PLATONIC SYMMETRY

N. J. Hitchin

Department of Pure Mathematics and Mathematical Statistics, 16 Mill Lane, Cambridge CB2 1SB, England

1. Introduction

The study of $SU(2)$ magnetic monopoles in $\mathbf{R}^3$ translates, via the twistor construction, into the language of rank 2 holomorphic vector bundles on the total space of the tangent bundle to the projective line. In this paper we approach from this point of view the question of finding monopoles which are invariant under the action of the symmetry group of a Platonic solid. The problem equates to one of finding some rather special symmetrical algebraic curves. We make use of essentially classical techniques to show that there exists a tetrahedrally symmetric monopole of charge 3 and an octahedrally symmetric monopole of charge 4.

There is an alternative construction of monopoles, using solutions to a system of ordinary differential equations – Nahm's equations – and in [4] this method is used to give more explicit information about the monopoles whose existence is proved here algebraically. It is also shown in that paper that there exists no icosahedrally-symmetric monopole of charge 6, even though there seems to be a good candidate for the corresponding algebraic curve. We explain this fact here from the twistor point of view.

The holomorphic bundle of a k-monopole is expressed as an extension. Restricting to the zero section of the tangent bundle of P^1 this bundle is trivial, and its global sections define a 2-dimensional subspace of $H^0(P^1, \mathcal{O}(k))$. An invariant monopole defines an invariant subspace. To find such subspaces, we regard $H^0(P^1, \mathcal{O}(k))$ as an irreducible representation space of $SU(2)$, and break it up under the action of the binary group $\tilde{G} \subset SU(2)$ covering the symmetry group G of a Platonic solid. The McKay correspondence, which underlies much of this paper, enables us easily to do this. We show this way that the first few admissible charges for an icosahedrally symmetric monopole are $k = 1, 7, 11, 13$. It is interesting to note that these are the first four exponents of the Lie algebra E_8.

The author wishes to thank M.F.Atiyah for many useful conversations, and in particular for suggesting an approach to show that the charge 6 icosahedrally symmetric monopole does not exist.

2. Monopoles, vector bundles and spectral curves

An $SU(2)$-monopole on $\mathbf{R}^3$ is a solution to the Bogomolny equations

$$\nabla\phi = *F \tag{1}$$

satisfying suitable boundary conditions (see for instance [2]). These conditions imply in particular that each solution determines a positive integer invariant, its charge k, and then

$$\int_{\mathbf{R}^3} |F|^2 = 4\pi k$$

Since the Bogomolny equations are invariant under oriented Euclidean transformations, we may consider solutions invariant by any subgroup of $SO(3)$.

From the holomorphic point of view $\mathbf{R}^3$ is replaced by its twistor space, the space of oriented straight lines. This is a 2-dimensional complex manifold, which has the structure of the tangent bundle TP^1, with projection $p : TP^1 \to P^1$ assigning to the line its direction. Reversing orientation of the line defines a real structure on TP^1, and the points of $\mathbf{R}^3$ correspond to real sections of the tangent bundle. Each such 2-sphere represents the oriented lines passing through the corresponding point of $\mathbf{R}^3$.

An arbitrary solution to (1) corresponds in twistor language to a holomorphic rank 2 bundle on TP^1, trivial on each real section, and with an appropriate real structure. If the boundary conditions are imposed, then a charge k solution is generated by a curve S in TP^1, the spectral curve of the monopole. The spectral curve is the divisor of a real section of $p^*\mathcal{O}(2k)$ (which we continue to denote by $\mathcal{O}(2k)$). If $\eta \in H^0(TP^1, \mathcal{O}(2))$ is the tautological section on $TP^1 (\cong \mathcal{O}(2))$, the equation of such a curve is of the form $\eta^k + a_1\eta^{k-1} + \cdots + a_k = 0$ where $a_i \in H^0(P^1, \mathcal{O}(2i))$ are real sections. Necessary and sufficient conditions for S to be a spectral curve are:

- L^2 is trivial on S
- $H^0(S, L^s(k-2)) = 0$ for $0 < s < 2$

Here L^s is an $SO(3)$-invariant line bundle on TP^1, restricted to S. It is invariantly defined in $H^1(S, \mathcal{O}^*)$ as $\exp(s\eta x)$ where $x \in H^1(P^1, \mathcal{O}(-2))$ is the canonical generator. The vector bundle E is obtained as an extension

$$L(-k) \to E \to L^*(k) \tag{2}$$

from the trivialization $s \in H^0(S, L^2)$ in the long exact sequence

$$\cdots \to H^0(TP^1, L^2) \to H^0(S, L^2) \to H^1(TP^1, L^2(-2k)) \to \ldots$$

Let $G \subset SO(3)$ be the symmetry group of a regular solid and $\tilde{G} \subset SU(2)$ the corresponding binary group. Let V be the 2-dimensional defining representation of $SU(2)$. The n-th symmetric power S^nV may be considered as the action of $SU(2)$ on homogeneous polynomials of degree n, or more geometrically as the space of holomorphic sections of the line bundle $\mathcal{O}(n)$ on P^1. The tautological section η of $\mathcal{O}(2)$ over TP^1 is $SO(3)$-invariant, so an invariant section of $\mathcal{O}(2k)$, such as defines a symmetrical spectral curve, has the form $\eta^k + a_1\eta^{k-1} + \cdots + a_k$ where each $a_i \in H^0(P^1, \mathcal{O}(2i))$ is invariant and thus corresponds to an invariant in $S^{2i}V$.

Let n be the smallest degree for which S^nV has an invariant under $\tilde{G}$. For the tetrahedral, octahedral and icosahedral group the value of n is $2k$ (see [4] or Section 6) where k is respectively 3,4,6. The invariant sections of $\mathcal{O}(2k)$ are given in terms of an affine parameter ζ by

- $\zeta(\zeta^4 - 1)$
- $\zeta^8 + 14\zeta^4 + 1$
- $\zeta(\zeta^{10} + 11\zeta^5 - 1)$

These polynomials have a straightforward geometrical interpretation in terms of their roots, which are points in $P^1 \cong S^2$. In the first case the roots (including ∞) are the mid-points of the six edges of a tetrahedron, in the second the vertices of a cube, and in the third the vertices of an icosahedron. Note that in all cases $|G| = 2k(k-1)$.

We define curves in TP^1 by

$$\eta^3 + a(\zeta(\zeta^4 - 1)) = 0 \tag{3}$$

$$\eta^4 + a(\zeta^8 + 14\zeta^4 + 1) = 0 \tag{4}$$

$$\eta^6 + a(\zeta(\zeta^{10} + 11\zeta^5 - 1)) = 0 \tag{5}$$

where a is real. Each such curve S is invariant by the corresponding group G and satisfies the reality conditions for a spectral curve. The question we ask is: can we choose a so that each is the spectral curve of a monopole?

We first prove some basic facts about S.

Lemma 1. *Each orbit of G on S is of length $|G|$ or $2k$. There is a unique orbit of length $2k$.*

Proof: Let m be a fixed point of an element of G. Consider its image in P^1. The group G acts on S through the natural action on the tangent bundle TP^1, but for isometries the action of an isotropy group on the tangent space of the point is faithful. Thus the zero vector is the only one fixed. But this is given by $\eta = 0$ and from the equation of the curve these points are the $2k$ zeros of the polynomial.

Lemma 2. *The curve S is smooth and of genus $(k-1)^2$.*

Proof: Smoothness is immediate since the polynomials by inspection have distinct roots. The genus formula is standard [3],[2].

Lemma 3. *The curve S is non-hyperelliptic.*

Proof: The hyperelliptic involution σ, if it exists, is unique and thus commutes with the action of G. The $2g+2$ fixed points of σ must therefore be a union of orbits of G. But $2g+2 = 2(k^2 - 2k + 2)$, which is less than $|G| = 2k(k-1)$ and bigger than $2k$. From Lemma 1, this is impossible.

Lemma 4. *The fixed point set of G acting on $H^1(S, \mathcal{O})$ is one-dimensional.*

Proof: By Serre duality, we can consider $H^0(S, K)$. Since S is the divisor of a section of $\mathcal{O}(2k)$ on TP^1, we have a canonical isomorphism $K \cong \mathcal{O}(2k-4)$. But from [3], every section s of $\mathcal{O}(2k-4)$ can be written uniquely as

$$s = \eta^{(k-2)} + a_1\eta^{(k-3)} + \cdots + a_{k-2}$$

where a_i is the pull-back of a section of $\mathcal{O}(2i)$ on P^1. Since η is G-invariant, this decomposition is respected by the action, and by the definition of k, there are no invariant sections of $\mathcal{O}(2i)$ for $i < k$. Hence the fixed point set consists of the scalar multiples of $\eta^{(k-2)}$, which is one-dimensional. Dually, the fixed points in $H^1(S, \mathcal{O})$ are multiples of ηx.

Lemma 5. *The identity component of the fixed point set of G in $H^1(S, \mathcal{O}^*)$ is an elliptic curve.*

Proof: Since the action of G is induced by a geometrical action on S, the integral lattice in $H^1(S, \mathcal{O})$ is preserved, and so the fixed point set is a union of abelian subvarieties. From Lemma 4, these are one-dimensional and hence elliptic curves.

With this data we can establish the first property of a spectral curve for S.

Proposition 1. *For each $k = 3, 4, 6$, there exists a real number a such that the corresponding curve S with equation (3), (4), (5) satisfies $L^2 \cong \mathcal{O}$. The number a is unique for $k = 4, 6$ and unique up to sign for $k = 3$.*

Proof: Let S_0 be the curve with $a = 1$. The line bundle L^s for $s \in \mathbf{R}$ restricted to S_0 lies in the identity component of the space of G-invariant line bundles. Moreover, L^s is imaginary with respect to the real structure τ in the sense that $\tau^* L^s = L^{-s}$. Since the identity component of the space of invariant bundles is an elliptic curve, the identity component of the space of imaginary bundles is a circle, thus there is a real number s, unique up to sign, such that L^{2s} is trivial, but L^s is not.

Now $L^s = \exp(s\eta x)$, so $L^2 \cong \mathcal{O}$ on the curve with $a = s^{-k}$. When k is even a is uniquely defined and when odd, defined up to sign.

The second, non-degeneracy, condition necessitates consideration of the representations of the group G. We have the following lemma:

Lemma 6. *Let U be an invariant line bundle such that $H^0(S, U)$ contains a one-dimensional representation of $\tilde{G}$. Then either $U \cong \mathcal{O}(2m)$ for some m, or $\deg U \geq 2k(k-1)$.*

Proof: If the section u generates the 1-dimensional space, then its divisor is a union of orbits. If the degree of the divisor is less than $2k(k-1) = |G|$, then by Lemma 1 it is the divisor η^m for some m. This however means that $U \cong \mathcal{O}(2m)$.

With a piece of classical algebraic geometry, we can now deal with the case $k = 3$.

3. Tetrahedral symmetry

Proposition 2. *There is a real number a, which is unique up to sign, such that $\eta^3 + a(\zeta(\zeta^4 - 1)) = 0$ is the spectral curve of a monopole with tetrahedral symmetry.*

Proof: Taking the value of a from the previous lemma, S is a curve of genus 4 satisfying the constraint $L^2 \cong \mathcal{O}$. We need to show that $H^0(S, L^s(1)) = 0$ for $0 < s < 2$.

Now the line bundle $L^s(1)$ on S has degree 3. From Lemma 3, S is non-hyperelliptic, so by Clifford's theorem,

$$\dim H^0(S, L^s(1)) \leq [3/2] + 1 = 2.$$

The dimension cannot be one from Lemma 6. On the other hand, if it is two-dimensional, we have a g_3^1 – a linear system of degree 3 and dimension 1. A non-hyperelliptic curve of genus 4 is, in its canonical embedding, the intersection of a cubic and a quadric, and has two g_3^1's if the quadric is nonsingular, and one if it is singular [1].

However, our curve S lies in TP^1. The four linearly independent sections of $\mathcal{O}(2)$ on TP^1 given by a basis of $H^0(P^1, \mathcal{O}(2))$ and the tautological section η embed TP^1 in P^3 as the complement of the vertex of a cone – a singular quadric. Thus since $K_S \cong \mathcal{O}(2k-4) \cong \mathcal{O}(2)$, the embedding $S \subset TP^1 \subset P^3$ is the canonical embedding and so S has a unique g_3^1, namely $\mathcal{O}(1)$. Thus if $s \neq 0, 2$ the line bundle $L^s(1)$ has no sections.

Since the conditions for a spectral curve are now satisfied, we deduce the existence of a corresponding monopole.

4. The McKay correspondence

To deal with the other cases, we need to consider irreducible representations of G or $\tilde{G}$ on sections of invariant line bundles on S. Depending on the degree, it is $\tilde{G}$ rather than G which acts.

The McKay correspondence relates the irreducible representations R_i of $\tilde{G}$ to the vertices of an extended Dynkin diagram. For the tetrahedral, octahedral and icosahedral groups these are the diagrams E_6, E_7, E_8 respectively:

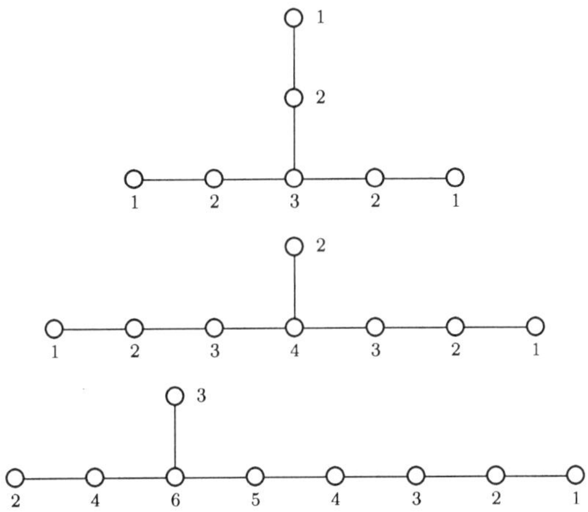

The two salient features of the correspondence are:

- The numbers attached to the vertices (by considering the coefficients of the highest root expanded in terms of the simple roots) are the dimensions of the representation spaces.
- The irreducible factors of $R_i \otimes V$, where V is the defining 2-dimensional representation, are given by the representations attached to the neighbouring vertices.

From the second remark above, since the graph is connected, it is clear that every irreducible is contained in some representation $\otimes^n V$, and hence from the Clebsch-Gordon decomposition in some $S^n V$. An equivalent way of determining the decomposition of the tensor product $R_i \otimes V$ is to use the incidence matrix C_{ij} of the diagram:

$$R_i \otimes V = \sum_j C_{ij} R_j \tag{6}$$

5. Octahedral symmetry

We begin with the following general lemma, which is a tool for induction. We shall apply it to the curve S given by (4).

Lemma 7. *Let U be an invariant line bundle on $\eta^k + a_1\eta^{k-1} + \ldots a_k = 0$ and suppose that $H^0(S, U(-1)) = 0$. Let R be an irreducible representation of $\tilde{G}$ in $H^0(S, U)$ and suppose R is contained in S^nV. Then $\deg U \geq 2k(k-1) - nk$.*

Proof: Since $V \cong V^*$ as an $SU(2)$ representation, if R is contained in S^nV, so is R^*. Thus

$$R \otimes R^* \subseteq H^0(S, U) \otimes H^0(P^1, \mathcal{O}(n))$$

and the multiples of the identity in $\operatorname{End} R \cong R \otimes R^*$ provide an invariant 1-dimensional subspace. Pulling back the sections of $\mathcal{O}(n)$ to S, and multiplying by those of U gives an invariant section of $U(n)$. But from Lemma 6, this is zero if $\deg U < 2k(k-1) - nk$.

Let $s_1, \ldots, s_r$ be a basis of $R \subseteq H^0(S, U)$, and

$$p_i(\zeta) = \sum_0^n a_{ij}\zeta^j$$

polynomials which form the dual basis for $R^* \subseteq H^0(P^1, \mathcal{O}(n))$. Then the vanishing of the invariant section of $U(n)$ means that

$$\sum_{i=1,j=0}^{i=r,j=n} a_{ij}\zeta^j s_i = 0 \tag{7}$$

But this implies that $s = \sum_1^r a_{i0}s_i$ vanishes where $\zeta = 0$. This is the divisor of a section of $\mathcal{O}(1)$, and so $s = \zeta t$ for some section t of $U(-1)$. By hypothesis $t = 0$, and so $s = 0$, and since the s_i form a basis,

$$a_{i0} = 0.$$

From this, we can rewrite (7) as

$$\sum_{i=1,j=1}^{i=r,j=n} a_{ij}\zeta^{j-1} s_i = 0$$

and repeat the argument, obtaining finally the vanishing of each coefficient a_{ij}, which gives $R = 0$, a contradiction.

We now turn to the case of the octahedral group.

Proposition 3. *There is a unique real number a such that $\eta^4 + a(\zeta^8 + 14\zeta^4 + 1) = 0$ is the spectral curve of a monopole with octahedral symmetry.*

Proof: We need to prove that $H^0(S, L^s(2)) = 0$ for $0 < s < 2$.

Consider the extended Dynkin diagram for E_7:

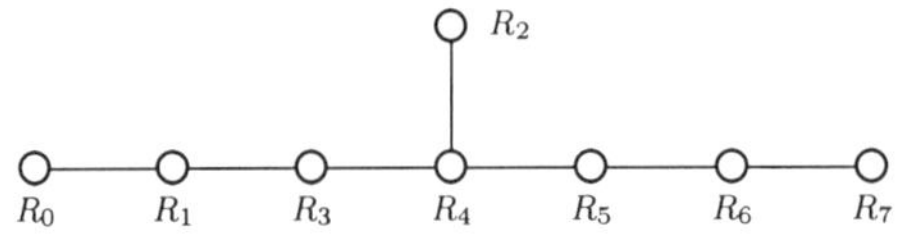

The octahedral group is isomorphic to the symmetric group S_4, which has the non-trivial 1-dimensional representation $g \mapsto \operatorname{sgn} g$. This is R_7 in the diagram. The symmetry between the left and right branches of the diagram can be seen in the McKay correspondence as the effect of tensoring a representation R with R_7. Now the proof of Lemma 7 can be easily adapted to the condition that $R \otimes R_7$ is contained in S^nV, since it depends on Lemma 6, which only requires a 1-dimensional invariant subspace, and not an invariant vector.

From the diagram, R or $R \otimes R_7$ is contained in S^nV for $n \leq 4$. Since $2(k-1)-1 = 6-1 = 5 > 4$, we deduce from Lemma 7 that $H^0(S, L^s(1)) = 0$ since L^s is non-trivial for $0 < s < 2$. Applying Lemma 7 again, we see that $H^0(S, L^s(2))$ contains no representations in S^nV for $n \leq 3$. There remains the question as to whether the 2-dimensional representation R_2, which lies in S^4V, is contained in the space.

This representation is the permutation representation on 3 objects obtained from the homomorphism $S_4 \to S_3$. In particular, it is real and carries a positive definite invariant inner product. Suppose this representation is contained in $H^0(S, L^s(2))$, and let s_1, s_2 be an orthonormal basis with respect to the inner product, then $s_1^2 + s_2^2 \in H^0(S, L^{2s}(4))$ is invariant, and by Lemma 6 vanishes unless L^{2s} is trivial. But $s_1^2 + s_2^2 = 0$ implies $s_1 = \pm i s_2$, contradicting linear independence. Thus $H^0(S, L^s(2)) = 0$ except possibly for $s = 1$.

In this latter case we must have

$$s_1^2 + s_2^2 = \lambda \eta^2 \tag{8}$$

for as in Lemma 4, η^2 is the only invariant section of $\mathcal{O}(4)$. Now in this particular representation, a 3-cycle in S_3 acts on $s_1 - is_2$ by multiplication by a cube root of 1. Thus the divisor of this section is A_4-invariant, and similarly for $s_1 + is_2$. But A_4 acts transitively on the vertices of the cube, which are the zeros of η. Thus (8) implies that either $\lambda = 0$, or $s_1 - is_2$, $s_1 + is_2$ and η have the same divisor which implies that L is trivial. Either possibility gives a contradiction, and so $H^0(S, L^s(2)) = 0$ for all $0 < s < 2$, thus proving the existence of the monopole.

One may use these methods to attack the icosahedrally symmetric curve (5), and in fact prove that $H^0(S, L^s(4)) = 0$ for $s \neq 1$. The explicit form of the solution to Nahm's equations in [4] shows, however that for $s = 1$ this is non-zero. In the next section, we shall see more directly that this curve cannot be the spectral curve of a monopole.

6. Invariant extensions

In Section 2 we saw that the holomorphic bundle E on TP^1 describing a monopole in twistor terms is an extension (2) of $L(k)$ by its dual. Restricting to a real section, the bundle must be trivial to describe a non-singular monopole, indeed this is a necessary condition for any solution of the Bogomolny equations. We therefore choose a point $x \in \mathbf{R}^3$ corresponding to a real section P_x of TP^1. The degree zero line bundle L is also trivial on the curve $P_x \cong P^1$, trivialized in an $SO(3)$-invariant manner. Since now $L(k) \cong \mathcal{O}(k)$ on P_x, we have an extension

$$\mathcal{O}(-k) \to E \to \mathcal{O}(k)$$

and from the exact cohomology sequence

$$0 \to H^0(P^1, E) \to H^0(P^1, \mathcal{O}(k))$$

the global sections of the trivial rank 2 bundle E define a 2-dimensional subspace of $H^0(P^1, \mathcal{O}(k))$.

Choosing the point x as the origin, a monopole which is symmetric under the action of $G \subset SO(3)$ defines a $\tilde{G}$-invariant subspace. We use this now to provide a restriction on the charge of a symmetric monopole. To proceed, we need to find invariant subspaces by decomposing $S^kV \cong H^0(P^1, \mathcal{O}(k))$ into irreducible representations under the action of $\tilde{G}$.

For any representation of $\tilde{G}$, the multiplicities of the irreducible factors define, according to the McKay correspondence, a function on the vertices of the corresponding Dynkin diagram, with non-negative integer values. For the representation S^kV, we denote this function v_k. Graphically, we can describe v_k by a labelling of the vertices of the Dynkin diagram. In fact the labelling in Section 4 is the regular representation.

Now recall that as a representation of $SU(2)$,

$$S^kV \otimes V = S^{k+1}V \oplus S^{k-1}V$$

so if $v_k = \sum_i a_i R_i$, then by the McKay correspondence

$$v_{k+1} + v_{k-1} = \sum_i a_i R_i \otimes V = \sum_{ij} C_{ij} a_i R_j$$

so that the vector v_k satisfies the recurrence relation

$$v_{k+1} - Cv_k + v_{k-1} = 0 \tag{9}$$

This is a straightforward relation to carry out by hand on the Dynkin diagram for small values of k: replace the value at a vertex by the sum of its neighbours and subtract the previous value.

6.1. Icosahedral symmetry. The basic extended Dynkin diagram for E_8

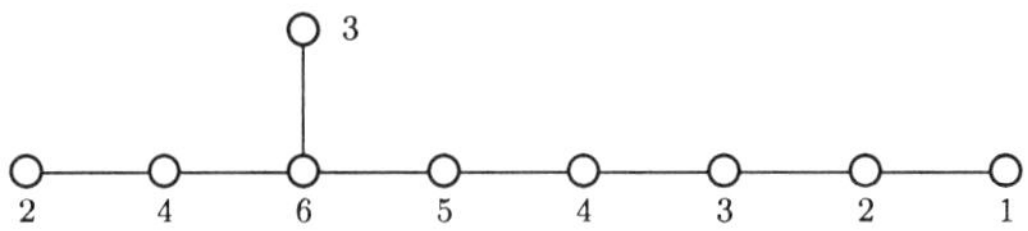

shows the dimensions of the irreducible representations of the binary icosahedral group. Other than the defining 2-dimensional representation, the only other one is at the far left of the diagram, 7 vertices away from the trivial representation at the far right. From the McKay correspondence it lies in S^7V. There are thus no 2-dimensional representations in S^kV for $1 < k < 7$ and hence no icosahedrally symmetric monopole for $k = 6$.

The charges k which are admissible under this criterion correspond to those for which S^kV contains a 2-dimensional representation. Working out by hand the vectors v_k for low values gives the following cases. Here we use a labelling of the Dynkin diagram to represent v_k.

The representation v_7 is

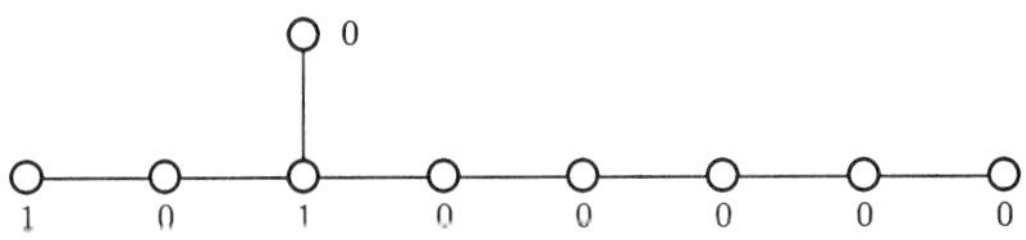

and v_{11}

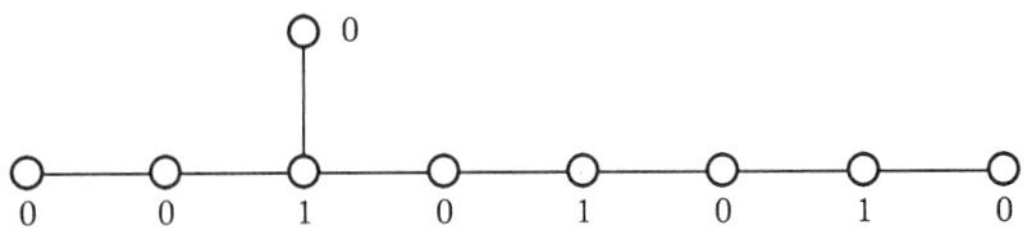

and v_{13}

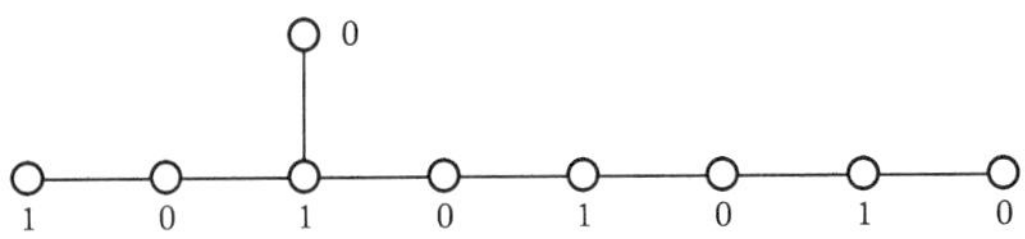

Each diagram has a non-zero label on a 2-dimensional vertex, and so the numbers $k = 1, 7, 11, 13$ are all therefore possible charges.

REFERENCES

[1] E.Arbarello, M.Cornalba, P.A.Griffiths & J.Harris, *Geometry of Algebraic Curves Vol 1*, Grundlehren No.267, Springer Verlag, New York (1985).

[2] M.F.Atiyah & N.J.Hitchin, *The Geometry and Dynamics of Magnetic Monopoles*, Princeton University Press, Princeton (1988).

[3] N.J.Hitchin, On the construction of monopoles, *Commun. Math. Phys* **89** (1983), 145-190.

[4] N.J.Hitchin, N.S.Manton & M.K.Murray, Symmetric Monopoles, Preprint, Cambridge DAMTP 95-17 (1995)

SEMI-UNIVERSAL FAMILY OF REFLEXIVE MODULES OVER A RATIONAL DOUBLE POINT OF TYPE A

Akira Ishii

Department of Mathematics, Kyoto University, Kyoto 606-01, Japan
e-mail: akira@kusm.kyoto-u.ac.jp

1. Introduction

In this paper, we shall summarize the results in [I2], give some examples and then do an explicit calculation in the case of type A.

For a reflexive module E over a rational double point, we denote by Def_E the semi-universal deformation space of E (see section 2). This space has a stratification such that each stratum corresponds to the isomorphism class of a reflexive module. We can determine which module appears as an stratum in Def_E (Theorem 4.3) and what are the singularities of (the reduced parts) of the closures of the minimal strata (Theorem 4.6). By using these results in [I2], we can draw a graph describing the stratification as we shall see in examples in section 4.3. It turns out that the graph for a module over the A_n-singularity coincides with a subgraph of the graph which describes the stratification of the nilpotent variety of type A_m stratified by conjugacy classes, studied in [KP1]. In section 5, we shall construct a versal family of a reflexive module over A_n-singularity. We can follow Faltings' construction of the semi-universal family of a module over the node of a curve [F]. Since A_n-singularity is a deformation of the node, we can use the construction in [F]. We shall obtain the nilpotent variety and its subvarieties as the reduced parts of Def_E (which is non-reduced).

Our results suggest that the reduced part of the deformation space Def_E should be isomorphic to the quiver variety ([N]) for a Dynkin diagram. The quiver variety for an affine Dynkin diagram is the moduli of ASD-connections on the ALE space (the minimal resolution of the rational double point $\mathbb{C}^2/\Gamma$ with 'ALE' metric) ([KN]) and Def_E should be related to the moduli of connections.

2. Deformation of modules – definitions

In this section, we define some basic notions on deformation of modules. Let $\mathcal{O}$ be a henselian local ring over an algebraically closed field k such that $\mathcal{O}/m_{\mathcal{O}} = k$. Here we use

Notation. For a local ring R, we denote by m_R the maximal ideal of R.

We denote by C the category of noetherian henselian local k-algebras with residue field k.

Notation. For an object R in C we denote by $\mathcal{O}_R$ the henselization of $\mathcal{O}\otimes_k R$ at the maximal ideal generated by $m_{\mathcal{O}}$ and m_R.

Let E be a finitely generated $\mathcal{O}$-module such that E is locally free over $\operatorname{Spec}(\mathcal{O}) - \{m_{\mathcal{O}}\}$. We define the deformation functor

$$\mathcal{D}ef_E : C \to \text{(Sets)}$$

of E as follows.

$$\mathcal{D}ef_E(R) = \left\{ \begin{array}{l} \text{finitely generated } \mathcal{O}_R\text{-module } \mathcal{E}_R \text{ flat over } R \\ \text{together with an isomorphism } \mathcal{E} \otimes_R k \cong E \end{array} \right\} \Big/ \sim$$

where $\sim$ denotes an isomorphism which induces the identity on E. Then $\mathcal{D}ef_E$ is a contravariant functor.

As in [A], We say a deformation $[\mathcal{E}_R] \in \mathcal{D}ef_E(R)$ is *algebraic* if and only if R is the henselization of a k-algebra of finite type and we make the following definition.

Definition 2.1. An algebraic deformation $\mathcal{E}_R$ is called *versal* if and only if the following condition holds:

Assume we are given a surjection $A \to \bar{A}$ in C, a deformation $\mathcal{E}_A \in \mathcal{D}ef_E(A)$ parametrized by A and a morphism $\bar{\phi} : R \to \bar{A}$ in C such that there is an isomorphism

$$\mathcal{E}_A \otimes_{\mathcal{O}_A} \mathcal{O}_{\bar{A}} \cong \mathcal{E}_R \otimes_{\mathcal{O}_R} \mathcal{O}_{\bar{A}}.$$

Then there are a lift $\phi : R \to A$ of $\bar{\phi}$ and an isomorphism $\mathcal{E}_A \cong \mathcal{E}_R \otimes_{\mathcal{O}_R} \mathcal{O}_A$ which induces the given isomorphism over $\mathcal{O}_{\bar{A}}$.

Furthermore, $\mathcal{E}_R$ is called *semi-universal* if and only if it is versal and the tangent map of ϕ (the dual of $m_R/m_R^2 \to m_A/m_A^2$) in the above definition is uniquely determined.

If there exists a semi-universal deformation, it is unique up to non-canonical isomorphism and any versal deformation is the pull-back of it by a smooth morphism.

Remark 2.2. We can also define the *universal deformation* by requiring the uniqueness of ϕ in the above condition. In our case, the object has many automorphisms and we cannot expect the existence of a deformation with such a good property.

Definition 2.3. If there is an algebraic semi-universal deformation $\mathcal{E}_R \in \mathcal{D}ef_E(R)$, we put

$$\operatorname{Def}_E = \operatorname{Spec}(R)$$

and call this the semi-universal deformation space of E.

We have defined a deformation of a module over a fixed local ring $\mathcal{O}$. We can also define a semi-universal deformation of a module over a given deformation of the singularity. Later we shall treat the case where $\mathcal{O}$ also deforms.

3. Reflexive modules over rational double points

Let $X = \operatorname{Spec}(\mathcal{O})$ be a rational double point with the origin 0, where $\mathcal{O}$ is the henselization of a finitely generated k-algebra. We denote by $\pi : \widetilde{X} \to X$ the minimal desingularization and by Z the fundamental cycle $\pi^{-1}(0)$. Then the dual graph of Z is the Dynkin diagram of type A, D or E. Furthermore, if we decompose Z as

$$Z = \sum_i m_i C_i \tag{3.1}$$

into irreducible components, the intersection matrix $(C_i \cdot C_j)$ is the negative of the Cartan matrix and the coefficients m_i are the same as those of the highest root of the corresponding root system.

There is another connection of this singularity with the Dynkin diagram, which is called the McKay correspondence. If the characteristic of k is zero, X can be written as $\mathbb{A}^2/\Gamma$ where Γ is a finite subgroup of $\mathrm{SL}_2(k)$. McKay [M] found the following relation between Γ and the Dynkin diagram.

Let $\rho_0, \ldots, \rho_n$ be the irreducible representations of Γ. There is a canonical (not necessarily irreducible) 2-dimensional representation c of Γ determined by the given inclusion $\Gamma \subset \mathrm{SL}_2(k)$. Define integers $a_{ij} (0 \leq i, j \leq n)$ as the coefficients of the decompositions

$$\rho_i \otimes c = \sum_j a_{ij} \rho_j.$$

Then $2I_{n+1} - (a_{ij})$ is, by calculation, the Cartan matrix associated with the affine Dynkin diagram of the same type as X where the extra vertex corresponds to the trivial representation. Moreover, the dimension of the representation ρ_i coincides with the coefficient m_i of the corresponding root in the highest root.

Artin and Verdier [AV] gave a geometric explanation of this fact using reflexive modules, in arbitrary characteristic.

We say an $\mathcal{O}$-module E is *reflexive* if it is finitely generated and the canonical morphism of E to its bidual $E^{\vee\vee}$ is an isomorphism. If (in char 0) there is a representation of the group Γ above, Γ acts on a trivial vector bundle on $\mathbb{A}^2$. Taking the invariant sections of the bundle with respect to the action (and localizing it), we obtain a reflexive $\mathcal{O}$-module E. Since any reflexive module over a regular local ring of dimension 2 is free, this gives a bijection of representations of Γ with reflexive $\mathcal{O}$-modules.

Artin and Verdier made a bijection of indecomposable reflexive modules with vertices of the Dynkin diagram in any characteristic in the following way. For a reflexive module E, put

$$\widetilde{E} = \pi^* E/\text{torsion}.$$

Then $\widetilde{E}$ is a locally free coherent sheaf on $\widetilde{X}$. The first Chern class $c_1(\widetilde{E})$ lies in

$$\mathrm{Pic}(\widetilde{X}) = (\bigoplus \mathbb{Z} C_i)^{\vee}$$

where we identify these modules via the intersection product.

Theorem 3.1 ([AV]). *The map* $E \mapsto (\mathrm{rank}(E), c_1(\widetilde{E}))$ *gives a bijection of the set of isomorphism classes of reflexive $\mathcal{O}$-modules with the set*

$$\{(r, D) \in \mathbb{Z}_{\geq 0} \times \mathrm{Pic}(\widetilde{X}) \mid D \cdot C_i \geq 0 \textit{ for all } i \textit{ and } D \cdot Z \leq r\}.$$

By this theorem, E is indecomposable if and only if corresponding (r, D) cannot be written as $(r, D) = (r_1, D_1) + (r_2, D_2)$ with $r_i \geq 1$ in the above set. Therefore, for a non-trivial indecomposable reflexive module E, there is a unique i such that $c_1(\widetilde{E}) \cdot C_j = \delta_{ij}$ and $\mathrm{rank}(E) = c_1(\widetilde{E}) \cdot Z = m_i$. Thus we can obtain a correspondence between non-trivial indecomposable reflexive modules and vertices of the Dynkin diagram.

Example 3.2. Let X be of type A_n. The Dynkin diagram is as follows.

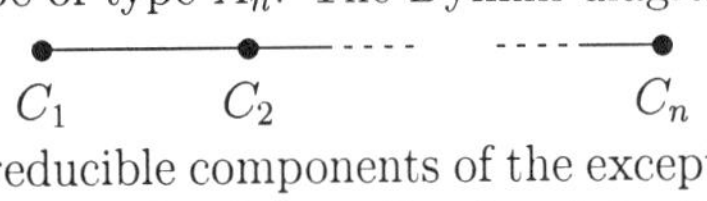

$C_1, \ldots, C_n$ denote the irreducible components of the exceptional locus and $Z = C_1 + \cdots + C_n$. Then there are n reflexive modules $L_1, \ldots, L_n$ of rank 1 with $c_1(\widetilde{L_i}) \cdot C_j = \delta_{ij}$ and any reflexive module is a direct sum of copies of these and $\mathcal{O}$'s.

If we put $L_0 = \mathcal{O}$, then we can use the notation L_i for $i \in \mathbb{Z}/(n+1)\mathbb{Z}$. Using this notation, we have

$$(L_i \otimes L_j)^{\vee\vee} \cong L_{i+j}.$$

This is easily deduced from the fact that $c_1(\widetilde{L}_i)+c_1(\widetilde{L}_j)-c_1(\widetilde{L}_{i+j})$ lies in $\bigoplus \mathbb{Z}C_i$. If char$(k) = 0$, L_i's correspond to characters of the group $\Gamma = \mathbb{Z}/(n+1)Z$ and the isomorphism above follows also from this fact.

4. Results on the stratification of Def_E

In this section, we assume char$(k) = 0$ (this assumption is not so essential) and we shall summarize the results in [I2]. We can obtain the existence of algebraic semi-universal deformation of a reflexive module E over the rational double point X by using Artin's criterion [A]. "The effectivity of formal versal deformation" is easily deduced from that for a sheaf on a projective surface by using the existence theorem in [I1].

4.1. Deformation of $(\Omega^1)^{\vee\vee}$.

The most fundamental reflexive module over the rational double point X (when we consider deformation) is

$$\Omega := (\Omega^1_X)^{\vee\vee}$$

where Ω^1_X is the module of 1-forms on X. Then Ω corresponds to the canonical representation c of Γ in the previous section and we have

$$c_1(\widetilde{\Omega}) = -Z \text{ in } \mathrm{Pic}(\widetilde{X}).$$

by McKay correspondence.

Theorem 4.1 ([I2]). *For the semi-universal deformation space of Ω, the following hold.*

(1) *The reduced part of* Def_Ω *is isomorphic to X.*
(2) Def_Ω *is non-reduced only at the origin.*

Furthermore, we can define the subfunctor of $\mathcal{D}ef_\Omega$ whose semi-universal deformation space is the reduced part of Def_Ω. Let C be the category in section 2 and define a functor

$$\mathcal{D}ef'_\Omega : C \to (\mathrm{Sets})$$

such that for an object R in C,

$$\mathcal{D}ef'_\Omega(R) = \left\{ \mathcal{E}_R \in \mathcal{D}ef_\Omega(R) \,\middle|\, \begin{array}{l} \text{the determinant line bundle of} \\ (\mathcal{E}_R \otimes_R R/m_R^n)|_{X_{\mathrm{reg}}} \text{ is trivial for any } n \end{array} \right\}$$

where X_{reg} is the regular locus of X and the determinant is taken as an $(\mathcal{O}_R/m_R^n)|_{X_{\mathrm{reg}}}$-module. Then

Theorem 4.2 ([I2]). *The reduced part of* Def_Ω *is the semi-universal deformation space for the functor* $\mathcal{D}ef'_\Omega$.

4.2. Stratification of the semi-universal deformation space.

By virtue of the fact $\mathrm{Ext}^1_{\mathcal{O}}(E,\mathcal{O}) = \mathrm{Ext}^1_{\mathcal{O}}(\mathcal{O},E) = 0$ for a reflexive module E, any deformation of $E \oplus \mathcal{O}$ is obtained from one of E by taking direct sum with the trivial deformation of $\mathcal{O}$. Especially, the semi-universal deformation spaces $\mathrm{Def}_{E\oplus\mathcal{O}}$ and Def_E are the same. Therefore, it is natural to ignore the trivial direct summands of a reflexive module when we consider deformation. Let $\sim$ be the equivalence relation among (the isomorphism classes of) reflexive $\mathcal{O}$-modules such that $\sim$ is generated by relations

$$E \sim E \oplus \mathcal{O}. \tag{4.1}$$

From now on, when we say an equivalence class of a reflexive $\mathcal{O}$-module, the class means the one with respect to this relation.

Let us translate Theorem 3.1 of Artin-Verdier using this equivalence relation. To ignore the trivial direct summands is to forget the rank in Theorem 3.1. Thus, we have a bijection

$$\text{(4.2)} \qquad \{\text{reflexive modules } E\}/\sim \quad \rightarrow \quad \{D \in \operatorname{Pic}(\widetilde{X}) \mid D \cdot C_i \geq 0 \text{ for all } i\}$$

which is given by $E \mapsto c_1(\widetilde{E})$. Note that the right hand side is identified with the set of the dominant weights of the corresponding root system.

Theorem 4.3 ([I2]). *Let E be a reflexive $\mathcal{O}$-module.*

(1) *There is a stratification of* Def_E *such that each stratum corresponds to an isomorphism class of a reflexive module.*

(2) *For a reflexive module F, the equivalence class of F appears as a stratum in* Def_E *if and only if $c_1(\widetilde{F}) - c_1(\widetilde{E})$ is represented by an effective divisor supported by the exceptional locus.*

Remark 4.4. If we identify the equivalence classes of reflexive modules with dominant weights, (2) above says that the weight corresponding to E is greater than or equal to that corresponding to F.

Corollary 4.5. *If reflexive $\mathcal{O}$-modules E and F are in a connected flat family of $\mathcal{O}$-modules, then the rank 1 modules $(\overset{r}{\wedge}E)^{\vee\vee}$ and $(\overset{r}{\wedge}F)^{\vee\vee}$ are isomorphic.*

By (2) of the theorem above, we can deduce that if the stratum of F is minimal, then $c_1(\widetilde{F}) - c_1(\widetilde{E})$ is either

(i) the fundamental cycle corresponding to a Dynkin subdiagram of the dual graph of Z or

(ii) one irreducible component C_i.

We can describe the singularity of the closure of a minimal stratum with the reduced induced structure.

Theorem 4.6. *In the case (i) above, the closure of the stratum of F is 2-dimensional and it is the rational double point of the type of the subdiagram. In the case (ii), put $m = (C_i, c_1(\widetilde{E})) - 1$. Then the closure is an isolated singularity of dimension $2m$ and there is a resolution of the singularity whose exceptional locus is the projective space $\mathbb{P}^m$ with normal bundle $T^*\mathbb{P}^m$ (the cotangent bundle).*

The normal singularity as in the case (ii) above is said to be of type a_m [KP1].

4.3. Examples.

Let us see how the semi-universal deformation spaces are stratified.

4.3.1. *The case of type A_n.*

There are n reflexive modules $L_1, \ldots, L_n$ as in Example 3.2. Consider a module of the form

$$E = E(m_1, \ldots, m_n) := L_1^{\oplus m_1} \oplus \cdots \oplus L_n^{\oplus m_n}$$

According to Theorem 4.3, the minimal strata in Def_E correspond to the following modules.

(1) $F = E(\ldots, m_{i-1}+1, m_i - 2, m_{i+1}+1, \ldots)$ for $1 \leq i \leq n$ with $m_i \geq 2$. In this case, $c_1(\widetilde{F}) - c_1(\widetilde{E}) = C_i$.

(2) $F = E(\ldots, m_{i-1}+1, 0, \ldots, 0, m_{j+1}+1, \ldots)$ for $1 \leq i < j \leq n$ such that $m_i = 1$, $m_{i+1} = \cdots = m_{j-1} = 0$ and $m_j = 1$. In this case, $c_1(\widetilde{F}) - c_1(\widetilde{E}) = C_i + \cdots + C_j$.

Theorem 4.6 says that (the normalization of) the closure of the stratum of F above has a singularity of type $a_{m_i - 1}$ in the case (i) and of type A_{j-i+1} in the case (ii). (We do not have to take normalization by the description in section 5.)

Example 4.7. Let X be of type A_5 and let $E = L_1^{\oplus 6} = E(6,0,0,0,0)$. Then the stratification of Def_E is as follows.

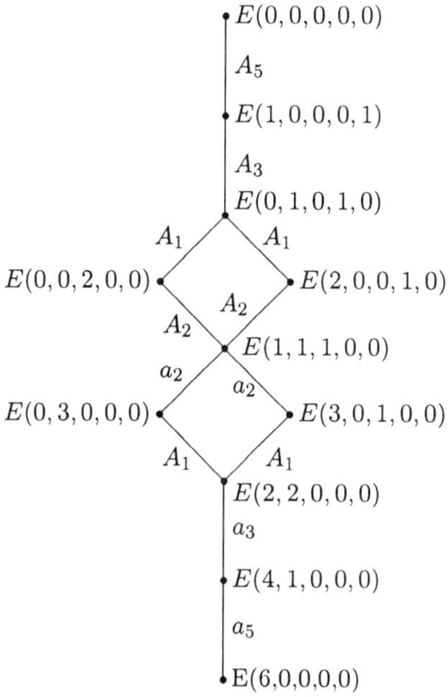

Here a vertex corresponds to the stratum of the module of the indicated class. The stratum of the lower vertex of an edge is contained in the closure of the stratum of the upper vertex of the edge and there are no other strata between them. The type attached to each edge is, modulo smooth equivalence, the type of the singularity of the closure of the stratum (with reduced structure) corresponding to the upper vertex along the stratum of the lower vertex.

The diagram above is the same as the diagram of the stratification of the nilpotent variety of type A_5 stratified with respect to the adjoint action of SL(6) [KP1].

For general $E(m_1, \ldots, m_n)$ over X of type A_n, we can see the following by comparing with [KP1]. Put $m = \sum_i i m_i$ and write $m = q(n+1) + d$ with $0 \leq d \leq n+1$. Let

$$J_i = \begin{pmatrix} 0 & 1 & & 0 \\ & \ddots & \ddots & \\ & & \ddots & 1 \\ 0 & & & 0 \end{pmatrix} \tag{4.3}$$

be the nilpotent Jordan cell of size i. Then the graph of the stratification of Def_E is the same as the subgraph of that of the nilpotent variety of type A_{m-1} consisting of vertices such that the corresponding orbits are contained in the closure of the orbit of

$$J_{n+1}^{\oplus q} \oplus J_d$$

and such that the closures of the corresponding orbits contain the orbit of

$$\bigoplus_{i=1}^{n} J_i^{\oplus m_i}.$$

In section 5, we shall construct a versal family of modules parametrized by a subscheme of the vector space $M_r(k)$ of matrices whose reduced part consists of the conjugacy classes above.

4.3.2. *Other cases.*
In the case of type D and E, the graph is different from that of the nilpotent variety. In the diagram of the stratification of the nilpotent variety of type D in [KP2], there appear singularities of type b, c and d which do not appear here.

Example 4.8. Let X be of type D_4. In this case, the Dynkin diagram is as follows.

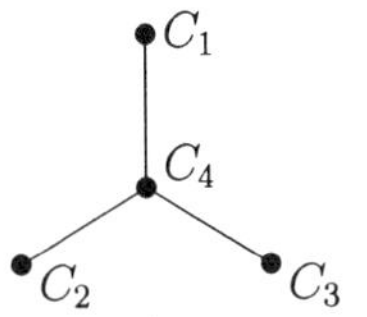

Then we have $Z = C_1 + C_2 + C_3 + 2C_4$. There are non-trivial indecomposable reflexive modules L_1, L_2, L_3 of rank 1 corresponding to C_1, C_2, C_3 respectively and C_4 corresponds to Ω which is of rank 2. Let $E = \Omega^{\oplus 2}$. Then the stratification of Def_E is as follows.

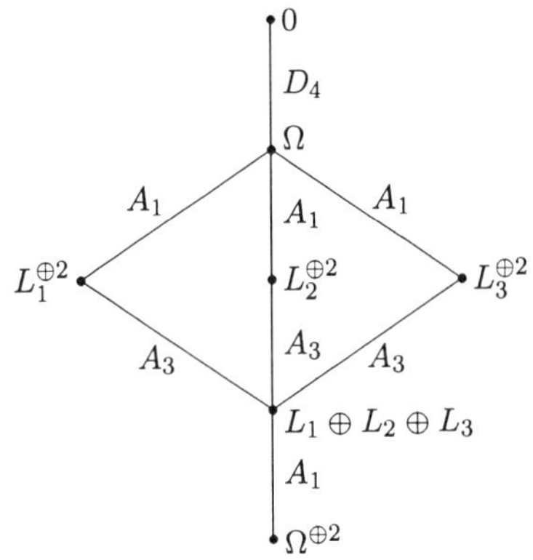

This is a part of the diagram of the stratification of the nilpotent variety of type D_4 [KP2](where appear singularities of type b_2 and d_4). The author doesn't know whether this fact has any meaning.

5. Construction of a semi-universal family in the case of type A

In the case of the node (dim $= 1$), Seshadri constructed a semi-universal deformation of a reflexive module [S]. Faltings treated the case where the underlying node deforms in [F]. We shall follow [F] and construct a semi-universal family of a reflexive module in the case of A_n-singularity. The method is completely the same as that in [F] and in fact we can use a result in [F] since A_n-singularity is the total space of a deformation of the node.

5.1. The case of the node. ([F])
Let R be a noetherian complete local ring with residue field k and let π be an element of the maximal ideal m_R.

$$A := R[[x, y]]/(xy - \pi)$$

is a deformation of the node $A_0 := k[[x, y]]/(xy)$. Faltings constructed a deformation over A of the reflexive A_0- module

$$E := (k[[x,y]]/(x) \oplus k[[x,y]]/(y))^{\oplus r}$$

as follows.

Construction 5.1 ([F]Construction 3.2). Let $P, Q \in M_r(R)$ be $r \times r$ matrices such that $PQ = QP = \pi \cdot 1$. Define $2r \times 2r$ matrices α and β with entries in A as

$$\alpha = \begin{pmatrix} x & P \\ Q & y \end{pmatrix}, \quad \beta = \begin{pmatrix} y & -P \\ -Q & x \end{pmatrix}.$$

Then, $\alpha\beta = \beta\alpha = 0$ and we obtain a complex

$$L(P,Q) : \cdots \xrightarrow{\beta} A^{2r} \xrightarrow{\alpha} A^{2r} \xrightarrow{\beta} A^{2r}.$$

We have the following facts on this complex ([F]).

(1) It is exact.
(2) $\mathcal{E}(P,Q) := \mathrm{Im}(\alpha) = \mathrm{Ker}(\beta)$ is flat over R.
(3) $\mathcal{E}(P,Q) \otimes_R k \cong E$ if $P \equiv Q \equiv 0 \mod m_R$.

Faltings expressed the endomorphism ring of $\mathcal{E}(P,Q)$ in terms of P and Q, and proved the following in [F] Proposition 3.3 and Remark 3.4.

Proposition 5.2 ([F]). *Let (P,Q) and (P',Q') be as above. Assume $P \equiv Q \equiv P' \equiv Q' \equiv 0 \mod m_R$. Then, $\mathcal{E}(P,Q) \cong \mathcal{E}(P',Q')$ if and only if there exist $U, V \in \mathrm{GL}(r,R)$ such that*

$$P' = UPV^{-1},\ Q' = VQU^{-1}.$$

Furthermore, if $P \equiv P', Q \equiv Q'$ modulo an ideal $I \subset R$ and if the given isomorphism is the identity modulo I, then we can choose U, V such that $U \equiv V \equiv 1 \mod I$.

Using this proposition, Faltings proves

Theorem 5.3. *Let $\mathcal{E}$ be a finitely generated A-module flat over R such that $\mathcal{E} \otimes_R R/m_R \cong E$. Then there exist $P, Q \in M_r(R)$ with $PQ = QP = \pi$ and $P \equiv Q \equiv 0$ modulo m_R such that $\mathcal{E} \cong \mathcal{E}(P,Q)$.*

This shows that Construction 5.1 gives rise to a formal semi-universal deformation of E. In the next subsection, we shall use the arguments used in the proof of this theorem in [F].

5.2. Construction of a formal family.

Here we shall construct a semi-universal deformation of a reflexive module in the case of A_{n-1}-singularity. As in Example 3.2, there are $n-1$ reflexive modules $L_1, \ldots, L_{n-1}$ of rank 1 corresponding to the vertices of the Dynkin diagram and any reflexive module is a direct sum of copies of these modules. Let r be a non-negative integer and we shall consider deformation of

$$E := L_1^{\oplus r}.$$

If we can describe the semi-universal deformation of E above, then we can derive the semi-universal deformation of any reflexive module by Theorem 4.3 and by the "openness of the versality" as we shall see in section 5.4.

As an defining equation of the rational double point $X = \mathrm{Spec}(\mathcal{O})$ of type A_{n-1}, we take

$$z^n + xy = 0$$

and we put

$$A_0 = k[[x,y,z]]/(xy + z^n) \cong \hat{\mathcal{O}}.$$

We use the same notation for $\mathcal{O}$-modules and A_0-modules. First, we shall construct a locally free resolution of L_1.

L_1 is isomorphic to the ideal I_D of the Weil divisor $D = \{y = z = 0\}$ in X. (It depends on the choice of the orientation of the Dynkin diagram whether this is L_1 or L_{n-1}.) Hence we can take a surjection

$$\mathcal{O}^{\oplus 2} \longrightarrow I_D \cong L_1$$

such that

$$\begin{pmatrix}1\\0\end{pmatrix} \mapsto z \text{ and } \begin{pmatrix}0\\1\end{pmatrix} \mapsto -y.$$

The kernel of this map is generated by

$$\begin{pmatrix}y\\z\end{pmatrix} \text{ and } \begin{pmatrix}-z^{n-1}\\x\end{pmatrix} = \frac{x}{z}\begin{pmatrix}y\\z\end{pmatrix}.$$

In this way, we obtain a resolution

$$\cdots \xrightarrow{\alpha_0} \mathcal{O}^{\oplus 2} \xrightarrow{\beta_0} \mathcal{O}^{\oplus 2} \xrightarrow{\alpha_0} \mathcal{O}^{\oplus 2} \xrightarrow{\beta_0} \mathcal{O}^{\oplus 2} \longrightarrow L_1 \longrightarrow 0 \tag{5.1}$$

where

$$\alpha_0 = \begin{pmatrix}x & z^{n-1}\\ -z & y\end{pmatrix} \text{ and } \beta_0 = \begin{pmatrix}y & -z^{n-1}\\ z & x\end{pmatrix}.$$

From this, we obtain

$$\dim_k \operatorname{Ext}^i(L_1, L_1) = 1 \text{ for } i \geq 1. \tag{5.2}$$

Let R be a noetherian complete local ring and let $t_1, \ldots, t_n$ be elements of the maximal ideal m_R. Define polynomials $f_i(z) (0 \leq i \leq n)$ in $R[z]$ as

$$f_i(z) = z^i + t_1 z^{i-1} + \cdots + t_i. \tag{5.3}$$

Then

$$A = R[[x, y, z]]/(xy + f_n(z))$$

is a deformation of A_0 parametrized by $\operatorname{Spec}(R)$. First we shall give a construction of a flat family which will give rise to a formal semi-universal family.

Construction 5.4. Let $M \in M_r(R)$ be an $r \times r$ matrices such that $f_n(M) = 0$. Define matrices $P, Q \in M_r(R[[z]])$ as

$$P = M^{n-1} + f_1(z)M^{n-2} + \cdots + f_{n-1}(z) \cdot 1$$
$$Q = M - z \cdot 1.$$

Then we have $PQ = QP = -f_n(z) \cdot 1$ and we can define a complex $L(M) := L(P, Q)$ as in Construction 5.1. Put

$$\mathcal{E}(M) := \mathcal{E}(P, Q).$$

Then

(1) $\mathcal{E}(M)$ is flat over R.

(2) $\mathcal{E}(M) \otimes_R k \cong E = L_1^{\oplus r}$ if $M \equiv 0 \mod m_R$.

If $\mathcal{E}(M) \cong \mathcal{E}(M')$ for two such matrices M and M' with $M \equiv M' \equiv 0 \mod m_R$, then there exist $U(z)$ and $V(z)$ in $\mathrm{GL}(r, R[[z]])$ such that

$$M' - z1 = V(z)(M - z1)U(z)^{-1}$$

by Proposition 5.2. Letting $z = 0$, we obtain

$$M' = V(0)MU(0)^{-1}. \tag{5.4}$$

Furthermore, if $M \equiv M'$ modulo an ideal $I \subset R$ and the given isomorphism is the identity modulo I, then we can choose U, V such that $U \equiv V \equiv 1 \mod I$.

Construction 5.4 gives rise to a formal versal deformation of the A_0-module E.

Theorem 5.5. *Let $I \subset R$ be an ideal with $m_R \cdot I = 0$. We put $\bar{R} = R/I$, $\bar{A} = A/IA$ and so on. Suppose we are given*

(1) *a finitely generated A-module $\mathcal{E}$ flat over R and*

(2) *a matrix $\bar{M} \in M_r(\bar{R})$ which satisfies $\bar{M} \equiv 0 \mod m_{\bar{R}}$ and $\bar{f}_n(\bar{M}) = 0$*

such that there is an isomorphism $\bar{\phi} : \mathcal{E} \otimes_A A_{\bar{R}} \xrightarrow{\sim} \mathcal{E}(\bar{M})$. Then there is an $M \in M_r(R)$ with $\bar{M} = M \bmod I$ and $f_n(M) = 0$ such that $\bar{\phi}$ can be lifted to $\phi : \mathcal{E} \xrightarrow{\sim} \mathcal{E}(M)$.

Proof. We shall prove this by following the proof of Theorem 5.3 by Faltings. Since E has a minimal free resolution (5.1) by free A_0-modules of rank $2r$, $\mathcal{E}$ also has a free resolution by a complex

$$L : \cdots \longrightarrow A^{2r} \longrightarrow A^{2r} \longrightarrow A^{2r}.$$

A minimal free resolution is unique up to isomorphism and hence $\bar{L} := L/IL$ is isomorphic to $L(\bar{M})$. We shall show that there is a lift $M \in M_r(R)$ of $\bar{M}$ such that $L \cong L(M)$.

We may assume $\dim_k I = 1$. The obstruction to lift $\bar{L}$ to L lies in

$$\mathrm{Ext}^2_{A_0}(E, E) \otimes_k I$$

and the choice of lifts are parametrized by

$$\mathrm{Ext}^1_{A_0}(E, E) \otimes_k I.$$

Both the vector spaces are of dimension r^2 by (5.2). The construction $M \mapsto L(M)$ gives rise to maps

$$\{\text{obstructions to lift } \bar{M}\} \longrightarrow \mathrm{Ext}^2_{A_0}(E, E) \otimes_k I$$
$$\{\text{choice of lifts of } \bar{M}\} \longrightarrow \mathrm{Ext}^1_{A_0}(E, E) \otimes_k I$$

We shall call the first map the obstruction map and the second one the tangent map. We shall show that

(1) the obstruction map is injective

(2) the tangent map is an isomorphism.

Then the assertion will follow: (1) shows that there is at least one lift of $\bar{M}$ and then we can choose a desired lift by (2).

(1) For an element $p = p(x, y, z)$ in A, we write

$$p = p_0(z) + p_{10}(z)x + p_{20}(z)x^2 + \cdots$$
$$+ p_{01}(z)y + p_{02}(z)y^2 + \cdots$$

and for an element $e \in R$ (or A), $\bar{e}$ denotes $e \bmod I$. Since $\bar{L} \cong L(\bar{M})$, $\bar{L}$ is a complex with maps

$$\bar{\alpha} = \begin{pmatrix} \bar{P} & x \\ y & \bar{Q} \end{pmatrix} \quad \bar{\beta} = \begin{pmatrix} \bar{Q} & -x \\ -y & \bar{P} \end{pmatrix}$$

where $\bar{P}$ and $\bar{Q}$ are matrices defined as in Construction 5.4 using $\bar{M}$. Then we have

$$\bar{\alpha}_{i0} = \bar{\alpha}_{0i} = \bar{\beta}_{i0} = \bar{\beta}_{0i} = 0 \text{ for } i \geq 2. \tag{5.5}$$

Assume the obstruction $\mathrm{ob}(\bar{L})$ to lift $\bar{L}$ vanishes. This means that there are α and β in $M_{2r}(A)$ such that $\alpha \bmod I = \bar{\alpha}$, $\beta \bmod I = \bar{\beta}$ and such that $\alpha\beta = 0$. Then by (5.5) and by the fact $m_R \cdot I = 0$, we obtain

$$0 = (\alpha\beta)_0(z) = \alpha_0(z)\beta_0(z) - (\alpha_{10}(z)\beta_{01}(z) + \alpha_{01}(z)\beta_{10}(z))f_n(z).$$

This implies

$$0 = \alpha_0(0)\beta_0(0) - (\alpha_{10}(0)\beta_{01}(0) + \alpha_{01}(0)\beta_{10}(0))t_n. \tag{5.6}$$

Let $M \in M_r(R)$ be the $(2,2)$-block of $2r \times 2r$ matrix $\alpha_0(0)$ divided into 4 matrices of size $r \times r$. Then we have $M \bmod I = \bar{M}$. Since

$$\bar{\alpha_0}(0) = \begin{pmatrix} \bar{P}(0) & 0 \\ 0 & \bar{Q}(0) \end{pmatrix} = \begin{pmatrix} \bar{f}_n(\bar{M}) & 0 \\ 0 & \bar{M} \end{pmatrix}$$

and

$$\bar{\beta_0}(0) = \begin{pmatrix} \bar{Q}(0) & 0 \\ 0 & \bar{P}(0) \end{pmatrix} = \begin{pmatrix} \bar{M} & 0 \\ 0 & \bar{f}_n(\bar{M}) \end{pmatrix}$$

by definition, and since all the entries of $\alpha_0(0)$ and $\beta_0(0)$ are in m_R, we obtain

$$\alpha_0(0)\beta(0) = \begin{pmatrix} f_{n-1}(M)M & 0 \\ 0 & Mf_{n-1}(M) \end{pmatrix}.$$

Similar calculation shows that

$$(\alpha_{10}(0)\beta_{01}(0) + \alpha_{01}(0)\beta_{10}(0))t_n = -t_n \cdot 1.$$

Putting these into (5.6), we obtain

$$0 = f_{n-1}(M)M + t_n \cdot 1 = f_n(M)$$

and this implies $\mathrm{ob}(\bar{M}) = 0$.

(2) Since both the vector spaces are of dimension r^2, it is sufficient to show that the map is injective. Let M and M' be two lifts of $\bar{M}$ such that there is an isomorphism

$$L(M) \cong L(M')$$

which induces the identity on $L(\bar{M})$. Then by (5.4), there exist $U(0), V(0) \in \mathrm{GL}(r, R)$ with $U(0) \equiv V(0) \equiv 1 \mod I$ such that

$$M' = V(0)MU(0)^{-1}.$$

Now since all the entries of M are in m_R and since $U(0) \equiv V(0) \equiv 1 \mod I$, we obtain

$$M' = M.$$

This proves (2) and hence the theorem. □

5.3. Algebraic semi-universal family.

Let R be the henselization of a local ring essentially of finite type over a field (or an excellent Dedekind domain) with residue field k. For elements $t_1, \ldots, t_n$ in m_R, we define $f_i(z)$ as in (5.3). We denote by A the henselization at (m_R, x, y, z) of the ring

$$R[x, y, z]/(xy + f_n(z)).$$

For a matrix $M \in M_r(R)$ with $f_n(M) = 0$, we can construct an A-module $\mathcal{E}(M)$ as in Construction 5.4.

Fix the parameter space $\mathrm{Spec}(R)$ and let V be the henselization at 0 of

$$R[s_{ij}]/(\text{entries of } f_n(S)) \tag{5.7}$$

where $R[s_{ij}]$ is the polynomial ring with r^2 variables s_{ij} $(1 \le i, j \le r)$ and S denotes the $r \times r$ matrix (s_{ij}). We have a 'universal matrix' $\mathcal{M} \in M_r(V)$ such that $f_n(\mathcal{M}) = 0$, which is the image of S.

For a local R-algebra T, we denote by A_T the henselization of $A \otimes_R T$.

Theorem 5.6. *$\mathcal{E}(\mathcal{M})$ is a semi-universal family of $L_1^{\oplus r}$ over the deformation*

$$\operatorname{Spec}(A) \to \operatorname{Spec}(R)$$

of X. Precisely, the following holds.

Let $T \to T'$ be a surjection of henselian local R-algebras with residue field k. Suppose we are given an A_T-module $\mathcal{E}_T$ flat over T and a map $\bar{\phi} : V \to \bar{T}$ such that there is an isomorphism $\mathcal{E}_T \otimes_{A_T} A_{\bar{T}} \cong \mathcal{E}(\mathcal{M}) \otimes_{A_V} A_{\bar{T}}$. Then there is a lift $\phi : V \to T$ of $\bar{\phi}$ such that the given isomorphism extends to $\mathcal{E}_T \cong \mathcal{E}(\mathcal{M}) \otimes_{A_V} A_T$.

Proof. Theorem 5.5 says that $\mathcal{E}(\mathcal{M})$ is formally versal in the sense of [A]. Then it is versal by virtue of [A] Theorem (3.3). □

5.4. Deformation of general reflexive modules.

Put $R = k$ in the previous subsection (this is the case where X does not deform). We have a flat family of reflexive $\mathcal{O}$-modules parametrized by the scheme

$$N(r, n; k) := \{M \in M_r(k) \mid M^n = 0\} = \operatorname{Spec} V$$

of finite type over k, where V is defined as in (5.7) with $R = k$. The henselization at 0 of the above scheme is the semi-universal deformation space of $L_1^{\oplus r}$. Thanks to the openness of the versality ([A]), the family at a point near 0 (and hence at any point by the group action) gives rise to a versal family of the corresponding module.

Let us determine which point corresponds to which module. Since $\mathrm{GL}(r, k)$ acts on $N(r, n; k)$ and on the family, the isomorphism class depends only on the conjugacy class of the matrix. Therefore we may assume that the matrix is a direct sum of Jordan cells J_i (4.3). Let

$$M = \bigoplus_{i=1}^{n} J_i^{\oplus n_i}$$

be a matrix of size $r = \sum i m_i$. Then the module corresponding to the matrix $M \in N(r, n; k)$ is

$$\mathcal{E}(M) = \bigoplus_{i=1}^{n} \mathcal{E}(J_i)^{\oplus m_i}.$$

Lemma 5.7. *We have isomorphisms*

$$\mathcal{E}(J_i) \cong L_i \oplus \mathcal{O}^{i-1} \text{ for } 1 \le i \le n-1 \text{ and } \mathcal{E}(J_n) \cong \mathcal{O}^{\oplus n}.$$

Proof. We put $L_n = \mathcal{O}$. Consider the versal family of reflexive modules parametrized by $N(i, n; k)$. The orbit of J_i is open dense in $N(i, n; k)$ and hence the corresponding reflexive module $\mathcal{E}(J_i)$ does not deform to other modules. This implies that the class of $\mathcal{E}(J_i)$ is minimal with respect to the order in Remark 4.4. Therefore, the class of $\mathcal{E}(J_i)$ is the class of a rank 1 module. In other words, there is j such that

$$\mathcal{E}(J_i) \cong \mathcal{O}^{\oplus i-1} \oplus L_j.$$

On the other hand, the origin of $N(i, n; k)$ corresponds to $L_1^{\oplus i}$ and hence Corollary 4.5 implies

$$L_j \cong (\overset{i}{\wedge} F)^{\vee\vee} \cong (\overset{i}{\wedge}(L_1^{\oplus i}))^{\vee\vee} \cong L_i.$$

This proves the Lemma. □

Thus we obtained an isomorphism

$$\mathcal{E}(M) \cong \mathcal{O}^{\oplus\left(\sum_{i=1}^{n}(i-1)m_i + m_n\right)} \oplus \bigoplus_{i=1}^{n-1} L_i^{\oplus m_i}.$$

Since the semi-universal deformation space does not depend on the equivalence class with respect to the relation (4.1), the following is proved.

Theorem 5.8. *Let* $E = \bigoplus_{i=1}^{n-1} L_1^{\oplus m_i}$ *be a reflexive module and put* $r = \sum i m_i$. *Then the semi-universal deformation space* Def_E *of* E *is isomorphic to the transversal slice of the scheme* $N(r, n; k)$ *to the orbit of the matrix* $\bigoplus_i J_i^{\oplus m_i}$. *The intersection of* Def_E *with the orbit of the matrix* $\bigoplus_i J_i^{\oplus n_i}$ *is a stratum of the module equivalent to* $\bigoplus_{i=1}^{n-1} L_i^{\oplus n_i}$.

Thus, the identification of the graphs in Section 4.3.1 comes from the isomorphism of (the reduced parts of) the spaces.

Remark 5.9. There exist countably many equivalence classes of reflexive modules and the graph of the closure relation as in Section 4.3 can be extended to a graph with infinitely many vertices by virtue of the openness of the versality. There should exist some infinite dimensional space stratified by countably many finite codimensional subspaces such that the strata correspond to the equivalence classes of reflexive modules.

In the case of type A_{n-1}, there are n connected components of such a graph, corresponding to rank 1 reflexive modules $\mathcal{O}, L_1, \ldots, L_{n-1}$. For an integer p with $0 \leq p \leq n-1$, such an infinite dimensional space should be

$$\varinjlim_i N(p + ni, n; k) = \{M \in \varinjlim_i M_{p+ni}(k) | M^n = 0\}$$

where the inclusion $M_m(k) \to M_{m+n}(k)$ is given by $M \mapsto M \oplus J_n$.

References

[A] Artin, M.: Versal deformations and algebraic stacks, Invent. Math. **27** (1974), 165-189.

[AV] Artin, M. and Verdier, J.-L.: Reflexive modules over rational double points, Math. Ann. **270** (1985), 79-82.

[F] Faltings, G.: Moduli-Stacks for Bundles on Semistable Curves, preprint.

[I1] Ishii, A.: On the moduli of reflexive sheaves on a surface with rational double points, Math. Ann. **294** (1992) 125-150.

[I2] ———: Deformation of reflexive modules over rational double points, in preparation.

[KP1] Kraft, H. and Procesi, C.: Minimal singularities in GL_n, Invent. Math. **62** (1981) 503-515.

[KP2] ———: On the geometry of conjugacy classes in classical groups, Comment. Math. Helvetici **57** (1982) 539-602.

[KN] Kronheimer, P. and Nakajima, H.: Yang-Mills instantons on ALE gravitational instantons, Math. Ann. **288** (1990) 267-307.

[M] McKay, J.: Graphs, singularities and finite groups, Proc. Symp. Pure Math., **37**, Amer. Math. Soc. (1980) 183-186.

[N] Nakajima, H.: Instantons on ALE spaces, quiver varieties, and Kac-Moody algebras, Duke. Math. **76** (1995) 365-416.

[S] Seshadri, C. S.: fibrés vectoriels sur les courbes algébriques, astérisque **96** (1982).

A REMARK ON DONALDSON'S POLYNOMIAL INVARIANTS FOR A SURFACE OF GENERAL TYPE WITH $p_g = 2$ AND $K^2 = 1$

Yukio Kametani

Department of Mathematics, Saga University, SAGA 840, JAPAN
e-mail:kametani@ms.saga-u.ac.jp

1. Introduction

Recently Kronheimer and Mrowka ([KM], [KM2]) have given a recurrence relation of Donaldson's polynomial invariants [D], which indicates that there is a universal relation between all the Donaldson invariants for a 4-manifold of simple type. From some results they conjectured a formula of all the Donaldson invariants for all minimal surfaces of general type with $p_g > 0$ [KM2], which was interpreted from a point of mathematical physics by Witten [W]. These results interest us in calculating the Donaldson invariant for more surfaces. In this paper we calculate one coefficient of the polynomial invariant for a surface of general type with $p_g = 2$ and $K^2 = 1$: Let $q_{d,w}$ be the polynomial invariant of degree d defined by the moduli space of anti-self-dual connections on a rank-two vector bundle E over a simply-connected smooth 4-manifold with $\det(E) = w$ and $d = 4c_2(E) - c_1(E)^2 - (3/2)(1 + b^+)$. Let Q the intersection form of X and Q^d the d-th power of Q by the rule of [DK, (10.4.2)].

Theorem *The polynomial invariant $q_{10,K}$ satisfies*

$$q_{10,K}(\Sigma_1, \cdots, \Sigma_{10}) = \frac{2}{5!} Q^5(\Sigma_1, \cdots, \Sigma_{10})$$

for surfaces $\Sigma_1, \cdots, \Sigma_{10}$ in X with $\langle c_1(K), \Sigma_i \rangle = 0$ $(1 \leq i \leq 10)$.

From some observation we show that Kronheimer-Mrowka's "basic classes" of X are only $\pm K$, by which we can determine the formal power series $q_w(h) = \sum_d q_{d,w}(h, \cdots, h)/d!$ for any w. For example, we can deduce that

Corollary *The formal expansion q_0 defined by Donaldson's $SU(2)$-polynomials is the form of*

$$q_0 = 2 \exp\left(\frac{Q}{2}\right) \sinh(K).$$

Our strategy of the proof depends on calculations by Donaldson and Kronheimer [DK]. They considered a surface of general type with $p_g = 3$ and $K^2 = 2$, which is obtained as the branched double cover over the projective space $\mathbb{CP}^2$ branched along a curve of degree 6. As to X, it was shown in [H] (see also [BPV, VII.10]) that the blow-up $\widehat{X}$ at the base point $Bs|K| = \{x_0\}$ is the branched double cover over the Hirzebruch surface Σ_2 branched along an effective divisor in the linear system $|6E + 10F|$, where E is a section with $E^2 = -2$ and F is a fiber. We take the branch locus to be a disjoint union of smooth curves; a quintic section and a section. Then $\widehat{X}$ has no (-2)-curve and so does X. We use some techniques in [DK] and [K] to calculate the polynomial invariant.

The author is grateful to Professor Shigeru Mukai and Professor Akira Fujiki for their advice.

2. Moduli spaces of vector bundles

Let L be an ample divisor on X. A holomorphic rank-two vector bundle $\mathcal{E}$ over X is L-stable if for any non-zero holomorphic map $s : \mathcal{O}(D) \to \mathcal{E}$, $D \cdot L < (1/2)c_1(\mathcal{E}) \cdot L$. We let M_k be the moduli space of rank-two L-stable bundles $\mathcal{E}$ over X with $c_1(\mathcal{E}) = K$ and $c_2(\mathcal{E}) = k$. By the Riemann-Roch theorem and the Serre duality, we have $\mathcal{X}(\mathcal{E}) = 2h^0(\mathcal{E}) - h^1(\mathcal{E}) = 6-k$. So if $k \leq 5$, there is an effective divisor D and a holomorphic map $\mathcal{O}(D) \to \mathcal{E}$ with a 0-dimensional scheme Z as the zero set. Then we have an exact sequence

$$0 \longrightarrow \mathcal{O}(D) \longrightarrow \mathcal{E} \longrightarrow \mathcal{O}(K-D) \otimes I_Z \longrightarrow 0. \tag{1}$$

By the calculation of the Chern character we get

$$|Z| = k - DK + D^2.$$

Since $\mathcal{E}$ is L-stable, $2DL < KL$. If L_0 is ample, so is $L = L_0 + nK$ for $n \geq 0$ by Nakai's criterion. Thus we obtain $KL_0 > n(2DK - 1) \geq n$ for all n, if $DK > 0$. This implies $DK = 0$ for large n. Then D is a empty set or a sum of (-2)-curves (cf. [BPV]), so $D = 0$ by the choice of our model. Hence there is an exact sequence:

$$0 \longrightarrow \mathcal{O} \longrightarrow \mathcal{E} \longrightarrow \mathcal{O}(K) \otimes I_Z \longrightarrow 0 \tag{2}$$

with $|Z| = k$. Conversely we have:

Lemma 1. *The bundle defined by (2) is stable.*

Proof. Suppose that there were a non-zero map $\mathcal{O}(C) \to \mathcal{E}$. From the exact cohomology sequence of (2), we get $h^0(-C) > 0$ or $h^0((K-C) \otimes I_Z) > 0$. When the former holds, $-C$ is linearly equivalent to an effective divisor, so $CK \leq 0$ and $CL_0 < 0$. When the latter holds, $K - C$ is linearly equivalent to an effective divisor, so we have $CK \leq 0$ and $CL_0 < KL_0$. This implies $2CL < KL$ for $L = L_0 + nK$ with n large. □

In [DK, Chapter 10] they investigated the moduli space of exact sequences rather than M_k as follows: The exact sequence (2) is uniquely determined by a triple $(\mathcal{E}, \psi, s)$, where $\mathcal{E}$ is a rank-two holomorphic bundle, ψ is an isomorphism from $\wedge^2\mathcal{E}$ to $\mathcal{O}(K)$ and s is an section of $\mathcal{E}$ with its zeros Z. We let $\widehat{N_k}$ be the moduli space of equivalence classes of triples $(\mathcal{E}, \psi, s)$. From the $\mathbb{C}^*$-action $\zeta \cdot (\mathcal{E}, \psi, s) = (\mathcal{E}, \psi, \zeta s)$ for $\zeta \in \mathbb{C}^*$, we obtain the quotient space N_k.

Then there is a natural forgetful map $\rho_k : N_k \to M_k$. The fiber of ρ_k at $[\mathcal{E}]$ is just the projective space $\mathbb{P}(H^0(\mathcal{E}))$. There is another obvious map from N_k to the k-fold symmetric product $s^k(X)$. The fiber at Z is an open set of the projective space $\mathbb{P}(\mathrm{Ext}^1(\mathcal{O}(K) \otimes I_Z, \mathcal{O}))$. Notice that there is an exact sequence

$$0 \longrightarrow \mathrm{Ext}^1(\mathcal{O}(K) \otimes I_Z, \mathcal{O}) \longrightarrow H^0(2K|_Z)^* \xrightarrow{e^T} H^0(2K)^*, \tag{3}$$

where $e : H^0(2K) \to H^0(2K|_Z)$ is the evaluation map and e^T is its transpose (cf. [DK, Chapter 10]). We evaluate the dimension of N_k using the fibration $N_k \to s^k(X)$.

Lemma 2. *For $k \leq 5$ the space N_k has at most dimension $2k - 1$ except when $k = 5$.*

Proof. Let N_k^i be the subspace of N_k in which $\dim \mathrm{Ker}(e) = i$ $(0 \leq i \leq h^0(2K) = 4)$. Then $N_k = \cup_{0 \leq i \leq 4} N_k^i$ defines a stratification. We show the lemma for each stratified space N_k^i: As to N_k^0, its transpose e^T is surjective. Using (3) we have

$$\dim N_k^0 \leq 2|Z| + \dim \mathrm{Ker}(e^T) - 1 = 3k - 5 \leq 2k - 1$$

except when $k = 5$. As to N_k^1, there is a non-zero section of $\mathcal{O}(2K)$ which vanishes on Z. Hence

$$\dim N_k^1 \leq h^0(2K) - 1 + |Z| + \dim \mathrm{Ker}(e^T) - 1 = 2k - 1.$$

As to N_k^2, there are two sections of $\mathcal{O}(2K)$ which vanish on Z. Hence we get a rough evaluation

$$\dim N_k^2 \leq h^0(2K) - 2 + |Z| + \dim \mathrm{Ker}(e^T) - 1 = 2k - 1.$$

As to N_k^3, we use the pluricanonical map $f_{2K} : X \setminus \{x_0\} \to \mathbb{C}P^3$, where $Bs|2K| = Bs|K| = \{x_0\}$. Then it was shown in [H] (see also [BPV, VII.10]) that f_{2K} is a branched double covering map. Hence $\dim \mathrm{Im}(e) = 1$ if and only if Z is contained in $Bs|2K| \cup f_{2K}^{-1}(\text{a point})$. Thus we get $\dim N_k^3 \leq \dim f_{2K}(X) + k - 2 \leq 2k - 1$. As to N_k^4, e is the zero map, so Z is supported on $\{x_0\}$. Thus we have $\dim N_k^4 \leq k - 1 \leq 2k - 1$. □

Lemma 3 *The moduli space M_5 is birational to the space $s^5(X)$, via the obvious map.*

Proof. For a scheme Z of generic five points, the evaluation map $e : \mathbb{C}^4 \to \mathbb{C}^5$ is injective. Hence the space N_5^0 is birationally isomorphic to $s^5(X)$. On the other hand, another evaluation map $e : H^0(K) \to H^0(K|_Z)$ is injective for a scheme Z of generic five points, so it follows from (2) that $h^0(\mathcal{E}) = 1$ and the fiber is one point. Thus the map ρ_5 is generically one-to-one. □

Now we can read off a value of the polynomial invariant $q_{10,K}$ by the same argument as in [DK, Chapter 10]: From the above results we can show that the value $q_{10,K}(\Sigma_1, \cdots, \Sigma_{10})$ is equal to twice the intersection number $\sharp(N_5 \cap W_{\Sigma_1} \cdots \cap W_{\Sigma_{10}})$ for surfaces Σ_i in X with $\langle c_1(K), \Sigma_i \rangle = 0$ $(1 \leq i \leq 10)$, where $W_{\Sigma_i} = \{(\mathcal{E}, \phi, s) \in N_5 \mid s_5^{-1}(0) \cap \Sigma_i \neq \emptyset\}$ is the Poincaré dual of the image of the μ-map $\mu([\Sigma_i])$. This number is equal to $(2/5!)Q^5(\Sigma_1, \cdots, \Sigma_{10})$ by Lemma 3. (We do not investigate the lower moduli spaces N_k $(k < 5)$ so precisely as in [DK, Chapter 10], however in proving that the lower moduli spaces N_k make no contribution to the value, the point in [DK, 10.4.3] was that for generic k points the map e^T is injective, which follows from the fact that the rank of e is full in our case too.)

3. Proof of Corollary

We first show that X has simple type [KM2]. By [KM2, Corollary 8.5] it suffices to find a sphere of square -2 meeting a torus of square 0 transversely in one point. To do this, we consider the fibration $\widehat{X} \to \Sigma_2$. A trace of vanishing cycles over a path connecting two singular fibers defines an embedded sphere of square -2. On the other hand, we choose a circle γ on the base Σ_2 to satisfy that γ meets the above path transversely in one point, that γ does not meet the base point of any singular fiber, and that the monodromy along γ is trivial. Then a trace of circles on the fibers over γ defines a torus of square 0. We can choose such a torus to meet the sphere in one point transversely.

Next we determine the basic classes. If one takes the branched cover $\widehat{X} \to \Sigma_2$ to have the branch locus along six sections and ten fibers, one can find many 2-spheres of square -2, in components of the ten singular fibers or by tracing vanishing circles over a path connecting two singular fibers. Then one can show that X has a big diffeomorphism group with respect to the canonical class K, as in the case of regular elliptic surfaces [FM].

Hence all the polynomial invariants $q_{d,K}$ are a polynomial of Q and K. So all the basic classes can be written as nK for some integer n. Because the canonical class K can be represented by a smooth embedded curve of genus 2 [BPV, Proposition 8.1.1], it follows from [KM2, Theorem 1.7 (b)] that the basic classes are only $\pm K$. Now the Corollary follows from [KM2, Theorem 1.7 (a)] and our main theorem.

References

[BPV] W. Barth, C. Peters and A. Van de Ven: *Compact Complex Surfaces*, Springer Verlag, Berlin-Heidelberg-New York, 1984.

[DP] S. K. Donaldson: Polynomial invariants for smooth four manifolds, Topology **29** (1990), 257–315.

[DK] S. K. Donaldson and P. B. Kronheimer: *The Geometry of Four-Manifolds*, Oxford Mathematical Monographs, 1990, Oxford Univ. Press.

[FM] R. Friedman and J. W. Morgan: *Geometry of Complex Surfaces*, 1994, Springer Verlag.

[H] E. Horikawa: Algebraic surfaces of general type with small c_1^2, I, Ann. of Math. **104** (1976), 357–387, II, Invent. Math. **37** (1976), 121–155, III, Invent. Math. **47** (1978), 209–248, IV, Invent. Math. **50** (1979), 103–128.

[Ko] D. Kotschick: On manifolds homeomorphic to $\mathbb{C}P^2\sharp 8\overline{\mathbb{C}P^2}$, Invent. Math. **95** (1989), 591–600.

[KM] P. B. Kronheimer and T. S. Mrowka: Recurrence relation and asymptotics for four-manifold invariants, Bull. Amer. Math. Soc. **30** (1994), 215–221.

[KMR] P. B. Kronheimer and T. S. Mrowka: Embedded surfaces and the structures of Donaldson's polynomial invariants, preprint.

[W] E. Witten: Supersymmetric Yang-Mills theory on a four manifold, preprint (1994).

MODULE DES FIBRÉS SEMI-STABLES ET FONCTIONS THÊTA

Joseph Le Potier

Université Paris 7, Institut de Mathématiques de Jussieu, Case Postale 7012, 2, Place Jussieu, 75251 Paris Cedex 05, France

e-mail: jlp@mathp7.jussieu.fr

INTRODUCTION

Cet article est inspiré d'un résultat de Faltings [3] dont il fournit une version effective. Soit $\mathrm{M}(r,d)$ l'espace de modules des classes d'équivalence de fibrés semi-stables de rang r et de degré d sur une courbe algébrique projective lisse X de genre $g \geq 1$. Une fois la courbe polarisée, on peut définir de manière canonique un fibré déterminant $\mathcal{D}$ sur $M(r,d)$; ce fibré inversible, dont nous rappelons la définition dans la première section est ample et est caractérisé par une propriété universelle décrite dans le corollaire 1.7. Soit $\delta = \mathrm{pgcd}(r,d)$. L'objet de cet article est de montrer que le fibré $\mathcal{D}^{\otimes k}$ est engendré par ses sections globales dès que

$$k > \mathrm{Max}(\delta, \frac{r^2}{4}(g-1)).$$

(*cf.* théorème 2.4). Plus précisément, on montre que ce fibré est engendré des sections particulières appelées fonctions thêtas. Considérons maintenant l'espace de modules $M(r,L)$ des classes d'équivalence de fibrés semi-stables F de déterminant L fixé. La restriction $\mathcal{D}_L = \mathcal{D}|_{M(r,L)}$ est un fibré inversible indépendant du choix de la polarisation de X. La même méthode permet d'obtenir l'énoncé suivant (*cf.* théorème 2.8): le fibré $\mathcal{D}_L^{\otimes k}$ est engendré par ses sections dès que

$$k > \frac{r^2}{4}(g-1).$$

Ces résultats sont combinés dans la section 3 avec le théorème de Flenner [4] sur la restriction des faisceaux semi-stables aux courbes pour obtenir sur les surfaces algébriques projectives lisses polarisées $(X, \mathcal{O}_X(1))$ des renseignements sur le fibré déterminant $\mathcal{D}$ de Donaldson associé à l'espace de modules $M = M(r, L, c_2)$ des faisceaux semi-stables de rang r, de déterminant L et classes de Chern c_2 fixés. On construit ainsi une suite d'entiers ℓ tels que le fibré $\mathcal{D}^{\otimes \ell}$ soit engendré par ses sections (*cf.* théorème 3.6), ce qui rend effectif le théorème 2 de [6]. Si on impose en plus que les fibrés L et $\mathcal{O}_X(1)$ soient numériquement liés, on peut en fait définir, ℓ étant convenablement choisi, un morphisme naturel

$$\sigma : M \to P(\mathrm{H}^0(\check{M}_\ell, \check{\mathcal{D}}_\ell))$$

de M dans un système linéaire complet sur un espace de modules adéquat $\check{M}_\ell$ de faisceaux semi-stables de dimension 1, tel que $\sigma^*(\mathcal{O}(1)) = \mathcal{D}^{\otimes \ell}$: le sous-schéma de $\check{M}_\ell$ des zéros de la section σ_F associée à un faisceau semi-stable F a pour support les points représentant les modules de saut de F. Comme conséquence, on obtient dans le théorème 3.7 une généralisation du théorème 6.11 de [6] relatif à l'hypersurface des droites de saut définie sous certaines conditions pour les faisceaux semi-stables de rang 2. Dans le cas des surfaces quelconques, on se heurte à des difficultés, plutôt psychologiques que réelles, dues au fait que les espaces de modules ne sont en général ni irréductibles, ni réduits et aux problèmes rencontrés pour définir les fibrés déterminants en toute généralité [7]. Nous avons préféré traiter d'abord le cas du plan projectif (*cf.* théorèmes 3.1 et 3.5) pour lequel les choses se comportent simplement.

Nous remercions C.S. Seshadri qui nous a expliqué l'article de Faltings lors de notre visite à Madras en 1992. La méthode que nous proposons ici dans la section 2, si elle s'inspire beaucoup des discussions que nous avons eues, conduit à des résultats sensiblement meilleurs que ceux qui pourraient être obtenus par la méthode que décrit C.S. Seshadri dans [8]. Une remarque arithmétique de G. Hein [5] nous a permis d'améliorer l'estimation proposée dans la première version. Nous sommes reconnaissant au CEFIPRA qui a subventionné notre voyage en Inde en 1992, et à la Fondation Taniguchi qui nous a invité à participer au Symposium Taniguchi en décembre 1994.

Notations

Par variété algébrique, on entend schéma de type fini et séparé sur $\mathbb{C}$. Les points sont les points fermés.

1. Le fibré déterminant sur les courbes

Cette section est consacrée à des rappels sans doute bien connus sur la construction du fibré déterminant $\mathcal{D}$ sur l'espace de modules $M(r, d)$ des classes d'équivalence de fibrés semi-stables de rang r et degré d sur une courbe projective lisse et irréductible X, et à ses propriétés. On gardera bien présent à l'esprit que ce fibré déterminant $\mathcal{D}$ dépend du choix d'une polarisation. Il résulte de sa construction que si on considère sa restriction $\mathcal{D}_L$ à l'espace de module $M(r, L)$ des fibrés semi-stables de déterminant fixé L, on obtient le générateur positif de Drézet et Narasimhan du groupe de Picard $\mathrm{Pic}(M(r, L)) \simeq \mathbb{Z}$ (*cf.* [1]). Ceci démontre que $\mathcal{D}_L$ est ample, et indépendant de tout choix de polarisation sur X. Par contre, l'amplitude de $\mathcal{D}$ n'est pas tout-à-fait évidente, et nous donnerons ici une démonstration de cette propriété. Dans [6] et [7] les diverses autres notions de fibrés déterminants sont introduites pour les variétés de modules de faisceaux semi-stables, qu'ils soient de torsion ou sans torsion, sur des variétés projectives lisses de dimension supérieure. Elles sont calquées sur le même principe; la seule difficulté qui se présente alors est que les espaces de modules ne sont en général ni lisses, ni même réduits, ce qui nécessite un peu de prudence. Ces fibrés déterminants seront examinés dans la section 3 pour les surfaces algébriques projectives.

1.1. Le complexe $Rf_*(\mathcal{F})$

Soit G un groupe algébrique. Soit $f : X \to S$ un morphisme de G-variétés algébriques, projectif et lisse de dimension relative 1 au-dessus de S, et muni d'un G-fibré inversible $\mathcal{O}_X(1)$ relativement ample. Pour $n > 0$, on a

$$f_*(\mathcal{O}_X(-n)) = 0.$$

Pour tout morphisme $\varphi : S' \to S$, G-équivariant, on désigne par $f' : X' \to S'$ la courbe relative obtenue à partir de $X \to S$ par changement de base :

$$\begin{array}{ccc} X' & \xrightarrow{\psi} & X \\ {\scriptstyle f'}\downarrow & & \downarrow{\scriptstyle f} \\ S' & \xrightarrow{\varphi} & S \end{array}$$

On considère un faisceau cohérent S-plat $\mathcal{F}$ sur X, muni d'une action de G. On désigne par $\mathcal{F}_{S'} = \psi^*(\mathcal{F})$ le faisceau cohérent sur X' obtenu à partir de $\mathcal{F}$ par le changement de base $\varphi : S' \to S$.

Lemme 1.1. *On peut associer fonctoriellement à $\mathcal{F}$ un complexe de longueur 1 de G-faisceaux localement libres $Rf_*(\mathcal{F})$ sur S, bien défini à quasi-isomorphismes près, et pour chaque i un isomorphisme fonctoriel*

$$\alpha_i : \mathrm{H}^i(Rf_*(\mathcal{F})) \simeq R^i f_*(\mathcal{F})$$

satisfaisant aux propriétés suivantes:

(i) *Pour tout $\mathcal{O}_S$-module localement libre A, muni d'une action de G, on a un isomorphisme*

$$Rf_*(\mathcal{F}) \otimes A \simeq Rf_*(\mathcal{F} \otimes f^*(A))$$

et cet isomorphisme commute avec les isomorphismes α_i;

(ii) *Pour tout changement de base G-équivariant $\varphi : S' \to S$, on a un isomorphisme*

$$\varphi^*(Rf_*(\mathcal{F})) \simeq Rf'_*(\mathcal{F}_{S'})$$

compatible avec les isomorphismes α_i.

Démonstration. On peut associer à $\mathcal{F}$ une résolution gauche $V_.$ de longueur ≤ 1 de $\mathcal{F}$ par un complexe de faisceaux cohérents V_i tels que $f_*(V_i) = 0$: il suffit de prendre, pour n assez grand, $V_0 = f^*f_*(\mathcal{F}(n)) \otimes \mathcal{O}_X(-n)$, et pour V_1 le noyau du morphisme canonique $V_0 \to F$. Alors le complexe $Rf_*(\mathcal{F})$ défini par

$$0 \to R^1 f_*(V_1) \to R^1 f_*(V_0) \to 0$$

est un complexe de faisceaux localement libres dont les faisceaux de cohomologie s'identifient aux images directes de $\mathcal{F}$. Il satisfait évidemment aux conditions du lemme. Si $W_.$ est une autre G-résolution de $\mathcal{F}$ satisfaisant aux mêmes conditions, et $\varphi : W_. \to V_.$ un morphisme de résolutions, on obtient un morphisme de complexes

$$\begin{array}{ccc} R^1 f_*(W_1) & \xrightarrow{\alpha_*} & R^1 f_*(W_0) \\ {\scriptstyle \varphi_1}\downarrow & & \downarrow{\scriptstyle \varphi_0} \\ R^1 f_*(V_1) & \xrightarrow{\beta_*} & R^1 f_*(V_0) \end{array}$$

qui est un quasi-isomorphisme. Etant données deux telles résolutions $V_.$ et $V'_.$ de $\mathcal{F}$ on peut les coiffer par une troisième résolution $W_.$ comme ci-dessus:

$$\begin{array}{ccc} & & V_. \\ & \nearrow{\scriptstyle \varphi} & \\ W_. & & \\ & \searrow{\scriptstyle \varphi'} & \\ & & V'_. \end{array}$$

où φ et φ' sont des quasi-isomorphismes. Alors les morphismes de complexes φ et φ' induisent sur les complexes associés par image directe des quasi-isomorphismes. Ainsi, le complexe ci-dessus est bien déterminé à quasi-isomorphisme près. □

1.2. Le morphisme $\lambda_{\mathcal{F}} : \mathrm{K}(X) \to \mathrm{Pic}^G(S)$

Soit X une courbe projective lisse de genre g, munie d'un point $a \in X$; le fibré inversible associé sera noté $\mathcal{O}_X(1)$. Etant donnée une variété algébrique S munie d'une action de G, on considère les projections

$$\begin{array}{ccc} S \times X & \xrightarrow{p_2} & X \\ {\scriptstyle p_1}\downarrow & & \\ S & & \end{array}$$

Soit $\mathrm{K}(X)$ l'algèbre de Grothendieck des classes de faisceaux algébriques cohérents sur la courbe X. Elle est équipée de la forme quadratique $u \mapsto \chi(u^2)$. Puisque X est lisse, on a un isomorphisme de groupes $\mathrm{K}(X) \simeq \mathbb{Z} \times \mathrm{Pic}(X)$ qui associe à la classe d'un fibré E le couple $(r(E), \det(E))$ formé du rang $r(E)$ et du déterminant $\det(E)$. On définit un homomorphisme de groupes $u \mapsto \lambda_{\mathcal{F}}(u)$

$$\mathrm{K}(X) \to \mathrm{Pic}^G(S)$$

de la manière suivante: on considère l'élément du groupe de Grothendieck $\mathrm{K}_G(S)$ des classes de G-faisceaux cohérents défini par $p_{1!}(\mathcal{F}.p_2^*(u))$. Cet élément provient d'un élément bien déterminé de $\mathrm{K}_G^0(S)$, groupe de Grothendieck des classes de G-faisceaux localement libres de type fini, pour lequel les notions de rang et de déterminant ont un sens; on a

$$\mathrm{rang}(p_{1!}(\mathcal{F}.p_2^*(u))) = dr(u) + r\chi(u)$$

et on pose dans $\mathrm{Pic}^G(S)$

$$\lambda_{\mathcal{F}}(u) = \det(p_{1!}(\mathcal{F}.p_2^*(u)))$$

Lemme 1.2.

(i) *La formation de $\lambda_{\mathcal{F}}(u)$ est compatible aux changements de base G-équivariants.*

(ii) *Etant donnée une suite exacte de G-faisceaux cohérents plats sur S*

$$0 \to \mathcal{F}' \to \mathcal{F} \to \mathcal{F}'' \to 0$$

on a dans $\mathrm{Pic}^G(S)$

$$\lambda_{\mathcal{F}}(u) = \lambda_{\mathcal{F}'}(u) \otimes \lambda_{\mathcal{F}''}(u).$$

(iii) *Pour tout G-fibré inversible A sur S on a*

$$\lambda_{\mathcal{F}\otimes p_1^*(A)}(u) = \lambda_{\mathcal{F}}(u) \otimes A^{\otimes(dr(u)+r\chi(u))}.$$

Démonstration. On peut supposer que u est la classe d'un fibré vectoriel E sur X. On applique alors le lemme ci-dessus à la famille $\mathcal{F} \otimes p_2^*(E)$: on a par définition

$$\lambda_{\mathcal{F}}(E) = \otimes_i \det R^i p_{1*}(\mathcal{F} \otimes p_2^*(E))^{(-1)^i}$$

et d'après le lemme ci-dessus, la formation du complexe $Rp_{1*}(\mathcal{F} \otimes p_2^*(E))$ commute aux changements de base. Les assertions (ii) et (iii) résultent trivialement du lemme 1.1. □

1.3. Application au schéma de Hilbert-Grothendieck

Soit E un fibré vectoriel fixé sur X. Désignons par P le polynôme $P(m) = mr + \chi$, où $\chi = d + r(1-g)$. La construction ci-dessus s'applique en particulier au cas où on prend pour variété de paramètres le schéma de Hilbert-Grothendieck $R = \mathrm{Hilb}(E, P)$ des faisceaux algébriques cohérents quotients de E, de polynôme de Hilbert P, équipé du faisceau universel quotient $\mathcal{F}$ sur $R \times X$. Considérons le diagramme

$$\begin{array}{ccc} R \times X & \xrightarrow{p_2} & X \\ {\scriptstyle p_1}\downarrow & & \\ R & & \end{array}$$

On sait que le schéma de Hilbert R est une variété projective; pour ℓ assez grand, l'image directe $R^1p_{1*}(\mathcal{F}(\ell))$ est nulle, et l'image directe $p_{1*}(\mathcal{F}(\ell))$ est un faisceau localement libre de rang $P(\ell)$ quotient du fibré trivial de fibre $\mathrm{H}^0(E(\ell))$. La propriété universelle de la grassmannienne $\mathrm{Grass}(\mathrm{H}^0(E(\ell)), P(\ell))$ des espaces vectoriels quotients de dimension $P(\ell)$ fournit un morphisme

$$\varphi_\ell : R \to \mathrm{Grass}(\mathrm{H}^0(E(\ell)), P(\ell))$$

qui associe au point q de R l'espace vectoriel quotient $\mathrm{H}^0(\mathcal{F}_q(\ell))$ de l'espace vectoriel des sections $\mathrm{H}^0(E(\ell))$. Ce morphisme est, si ℓ est assez grand, un plongement fermé: ceci résulte de la construction de Grothendieck de ce schéma de Hilbert. On désigne par H_ℓ le fibré inversible image réciproque de $\mathcal{O}_{\mathrm{Grass}(\mathrm{H}^0(E(\ell)))}(1)$ par ce plongement.

Soit $GL(E)$ le groupe des automorphismes de E. Considérons l'action naturelle du groupe des automorphismes de E sur R. Il opère aussi sur la grassmannienne et le plongement φ_ℓ est équivariant pour cette action. En outre, l'action s'étend en une action linéaire sur le fibré $\mathcal{O}_{\mathrm{Grass}(\mathrm{H}^0(E(\ell)),P(\ell))}(1)$ et par suite sur H_ℓ.

Soient m un entier, et R_m l'ouvert de R des points $q \in R$ satisfaisant aux conditions suivantes

(i) $\mathrm{H}^1(\mathcal{F}_q(m)) = 0$

(ii) L'application linéaire induite par la projection $E \to \mathcal{F}_q$:

$$\mathrm{H}^0(E(m)) \to \mathrm{H}^0(\mathcal{F}_q(m))$$

est inversible.

L'ouvert R_m est invariant par $GL(E)$ et on peut considérer le groupe de Picard $\mathrm{Pic}^{GL(E)}(R_m)$ des classes d'isomorphisme de fibrés inversibles muni d'une action de $GL(E)$.

Proposition 1.3. *Soit a un point de X, et $h \in \mathrm{K}(X)$ défini par le faisceau structural du point a. Soient r' et d' deux entiers. Considérons l'élément $u = r' + d'h$ de $\mathrm{K}(X)$. Alors, dans $\mathrm{Pic}(R_m)$, noté additivement, on a*

$$(\ell - m)\lambda_{\mathcal{F}}(r' + d'h) = (-mr' + d')\,\mathrm{H}_\ell\,.$$

Démonstration. On a évidemment, dans $K(X)$,

$$\mathcal{O}_X(i) = 1 + ih.$$

On applique ceci en remplaçant successivement i par ℓ et m. Du fait que $\lambda_{\mathcal{F}}(\mathcal{O}_X(\ell)) = \det(p_{1*}(\mathcal{F}(\ell))) = \mathrm{H}_\ell$ on obtient alors dans $\mathrm{Pic}(R)$ la relation

$$\mathrm{H}_\ell = \lambda_{\mathcal{F}}(1) + \ell\lambda_{\mathcal{F}}(h) \tag{1.1}$$

Sur l'ouvert R_m on a évidemment $R^1p_{1*}(\mathcal{F}(m)) = 0$, et le faisceau $p_{1*}(\mathcal{F}(m))$ est le faisceau trivial. Il en résulte la relation dans $\mathrm{Pic}(R_m)$

$$0 = \lambda_{\mathcal{F}}(1) + m\lambda_{\mathcal{F}}(h) \tag{1.2}$$

Des deux équations (1.1) et (1.2) on tire dans $\mathrm{Pic}(R_m)$ les relations

$$(\ell - m)\lambda_{\mathcal{F}}(h) = \mathrm{H}_\ell \quad \text{et} \quad (\ell - m)\lambda_{\mathcal{F}}(1) = -m\,\mathrm{H}_\ell\,.$$

La proposition résulte alors de la linéarité de $\lambda_{\mathcal{F}}(u)$ par rapport à u. □

Remarque 1.4. La relation ci-dessus n'est pas vraie dans le groupe $\mathrm{Pic}^{GL(E)}(R_m)$: en fait, dans ce groupe les deux membres diffèrent par une puissance $GL(E)$-fibré inversible $\det(p_{1*}(\mathcal{F}(m)))$ (lequel est trivial dans $\mathrm{Pic}(R_m)$).

Corollaire 1.5. *On considère le $GL(E)$-fibré inversible sur R défini par*

$$\mathcal{D}_R = \frac{1}{\mathrm{pgcd}(r,d)} \lambda_{\mathcal{F}}(-r + \chi h).$$

Si m est assez grand, le fibré inversible $\mathcal{D}_R$ est isomorphe, sur l'ouvert R_m, à la restriction d'un fibré inversible ample.

Démonstration. Il suffit de choisir $\ell > m$ et assez grand. □

1.4. Le fibré déterminant

Soient m un entier assez grand, et H un espace vectoriel de dimension $P(m)$. On considère le fibré vectoriel $E = \mathrm{H} \otimes \mathcal{O}_X(-m)$, et le schéma de Hilbert-Grothendieck $R = \mathrm{Hilb}(E, P)$ des faisceaux cohérents quotients de E de polynôme de Hilbert P, polarisé comme ci-dessus par le plongement de Grothendieck, et muni de la famille universelle quotient $\mathcal{F}$ sur $R \times X$. Le groupe des automorphismes $GL(E)$ est isomorphe au groupe $GL(\mathrm{H})$. On sait que l'ouvert R^{ss} des points $SL(\mathrm{H})$-semi-stables coïncide avec l'ouvert des points $q \in R$ satisfaisant à la condition (ii) ci-dessus et à la condition suivante:

(iii) Le faisceau $\mathcal{F}_q$ est semi-stable.

En particulier, $R^{ss} \subset R_m$.

L'espace de modules $M(r,d)$ est le quotient de Mumford $R^{ss}/GL(\mathrm{H})$. C'est une variété projective: elle se plonge dans une variété projective par le système linéaire des sections invariantes associé à une puissance convenable du fibré $\mathcal{O}_R(1)$.

Soit $u \in \mathrm{K}(X)$, tel que le rang $r(u)$ et la caractéristique d'Euler-Poincaré $\chi(u)$ de u soient liés par la relation suivante:

$$dr(u) + r\chi(u) = 0 \qquad (*)$$

Ceci entraîne que l'élément $p_{1!}(\mathcal{F}.p_2^*(u))$ de $\mathrm{K}^0_G(R)$ est de rang 0. Alors le $GL(E)$-fibré inversible $\lambda_{\mathcal{F}}(u)$ sur R^{ss} provient d'un fibré inversible sur $M(r,d)$: d'après un lemme dû à Kempf [1], il suffit de vérifier qu'en un point d'orbite fermée, le stabilisateur opère trivialement sur $\lambda_{\mathcal{F}}(u)$; ceci résulte facilement des sorites énoncés dans le lemme 1.2. Ceci s'applique en particulier au $GL(E)$-fibré inversible $\mathcal{D}_R$, qui provient d'un fibré inversible $\mathcal{D}$ sur $M(r,d)$, qu'on appellera le *fibré déterminant.*

Corollaire 1.6. *Le fibré $\mathcal{D}$ est ample.*

Démonstration. On sait que le morphisme image réciproque

$$\mathrm{Pic}(M(r,d)) \to \mathrm{Pic}(R^{ss})$$

est injectif. D'après la théorie invariante de Mumford, l'image réciproque dans $\mathrm{Pic}(R^{ss})$ du fibré $\mathcal{O}(1)$ sur l'espace de modules $M(r,d)$ est une puissance positive de H_ℓ. Ainsi, la relation entre obtenue $\mathcal{D}_R$ et H_ℓ dans la proposition 1.3 provient d'une relation (écrite en notation additive)

$$\alpha\mathcal{D} = \beta\mathcal{O}(1)$$

où α et β sont des entiers > 0. Par suite, $\mathcal{D}$ est ample. □

Corollaire 1.7. *Soit S une variété algébrique, $\mathcal{F}$ un faisceau cohérent sur $S \times X$, S-plat, de polynôme de Hilbert P. Considérons le fibré inversible $\mathcal{D}_{\mathcal{F}}$ sur S défini par*

$$\mathcal{D}_{\mathcal{F}} = \frac{1}{\mathrm{pgcd}(r,d)} \lambda_{\mathcal{F}}(-r + \chi h).$$

(1) *Le fermé de S des points $s \in S$ tels que $\mathcal{F}_s$ soit instable est intersection de fermés des zéros de sections de $\mathcal{D}_{\mathcal{F}}^{\otimes k}$, avec k assez grand.*
(2) *Supposons que $\mathcal{F}_s$ soit semi-stable pour tout $s \in S$; désignons par*

$$f_{\mathcal{F}} : S \to M(r,d)$$

le morphisme modulaire. Dans $\mathrm{Pic}(S)$, *on a la relation* $\mathcal{D}_{\mathcal{F}} = f_{\mathcal{F}}^{*}(\mathcal{D})$.

2. Les fonctions thêta

Soient S une variété algébrique, munie d'une action d'un groupe algébrique G, et $\mathcal{F}$ une G-famille plate de faisceaux cohérents sur X, de rang r et degré d, paramétrée par S. On considère ici des éléments $u \in \mathrm{K}(X)$, dont le rang $r(u)$ et la caractéristique d'Euler-Poincaré $\chi(u)$ de u sont liés comme ci-dessus par la relation :

$$dr(u) + r\chi(u) = 0 \qquad (*)$$

Si u peut se représenter par un fibré vectoriel algébrique V, cette hypothèse signifie que pour tout s on a $\chi(\mathcal{F}_s \otimes V) = 0$; autrement dit l'élément $p_{1!}(\mathcal{F}.p_2^{*}(u))$ de $\mathrm{K}_G^0(S)$ est de rang 0. On va associer à un tel fibré vectoriel une section G-invariante du fibré $\lambda_{\mathcal{F}}(-u)$, bien déterminée à inversibles près. En choisissant le déterminant $\det(V)$ multiple de $\mathcal{O}_X(1)$, et en appliquant ceci à la famille universelle $\mathcal{F}$ paramétrée par le schéma de Hilbert-Grothendieck, on obtiendra par passage au quotient une section d'une puissance du fibré déterminant $\mathcal{D}$ sur $M(r,d)$, bien déterminée à inversibles près, et qu'on appellera fonction thêta.

2.1. Les sections θ_V

Soit $\mathcal{F}$ une G-famille de faisceaux cohérents sur X comme ci-dessus, paramétrée par la G-variété S.

Lemme 2.1. *Soit V un fibré vectoriel algébrique sur X, de classe de Grothendieck u dans* $\mathrm{K}(X)$. *On suppose que le rang et la caractéristique d'Euler-Poincaré de V sont liées par la relation* $(*)$. *Alors le G-fibré inversible $\lambda_{\mathcal{F}}(u)^{-1}$ est muni d'une section G-invariante $\theta_{\mathcal{F},V}$, bien déterminée à éléments inversibles près, dont le schéma des zéros a pour support les points $s \in S$ tels que* $\mathrm{H}^0(\mathcal{F}_s \otimes V) \neq 0$.

Remarquons que comme $\chi(\mathcal{F}_s \otimes V) = 0$, la dernière condition est équivalente à $\mathrm{H}^1(\mathcal{F}_s \otimes V) \neq 0$. En termes de pentes, si on pose $\dfrac{d}{r} = \mu$ la condition $(*)$ est équivalente à

$$\mu + \mu(V) = g - 1.$$

Démonstration. Puisque $\mathcal{F}$ est un faisceau S-plat, on peut choisir pour le G-faisceau cohérent $\mathcal{G} := \mathcal{F} \otimes p_2^*(V)$ une résolution gauche localement libre

$$\begin{array}{ccccccc} 0 & \longrightarrow & V_1 & \xrightarrow{\alpha} & V_0 & \longrightarrow & 0 \\ & & & & \downarrow & & \\ & & & & \mathcal{G} & & \end{array}$$

telle que $p_{1*}(V_0) = 0$. Alors on a $p_{1*}(V_1) = 0$ ce qui prouve que l'élément de $\mathrm{K}^0_G(S)$ défini par $\lambda_{\mathcal{F}}(u)$ est représenté par le G-fibré inversible $\det(R^1p_{1*}(V_0)) \otimes \det(R^1p_{1*}(V_1))^{-1}$; le morphisme α induit un morphisme

$$\alpha_* : R^1p_{1*}(V_1) \to R^1p_{1*}(V_0)$$

G-équivariant dont le déterminant $\theta = \det(\alpha_*)$ définit une section G-invariante de $\lambda_{\mathcal{F}}(u)^{-1}$. Le sous-schéma de S défini par θ est la sous-variété de Fitting du faisceau $R^1p_{1*}(\mathcal{G})$. Ceci prouve qu'à inversibles près (dans l'algèbre des sections G-invariantes du faisceau structural) cette section θ ne dépend pas de la résolution choisie. □

Definition 2.2. *La section $\theta = \theta_{\mathcal{F},V}$ ainsi définie s'appelle fonction thêta associée au fibré vectoriel V et à la famille $\mathcal{F}$.*

Lemme 2.3.

(i) *La formation de la fonction $\theta_{\mathcal{F},V}$ est compatible aux changements de base G-équivariants.*

(ii) *Etant donnée une suite exacte de fibrés vectoriels $0 \to V' \to V \to V'' \to 0$ satisfaisant à la condition (*) on a, à inversibles près,*

$$\theta_{\mathcal{F},V} = \theta_{\mathcal{F},V'}\theta_{\mathcal{F},V''}.$$

(iii) *Etant donnée une suite exacte $0 \to \mathcal{F}' \to \mathcal{F} \to \mathcal{F}'' \to 0$ de G-faisceaux cohérents plats sur S dont le rang et le degré satisfont à la condition (*) on a, à inversibles près,*

$$\theta_{\mathcal{F},V} = \theta_{\mathcal{F}',V}\theta_{\mathcal{F}'',V}.$$

Lorsque V varie, à rang et déterminant constant, sa classe dans $\mathrm{K}(X)$ reste constante; ainsi la classe d'isomorphisme du fibré $\lambda_{\mathcal{F}}(V)$ ne change pas. Dans la suite, le rang et le déterminant de V seront choisis comme suit:

$$r(V) = \frac{kr}{\mathrm{pgcd}(r,d)} \quad ; \quad \det(V) = \mathcal{O}_X\left(-\frac{k\chi}{\mathrm{pgcd}(r,d)}\right).$$

Dans ce cas la section $\theta_{\mathcal{F}}(V)$ est section du fibré $\mathcal{D}_{\mathcal{F}}^{\otimes k}$. Ceci s'applique en particulier au cas où on prend pour S le schéma de Hilbert-Grothendieck R et pour $\mathcal{F}$ le faisceau quotient universel. Alors la section $\theta_{\mathcal{F},V}$ passe au quotient et définit une section de $\mathcal{D}^{\otimes k}$ sur $M = M(r,d)$, qui sera appelée fonction thêta associée au fibré vectoriel V, et notée θ_V.

2.2. Le principal résultat

Théorème 2.4. *Soient $\delta = \mathrm{pgcd}(r,d)$, et k un entier $> \delta$ et tel que*

$$k > \frac{r^2}{4}(g-1).$$

Alors le fibré $\mathcal{D}^{\otimes k}$ est engendré par les fonctions thêta.

Cet énoncé est la version effective d'un résultat dû à Faltings [3]; il signifie que pour tout fibré semi-stable F sur la courbe X, il existe un fibré vectoriel V dont le rang et le déterminant sont donnés par

$$r(V) = \frac{kr}{\delta} \quad ; \quad \det(V) = \mathcal{O}_X(-\frac{k\chi}{\delta}),$$

et tel que $\mathrm{H}^0(F \otimes V) = 0$, ou ce qui revient au-même $\mathrm{H}^1(F \otimes V) = 0$. Un tel k étant choisi, si $W \subset \mathrm{H}^0(M(r,d), \mathcal{D}^{\otimes k})$ désigne le système linéaire engendré par les fonctions thêta on obtient un morphisme fini

$$M(r,d) \to \mathbf{P}(W).$$

Ceci entraîne que l'algèbre graduée engendrée par les fonctions thêta du type précédent est une algèbre de type fini.

La démonstration que nous donnons ici est inspirée de celle qu'ont donnée Seshadri et Nori [8]. Pour $g = 0$, l'énoncé est trivial. Pour $r = 1$, l'espace de modules est isomorphe à la jacobienne des fibrés inversibles de degré $g - 1$ et il est bien connu que cet énoncé est vrai en fait en prenant $k \geq 2$: ainsi, cet énoncé peut sans aucun doute être amélioré. Commençons par une remarque qui précise comment on doit choisir V:

Lemme 2.5. *Soient F un faisceau cohérent sur X et V un fibré vectoriel sur X tels que $\chi(F \otimes V) = 0$. Si $h^0(F \otimes V) = 0$, le faisceau F est localement libre et F et V sont semi-stables.*

Démonstration. Il est clair que si F avait de la torsion $F \otimes V$ devrait avoir des sections non nulles. On est ainsi ramené à prouver que si F et V sont deux fibrés vectoriels satisfaisant aux conditions du lemme, ils sont semi-stables. Montrons par exemple que V est semi-stable. Soit $V' \subset V$ un sous-fibré non nul et de pente $\mu(V')$. L'hypothèse $h^0(F \otimes V) = 0$ entraîne alors $h^0(F \otimes V') = 0$, et donc $\chi(F \otimes V') \leq 0$. Mais alors $\chi(F \otimes V') \leq \chi(F \otimes V)$, ce qui implique $\mu(V') \leq \mu(V)$. □

En raison de l'assertion (iii) du lemme 2.3 on peut supposer que F est non seulement semi-stable, mais stable. En effet, l'espace de modules des fibrés stables V, dont le rang et le déterminant sont fixés est irréductible, et les points représentant les fibrés V tels que $h^0(F \otimes V) = 0$ forment un ouvert de Zariski d'après le théorème de semi-continuité.

2.3. Sections d'un schéma de Grothendieck

Soit $\mathcal{E} \to S \times X$ une famille de fibrés vectoriels de rang r et de pente μ sur la courbe X, paramétrée par une variété algébrique lisse S; on suppose que la famille de fibrés inversibles $\det \mathcal{E}_s$ est constante. On considère le schéma de Hilbert-Grothendieck relatif

$$\pi : \mathrm{Hilb}(\mathcal{E}/S, P) \to S$$

dont les points sont les paires $t = (s, E'')$, où s est un point de S, et $E'' = \mathcal{E}_s/E'$ un faisceau cohérent quotient de $\mathcal{E}_s$ de polynôme de Hilbert P, par un sous-faisceau E'. Quand $\mathcal{E}$ est fixé, le polynôme de Hilbert P n'est pas quelconque: on sait déjà que la pente de E'' est minorée; sous des hypothèses supplémentaires sur la famille $\mathcal{E}$ on peut préciser cette estimation. C'est l'objet de ce paragraphe.

Soit $\mathrm{Ext}^1_0(\mathcal{E}_s, \mathcal{E}_s)$ le noyau du morphisme trace

$$\mathrm{trace} \ : \mathrm{Ext}^1(\mathcal{E}_s, \mathcal{E}_s) \to \mathrm{H}^1(\mathcal{O}_X).$$

On a une suite exacte naturelle ([2], proposition 1.5)

$$T_t \operatorname{Hilb}(\mathcal{E}/S, P) \xrightarrow{T\pi} T_s S \xrightarrow{\nu} \operatorname{Ext}^1(E', E'')$$

dans laquelle l'application linéaire ν est la composée de l'application de déformation infinitésimale de Kodaira-Spencer $\kappa_s : T_s S \to \operatorname{Ext}^1_0(\mathcal{E}_s, \mathcal{E}_s)$ au point $s \in S$, et de la projection canonique $\operatorname{Ext}^1_0(\mathcal{E}_s, \mathcal{E}_s) \to \operatorname{Ext}^1(E', E'')$.

Lemme 2.6. *Soit* $t = (s, E'')$ *un point de* $\operatorname{Hilb}(\mathcal{E}/S, P)$. *On suppose que* E'' *est de rang* $r'' > 0$ *et on désigne par* μ'' *sa pente. On se place sous les hypothèses suivantes:*

(i) *l'application* κ_s *est surjective.*

(ii) *il existe, au voisinage* U *du point* s, *une section locale* $\sigma : U \to \operatorname{Hilb}(\mathcal{E}/S, P)$ *passant par le point* t.

On a alors

$$\mu'' - \mu \geq (1 - \frac{r''}{r})(g - 1).$$

Démonstration. Rappelons d'abord la notation suivante: étant donnés deux faisceaux algébriques cohérents A et B sur une variété algébrique lisse, on pose

$$\chi(A, B) = \sum_q (-1)^q \dim \operatorname{Ext}^q(A, B).$$

Puisque κ_s est par hypothèse une application linéaire surjective, il en résulte que l'application linéaire ν est surjective. Puisque π a une section locale au voisinage du point s, son application linéaire tangente est aussi surjective. Il en résulte que $\operatorname{Ext}^1(E', E'') = 0$. Ainsi, on obtient $\chi(E', E'') \geq 0$, ce qui est équivalent à $\chi(\mathcal{E}_s, E'') \geq \chi(E'', E'')$. Compte-tenu de la formule de Riemann-Roch, ceci s'écrit encore

$$rr''(\mu'' - \mu + 1 - g) \geq r''^2(1 - g)$$

ce qui entraîne évidemment le résultat. □

2.4. Démonstration du théorème 2.4

Soit F un fibré stable de rang r, de caractéristique d'Euler-Poincaré χ; on désigne par d le degré, et par $\mu = \dfrac{d}{r}$ la pente de F. On se propose de montrer qu'il existe un fibré stable E de rang et déterminant

$$r(E) = k\frac{r}{\delta} \quad ; \det(E) = L = \mathcal{O}(\frac{k\chi}{\delta})$$

et tel que $\operatorname{Hom}(E, F) = 0$. Ceci signifie que la fonction θ_{E^*} ne s'annule pas au point de $M(r, d)$ défini par F, ce qui démontrera le théorème.

Commençons par paramétrer ces fibrés stables E: soit Q le polynôme défini par

$$Q(m) = \frac{k}{\delta}(rm + \chi + r(1 - g)).$$

On considère, pour m assez grand, un espace vectoriel H de dimension $Q(m)$, et le fibré vectoriel $W = \mathrm{H} \otimes \mathcal{O}_X(-m)$. Dans le schéma de Hilbert-Grothendieck $R = \operatorname{Hilb}(W, Q)$ des faisceaux cohérents quotients de W de polynôme de Hilbert Q on considère l'ouvert R^s qui paramètre les fibrés quotients stables E tels que l'application linéaire induite $\mathrm{H} \to H^0(E)$ soit bijective. On désigne par S le sous-schéma de R^s défini par l'image réciproque de L par le morphisme déterminant:

$$\det : R^s \to \operatorname{Pic}(X).$$

Ce sous-schéma est lisse, irréductible et est muni d'un fibré quotient universel $\mathcal{E} \to S \times X$; de plus l'application linéaire de déformation infinitésimale de Kodaira-Spencer associée à cette famille en un point $s \in S$ correspondant au fibré quotient $E = \mathcal{E}_s$

$$\kappa_s : T_sS \to \mathrm{Ext}^1_0(E, E)$$

est surjective. Le théorème sera démontré si on montre qu'il existe un point $s \in S$ tel que $\mathrm{Hom}(E, F) = 0$, ou ce qui revient au même $\mathrm{Ext}^1(E, F) = 0$.

L'accouplement de Yoneda

Soit ℓ le plus petit des nombres $\dim \mathrm{Ext}^1(\mathcal{E}_s, F)$ quand s parcourt S, et supposons ce rang non nul. La variété déterminantielle Σ_ρ définie par les mineurs traduisant les conditions de rang $\dim \mathrm{Ext}^1(\mathcal{E}_s, F) \geq \rho$ a pour espace tangent de Zariski en un point s où cette dimension est ρ le noyau du morphisme de Petri

$$\tau : T_sS \to \mathrm{L}(\mathrm{Hom}(E, F), \mathrm{Ext}^1(E, F))$$

où l'on a posé comme ci-dessus $E = \mathcal{E}_s$. Si on prend $\rho = \ell$, cette sous-variété déterminantielle coïncide avec S, par conséquent son espace de tangent de Zariski en un point $s \in S$ est l'espace vectoriel T_sS; ainsi, l'application linéaire τ est nulle en tout point de l'ouvert U des points s tel que $\dim \mathrm{Hom}(E, F) = \ell$. Or le morphisme de Petri est bien entendu l'application linéaire composée

$$T_sS \xrightarrow{\kappa_s} \mathrm{Ext}^1_0(E, E) \to \mathrm{L}(\mathrm{Hom}(E, F), \mathrm{Ext}^1(E, F))$$

où la deuxième flèche est, au signe près, l'accouplement de Yoneda. Ainsi, cet accouplement est nul: autrement dit l'application bilinéaire

$$\mathrm{Ext}^1_0(E, E) \otimes \mathrm{Hom}(E, F) \to \mathrm{Ext}^1(E, F)$$

de Yoneda est nulle. Par dualité de Serre, on obtient que l'accouplement naturel

$$\mathrm{Hom}(E, F) \otimes \mathrm{Hom}(F, E \otimes \omega_X) \to \mathrm{Hom}(E, E \otimes \omega_X)/\,\mathrm{H}^0(\omega_X)$$

donné par la composition des morphismes est nul. Ainsi, la composition des homomorphismes fournit une homothétie dont le rapport est une forme différentielle.

La famille $\mathcal{F}'_s$

Considérons le morphisme canonique au-dessus de U

$$\underline{\mathrm{Hom}}_{p_1}(\mathcal{E}, p_2^*(F)) \otimes \mathcal{E} \to p_2^*(F)$$

dont le rang est non nul; quitte à diminuer au besoin U, on peut supposer que le fibré $\underline{\mathrm{Hom}}_{p_1}(\mathcal{E}, p_2^*(F))$ est muni d'une section non nulle u, ce qui définit une famille de morphimes non nuls

$$u_s : \mathcal{E}_s \to F$$

paramétrée par U. On peut supposer en outre que le conoyau du morphisme u_s définit une famille plate de faisceaux cohérents au-dessus de U. On désigne par $\mathcal{F}'_s$ l'image de ce morphisme: c'est une famille plate de faisceaux localement libres; on a évidemment par semi-stabilité

$$\mu(\mathcal{E}_s) \leq \mu(\mathcal{F}'_s) \leq \mu.$$

On va voir en fait que si k augmente, la pente de $\mathcal{F}'_s$ a tendance à se rapprocher de μ. En effet la famille $\mathcal{F}'_s$ est une famille plate quotient de $\mathcal{E}_s$, et la propriété universelle du schéma de Grothendieck relatif associe à cette famille une section du schéma de Hilbert relatif correspondant: les hypothèses du lemme 2.6 ci-dessus sont donc satisfaites, et on a donc d'après ce lemme

$$\mu(\mathcal{F}'_s) - \mu(\mathcal{E}_s) \geq (1 - \frac{r(\mathcal{F}'_s)}{r(\mathcal{E}_s)})(g-1)$$

ce qui s'écrit encore, compte-tenu de la relation $\mu - \mu(\mathcal{E}_s) = g - 1$

$$0 \leq \mu - \mu(\mathcal{F}'_s) \leq \frac{r(\mathcal{F}'_s)}{r(\mathcal{E}_s)}(g-1) \tag{2.1}$$

La famille $\mathcal{F}''_s$

Considérons le morphisme v_s obtenu en choisissant une section non nulle du fibré $\underline{\mathrm{Hom}}_{p_1}(p_2^*(F), \mathcal{E} \otimes \omega_X)$

$$v_s : F \longrightarrow \mathcal{E}_s \otimes \omega_X$$

et désignons par $\mathcal{F}''_s$ le faisceau image de ce morphisme. Ce faisceau est pour chaque point $s \in U$ localement libre et non nul, et ici encore, quitte à diminuer éventuellement l'ouvert U, on peut supposer que la famille des conoyaux est plate, ce qui entraîne que la famille des faisceaux $\mathcal{F}''_s$ est elle aussi plate. Le lemme ci-dessus appliqué à cette famille de quotients donne cette fois

$$0 \leq \mu(\mathcal{F}''_s) - \mu \leq \frac{r(\mathcal{F}''_s)}{r(\mathcal{E}_s)}(g-1) \tag{2.2}$$

Le bilan

Considérons, pour $s \in U$ le diagramme

$$\begin{array}{ccccc} \mathcal{E}_s & \longrightarrow & \mathcal{F}'_s & \longrightarrow & 0 \\ & & \downarrow & & \\ & & F & & \\ & & \downarrow & & \\ 0 & \longrightarrow & \mathcal{F}''_s & \longrightarrow & \mathcal{E}_s \otimes \omega_X \end{array}$$

dans lequel les lignes sont des suites exactes. D'après la propriété de l'accouplement de Yoneda énoncée ci-dessus, le morphisme composé

$$\varphi : \mathcal{E}_s \longrightarrow \mathcal{E}_s \otimes \omega_X$$

est une homothétie dont le rapport au-dessus de chaque point s est la multiplication par une forme différentielle. Si cette forme différentielle n'est pas nulle, le morphisme correspondant devrait être génériquement un isomorphisme. Mais ce morphisme se factorise à travers F, et l'hypothèse $k > \delta$ entraîne que le rang $r(\mathcal{E}_s)$ est strictement supérieur à r. Ainsi, il n'est pas possible que φ soit génériquement un isomorphisme. Par suite $\varphi = 0$. Ceci implique que dans le diagramme ci-dessus la ligne verticale est une 0-suite. On a alors,

$$r(\mathcal{F}'_s) + r(\mathcal{F}''_s) \leq r \tag{2.3}$$

Ceci implique en particulier $r(\mathcal{F}'_s) \leq r - 1$ et $r(\mathcal{F}''_s) \leq r - 1$. En ajoutant (2.1) et (2.2), on obtient, compte-tenu de (2.3),

$$\begin{aligned} 0 \leq \mu(\mathcal{F}''_s) - \mu(\mathcal{F}'_s) &\leq \frac{r}{r(\mathcal{E}_s)}(g-1) \\ &\leq \frac{\delta}{k}(g-1). \end{aligned}$$

On a obligatoirement $\mu(\mathcal{F}''_s) > \mu > \mu(\mathcal{F}'_s)$, par la condition de stabilité. Mais alors, on a la minoration

$$\mu(\mathcal{F}''_s) - \mu(\mathcal{F}'_s) \geq \frac{\delta}{r}(\frac{1}{r(\mathcal{F}'_s)} + \frac{1}{r(\mathcal{F}''_s)})$$
$$\geq \frac{4\delta}{r^2}.$$

Il en résulte $k \leq \frac{r^2}{4}(g-1)$, ce qui est contraire à l'hypothèse. Ainsi, $\ell = 0$, ce qui prouve le théorème.

Corollaire 2.7. *Soit $\mathcal{F}$ une famille plate de faisceaux semi-stables de rang r et de degré d sur la courbe C, paramétrée par une variété algébrique S. Soient $\delta = \mathrm{pgcd}(r, d)$, et k un entier $> \delta$ et tel que*

$$k > \frac{r^2}{4}(g-1).$$

Alors l'idéal du fermé des points $s \in S$ tels que $\mathcal{F}_s$ soit instable est le radical de l'idéal engendré par les fonctions $\theta \in \mathrm{H}^0(\mathcal{D}_{\mathcal{F}}^{\otimes k})$.

2.5. Une variante

L'énoncé ci-dessous est apparenté au théorème 2.4, mais en fait beaucoup plus facile; il est un peu plus précis que le résultat obtenu par Georg Hein dans [5]. Lorsque $r^2(g-1) \geq 4\delta$, cet énoncé est en fait une conséquence du théorème 2.4. Soit L un fibré inversible de degré d sur X, et $M(r, L)$ l'espace de modules des classes d'équivalence de fibrés semi-stables sur X de rang r et de déterminant L. Soit $\mathcal{D}_L$ le fibré induit sur $M(r, L)$ par $\mathcal{D}$. Ce fibré inversible est le générateur positif du groupe cyclique $\mathrm{Pic}(M(r, L))$ et ne dépend pas du choix de la polarisation.

Théorème 2.8. *Le fibré inversible $\mathcal{D}_L^{\otimes k}$ est engendré par ses sections dès que*

$$k > \frac{r^2}{4}(g-1).$$

Démonstration. La restriction du fibré inversible $\mathcal{D}^{\otimes k}$ à $M(r, L)$ est le fibré inversible construit par passage au quotient à partir du fibré $\lambda_{\mathcal{F}}(-V)$, où V est un fibré vectoriel dont le rang et le degré sont donnés par

$$r(V) = k\frac{r}{\delta} \quad ; d(V) = -\frac{k\chi}{\delta},$$

et où $\mathcal{F}$ est une famille de fibrés semi-stables de rang r et de déterminant L fixé, comme au §2.1. L'énoncé signifie donc que pour tout fibré semi-stable F de rang r et degré d il existe un tel fibré vectoriel V satisfaisant à la condition $\mathrm{H}^0(F \otimes V) = 0$. Cet énoncé est plus facile que le précédent, car on n'impose pas ici que le déterminant de V soit fixé. Alors on peut remplacer dans la démonstration ci-dessus $\mathrm{Ext}^1_0(E, E)$ par $\mathrm{Ext}^1(E, E)$ et la condition $k > \delta$ n'est plus nécessaire pour obtenir que la suite $\mathcal{F}'_s \to F \to \mathcal{F}''_s$ soit une 0-suite. □

3. Fibré déterminant sur les surfaces

Les constructions du fibré déterminant données dans le premier paragraphe peuvent s'étendre à des variétés de dimension supérieure: sur les espaces de modules des classes de faisceaux semi-stables sur les surfaces algébriques projectives et lisses, on peut en particulier définir le fibré déterminant de Donaldson $\mathcal{D}$, dont on sait qu'une puissance est engendré par les sections ([6], théorème 2). On donne ici, en s'appuyant sur le théorème 2.8 une version effective de cet énoncé. En fait, le théorème 2.4 permet d'étendre aux surfaces la notion de morphisme de droites de saut introduite sur l'espace projectif pour les fibrés stables de rang 2 et de classe de Chern $c_1 = 0$. Nous commençons par le cas de l'espace projectif qui est plus facile à traiter.

3.1. Le cas du plan projectif

Considérons l'algèbre de Grothendieck $K(P_2)$ associée aux faisceaux algébriques cohérents sur le plan projectif. La classe de Grothendieck d'un faisceau algébrique cohérent F dans $K(P_2)$ est notée $c(F)$. Un élément $c \in K(P_2)$ est déterminé par son rang r, sa classe de Chern c_1 et sa caractérique d'Euler Poincaré χ. Le groupe de Grothendieck $\mathrm{K}(P_2)$ est muni de la forme quadratique entière $c \mapsto \chi(c^2)$, dont la forme polaire associée est notée $< \, , \, >$. La forme quadratique ci-dessus est alors donnée par

$$\chi(c^2) = 2r\chi + c_1^2 - r^2.$$

On désigne par h la classe de Grothendieck du faisceau structural d'une droite dans l'algèbre $K(P_2)$.

Soit $M = M(c)$ l'espace de modules des classes d'équivalence de faisceaux semi-stables de classe de Grothendieck $c \in K(P_2)$. On sait que c'est une variété algébrique projective irréductible et normale, localement factorielle. Soit $c^{\perp}$ l'orthogonal de c dans $K(P_2)$. On a alors un homomorphisme canonique

$$\lambda_M : c^{\perp} \to \mathrm{Pic}(M)$$

dont la construction est semblable à celle qui a été donnée dans la section 1: avec les notations de cette section, pour $v \in c^{\perp}$ le fibré inversible $\lambda_M(v)$ provient, par passage au quotient, du fibré inversible $\lambda_{\mathcal{F}}(v)$ sur le schéma de Hilbert-Grothendieck adéquat, $\mathcal{F}$ désignant le faisceau quotient universel paramétré par ce schéma de Hilbert-Grothendieck [7]. Cette construction est valable aussi pour les espaces de modules de faisceaux de dimension 1. On note r, c_1, et χ, le rang, la classe de Chern, et la caractéristique d'Euler-Poincaré de c; on suppose $r \geq 1$, et on pose

$$\delta = \mathrm{pgcd}(r, c_1).$$

On prend pour v l'opposé de la classe de Grothendieck $c(G) \in K(P_2)$ des faisceaux algébriques cohérents de dimension 1, de multiplicité $\frac{r}{\delta}$ et de caractéristique d'Euler-Poincaré $-\frac{c_1}{\delta}$: la classe $c(G)$ est donnée par la formule

$$c(G) = \frac{r}{\delta}h - \frac{c_1 + r}{\delta}h^2.$$

La relation d'orthogonalité $< c, v >= 0$ est satisfaite, de sorte que le fibré inversible $\mathcal{D} = \lambda_M(v)$ a un sens. C'est par définition le fibré déterminant de Donaldson sur M.

Théorème 3.1. *Soit $d = [\frac{r^2}{2}]$ la partie entière de $\frac{r^2}{2}$. Soit ℓ un entier strictement positif divisible par d et tel que $\ell > \frac{d^2 r^2}{8}(d-3)$. Le fibré inversible $\mathcal{D}^{\otimes \ell}$ est engendré par ses sections.*

En fait dans le cas où r est pair, on peut donner un énoncé légèrement meilleur: on peut affirmer que le fibré $\mathcal{D}^{\otimes \ell}$ est engendré par ses sections dès que ℓ est divisible par d et $\ell > \frac{d^2 r\delta}{4}(d-3)$.

L'énoncé ci-dessus est en fait une conséquence d'un énoncé plus précis, qui va nous permettre de définir pour chaque faisceau semi-stable F sur le plan projectif une hypersurface jouant le même rôle que la courbe des droites de saut pour les faisceaux semi-stables de rang 2. On considère pour ceci pour ℓ entier positif les faisceaux semi-stables G de dimension 1, dont la classe de Chern $\check{c}_1 = c_1(G)$ et la caractéristique d'Euler-Poincaré $\check{\chi} = \chi(G)$ sont données par

$$\check{c}_1 = \ell\frac{r}{\delta} \quad ; \quad \check{\chi} = -\ell\frac{c_1}{\delta}.$$

La classe de Grothendieck de tels faisceaux dans $K(P_2)$ est notée $\check{c}$; cette classe est orthogonale à c. On désigne par $\check{M}_\ell = M(\check{c})$ l'espace de modules des classes de faisceaux semi-stables de classe $\check{c}$. Sur cet espace de modules, le faisceau inversible $\check{\mathcal{D}}_\ell = \lambda_{\check{M}_\ell}(-c)$ a un sens.

Definition 3.2. *Soient v une classe de $K(P_2)$ orthogonale à c et F un faisceau semi-stable de classe c dans $K(X)$. On dit qu'un faisceau algébrique cohérent G de classe de Grothendieck v, transverse à F, est un module de saut pour F si*

$$\mathrm{H}^1(F \otimes G) \neq 0.$$

La condition de transversalité signifie que $\underline{\mathrm{Tor}}_q(F, G) = 0$ pour $q > 0$. On se limite ici à l'étude des modules de saut G de dimension 1 et semi-stables: ils sont par définition de Cohen-Macauley. Dans ce cas, le faisceau G est tranverse à F et la condition ci-dessus est équivalente à $\mathrm{H}^0(F \otimes G) \neq 0$.

Proposition 3.3. *Soit ℓ un entier strictement positif. On suppose $\ell > \frac{d^2 r^2}{8}(d-3)$ et divisible par d. Soit F un faisceau semi-stable de classe de Grothendieck c dans $K(P_2)$. Dans l'espace de modules $\check{M}_\ell$ les classes d'équivalence des faisceaux semi-stables G qui sont des modules de saut pour F constituent une hypersurface σ_F obtenue comme schéma des zéros d'une section θ_F du fibré inversible $\check{\mathcal{D}}_\ell$ sur $\check{M}_\ell$.*

Démonstration. Cet énoncé est une combinaison du théorème de Flenner sur la restriction des faisceaux semi-stables sur une surface algébrique à une courbe, et du théorème 2.8. Soit F un faisceau semi-stable de classe de Grothendieck c. On sait d'après le théorème de Flenner [4] qu'il existe une courbe lisse Y de degré $d = [\frac{r^2}{2}]$ telle que le faisceau $F|_Y$ soit semi-stable. Notons qu'avec notre choix de d, on a $\mathrm{pgcd}(r, dc_1) = r$ si r est pair, et δ si r est impair. Le genre d'une telle courbe Y est donné par la formule

$$g - 1 = \frac{d(d-3)}{2}.$$

Si $k > \frac{r^2}{4}(g-1)$, on peut alors trouver d'après le théorème 2.8 un fibré vectoriel V stable de rang $k\frac{r}{\delta}$ sur la courbe Y tel que $\chi(F \otimes_Y V) = 0$ et $\mathrm{H}^1(F \otimes_Y V) = 0$. On prend $\ell = kd$. Alors le faisceau $G = \mathcal{O}_Y(V)$ définit un faisceau stable de dimension 1 sur P_2 de multiplicité $\ell\frac{r}{\delta}$, de caractéristique d'Euler-Poincaré $\check{\chi} = -\ell\frac{c_1}{\delta}$; ce faisceau stable n'est pas de saut pour F. Comme dans la section 2, le faisceau F définit, à inversibles près, une section θ_F du fibré inversible $\check{\mathcal{D}}_\ell$, dont le schéma des zéros correspond à l'idéal de Fitting définissant la variété déterminantielle donnée par la condition $\mathrm{H}^1(F \otimes G) \neq 0$. Les points de ce sous-schéma représentent les modules de saut pour le faisceau semi-stable F. La section θ_F n'est pas identiquement nulle; ceci entraîne que s'il existe des modules de saut dans $\check{M}_\ell$, ces modules de saut constituent une hypersurface de $\check{M}_\ell$.

Dans le cas où r est pair, il suffit en fait d'imposer $k > \frac{r\delta}{4}(g-1)$ pour obtenir les mêmes conclusions. □

Remarque 3.4. (1) Quand $r = 2$ et $c_1 = 0$, le théorème de Grauert-Mülich est un peu meilleur que l'énoncé de Flenner que nous avons utilisé. On peut prendre $d = 1$, et pour $\check{M}$ l'espace de modules les faisceaux semi-stables de multiplicité 1, et de caractéristique d'Euler-Poincaré $\chi = 0$. Ces faisceaux sont les faisceaux $\mathcal{O}_\ell(-1)$ où ℓ est une droite. Ces modules sont de saut si $h^1(F(-1)|_\ell) \neq 0$ ce qui signifie que ℓ est une droite de saut pour F.

(2) Quand $r = 2$ et $c_1 = -1$, on peut travailler avec l'espace de modules $\check{M}_1$ des faisceaux semi-stables de multiplicité 2 et caractéristique d'Euler-Poincaré $\chi = 1$. Cet espace de modules est isomorphe au système linéaire des coniques de P_2, et l'hypersurface θ_F est l'hypersurface des coniques de saut. On ne peut pas travailler avec les espaces de modules des faisceaux de dimension 1 et multiplicité 1, car on ne peut pas réaliser avec de tels faisceaux G la condition d'orthogonalité $< c, c(G) >= 0$.

(3) On sait que si $r = 3, c_1 = 0$, il peut arriver que toute droite soit de saut: c'est le cas par exemple pour le fibré $S^2T(-2)$. C'est pourquoi il est nécessaire d'introduire des faisceaux de dimension 1 de multiplicité supérieure.

Théorème 3.5. *On garde les notations ci-dessus, et on suppose encore* $\ell > \frac{d^2r^2}{4}(d-3)$, *strictement positif et divisible par d. Soit* $|\check{\mathcal{D}}_\ell|$ *le système linéaire complet associé à* $\check{\mathcal{D}}_\ell$ *dans* $\check{M}_\ell$.

(i) *On a un morphisme canonique*

$$\sigma : M \to |\check{\mathcal{D}}_\ell|$$

tel que $\sigma^*(\mathcal{O}(1)) = \mathcal{D}^{\otimes \ell}$. *Ce morphisme associe à un point représentant un faisceau semi-stable F l'hypersurface* σ_F.

(ii) *Soit U l'ouvert de M des points représentant des faisceaux localement libres* μ*-stables. Alors la restriction*

$$\sigma|_U : U \to |\check{\mathcal{D}}|$$

est quasi-finie. Autrement dit, les fibres de σ *rencontrent l'ouvert U suivant un nombre fini de points.*

Démonstration. La démonstration est une variante de celle qui a été donnée dans [6] pour les droites de saut. Dans $M \times \check{M}_\ell$ les points représentant les couples (F, G) de faisceaux semi-stables tels que $h^1(F \otimes G) = 0$ constituent une hypersurface obtenue comme schéma des zéros d'une section canonique (à inversibles près) du fibré $\mathcal{D}^{\otimes \ell} \boxtimes \check{\mathcal{D}}_\ell$. Cette section définit une section canonique θ du fibré $\mathcal{D}^{\otimes \ell} \otimes \mathrm{H}^0(\check{M}_\ell, \check{\mathcal{D}}_\ell)$, dont la valeur au point F est (à inversibles près) est θ_F. Cette section est partout non nulle d'après la proposition ci-dessus. On en déduit un morphisme $\sigma : M \to |\check{\mathcal{D}}_\ell|$ satisfaisant à la condition demandée en (i). L'assertion (ii) repose sur le fait qu'il existe des puissances $\mathcal{D}^{\otimes \nu}$ du fibré déterminant de Donaldson qui sont engendré par leurs sections et telles que le morphisme associé au système linéaire complet

$$M \to |\mathcal{D}^{\otimes \nu}|^*$$

induise un plongement de l'ouvert U sur une sous-variété localement fermée du second membre ([6], théorème 2). Alors la restriction de $\mathcal{D}$ à toute courbe projective irréductible rencontrant U est de degré > 0. Dès lors, s'il existait une courbe irréductible C contenue dans une fibre de $\sigma|_U$, le morphisme σ serait constant sur ses fibres. L'adhérence $\overline{C}$ serait alors contenue dans une fibre de σ, et le fibré $\mathcal{D}|_{\overline{C}}$ devrait être trivial, ce qui aboutit à une contradiction. □

3.2. Autres surfaces

Considérons maintenant une surface algébrique projective lisse polarisée $(X, \mathcal{O}_X(1))$. Les énoncés ci-dessus s'étendent, à condition de prendre quelques précautions, à toutes les surfaces projectives irréductibles et lisses. On désigne encore par $K(X)$ l'algèbre de Grothendieck de X, équipée de sa forme quadratique naturelle: en termes de rang r, de classe de Chern c_1 et de caractéristique d'Euler-Poincaré χ, cette forme quadratique est donnée par

$$c \mapsto \chi(c^2) = 2r\chi + c_1^2 - r^2\chi(\mathcal{O}_X).$$

On désigne par $K_{\text{num}}(X)$ le quotient de $K(X)$ par le noyau de cette forme quadratique. Ce noyau est un idéal de $K(X)$, et par conséquent $K_{\text{num}}(X)$ est une algèbre équipée de la forme quadratique obtenue par passage au quotient. Deux faisceaux algébriques cohérents sur X ont même classe de Grothendieck dans $K_{\text{num}}(X)$ si et seulement si ils ont même rang, même classe de Chern dans $A^1_{\text{num}}(X)$ et même caractéristique d'euler-Poincaré.

On considère ici une classe de Grothendieck $c \in K_{\text{num}}(X)$, dont on désigne par r le rang, c_1 la classe de Chern, et χ la caractéristique d'Euler-Poincaré, et un fibré inversible L de classe de Chern $c_1(L) = c_1$. On désigne par $M = M(c, L)$ l'espace de modules des classes d'équivalence de faisceaux semi-stables de classe de Grothendieck c et de déterminant $\det F = L$. On sait que c'est une variété projective. Pour étendre les énoncés ci-dessus au cas général, on se heurte à plusieurs difficultés, dont la première est que le module $M(c, L)$ n'a aucune raison d'être ni irréductible ni même réduit. De plus, une classe $v \in K(X)$, orthogonale à c, détermine un fibré inversible $\lambda_{M^s}(v)$ sur l'ouvert M^s des faisceaux stables, mais ce fibré inversible ne s'étend pas en général à $M(c, L)$. C'est pourtant le cas si on impose à v d'avoir son déterminant numériquement lié à $\mathcal{O}_X(1)$ (*cf.* [6], [7]).

Soient $\eta = c_1(\mathcal{O}_X(1))$ la classe de Chern du fibré $\mathcal{O}_X(1)$, et $\omega = c_1(\Omega^2_X)$ la classe canonique. On suppose que r un entier ≥ 1. On pose

$$\begin{aligned} n &= \eta^2 \\ m &= \omega\eta \\ d &= n[\frac{r^2}{2}] \\ g - 1 &= \frac{1}{2}(d^2 n + dm) \\ \delta &= \text{pgcd}(r, c_1.\eta). \end{aligned}$$

On désigne par h la classe de Grothendieck du faisceau structural d'une section hyperplane dans $K(X)$, et par $H \subset K(X)$ la sous-algèbre unitaire engendrée par h. Soit $\mathcal{D} = \lambda_M(v)$ le fibré déterminant de Donaldson obtenu en prenant pour v l'opposé de la classe dans $K(X)$ des faisceaux algébriques cohérents G de dimension 1, de déterminant $\det G = \mathcal{O}_X(\frac{r}{\delta})$ et de caractéristique d'Euler-Poincaré $\chi(G) = -\frac{c_1.\eta}{\delta}$. Ce fibré a bien un sens, du fait que la relation d'orthogonalité $< c, v >= 0$ est satisfaite et qu'une telle classe définit un point du biorthogonal $H^{\perp\perp}$ de H (*cf.* [7]).

Théorème 3.6. *Si ℓ un entier > 0 divisible par d et tel que*

$$\ell > \frac{d^2r^2}{8}(dn+m).$$

Alors le fibré $\mathcal{D}^{\otimes\ell}$ sur $M(c, L)$ est engendré par ses sections globales.

Démonstration. Pour établir ce théorème, on construit pour ℓ entier naturel assez grand et divisible par d un faisceau semi-stable G de dimension 1 dont le déterminant et la caractéristique d'Euler-Poincaré sont données par

$$\det G = \check{L} = \mathcal{O}_X(\frac{\ell r}{\delta}) \quad ; \quad \check{\chi} = -\frac{\ell c_1.\eta}{\delta},$$

et tel que $h^1(F \otimes G) = 0$. Si F est un faisceau semi-stable de classe de Grothendieck c, le théorème de Flenner [4] dit cette fois qu'il existe dans le système linéaire $|\mathcal{O}_X(d)|$ une courbe lisse Y telle que $F|_Y$ soit un fibré semi-stable. Le nombre g défini ci-dessus est le genre d'une telle courbe. Sur une telle courbe, on peut trouver, d'après le théorème 2.8, un fibré vectoriel stable V de rang $k\frac{r}{\delta}$ tel que $\chi(F \otimes_Y V) = 0$ et $h^0(F \otimes V) = 0$ dès que $k > \frac{r^2}{4}(g-1)$. On prend $\ell = dk$, k étant ainsi choisi. Ce fibré vectoriel définit un faisceau semi-stable $G = \mathcal{O}_Y(V)$ de dimension 1, dont le déterminant est $\check{L}$ et la caractéristique d'Euler-Poincaré $\check{\chi}$. La classe de Grothendieck $c(G) \in K(X)$ d'un tel faisceau est orthogonale à c et appartient à $H^{\perp\perp}$; le fibré inversible associé $\lambda_M(-c(G))$ est $\mathcal{D}^{\otimes\ell}$. Le faisceau G ainsi construit définit, à inversibles près, une section de $\mathcal{D}^{\otimes\ell}$ qui ne s'annule pas au point F. D'où l'énoncé. □

Comme pour le plan projectif, on peut en fait obtenir un morphisme naturel σ de M dans un espace projectif associant à un faisceau semi-stable F un sous-schéma de modules de saut, à condition de supposer que L et $\mathcal{O}_X(1)$ numériquement liés. Soit $\check{c}$ la classe de Grothendieck dans $K_{\text{num}}(X)$ des faisceaux algébriques cohérents G de dimension 1 dont la classe de Chern et la caractéristique d'Euler Poincaré sont données par

$$\check{c}_1 = \frac{\ell r}{\delta}\eta \quad ; \quad \check{\chi} = -\frac{\ell c_1.\eta}{\delta}.$$

On désigne par $M(\check{c})$ l'espace de modules des classes d'équivalence de faisceaux semi-stables de classe de Grothendieck $\check{c}$. Considérons le morphisme canonique, à valeurs dans la composante du groupe de Picard de X qui paramètre les fibrés inversibles de classe de Chern $\check{c}_1$,

$$M(\check{c}) \to \mathrm{Pic}^{\check{c}_1}(X)$$

qui associe à un faisceau semi-stable G son déterminant $\det(G)$. On note $\check{M}_\ell = M(\check{c}, \check{L})$ la fibre au-dessus du point $\check{L}$. Les faisceaux semi-stables F de classe de Grothendieck c dans $K_{\text{num}}(X)$ et de déterminant L définissent un élément bien déterminé de $K(X)$, modulo les cycles de codimension 2 et de degré 0. Le fibré inversible $\check{\mathcal{D}}_\ell = \lambda_{\check{M}_\ell}(-c(F))$ associé à la classe de Grothendieck $c(F) \in K(X)$ d'un faisceau semi-stable F représentant un point de $M(c, L)$ est alors indépendant de F : il suffit en effet de constater que pour un cycle u de codimension 2 et de degré 0 dans X, le fibré $\lambda_{\check{M}}(u)$ est trivial. Mais ceci résulte facilement de la proposition 2.8 de [7]. Si on suppose

$$\ell > \frac{d^2r^2}{8}(dn+m),$$

dans l'espace de modules $\check{M}_\ell$ les points représentant les faisceaux semi-stables G qui sont des modules de saut pour F, c'est-à-dire tels que $\mathrm{H}^1(F \otimes G) \neq 0$ sont les points d'un fermé σ_F obtenu comme ensemble des zéros d'une section non nulle θ_F du faisceau inversible $\check{\mathcal{D}}_\ell$. On désigne par $|\check{\mathcal{D}}_\ell|$ l'espace projectif $P(\mathrm{H}^0(\check{\mathcal{D}}_\ell))$ des droites de l'espace vectoriel $\mathrm{H}^0(\check{\mathcal{D}}_\ell)$. En s'appuyant sur le théorème 2 de [6], on obtient alors, comme pour le plan projectif, le résultat suivant qui précise le théorème 3.6 ci-dessus:

Théorème 3.7. *On suppose que L et $\mathcal{O}_X(1)$ sont numériquement liés; soit ℓ un entier > 0 divisible par d et tel que*

$$\ell > \frac{d^2 r^2}{8}(dn + m).$$

(i) *Il existe un morphisme naturel*

$$\sigma : M \to |\check{\mathcal{D}}_\ell|$$

tel que $\sigma^*(\mathcal{O}(1)) = \mathcal{D}^{\otimes \ell}$. Ce morphisme σ associe au point F la classe de la section θ_F dans l'espace projectif $|\check{\mathcal{D}}_\ell|$.

(ii) *Soit U l'ouvert de $M = M(c, L)$ des classes d'isomorphisme de faisceaux μ-stables localement libres. La restriction $\sigma|_U$ est quasi-finie.*

Références

[1] J. M. Drézet et M. S. Narasimhan: Groupe de Picard des variétés de modules de fibrés semi-stables sur les courbes algébriques, Invent. math. **97** (1989), 53–94.

[2] J.-M. Drézet et J. Le Potier: Fibrés stables et fibrés exeptionnels sur le plan projectif, Ann. scient. Ec. Norm. Sup. 4e série **18** (1985), 193–244.

[3] G. Faltings: Stable G-bundles and projective connections, J. Alg. Geometry **2** (1993), 507–568.

[4] H. Flenner: Restrictions of semistable bundles on projective varieties, Comment. Math. Helvitici **59** (1984), 635–650.

[5] G. Hein: On the generalized theta divisor, Preprint Humbolt-Universität, Berlin (1994).

[6] J. Le Potier: Fibré déterminant et courbes de saut sur les surfaces algébriques, Complex Projective Geometry (Bergen 1989) London Mathematical Society, (1992), 213–240.

[7] J. Le Potier: Faisceaux semi-stables et systèmes cohérents, Proceedings de la Conference de Durham on Vector bundles in Algebraic Geometry (1993), Cambridge University Press (1995), 179–239.

[8] C.S. Seshadri: Vector bundles on curves, Preprint, School of Mathematics, SPIC Foundation, Madras.

SYSTÈMES COHÉRENTS ET POLYNÔMES DE DONALDSON

Joseph Le Potier

Université Paris 7, Institut de Mathématiques de Jussieu, Case Postale 7012, 2, Place Jussieu, 75251 Paris Cedex 05, France

e-mail: jlp@mathp7.jussieu.fr

INTRODUCTION

Soit M_n l'espace de modules des classes d'équivalence de faisceaux semi-stables de rang 2, de classes de Chern $(0, n)$ sur le plan projectif complexe; c'est une variété projective irréductible et normale de dimension $d = 4n - 3$. Le polynôme de Donaldson de degré d pour le plan projectif P_2 est donné par le coefficient

$$q_d = \int_{M_n} c_1(\mathcal{D}_n)^d$$

qu'on appellera nombre de Donaldson de degré d; dans cette intégrale, $\mathcal{D}_n$ est le fibré déterminant de Donaldson (*cf.* section 5) sur M_n. Le but de cet article est de montrer comment on peut obtenir des relations entre ces nombres et certains nombres d'intersection sur le schéma de Hilbert $\mathrm{Hilb}^{n+\ell^2}(P_2)$ des sous-schémas finis de longueur $n + \ell^2$, pour $n \leq (\ell+1)(\ell+2)$. Pour ceci on doit mettre en relation l'espace de modules M_n et le schéma de Hilbert $\mathrm{Hilb}^{n+\ell^2}(P_2)$. L'idée de la démonstration est la suivante: si F est un faisceau semi-stable de rang 2 et de classe de Chern $(0, n)$, on a $\chi(F(\ell)) = (\ell+1)(\ell+2) - n$, et le faisceau $F(\ell)$ possède au moins une section non nulle si $n < (\ell+1)(\ell+2)$. Quand $n = (\ell+1)(\ell+2)$, les faisceaux semi-stables F tels que $F(\ell)$ ait au moins une section non nulle définissent une hypersurface de M_n. On introduit alors l'espace de modules $\mathrm{Syst}_\alpha(c(\ell), 1)$ des classes d'équivalence de systèmes cohérents de type $(c(\ell), 1)$ et α-semi-stables: il s'agit des paires $(\Gamma, F(\ell))$ où F est un faisceau de rang 2, de classes de Chern $(0, n)$, et Γ est un sous-espace vectoriel de dimension 1 de l'espace des sections $\mathrm{H}^0(F(\ell))$, soumises à une condition de semi-stabilité expliquée dans la section 1. Ici le paramètre α est un polynôme positif, à coefficients rationnels, de degré ≤ 1. Ces espaces de modules sont des variétés projectives. Quand α est une constante ϵ assez petite, on peut construire un morphisme $(\Gamma, F(\ell)) \mapsto F$

$$f : \mathrm{Syst}_\epsilon(c(\ell), 1) \to M_n$$

dont la fibre générique est un espace projectif. Le calcul de l'intégrale de Donaldson se ramène, pour $n \leq (\ell+1)(\ell+2)$ au calcul de certaines intégrales $J_{\epsilon,n}$ sur la variété projective $\mathrm{Syst}_\epsilon(c(\ell), 1)$.

Quand α varie, le comportement de l'espace de modules $\mathrm{Syst}_\alpha(c(\ell), 1)$ change au passage de certaines valeurs appelées valeurs critiques, suivant un modèle proposé sur les courbes par Thaddeus [22]. On peut encore définir sur $\mathrm{Syst}_\alpha(c(\ell), 1)$ des intégrales $J_{\alpha,n}$ jouant le même rôle que $J_{\epsilon,n}$. L'examen du comportement de ces intégrales au passage des valeurs critiques permet d'envisager des relations entre ces intégrales. Quand α prend une valeur

β_- suffisamment proche de la plus grande valeur β pour lequel l'espace de modules est non vide (c'est le polynôme $\beta(m) = 2\ell m + \ell^2 + 3\ell - n$) il existe un morphisme surjectif

$$\mathrm{Syst}_{\beta_-}(c(\ell), 1) \to \mathrm{Hilb}^{n+\ell^2}(P_2)$$

dont la fibre générique est un espace projectif. Ceci permet d'espérer une relation entre $J_{\epsilon,n}$ et certains nombres d'intersection sur $\mathrm{Hilb}^{n+\ell^2}(P_2)$; par suite, ceci devrait donner les nombres de Donaldson q_d en termes de nombres d'intersection sur le schéma de Hilbert $\mathrm{Hilb}^{n+\ell^2}(P_2)$. Par exemple, une telle relation est décrite dans le cas $\ell = 1$ dans le théorème 5.1.

Lorsque $\ell > 1$, ce programme se heurte à plusieurs difficultés, dont la plus importante est qu'en général l'espace de modules $\mathrm{Syst}_\alpha(c(\ell), \ell)$ n'a aucune raison d'être lisse même si α n'est pas valeur critique, ce qui rend plus délicate l'adaptation des résultats de Thaddeus. On explique ici en détail la méthode pour $\ell = 1$, qui permet le calcul des nombres de Donaldson q_{4n-3} pour $2 \leq n \leq 6$. Dans ce cas, l'espace de module $S_{\alpha,n} := \mathrm{Syst}_\alpha(c(1), 1)$ est lisse si α n'est pas valeur critique, et la situation est aussi belle que celle que rencontrait Thaddeus sur les courbes pour l'espace de modules des paires semi-stables. Il se trouve que les nombres de Donaldson ont une interprétation géométrique simple en termes du morphisme

$$\gamma : M_n \to |\mathcal{O}_{P_2^*}(n)|$$

qui associe à la classe d'un faisceau semi-stable F la courbe γ_F du plan projectif dual P_2^* de ses droites de saut (*cf.* section 5). Le fibré déterminant $\mathcal{D}_n$ est alors l'image réciproque de $\mathcal{O}(1)$ par ce morphisme γ. Ce morphisme est génériquement fini sur son image, et le nombre q_{4n-3} n'est autre que le produit du degré de γ par le degré de l'image. Quand $\ell = 1$, le faisceau F sous-jacent à un système cohérent α-semi-stable $(\Gamma, F(1))$ est μ-semi-stable (sauf pour la valeur critique β), ce qui permet encore de définir la courbe des droites de saut de F. On obtient ainsi un morphisme

$$\gamma_\alpha : S_{\alpha,n} \to |\mathcal{O}_{P_2^*}(n)|$$

qui associe à la classe de $(\Gamma, F(1))$ la courbe des droites de saut de F. Ceci conduit à une démontration géométrique rapide du fait que les intégrales $J_{\alpha,n}$ restent inchangées au passage des points critiques: c'est essentiellement la méthode qu'avec Ellingsrud et Strømme nous avons choisie d'expliquer dans [8]. Les résultats déjà connus fournissent les nombres q_{4n-3} pour $n = 2, n = 3$, et $n = 4$ ([2], [19], [24] mais le résultat est nouveau pour $n = 5$ et $n = 6$). Le calcul des nombres d'intersection sur le schéma de Hilbert $\mathrm{Hilb}^{n+1}(P_2)$, mené à bien par G. Ellingsrud et S. A. Strømme grâce à l'usage de la formule des résidus de Bott et d'un programme sur ordinateur [8] donne finalement $q_{17} = 2540$ et $q_{21} = 233208$.

L'étude du cas $\ell = 2$ devrait fournir les nombres de Donaldson q_{4n-3} pour $n \leq 12$. Dans ce cas, le faisceau F sous-jacent à un système cohérent α-semi-stable $(\Gamma, F(2))$ n'est plus obligatoirement μ-semi-stable, et il n'est plus possible de définir un morphisme γ_α comme ci-dessus. On doit obligatoirement comprendre comment varie les intégrales $J_{\alpha,n}$ au passage des valeurs critiques [11].

Signalons que G. Ellingsrud et L. Göttsche viennent de donner une méthode, généralisant celle de [19], qui permet théoriquement de calculer tous les nombres de Donaldson du plan projectif. Ceci fait encore appel naturellement à des programmes sur ordinateur; il se pose maintenant la question de l'existence de formules acceptables pour calculer de tels nombres.

Cet article est organisé de la manière suivante: dans la section 1, nous rappelons la définition des espaces de modules de systèmes cohérents, et dans la section 2, nous donnons l'énoncé du critère de lissité (*cf.* théorème 2.6) comme l'a obtenu Min He [11]. La section 3 est consacrée à la généralisation des résultats de [6] : sous des hypothèses adéquates, on peut calculer la codimension des strates de Harder-Narasimhan pour une famille complète de systèmes cohérents (*cf.* proposition 3.2). Ces résultats théoriques trouvent leur application au chapitre 4 dans l'étude de l'espace de modules $S_{\alpha,n} := \mathrm{Syst}_\alpha(c(1), 1)$. La lissité si α n'est pas valeur critique obtenue dans le théorème 4.1 par application du critère de Min He nécessite la compréhension du passage des valeurs critiques: il s'agit d'un phénomène de flip, décrite dans le théorème 4.4, analogue à la situation décrite par Thaddeus [22] sur les courbes pour les espaces de modules de paires semi-stables. Les relations entre les espaces de modules $S_{\alpha,n}$ pour les grandes valeurs de α et le schéma de Hilbert $\mathrm{Hilb}^{n+1}(P_2)$ sont décrites dans le corollaire 4.3. L'étude du morphisme $f : S_{\epsilon,n} \to M_n$ (pour ϵ rationnel > 0 suffisamment petit) est donnée dans le théorème 4.7; pour $n \geq 6$, ce morphisme permet l'étude du sous-schéma de Brill-Noether B de M_n défini par les classes de faisceaux semi-stables F satisfaisant satisfaisant à la condition déterminantielle $h^0(F(1)) \geq 1$: on obtient ainsi que B est une sous-variété irréductible et normale. L'espace de module $S_{\epsilon,n}$ est en fait une désingularisation de ce sous-schéma. Ce résultat, intéressant en lui-même, ne sera en fait utilisé que pour $n = 6$. Enfin, on utilise ces résultats dans la section 5 pour le calcul des nombres de Donaldson: les propositions 5.3 et 5.7 constituent l'essentiel de la démonstration; elles montrent par un calcul direct que certaines intégrales $I_{\alpha,n}$ et $J_{\alpha,n}$ ne changent pas au passage des valeurs critiques.

Notations

Par variété algébrique, on entend schéma de type fini et séparé sur $\mathbb{C}$. Les points sont les points fermés.

1. Espaces de modules de systèmes cohérents.

Soit X une variété algébrique projective lisse de dimension n, munie d'une polarisation $\mathcal{O}_X(1)$. On désigne par $K(X)$ l'algèbre de Grothendieck des faisceaux algébriques cohérents équipée de la forme quadratique entière $u \mapsto \chi(u^2)$. D'après le théorème de Riemann-Roch, cette forme quadratique se calcule en termes de rang et de classes de Chern. Le noyau $\ker q$ est un idéal de $K(X)$; le quotient

$$K_{\mathrm{num}}(X) = K(X)/\ker q.$$

est muni d'une forme quadratique canonique, dont on désigne par $< \ , \ >$ la forme polaire associée. Deux faisceaux algébriques cohérents qui ont même rang et mêmes classes de Chern ont même classe dans $K_{\mathrm{num}}(X)$. La classe d'un faisceau algébrique cohérent F dans $K_{\mathrm{num}}(X)$ sera notée c(F); de telles classes sont dites effectives. On désigne par h la classe dans $K_{\mathrm{num}}(X)$ du faisceau structural d'une section hyperplane.

Soit F un faisceau algébrique cohérent de dimension d, de classe c dans $K_{\mathrm{num}}(X)$. Le polynôme de Hilbert de F s'écrit

$$P_F(m) = < c, (1-h)^{-m} > = \sum_{0 \leq i \leq d} C^i_{m+i-1} < c, h^i >$$

L'entier $r = < c, h^d >$ s'appelle multiplicité de F; le polynôme $p_F = \dfrac{P_F}{r}$ s'appelle le polynôme de Hilbert réduit de F.

Definition 1.1. *Soit $c \in K_{\mathrm{num}}(X)$ et k un entier ≥ 0. On appelle système cohérent de type (c, k) sur X une paire $\Lambda = (\Gamma, F)$ où F est un faisceau algébrique cohérent de classe c sur X et $\Gamma \subset \mathrm{H}^0(F)$ un sous-espace vectoriel de dimension k de l'espace vectoriel des sections de F.*

La dimension k de Γ sera appelée l'ordre du système cohérent (Γ, F). Les notions de dimension et multiplicité d'un système cohérent sont relatives au faisceau algébrique cohérent sous-jacent. Un morphisme de systèmes cohérents $(\Gamma, F) \to (\Gamma', F')$ est un morphisme de faisceaux algébriques cohérents $f : F \to F'$ tel que $f(\Gamma) \subset \Gamma'$.

On se fixe un polynôme $\alpha > 0$ à coefficients rationnels. Si $\Lambda = (\Gamma, F)$ est un système cohérent, on pose

$$P_\Lambda = \alpha \dim \Gamma + P_F$$

C'est le polynôme de Hilbert du système cohérent Λ. Le quotient $p_\Lambda = \dfrac{P_\Lambda}{r}$, où r est la multiplicité de F, s'appelle le polynôme de Hilbert réduit de Λ.

Definition 1.2. *Soit $\Lambda = (\Gamma, F)$ un système cohérent sur X, dont le faisceau F sous-jacent est de dimension d et de multiplicité r. On dit que Λ est α-semi-stable (resp. stable) si*
(i) *le faisceau cohérent F est pur de dimension d;*
(ii) *pour tout sous-faisceau cohérent F' de F de multiplicité $0 < r' < r$ on a*

$$p_{(\Gamma', F')} \leq p_{(\Gamma, F)} \quad (\text{resp.} <)$$

où $\Gamma' = \Gamma \cap \mathrm{H}^0(F')$.

Les systèmes cohérents α-semi-stables de polynôme de Hilbert réduit fixé constituent une catégorie abélienne. Il en résulte qu'il existe dans cette catégorie des filtrations de Jordan-Hölder et par suite la notion de S-équivalence a un sens: deux systèmes cohérents α-semi-stables Λ et Λ' sont S-équivalents si les gradués de Jordan-Hölder $gr(\Lambda)$ et $gr(\Lambda')$ sont isomorphes. Les objets simples de cette catégorie sont les systèmes cohérents stables.

Soient S une variété algébrique, $p_1 : S \times X \to S$ la première projection. Une famille de systèmes cohérents paramétrée par S est la donnée d'une paire $\Lambda = (\mathcal{V}, \mathcal{F})$ où $\mathcal{F}$ est un faisceau algébrique cohérent sur $S \times X$ et $\mathcal{V}$ un sous-faisceau localement libre de l'image directe $p_{1*}(\mathcal{F})$ satisfaisant à la propriété suivante: pour tout $s \in S$ l'application linéaire canonique induite au-dessus du point s

$$\Gamma_s \to \mathrm{H}^0(\mathcal{F}_s)$$

est injective. Ainsi, pour une telle famille, $\Lambda_s = (\Gamma_s, \mathcal{F}_s)$ définit un système cohérent. Si le faisceau $\mathcal{F}$ est S-plat, on dit que la famille est S-plate. On désigne par $\mathbf{Syst}_\alpha(c, k)$ le foncteur contravariant associant à la variété algébrique S l'ensemble des classes d'isomorphisme des famille plates paramétrées par S de systèmes cohérents α-semi-stables $(\Gamma_s, \mathcal{F}_s)$ de type (c, k).

Théorème 1.3. *Soit $c \in K_{\mathrm{num}}(X)$ une classe effective, et k un entier ≥ 0.*
(i) *Il existe pour le foncteur $\mathbf{Syst}_\alpha(c, k)$ un espace de modules grossier $\mathrm{Syst}_\alpha(c, k)$.*
(ii) *Cet espace de modules est une variété projective; ses points sont les classes de S-équivalence de systèmes cohérents α-semi-stables.*

Dans le cas particulier où on prend pour α le polynôme de Hilbert du faisceau structural de X cet énoncé est démontré dans [16]. La version ci-dessus, valable pour α quelconque, est une variante obtenue par Min He [11].

2. Etude infinitésimale

L'essentiel de cette section repose sur le travail de Min He [11]. La catégorie des systèmes cohérents n'est pas une catégorie abélienne. Pour pallier à cet inconvénient, on introduit la notion plus générale de système algébrique.

Definition 2.1. *Un système algébrique sur X est un triplet $\Lambda = (\Gamma, i, F)$ où F est un faisceau algébrique sur X, Γ un espace vectoriel, et $i : \Gamma \to \mathrm{H}^0(F)$ une application $\mathbb{C}$-linéaire.*

Un morphisme de systèmes algébriques $\Lambda = (\Gamma, i, F) \to \Lambda' = (\Gamma', i', F')$ est un couple (f, g) où $g : F \to F'$ est un morphisme de faisceau algébriques, et $f = \Gamma \to \Gamma'$ une application $\mathbb{C}$-linéaire, rendant commutatif le diagramme

$$\begin{array}{ccc} \Gamma & \xrightarrow{f} & \Gamma' \\ \downarrow & & \downarrow \\ \mathrm{H}^0(F) & \xrightarrow{g} & \mathrm{H}^0(F') \end{array}$$

Lemme 2.2. *La catégorie des systèmes algébriques est une catégorie abélienne qui a assez d'injectifs.*

On vérifie en effet qu'un système algébrique de la forme (I, φ, J) où J est un faisceau algébrique injectif, et $\varphi : I \to \mathrm{H}^0(J)$ une application linéaire surjective est un objet injectif de cette catégorie. Il est clair que tout système algébrique se plonge dans un tel objet injectif.

Ceci donne un sens aux foncteurs dérivés $\mathrm{Ext}^q(\Lambda, \Lambda')$.

Proposition 2.3. *Soient $\Lambda = (\Gamma, i, F)$ et $\Lambda' = (\Gamma', i', F')$ deux systèmes algébriques. On a une suite exacte longue*

$$\begin{aligned} 0 \to \mathrm{Hom}(\Lambda, \Lambda') &\to \mathrm{Hom}(\Gamma, \Gamma') \oplus \mathrm{Hom}(F, F') \to \mathrm{Hom}(\Gamma, \mathrm{H}^0(F')) \\ &\to \mathrm{Ext}^1(\Lambda, \Lambda') \to \mathrm{Ext}^1(F, F') \to \mathrm{Hom}(\Gamma, \mathrm{H}^1(F')) \to \dots \end{aligned}$$

Dans le cas particulier où $i' : \Lambda' \to \mathrm{H}^0(F')$ est injectif, cette suite exacte peut s'écrire sous une forme légèrement différente. Ceci s'appliquera en particulier aux systèmes cohérents:

Proposition 2.4. *Soient $\Lambda = (\Gamma, i, F)$ et $\Lambda' = (\Gamma', i', F')$ deux systèmes algébriques; supposons que l'application linéaire $i' : \Gamma' \to \mathrm{H}^0(F')$ soit injective. Alors on a une suite exacte longue*

$$\begin{aligned} 0 \to \mathrm{Hom}(\Lambda, \Lambda') &\to \mathrm{Hom}(F, F') \to \mathrm{Hom}(\Gamma, \mathrm{H}^0(F')/\Gamma') \\ &\to \mathrm{Ext}^1(\Lambda, \Lambda') \to \mathrm{Ext}^1(F, F') \to \mathrm{Hom}(\Gamma, \mathrm{H}^1(F')) \to \dots \end{aligned}$$

La proposition qui suit est utile pour évaluer $\mathrm{Ext}^q(\Lambda, \Lambda')$ pour $q \geq 2$ dans le cas où Λ et Λ' sont des systèmes cohérents.

Proposition 2.5. *Soient $\Lambda = (\Gamma, i, F)$ et $\Lambda' = (\Gamma', i', F')$ deux systèmes algébriques; on suppose que le morphisme d'évaluation $\Gamma \otimes \mathcal{O}_X \to F$ est injectif, de conoyau Θ. Alors il existe un homomorphisme canonique*

$$\mathrm{Ext}^q(\Theta, F') \to \mathrm{Ext}^q(\Lambda, \Lambda')$$

qui est un isomorphisme pour $q \geq 2$, et un épimorphisme dans le cas $q = 1$.

Démonstration. Soit $p : F \to \Theta$ la projection canonique. Considérons une résolution injective $R^\cdot = (I^\cdot, i^\cdot, J^\cdot)$ de Λ', avec J^k injectif. On a alors un morphisme de complexes

$$\mathrm{Hom}(\Theta, J^\cdot) \to \mathrm{Hom}(\Lambda, R^\cdot)$$

On obtient un homomorphisme

$$\mathrm{Ext}^q(\Theta, F') \to \mathrm{Ext}^q(\Lambda, \Lambda')$$

indépendant des résolutions choisies. Le diagramme commutatif, dans lequel $\mathrm{Hom}(\Gamma, \Gamma')$ désigne le complexe réduit au seul terme $\mathrm{Hom}(\Gamma, \Gamma')$ placé en degré 0

$$\begin{array}{ccccccccc} 0 & \longrightarrow & \mathrm{Hom}(\Theta, J^{\cdot}) & \longrightarrow & \mathrm{Hom}(F, J^{\cdot})) & \longrightarrow & \mathrm{Hom}(\Gamma, \mathrm{H}^0(J^{\cdot}) & \longrightarrow & 0 \\ & & \downarrow & & \downarrow & & \downarrow \wr & & \\ 0 & \longrightarrow & \mathrm{Hom}(\Lambda, R^{\cdot}) & \longrightarrow & \mathrm{Hom}(F, J^{\cdot}) \oplus \mathrm{Hom}(\Gamma, \Gamma') & \longrightarrow & \mathrm{Hom}(\Gamma, \mathrm{H}^0(J^{\cdot})) & \longrightarrow & 0 \end{array}$$

fournit un morphisme des suites exactes longues associées, et la proposition s'obtient par application du lemme des cinq. □

Théorème 2.6. *(Min He) Soit Λ un système cohérent stable sur X définissant un point p de l'espace de modules* $\mathrm{Syst}_\alpha(c, k)$.

(i) *L'espace tangent de Zariski $T_p \mathrm{Syst}_\alpha(c, k)$ est canoniquement isomorphe à* $\mathrm{Ext}^1(\Lambda, \Lambda)$.
(ii) *Si* $\mathrm{Ext}^2(\Lambda, \Lambda) = 0$ *l'espace de modules $\mathrm{Syst}_\alpha(c, k)$ est lisse au voisinage du point p.*

3. Filtrations de Harder-Narasimhan

Considérons un système cohérent $\Lambda = (\Gamma, F)$ sur X, où F est un faisceau cohérent pur de dimension d. On peut trouver une filtration croissante finie de longueur ℓ de Λ par des sous-systèmes cohérents

$$0 \subset F_1\Lambda \subset \cdots \subset F_\ell\Lambda = \Lambda$$

satisfaisant aux conditions suivantes:

(i) le système algébrique $gr_i(\Lambda) = F_i\Lambda / F_{i-1}\Lambda$ est un système cohérent α-semi-stable pour $1 \leq i \leq \ell$;
(ii) on a, pour les polynômes de Hilbert réduits associés

$$p_{gr_i} > p_{gr_{i+1}}$$

Cette filtration est unique et s'appelle la filtration de Harder-Narasimhan de Λ relative à α.

Etant donnée une famille plate de systèmes cohérents $\Lambda = (\mathcal{V}, \mathcal{F})$ paramétrée par S, avec $\mathcal{F}_s$ pur de dimension d, on peut considérer les points s pour lesquels la filtration de Harder-Narasimhan $F_i\Lambda_s$ est de longueur ℓ, et telle que le gradué $gr_i(\Lambda)$ soit de polynôme de Hilbert fixé. On obtient ainsi une sous-variété localement fermée de S qu'on appelle strate de Harder-Narasimhan de S. Comme dans le cas des familles de faisceaux cohérents, on peut sous certaines conditions évaluer la codimension de cette strate. Pour ceci on doit introduire la notion de déformation infinitésimale de Kodaira et Spencer.

3.1. Familles complètes de systèmes cohérents

Considérons une famille plate de systèmes cohérents $\Lambda = (\mathcal{V}, \mathcal{F})$ paramétrée par une variété algébrique lisse S. Soit s un point de S, T_s l'espace vectoriel tangent au point s à S. Quand on considère le voisinage infinitésimal $\{s^{(2)}\}$ d'ordre 1 de s, on obtient une suite exacte de familles de systèmes cohérents paramétrée par $s^{(2)}$

$$0 \to \Lambda_s \otimes T_s^* \to \Lambda_{s^{(2)}} \to \Lambda_s \to 0$$

qu'on peut considérer comme une suite exacte de systèmes cohérents sur $\{s\} \times X$ compte-tenu du morphisme canonique $\{s^{(2)}\} \to \{s\}$. Par suite, on obtient une application linéaire

$$T_s S \to \operatorname{Ext}^1(\Lambda_s, \Lambda_s)$$

qu'on appelle morphisme de déformation infinitésimale de Kodaira et Spencer de la famille au point s. Alors la notion de famille complète de systèmes cohérents paramétrée par S a un sens:

Definition 3.1. *Une famille plate Λ de systèmes cohérents paramétrée par une variété algébrique lisse S est dite complète si le morphisme de déformation infinitésimale de Kodaira et Spencer*

$$\kappa_s = T_s S \to \operatorname{Ext}^1(\Lambda_s, \Lambda_s)$$

est surjectif en tout point $s \in S$.

Considérons maintenant un système cohérent Λ sur X muni d'une filtration. On peut introduire comme dans toute catégorie abélienne ayant assez d'objets injectifs les foncteurs $\operatorname{Ext}^q_-(\Lambda, \Lambda)$ qui respectent la filtration, et les foncteurs $\operatorname{Ext}^q_+(\Lambda, \Lambda)$ associés: si $R^\cdot$ est une résolution injective de Λ, dans la catégorie des complexes filtrés, on a par définition

$$\operatorname{Ext}^q_-(\Lambda, \Lambda) = \mathrm{H}^q(\operatorname{Hom}_-(\Lambda, R^\cdot))$$

où $\operatorname{Hom}_-(\Lambda, R^\cdot)$ désigne le complexe des homomorphismes qui sont compatibles avec les filtrations. De même, $\operatorname{Ext}^q_+(\Lambda, \Lambda) = \mathrm{H}^q(\operatorname{Hom}_+(\Lambda, R^\cdot))$ où $\operatorname{Hom}_+(\Lambda, R^\cdot)$ est le complexe quotient

$$\operatorname{Hom}_+(\Lambda, R^\cdot) = \operatorname{Hom}(\Lambda, R^\cdot) / \operatorname{Hom}_-(\Lambda, R^\cdot)$$

On a alors une suite exacte canonique

$$\ldots \to \operatorname{Ext}^q_-(\Lambda, \Lambda) \to \operatorname{Ext}^q(\Lambda, \Lambda) \to \operatorname{Ext}^q_+(\Lambda, \Lambda) \to \ldots$$

On peut en particulier appliquer ceci au cas où la filtration de Λ est la filtration de Harder-Narasimhan relative à α. L'énoncé suivant est identique à celui de [6].

Théorème 3.2. *Soit $\Lambda = (\Lambda_s)_{s \in S}$ une famille plate de systèmes cohérents purs de dimension d, paramétrée par une variété algébrique lisse S. Soit $s \in S$. On suppose que conditions suivantes sont satisfaites*

(i) *la famille Λ est une famille complète;*

(ii) $\operatorname{Ext}^2_-(\Lambda_s, \Lambda_s) = 0$ *et* $\operatorname{Ext}^2_+(\Lambda_s, \Lambda_s) = 0$.

Alors la strate de Harder-Narasimhan est lisse en s, de codimension

$$\dim \operatorname{Ext}^1_+(\Lambda_s, \Lambda_s)$$

En fait, l'espace vectoriel $\operatorname{Ext}^1_+(\Lambda_s, \Lambda_s)$ s'identifie à l'espace normal en s à la strate de Harder-Narasimhan passant par le point s.

4. Systèmes cohérents sur le plan projectif

Pour le plan projectif $X = P_2$, on a $K(P_2) = K_{\text{num}}(P_2)$. Soit n un entier; les faisceaux cohérents de rang 2 et de classe de Chern $c_1 = 0$ et $c_2 = n$ sur le plan projectif ont pour classe $c = 2 - nh^2$ dans $K(P_2)$. L'espace de modules des classes d'équivalence de faisceaux semi-stables F de classe c sera noté M_n. On se propose de montrer comment les espaces de modules de systèmes cohérents permettent de relier cet espace de modules au schéma de Hilbert $\operatorname{Hilb}^{n+1}(P_2)$ des sous-schémas de longueur $n+1$.

On considère l'espace de modules $S_{\alpha,n} := \mathrm{Syst}_\alpha(c(1),1)$ des classes de systèmes cohérents α-semi-stables $\Lambda = (\Gamma, F)$ sur le plan projectif, où F faisceau cohérent de rang 2, de classes de Chern $c_1 = 2$ et $c_2 = n+1$, et $\dim \Gamma = 1$. Par définition, un tel système cohérent $\Lambda = (\Lambda, F)$ est α-semi-stable si F est sans torsion, et si pour tout sous-faisceau cohérent $F' \subset F$ de rang 1, on a

$$p_{F'} \leq p_F + \frac{\alpha}{2} \tag{A}$$

si $\Gamma \cap \mathrm{H}^0(F') = \{0\}$, et

$$p_{F'} \leq p_F - \frac{\alpha}{2} \tag{B}$$

si $\Gamma \subset \mathrm{H}^0(F')$. Remarquons que si cet espace de modules n'est pas vide, on a obligatoirement

$$0 < \alpha \leq \beta := 2(p_F - P_{\mathcal{O}})$$

où $P_{\mathcal{O}}$ est le polynôme de Hilbert du faisceau structural $\mathcal{O} = \mathcal{O}_{P_2}$ de P_2. Le membre de droite est le polynôme de degré 1 défini par $\beta(m) = 2m + 4 - n$.

Un point de $S_{\alpha,n}$ est dit strictement semi-stable si c'est la classe d'un faisceau α-semi-stable qui n'est pas stable.

Théorème 4.1. *Soit $0 < \alpha \leq \beta$. On suppose $n \geq 2$.*

(i) *Les variétés $S_{\alpha,n}$ sont irréductibles et normales de dimension $3n+2$.*

(ii) *La variété $S_{\alpha,n}$ n'a pas de point strictement semi-stable, sauf si $\alpha \equiv n \mod 2$, et $0 < \alpha \leq n$. Ces valeurs sont appelées valeurs critiques. Entre deux valeurs critiques, l'espace de modules $S_{\alpha,n}$ est lisse et reste constant.*

(iii) *Si α n'est pas valeur critique, il existe une famille universelle de systèmes cohérents paramétrée par $S_{\alpha,n}$.*

(iv) *Si α est valeur critique $< \beta$, et $n \geq 3$, la variété $S_{\alpha,n}$ est singulière aux points strictement semi-stables, et lisse en dehors.*

L'essentiel de cette section est consacré à la démonstration de ce théorème. Le point essentiel de la démonstration est la vérification du critère de lissité de Min He: cette vérification se fera en utilisant la description du saut au passage des points critiques.

4.1. Les espaces de modules $S_{\beta,n}$ et $S_{\beta_-,n}$

Commençons par la description de $S_{\alpha,n}$ pour $\alpha = \beta$ et α proche de β.

Proposition 4.2. *Soit $\alpha \leq \beta$ un polynôme de degré 1.*

(i) *Si $\Lambda = (\Gamma, F)$ est un système cohérent α-semi-stable, on a une extension*

$$0 \to \Lambda' \to \Lambda \to \Lambda'' \to 0$$

où $\Lambda' = (\mathbb{C}, \mathcal{O}_{P_2})$ et $\Lambda'' = (0, I_Z(2))$, où I_Z est l'idéal d'un sous-schéma Z du plan projectif de longueur $n+1$.

(ii) *Supposons $\alpha < \beta$. Alors cette extension n'est pas scindée. Réciproquement, une telle extension, si elle est non scindée, définit un système cohérent $\Lambda = (\Gamma, F)$ α-semi-stable. De plus, le faisceau F est μ-semi-stable.*

Démonstration. (i) La condition (B) entraîne que le conoyau du morphisme d'évaluation $\Lambda \otimes \mathcal{O}_X \to F$ est un faisceau sans torsion, donc isomorphe à un faisceau $I_Z(2)$, avec Z sous-schéma de longueur $n+1$ du plan projectif. On obtient donc une extension

$$0 \to \Lambda' \to \Lambda \to \Lambda'' \to 0$$

où $\Lambda' = (\mathbb{C}, \mathcal{O}_X)$ et $\Lambda'' = (0, I_Z(2))$. D'où l'assertion (i).

(ii) Supposons $\alpha < \beta$. Si l'extension était scindée, le faisceau F sous-jacent à Λ aurait un sous-faisceau F', isomorphe à $I_Z(2)$; la condition (A) pour ce faisceau conduit à l'inégalité $\beta \leq \alpha$, ce qui est contraire à l'hypothèse. Réciproquement, considérons une extension non scindée

$$0 \longrightarrow \Lambda' \longrightarrow \Lambda \longrightarrow \Lambda'' \longrightarrow 0$$

où $\Lambda' = (\mathbb{C}, \mathcal{O}_X)$ et $\Lambda'' = (0, I_Z(2))$. Soit F le faisceau cohérent sous-jacent à Λ. Alors F est sans torsion. Si F' est un sous-faisceau de rang 1 de F tel que le quotient F/F' soit sans torsion, on a soit $F' = \mathcal{O}_X$, soit $\mathcal{O}_X \longrightarrow F''$ est injectif: ainsi $F' \cap \mathcal{O}_X = 0$ et par suite $F' \longrightarrow I_Z(2)$ serait lui aussi injectif; donc $c_1(F') \leq 2$. Si $c_1(F') = 2$, le fait que l'extension ne soit pas scindée implique $c_2(F') \geq n + 2$. C'est impossible, à cause de la relation

$$c_2(F') + c_2(F'') = n + 1.$$

Par suite, $c_1(F') \leq 1$, et par suite $p_{F'} < p_F + \dfrac{\alpha}{2}$. On vient de voir en fait que le faisceau F sous-jacent à Λ est obligatoirement μ-semi-stable. □

Corollaire 4.3.

(i) *Pour $\alpha = \beta$, tous les points sont strictement semi-stables, et* $S_{\beta,n} = \mathrm{Hilb}^{n+1}(P_2)$.

(ii) *Soit $\beta_- < \beta$ proche de β. L'espace de modules $S_{\beta_-,n}$ est l'espace des extensions non scindées:*

$$0 \longrightarrow (\mathbb{C}, \mathcal{O}_X) \longrightarrow \Lambda \longrightarrow (0, I_Z(2)) \longrightarrow 0$$

où I_Z est l'idéal d'un sous-schéma de longueur $n+1$ de P_2. Soient $\Xi \subset$ $\mathrm{Hilb}^{n+1}(P_2) \times P_2$ *le sous-schéma universel, p_1 la première projection, et $\mathcal{V}$ le fibré vectoriel de rang $n+1$ défini sur* $\mathrm{Hilb}^{n+1}(P_2)$ *par $\mathcal{V} = p_{1*}(\mathcal{O}_\Xi(0, -1))$. On a*

$$S_{\beta_-,n} \simeq P(\mathcal{V}).$$

Démonstration. La suite exacte $0 \longrightarrow \Lambda' \longrightarrow \Lambda \longrightarrow \Lambda'' \longrightarrow 0$ de la proposition ci-dessus, où $\Lambda' = (\mathbb{C}, \mathcal{O}_{P_2})$ et $\Lambda'' = (0, I_Z(2))$, définit dans le cas $\alpha = \beta$ une filtration de Jordan-Hölder. Par suite Λ est S-équivalent à $\Lambda' \oplus \Lambda''$. D'où l'assertion (i). Le faisceau algébrique cohérent sur $\mathrm{Hilb}^{n+1}(P_2)$ défini par $\underline{\mathrm{Ext}}^1_{p_1}(\mathcal{I}_\Xi(0, 2)), \mathcal{O})$ est le dual du faisceau

$$R^1 p_{1*}(\mathcal{I}_\Xi(0, -1)) \simeq p_{1*}(\mathcal{O}_\Xi(0, -1)).$$

La proposition ci-dessus montre que l'espace de modules $S_{\beta_-,n}$ est isomorphe au fibré en espaces projectifs (au sens de Grothendieck) $P(p_{1*}(\mathcal{O}_\Xi(0, -1)))$. □

4.2. Les flips de Thaddeus

On se préoccupe ici de ce qui se passe au passage des valeurs critiques. L'énoncé est analogue au résultat obtenu par Thaddeus [22] sur les courbes.

Théorème 4.4. *Soient α une valeur critique $< \beta$, et $\alpha_- < \alpha < \alpha_+$ des valeurs proches de α. Alors il existe des morphismes birationnels*

$$\begin{array}{ccccc} & & \widetilde{S}_{\alpha,n} & & \\ & \pi_- \swarrow & & \searrow \pi_+ & \\ S_{\alpha_-,n} & & & & S_{\alpha_+,n} \\ & p_- \searrow & & \swarrow p_+ & \\ & & S_{\alpha,n} & & \end{array}$$

Dans ce diagramme p_- et p_+ sont des éclatements du fermé Σ des points strictement semi-stables de $S_{\alpha,n}$. Le morphisme π_- est l'éclatement de la sous-variété Σ_- des points qui ne sont pas α_+-stable; le morphisme π_+ est l'éclatement de la sous-variété Σ_+ des points qui ne sont pas α_--stable. Ces sous-variétés Σ_- et Σ_+ sont lisses de codimension

$$\operatorname{codim}\Sigma_+ = n-1\ ; \operatorname{codim}\Sigma_- = \frac{n+\alpha}{2}+2.$$

Le diviseur exceptionnel dans $\widetilde{S}_{\alpha,n}$ est le même pour les deux éclatements.

La démonstration de cet énoncé repose sur la description des strates de Harder-Narasimhan, donnée dans la section 3, et l'étude des foncteurs $\operatorname{Ext}^q_+(\Lambda,\Lambda)$ et $\operatorname{Ext}^q_-(\Lambda,\Lambda)$.

Lemme 4.5. *Soient I' et I'' deux faisceaux de rang 1 sans torsion de classe $c(I') = 1-\ell' h^2$ et $c(I'') = 1-\ell'' h^2$ respectivement. Considérons deux systèmes cohérents $\Lambda' = (\Gamma', I'(1))$ et $\Lambda'' = (\Gamma'', I''(1))$ d'ordre 0 ou 1. Si $\ell' \geq \ell'' - 2$, on a*

$$\operatorname{Ext}^q(\Lambda',\Lambda'') = 0 \quad \text{pour} \quad q \geq 2.$$

Démonstration. On pose $F' = I'(1)$, et $F'' = I''(1)$. On a la suite exacte longue de la proposition 2.4

$$\ldots \to \operatorname{Hom}(\Gamma', \mathrm{H}^1(F'')) \to \operatorname{Ext}^2(\Lambda',\Lambda'') \to \operatorname{Ext}^2(F',F'') \to \ldots$$

et on a d'après le théorème de dualité de Serre $\operatorname{Ext}^2(F',F'') = 0$. Le résultat est donc évident si $\Gamma' = 0$. Supposons maintenant que Λ' soit d'ordre 1: on a donc une suite exacte

$$0 \to \Gamma' \otimes \mathcal{O}_X \to F' \to \Theta' \to 0$$

où $\Theta' = \mathcal{O}_{d'}(1-\ell')$ est un faisceau localement libre de degré $1-\ell'$ sur une droite d' du plan. D'après la proposition 2.5 il suffit de vérifier que $\operatorname{Ext}^2(\Theta', F'') = 0$. Mais cet espace vectoriel est isomorphe à $\mathrm{H}^1(F''|_{d'}(\ell'))$; le faisceau $F''|_{d'}$ est isomorphe à $\mathcal{O}_{d'}(x) \oplus T$ où T est un faisceau de dimension 0. Soit $\ell(T)$ la longueur de T. On a alors $x+1+\ell(T) = 2$ et par conséquent $x = 1-\ell(T)$. On a évidemment

$$\ell(T) = \ell((\mathcal{O}/I)|_{d'}) \leq \ell''.$$

Par conséquent, $x \geq 1-\ell''$. Ainsi, si $\ell' - \ell'' + 1 \geq -1$, *i.e.* si $\ell' \geq \ell'' - 2$ on aura $\mathrm{H}^1(F''|_{d'}(\ell')) = 0$. Ceci démontre le lemme. □

Lemme 4.6. *Soit n un entier ≥ 2. Soit α une valeur critique $\neq \beta$, et $\Lambda = (\Gamma, F)$ un système cohérent de type $(c(1), 1)$.*

(i) *On suppose que Λ est α_+-semi-stable et n'est pas α_--semi-stable. Alors, pour la filtration de Harder-Narasimhan relative à α_-, on a*

$$\operatorname{Ext}^2_-(\Lambda,\Lambda) = 0;$$

$$\operatorname{Ext}^i_+(\Lambda,\Lambda) = 0 \quad \text{si} \quad i \neq 1\ ;\ \dim \operatorname{Ext}^1_+(\Lambda,\Lambda) = n-1.$$

(ii) *On suppose que Λ est α_--semi-stable et n'est pas α_+-semi-stable. Alors, pour la filtration de Harder-Narasimhan relative à α_+,*

$$\operatorname{Ext}^2_-(\Lambda,\Lambda) = 0;$$

$$\operatorname{Ext}^i_+(\Lambda,\Lambda) = 0 \quad \text{si} \quad i \neq 1 \quad ; \quad \dim \operatorname{Ext}^1_+(\Lambda,\Lambda) = \frac{n+\alpha}{2}+2.$$

Démonstration. (i) La filtration de Harder-Narasimhan relative à α_- conduit à une suite exacte

$$0 \to \Lambda' \to \Lambda \to \Lambda'' \to 0,$$

avec $\Lambda' = (0, F')$ et $\Lambda'' = (\Gamma'', F'')$ d'ordre 1, où $F' = I'(1)$, et $F'' = I''(1)$, où I' et I'' sont les idéaux de sous-schémas finis de longueurs respectives $\ell' = \frac{1}{2}(n - \alpha)$ et $\ell'' = \frac{1}{2}(n + \alpha)$. D'après le lemme 4.5 ci-dessus, on a

$$\mathrm{Ext}^2(\Lambda'', \Lambda') = \mathrm{Ext}^2(\Lambda', \Lambda') = \mathrm{Ext}^2(\Lambda'', \Lambda'') = 0,$$

d'où découle $\mathrm{Ext}^2_-(\Lambda, \Lambda) = 0$. On a aussi, d'après la suite exacte de la proposition 2.4, $\mathrm{Ext}^q(\Lambda', \Lambda'') = \mathrm{Ext}^q(F', F'')$ pour tout entier q. Puisque $\alpha > 0, \mathrm{Hom}(F', F'') = 0$; d'autre part, par dualité de Serre, $\mathrm{Ext}^2(F', F'') = 0$. Le théorème de Riemann-Roch conduit à l'égalité suivante

$$\chi(F', F'') = \sum_i \dim \mathrm{Ext}^i(F', F'') = 1 - (\ell' + \ell'')$$

et donc $\dim \mathrm{Ext}^1(F', F'') = n - 1$.

(ii) De même, la filtration de Harder-Narasimhan relative à α_+ fournit une suite exacte

$$0 \to \Lambda' \to \Lambda \to \Lambda'' \to 0,$$

avec $\Lambda' = (\Gamma', F')$ d'ordre 1 et $\Lambda'' = (0, F'')$, où $F' = I'(1)$, et $F'' = I''(1)$ sont de classes de Chern $c_2(F') = \frac{1}{2}(n + \alpha)$ et $c_2(F'') = \frac{1}{2}(n - \alpha)$. On a alors évidemment $\mathrm{Ext}^2(\Lambda'', \Lambda') = 0$ et d'après le lemme 4.5 ci-dessus $\mathrm{Ext}^2(\Lambda', \Lambda') = \mathrm{Ext}^2(\Lambda'', \Lambda'') = 0$. Il en résulte encore $\mathrm{Ext}^2_-(\Lambda, \Lambda) = 0$. Le lemme 4.5 fournit encore $\mathrm{Ext}^q(\Lambda', \Lambda'') = 0$ pour $q \geq 2$ et $\mathrm{Hom}(\Lambda', \Lambda'') = 0$ parce qu'il s'agit d'une filtration de Harder-Narasimhan. Posons

$$\chi(\Lambda', \Lambda'') = \sum_i (-1)^i \dim \mathrm{Ext}^i(\Lambda', \Lambda'').$$

On a alors d'après la proposition 2.4

$$\begin{aligned}\chi(\Lambda', \Lambda'') &= \chi(F', F'') - \dim \Gamma'(\chi(F'') - \dim \Gamma'') \\ &= 1 - n - (3 - \frac{1}{2}(n - \alpha)) \\ &= -2 - \frac{1}{2}(n + \alpha).\end{aligned}$$

Ainsi, $\dim \mathrm{Ext}^1(\Lambda', \Lambda'') = \frac{1}{2}(n + \alpha) + 2$. □

4.3. Démonstration du théorème 4.1

En dehors des valeurs critiques, tous les points sont stables: pour obtenir la lissité en un tel point, il suffit d'après le critère de Min He de vérifier que $\mathrm{Ext}^2(\Lambda, \Lambda) = 0$ en un tel point. Pour $n < \alpha < \beta$, il suffit par semi-continuité de le vérifier pour le système cohérent $\Lambda = \Lambda' \oplus \Lambda''$, avec $\Lambda' = (\mathbb{C}, \mathcal{O}_{P_2})$ et $\Lambda'' = (0, I_Z(2))$ où I_Z est l'idéal d'un sous-schéma Z de P_2 de longueur $c_2(I_Z) = n + 1$. La vérification est analogue à celle du lemme 4.5; la seule chose à constater est $\mathrm{Ext}^2(\Lambda', \Lambda'') = 0$. Ceci résulte de la suite exacte longue 2.5: on a en effet $\mathrm{H}^2(I_Z(2)) = 0$, et le morphisme naturel $\mathrm{Ext}^1(\Lambda', \Lambda'') \to \mathrm{Hom}(\mathbb{C}, \mathrm{H}^1(I_Z(2)))$ est trivialement un isomorphisme.

Au passage d'une valeur critique α, (en décroissant), les extensions du type (i) dans le lemme 4.6 sont remplacées par des extensions $0 \to \Lambda'' \to \widetilde{\Lambda} \to \Lambda' \to 0$ de type (ii); ces extensions sont α_--stables si elles sont non scindées; de telles extensions, classées par l'espace projectif $P(\mathrm{Ext}^1(\Lambda', \Lambda''))$ existent sous l'hypothèse $n > 1$, car $\dim \mathrm{Ext}^1(\Lambda', \Lambda'') = n - 1 > 0$. Il résulte de la suite exacte

$$\mathrm{Ext}^2_-(\Lambda, \Lambda) \to \mathrm{Ext}^2(\Lambda, \Lambda) \to \mathrm{Ext}^2_+(\Lambda, \Lambda)$$

et du lemme 4.6 ci-dessus que $\mathrm{Ext}^2(\Lambda, \Lambda) = 0$ pour de tels systèmes cohérents. Les autres systèmes cohérents α_--semi-stables figuraient déjà dans l'espace de modules $S_{\alpha_+,n}$ et donc héritent de cette condition. Il résulte du critère de Min He que tous les espaces $S_{\alpha,n}$ sont lisses si α n'est pas valeur critique.

Pour α proche de β, l'espace de modules $S_{\alpha,n}$ est irréductible puique c'est un fibré en espaces projectifs associé à un fibré vectoriel. Après éclatement et contraction l'irréductibilité est conservée: tous les espaces de modules $S_{\alpha,n}$ sont donc irréductibles pour $n \geq 2$. Reste le calcul de la dimension: on a d'après la suite exacte de la proposition 2.4

$$\begin{aligned}\chi(\Lambda, \Lambda) &= 1 - \dim \mathrm{Ext}^1(\Lambda, \Lambda) \\ &= \chi(F, F) - (\chi(F(1)) - 1) \\ &= 4 - 4n - (5 - n) = -1 - 3n\end{aligned}$$

et par suite $\dim \mathrm{Ext}^1(\Lambda, \Lambda) = 3n + 2$.

La construction des espaces de modules de systèmes cohérents et une variante du critère de Min He permet en fait de vérifier que l'espace de modules $S_{\alpha,n}$ est toujours un bon quotient d'une variété lisse: ceci entraîne la normalité de $S_{\alpha,n}$ pour α valeur critique. Au voisinage d'un point $\Lambda = \Lambda' \oplus \Lambda''$, avec $\Lambda' = (\Gamma', I'(1))$ d'ordre 1 et $\Lambda'' = (0, I''(1))$ où I' et I'' idéaux de sous-schémas finis de longueurs $\ell' = \frac{1}{2}(n + \alpha)$ et $\ell'' = \frac{1}{2}(n - \alpha)$ l'espace de modules est localement isomorphe (au sens analytique) au produit d'une variété lisse par le cône de l'espace vectoriel

$$\mathrm{Ext}^1(\Lambda', \Lambda'') \otimes \mathrm{Ext}^1(\Lambda'', \Lambda')$$

des tenseurs de rang ≤ 1. Ce cône est singulier, sauf si $n = 2$. □

L'existence d'une famille universelle paramétrée par $S_{\alpha,n}$ si α n'est pas valeur critique résulte de la construction ([15], [11]) de $S_{\alpha,n}$: en effet, cette variété algébrique est un quotient géométrique d'un ouvert $G_{\alpha,n}$ d'une grassmannienne relative au-dessus d'un schéma de Hilbert $\mathrm{Hilb}(H \otimes \mathcal{O}_X(-m), c(1))$. Ici H est un espace vectoriel fixé de dimension $\chi(c(m+1))$, et m est un entier assez grand. Le groupe qui agit est $SL(H)$, et le stabilisateur de chaque point est le centre. Cet ouvert paramètre un système cohérent universel $(\mathcal{V}, \mathcal{F})$, muni d'une action de $SL(H)$. Alors le système cohérent $(\mathcal{O}, \mathcal{V}^{-1} \otimes \mathcal{F})$ passe au quotient d'après le lemme de Kempf [7] et définit un système cohérent universel sur $S_{\alpha,n}$.

4.4. Le morphisme $f : S_{\epsilon,n} \to M_n$ et le lieu de Brill-Noether

Reste l'étude des petites valeurs de α. On rappelle que M_n est l'espace de modules des faisceaux semi-stables sur le plan projectif, de classe $c = 2 - nh^2$ dans $K(P_2)$.

Théorème 4.7. *On suppose $n \geq 2$. Pour ϵ assez petit, on a un morphisme canonique $f : S_{\epsilon,n} \to M_n$ associant au système cohérent ϵ-semi-stable $\Lambda = (\Gamma, F)$ la classe du faisceau $F(-1)$.*

(i) *Pour $n \leq 5$, ce morphisme est surjectif, et sa fibre générique est P_{5-n}.*

(ii) *Pour $n \geq 6$, le sous-schéma de Brill-Noether B de M_n défini par les classes de faisceaux semi-stables F satisfaisant à la condition déterminantielle $h^0(F(1)) > 0$ est irréductible et normal de codimension $n - 5$. Le morphisme f est génériquement injectif et a B pour image.*

Démonstration. Le morphisme f est bien défini, car si $\Lambda = (\Gamma, F)$ est un système cohérent ϵ-semi-stable représentant un point de $S_{\epsilon,n}$, le faisceau F sous-jacent est évidemment semi-stable. La première assertion résulte du fait que si F est un faisceau semi-stable de rang 2 et classes de Chern $(0, n)$, on a $h^0(F(1)) = \chi(F(1)) = 6 - n$ pour $n = 2$ et $n = 3$ et en dehors d'un fermé de codimension 3 pour $n = 4$, et d'une hypersurface pour $n = 5$ ([12]). Si F est un faisceau stable générique de rang 2 et classes de Chern $(0, n)$, tout sous-espace vectoriel $\Gamma \subset \mathrm{H}^0(F(1))$ définit un système cohérent ϵ-stable. Ainsi, l'image réciproque du point de M_n défini par F s'identifie à l'espace projectif $P(\mathrm{H}^0(F(1)))$.

Pour vérifier l'assertion (ii), remarquons que l'image de f est évidemment contenue dans le lieu de Brill-Noether. Réciproquement, un faisceau semi-stable F tel que $h^0(F(1)) \geq 1$ définit un point de l'image de f: c'est évident si F est stable, pourvu que ϵ ait été choisi assez petit. Si F est strictement semi-stable, équivalent à $I' \oplus I''$, on a par exemple $h^0(I'') \geq 1$. Considérons les systèmes cohérents $\Lambda' = (0, I'(1))$, et $\Lambda'' = (\Gamma'', I''(1))$, avec Γ'' de dimension 1. Les extensions $0 \to \Lambda' \to \Lambda \to \Lambda'' \to 0$ sont classées par $\mathrm{Ext}^1(\Lambda'', \Lambda')$, et on a la suite exacte

$$\mathrm{Ext}^1(\Lambda'', \Lambda') \to \mathrm{Ext}^1(I'', I') \to L(\Gamma'', \mathrm{H}^1(I'(1))).$$

On a $\dim \mathrm{Ext}^1(I'', I') \geq 2n - 1$, et $\dim \mathrm{H}^1(I'(1)) \leq \dfrac{n}{2} - 2$. Par suite, il existe des extensions Λ qui fournissent une suite exacte non scindée $0 \to I' \to F \to I'' \to 0$. Une telle extension est ϵ-stable, ce qui montre que tout point du lieu de Brill-Noether est contenu dans l'image de f.

Il en résulte que la variété B est irréductible et que $\dim B \leq 3n + 2$. Ceci aura pour conséquence que B a la codimension attendue $n - 5$. Pour vérifier que B est normale, on va commencer par vérifier que l'ensemble singulier de B est de codimension ≥ 2. On vérifiera en même temps que le morphisme f est un plongement au-dessus d'un ouvert de M_n qui rencontre l'image de f. Considérons l'ouvert U des points de M_n représentant les faisceaux stables F tels que $h^0(F(1)) \leq 1$. Le complémentaire de $B \cap U$ dans B est la réunion du fermé des points satisfaisant à la condition déterminantielle $h^0(F(1)) \geq 2$ et du fermé des points de B qui sont strictement semi-stables.

Lemme 4.8. *L'intersection $B \cap U$ du schéma de Brill-Noether B avec U est une sous-variété lisse, de codimension $n - 5$, et au-dessus de U, le morphisme f est un plongement d'image $B \cap U$.*

Démonstration. Si F est un faisceau représentant un point de $B \cap U$, le système cohérent $\Lambda = (\Gamma, F(1))$ défini par $\Gamma = \mathrm{H}^0(F(1))$ est ϵ-stable, et c'est le seul point de $S_{n,\epsilon}$ au-dessus de a. Dans la suite exacte

$$0 \to \mathrm{Ext}^1(\Lambda, \Lambda) \to \mathrm{Ext}^1(F, F) \to L(\Gamma, \mathrm{H}^1(F(1)))$$

la seconde flèche est surjective du fait que $\mathrm{Ext}^2(\Lambda, \Lambda) = 0$; ceci implique que le morphisme de Pétri est surjectif, et donc que B est lisse au point défini par F. La première flèche représente l'application linéaire tangente à f au point Λ, qui est donc injective. □

Reste à évaluer la codimension du fermé $B \setminus U$. C'est l'objet des deux lemmes suivants:

Lemme 4.9. *La sous-variété de Brill-Noether B_2 de M_n définie par les classes de faisceaux semi-stables F satisfaisant à la condition déterminantielle $h^0(F(1)) \geq 2$ est irréductible de dimension $\leq 2n + 5$.*

Cet énoncé implique en particulier que cette sous-variété de Brill-Noether a la codimension attendue $2n - 8$. Cet énoncé sera démontré plus tard.

Lemme 4.10. *Le fermé de B des points strictement semi-stables est de codimension $\geq \frac{3n}{2}$ dans B.*

Démonstration. Le fermé en question est évidemment vide si n est impair. Si n est pair, les points de ce fermé correspondent aux faisceaux $F = I \oplus I''$ où I' et I'' sont les idéaux de sous-schémas finis de P_2 de même longueur $\frac{n}{2}$, et tels que soit $h^0(I'(1)) \geq 1$, soit $h^0(I''(1)) \geq 1$. Il en résulte que la dimension de ce fermé est $\leq 2 + \frac{n}{2} + n$. Par suite sa codimension dans B est $\frac{3n}{2}$. □

4.5. Fin de la démonstration du théorème 4.7

Il reste à vérifier que B est une variété normale. C'est plus facile dans le cas n impair, car alors M_n est lisse et paramètre un faisceau universel $\mathcal{F}$. La variété B est la variété déterminantielle définie par l'idéal de Fitting du faisceau $pr_{1*}(\mathcal{F}(0,1))$; sa dimension est la dimension attendue: ainsi, elle est de Cohen-Macaulay [1]. Puisque son ensemble singulier est de codimension ≥ 2, il résulte du critère de Serre qu'elle est normale. □

Dans le cas où n est pair, on choisit un entier m assez grand: désignons par H un espace vectoriel de dimension $N = \chi(c(m))$. On considère le fibré $E = H \otimes \mathcal{O}_{P_2}(-m)$ et l'ouvert $SL(H)$-invariant Ω des points du schéma de Hilbert Hilb(E, c) représentant les faisceaux quotients F de E semi-stables de classe c. L'ouvert Ω est une variété lisse de dimension $4n-3+N^2-1$, et l'espace de modules M_n est un bon quotient de Ω par $SL(H)$. Considérons la sous-variété de Brill-Noether $\widetilde{B}$ de Ω des points représentant les faisceaux quotients F satisfaisant à la condition déterminantielle $h^0(F(1)) \geq 1$. Ce sous-schéma est $SL(H)$-invariant, et on a un morphisme surjectif $\widetilde{B} \to B$ qui fait de B est un bon quotient de $\widetilde{B}$ par $SL(H)$. Il suffit de montrer que $\widetilde{B}$ est normale. Pour ceci, il suffit encore de vérifier que $\widetilde{B}$ a la dimension attendue, et que son ensemble singulier est de codimension ≥ 2. Mais le morphisme $\widetilde{B} \to B$ est lisse de dimension $N^2 - 1$ au-dessus de $B \cap U$. Il suffit donc de vérifier les assertions suivantes:

(1) le fermé de $\widetilde{B}$ correspondant aux faisceaux F strictement semi-stables est de dimension $\leq \frac{5n}{2} + 2 + N^2$.

Ceci résulte du fait que les points de ce fermé correspondent aux paires (F, φ) où F est une extension $0 \to I' \to F \to I'' \to 0$ où I' et I'' sont des idéaux de sous-schémas finis de même longueur $\frac{n}{2}$, tels que soit $h^0(I'(1)) \geq 1$, soit $h^0(I''(1)) \geq 1$, et φ un isomorphisme $H \simeq \mathrm{H}^0(F(\ell))$.

(2) la sous-variété de $\widetilde{B}$ des points correspondant aux faisceaux quotients F stables tels que $h^0(F(1)) \geq 2$ est de dimension $2n + 4 + N^2$. Ceci résulte du lemme 4.9.

Ainsi, $\widetilde{B}$ est de dimension $3n + 1 + N^2$: c'est la dimenion attendue, et donc $\widetilde{B}$ est de Cohen-Macaulay; de plus, son ensemble singulier est de codimension ≥ 2. Par suite $\widetilde{B}$ est normale, ainsi que B. □

Signalons que dans le cas $n = 6$, le fait que la sous-variété de Brill-Noether B ait la bonne dimension et soit irréductible et réduite est un cas particulier d'un résultat de Göttsche et Hirschowitz [10].

Remarque 4.11. (1) Pour $n = 0$, la seule valeur critique est β. L'espace de modules $S_{\alpha,0}$ est isomorphe au plan projectif P_2 et le morphisme f est le morphisme constant.

(2) Pour $n = 1$, les seules valeurs critiques sont 1 et β. Pour $\alpha < 1$ l'espace de module $S_{\alpha,1}$ est vide. Pour $1 < \alpha < \beta$ on a encore des morphismes

$$\begin{array}{ccc} & S_{\alpha,1} & \\ {}^{p_1}\swarrow & & \searrow^{p_\beta} \\ S_{1,1} & & S_{\beta,1} \end{array}$$

mais ces projections p_1 et p_β ne sont plus des morphismes birationnels. La projection p_1 s'interprète de la manière suivante: l'espace de modules $S_{1,1}$ est isomorphe à l'espace de modules des systèmes cohérents $\Lambda = (\Gamma, I_a(1))$ d'ordre 1, où I_a est l'idéal d'un point a. Cet espace de modules est isomorphe à la variété d'incidence des couples (droite, point): c'est donc une variété projective lisse de dimension 3. L'espace de modules $S_{\alpha,1}$ est, on l'a vu, un fibré en droites projectives au-dessus de $\mathrm{Hilb}^2(P_2)$. Pour tout tel système cohérent α-semi-stable peut s'écrire (de manière unique) comme extension $0 \to \Lambda' \to \Lambda \to \Lambda'' \to 0$ avec $\Lambda' = (0, \mathcal{O}_{P_2}(1))$ et $\Lambda'' = (\mathbb{C}, I_a(1))$. Cette extension correspond en fait à une filtration de Jordan-Hölder pour $\alpha = 1$. La projection p_1 est celle qui associe à la classe de Λ la classe de $\Lambda' \oplus \Lambda''$, qu'on identifie au point de $S_{1,1}$ défini par Λ''.

4.6. Etude de la sous-variété de Brill-Noether B_2

On suppose comme ci-dessus $n \geq 2$. Considérons, sur le plan projectif, l'espace de modules $\mathrm{Syst}_\alpha(c(1), 2)$ des systèmes cohérents α-semi-stables $\Lambda = (\Gamma, F)$ où F est un faisceau cohérent de rang 2, et de classes de Chern $c_1 = 2, c_2 = n + 1$ et $\Gamma \subset \mathrm{H}^0(F)$ un sous-espace vectoriel de dimension 2. Par définition, un tel système cohérent est α-semi-stable si pour tout sous-faisceau de rang 1 sans torsion $F' \subset F$ on a

$$p_{F'} \leq \begin{cases} p_F - \alpha & \text{si } \Gamma \subset \mathrm{H}^0(F') \qquad \text{(A)} \\ p_F & \text{si } \dim \Gamma \cap \mathrm{H}^0(F') = 1 \qquad \text{(B)} \\ p_F + \alpha & \text{si } \Gamma \cap \mathrm{H}^0(F') = 0 \qquad \text{(C)} \end{cases}$$

Pour α petit, le faisceau F sous-jacent à un système cohérent (Γ, F) α-semi-stable est encore semi-stable. Il en résulte que l'on a encore un morphisme d'oubli

$$g : \mathrm{Syst}_\alpha(c(1), 2) \to M_n$$

qui associe à la classe du système cohérent semi-stable $\Lambda = (\Gamma, F)$ la classe du faisceau semi-stable $F(-1)$. Le lieu de Brill-Noether B_2 coïncide avec l'image de g. Ainsi, le lemme 4.9 est une conséquence de l'énoncé suivant:

Lemme 4.12. *L'espace de modules* $\mathrm{Syst}_\alpha(c(1), 2)$ *est irréductible et génériquement réduit de dimension* $2n + 5$.

Démonstration. Il résulte de la définition ci-dessus que la situation (A) ne peut se présenter si α est un polynôme de degré > 0. Si α est une constante > 0, cette situation ne se présente que si $F' = I'(1)$, où I' est l'idéal d'un sous-schéma fini de longueur $\frac{n}{2} + \alpha$; pour $n \geq 2$, ceci implique $h^0(F') \leq 1$ ce qui montre que la situation (A) ne peut pas non plus se présenter. Il en résulte que si β est un polynôme > 0 de degré 2, un tel système cohérent est aussi β-semi-stable; on peut prendre par exemple le polynôme de Hilbert du faisceau structural du plan projectif. Réciproquement, tout système cohérent β-semi-stable est en fait α-semi-stable: il suffit en effet, de vérifier la condition (C). Si (Γ, F'') est un système cohérent quotient de

(Γ, F), avec F'' sans torsion de rang 1, et $\dim \Gamma = 2$, on doit avoir $c_1(F'') \geq 1$, et en cas d'égalité, $c_2(F'') \leq 1$. Par suite, $p_{F''} > p_F - \alpha$; ainsi, la condition (C) est bien satisfaite.

Ainsi, on a un isomorphisme

$$\mathrm{Syst}_\alpha(c(1), 2) \simeq \mathrm{Syst}_\beta(c(1), 2).$$

Il résulte du corollaire 5.12 de [15] que le membre de droite est isomorphe à l'espace de modules $\mathrm{Syst}_\beta(2h + nh^2, 2)$ des systèmes cohérents β-semi-stables $\Lambda = (\Gamma, \Theta)$, où Θ est un faisceau cohérent de dimension 1 de classe $2h+nh^2$ dans $K(P_2)$ et $\Gamma \subset \mathrm{H}^0(\Theta)$ un sous-espace de dimension 2. Le support schématique d'un tel faisceau est une conique.

L'espace de modules $\mathrm{Syst}_\alpha(2h+nh^2, 2)$ reste constant tant que α est un polynôme de degré ≥ 1; quand α est de degré 0, cet espace change au passage des valeurs α pour lesquelles il existe des systèmes cohérents strictement semi-stables du type $(\Gamma', \Theta') \oplus (0, \Theta'')$, avec Θ' et Θ'' de multiplicité 1 et

$$\chi(\Theta') = \frac{n}{2} + 1 - \alpha \quad ; \quad \chi(\Theta'') = \frac{n}{2} + 1 + \alpha$$

c'est-à-dire les nombres rationnels $\alpha > 0$ tels que $n - 2\alpha \in 2\mathbb{Z}$, et ≥ 2. Ces valeurs α seront appelées valeurs critiques. Pour $0 < \epsilon < \frac{1}{2}$, le faisceau Θ sous-jacent à un tel système cohérent (Γ, Θ) ϵ-semi-stable est lui-même semi-stable. Quand n est un entier impair $n = 2k + 1$, le faisceau Θ est en fait stable, et isomorphe à $\mathcal{O}_C(k)$, où C est une conique de P_2; réciproquement, n'importe quel sous-espace vectoriel $\Gamma \subset \mathrm{H}^0(\Theta)$ de dimension 2 définit un système cohérent ϵ-stable. Il en résulte que si $\Xi \subset P_5 \times P_2$ désigne la conique universelle, et $\mathcal{W}$ le fibré vectoriel $pr_{1*}(\mathcal{O}_\Xi(0, k+1))$ l'espace de modules des systèmes cohérents $\mathrm{Syst}_\epsilon(2h + nh^2, 2)$ s'identifie à la grassmannienne relative

$$\mathrm{Syst}_\epsilon(2h + nh^2, 2) \simeq \mathrm{Grass}(2, \mathcal{W}).$$

Ainsi, cet espace de modules est lisse de dimension $2n+5$. Quand n est un entier pair $n = 2k$, la situation est plus compliquée: Θ a une présentation comme conoyau d'un morphisme génériquement injectif

$$0 \to H(k-1) \xrightarrow{u} H(k) \to \Theta \to 0$$

où H désigne le fibré trivial de rang 2. Dans l'espace vectoriel de tous ces morphismes, désignons par Ω l'ouvert de l'espace vectoriel $\mathrm{Hom}(H(k-1), H(k))$ défini par les morphismes u génériquement injectifs pour lesquels le conoyau Θ est semi-stable; au-dessus de $\Omega \times P_2$, on dispose d'un faisceau universel, noté encore Θ; le faisceau $\mathcal{W} = pr_{1*}(\Theta)$ est alors localement libre de rang $n + 2$, et on peut considérer la grassmanienne relative

$$\mathcal{G} = \mathrm{Grass}(2, \mathcal{W}) \to \Omega.$$

Le groupe $G = GL(H) \times GL(H)$ agit de manière naturelle sur Ω, par conjugaison, et sur le faisceau Θ; il en résulte une action naturelle sur la grassmannienne relative $\mathcal{G}$. L'ouvert des points semi-stables $\mathcal{G}^{ss}$ a alors un bon quotient qui s'identifie à l'espace de modules $\mathrm{Syst}_\epsilon(2h + nh^2, 2)$. L'action de G se factorise en une action de $G/\mathbb{C}^*$ libre sur l'ouvert des points stables. Il en résulte que cet espace de modules est irréductible et normal de dimension $2n + 5$, lisse aux points stables.

Il resterait à voir ce qui se passe au passage des valeurs critiques. L'étude serait comparable, quoique que légèrement plus compliquée, à celle qui a été menée ci-dessus pour les systèmes cohérents d'ordre 1. Nous nous contenterons d'un résultat plus faible:

Lemme 4.13. *Soit β un polynôme > 0, de degré ≥ 1.*

(i) *Dans l'espace de modules $\mathrm{Syst}_\beta(2h+nh^2, 2)$ le fermé des points qui ne sont pas ϵ-semi-stables est de dimension $\leq 2n + 4$.*

(ii) *L'espace de modules* $\mathrm{Syst}_\beta(2h + nh^2, 2)$ *est irréductible et génériquement réduit de dimension* $2n + 5$.

Démonstration. Les points qui ne sont pas ϵ-semi-stables se représentent par des systèmes cohérents $\Lambda = (\Gamma, \Theta)$ qui s'écrivent comme extension $0 \to \Lambda'' \to \Lambda \to \Lambda' \to 0$ avec $\Lambda' = (\Gamma', \Theta')$ et $\Lambda'' = (0, \Theta'')$ de multiplicité 1, et de caractéristique d'Euler-Poincaré donnée par

$$\chi'' := \chi(\Theta'') > \frac{n}{2} + 1 \quad ; \quad 2 \leq \chi' := \chi(\Theta') < \frac{n}{2} + 1.$$

De telles extensions, pour Λ' et Λ'' fixés, sont classées par l'espace projectif $P(\mathrm{Ext}^1(\Lambda', \Lambda''))$. On a la suite exacte

$$L(\Gamma', \mathrm{H}^1(\Theta'')) \to \mathrm{Ext}^2(\Lambda', \Lambda'') \to \mathrm{Ext}^2(\Theta', \Theta'')$$

et on a $\mathrm{Ext}^2(\Theta', \Theta'') = 0$ par dualité de Serre; d'autre part, $\mathrm{H}^1(\Theta'') = 0$, et par suite $\mathrm{Ext}^2(\Lambda', \Lambda'') = 0$. De même, $\mathrm{Ext}^q(\Lambda', \Lambda'') = 0$ pour $q > 2$. D'autre part, $\mathrm{Hom}(\Lambda', \Lambda'') = 0$. On obtient

$$\begin{aligned}\chi(\Lambda', \Lambda'') &= \chi(\Theta', \Theta'') - 2\chi'' \\ &= -1 - 2\chi''.\end{aligned}$$

Il en résulte $\dim \mathrm{Ext}^1(\Lambda', \Lambda'') = 2\chi'' + 1$. Il en résulte que la strate de Harder-Narasimhan, relative à ϵ, passant par un tel point Λ est irréductible et lisse de dimension $4 + 2(\chi' - 2) + 2\chi'' = 2n + 4$. D'où l'assertion (i). Le complémentaire de la réunion de ces strates est un ouvert U qui s'identifie à un ouvert de l'espace de modules $\mathrm{Syst}_\epsilon(2h + nh^2, 2)$: cet ouvert est irréductible de dimension $2n + 5$. D'autre part, en un point Λ de la strate ci-dessus, on a $\mathrm{Ext}^2(\Lambda'', \Lambda') = 0$, sauf peut-être si le support de Θ' et le support de Θ'' coïncident: il en résulte qu'en un tel point $\mathrm{Ext}^2(\Lambda, \Lambda) = 0$. Le critère de Min He montre que l'espace de modules est lisse de dimension $2n+5$ en tout point de cette strate, sauf peut-être sur un fermé de dimension $\leq 2n + 2$. Par suite, toute strate de Harder-Narasimhan est contenue dans l'adhérence de U. Il en résulte que l'espace de modules $\mathrm{Syst}_\beta(2h + nh^2, 2)$ est irréductible de dimension $2n + 5$. D'où le lemme. □

5. Polynômes de Donaldson

5.1. Rappels sur les polynômes de Donaldson ([3], [4])

Soit X une surface algébrique projective lisse et simplement connexe, polarisée. On considère l'espace de modules M_n des faisceaux semi-stables sur X de rang 2, de classes de Chern $(0, n)$: autrement dit, il s'agit des faisceaux semi-stables de classe $c = 2 - nh^2$ dans $K_{\mathrm{num}}(X)$. On rappelle que $K_{\mathrm{num}}(X)$ est munie d'une forme quadratique naturelle. Considérons les couples (n, L) formés d'un entier n et d'une polarisation $L = \mathcal{O}_X(1)$ satisfaisant aux conditions suivantes:

(i) l'espace de modules M_n est une variété normale de dimension

$$d = 4n - 3(1 + p_g);$$

(ii) pour tout $i > 0$ on a $\dim M_{n-i} \times S^i X < d$;

(iii) il n'existe pas de fibré inversible sur X de classe de Chern $u \in \mathrm{H}^2(X)$ orthogonale à $\ell = c_1(L)$ pour la forme d'intersection et telle que

$$-n \leq u^2 < 0.$$

D'après Donaldson [3], les assertions concernant la dimension dans (i) et (ii) peuvent toujours être réalisées pour n assez grand, quitte au besoin à modifier la polarisation pour obtenir la condition (iii). La normalité est un résultat de J. Li [18]. La condition (iii) implique que s'il existe dans M_n des points strictement semi-stables représentés par des faisceaux F s'insérant dans une suite exacte

$$0 \longrightarrow F' \longrightarrow F \longrightarrow F'' \longrightarrow 0$$

telle que $p_{F'} = p_F$ alors la classe $F - 2F'$ dans $K_{\mathrm{num}}(X)$ est de dimension 0. Il en résulte en particulier que toute classe $v \in K(X)$ qui est orthogonale à F est aussi orthogonale à F' et F''.

On définit de manière naturelle une application $\mathbb{Z}$-linéaire

$$\mu : \mathrm{H}_2(X, \mathbb{Z}) \longrightarrow \mathrm{H}^2(M_n, \mathbb{Z}).$$

Supposons d'abord n impair; alors tous les points sont stables, et il existe un faisceau universel $\mathcal{F}$ paramétré par M_n. Alors on définit pour $u \in \mathrm{H}_2(X, \mathbb{Z})$

$$\mu(u) = (c_2 - \frac{c_1^2}{4})(\mathcal{F})/u$$

où le membre de droite est le slant-product. Cette classe $\mu(u)$ est indépendante du choix du faisceau universel. Le polynôme de Donaldson de degré d est alors défini par

$$q_{d,H}^{alg}(u) = \int_{M_n} \mu(u)^d$$

pour $u \in \mathrm{H}_2(X, \mathbb{Z})$.

Quand n est pair, il n'y a plus de faisceau universel. On introduit la variété algébrique $\widehat{X}$ éclatée de X en un point, convenablement polarisée: si E est le diviseur exceptionnel, on choisit la polarisation $\widehat{H} = mH - E$ pour $m >> 0$. On désigne par e la classe fondamentale du diviseur exceptionnel E; il y a lieu d'introduire l'espace de modules $\widehat{M}_{e,n}$ des classes de faisceaux semi-stables F sur $\widehat{X}$ de classes de Chern $c_1(F) = e, c_2(F) = n$, pour lequel il existe cette fois un faisceau universel. Cet espace de modules est de dimension $d+1$ et on peut définir un polynôme de Donaldson homogène de degré $d+1$ associé $\widehat{q}_{e,d+1}^{alg}$, en utilisant la même formule que ci-dessus pour définir μ, relative à l'espace de modules $\widehat{M}_{n,e}$. L'espace $\mathrm{H}_2(X)$ est considéré comme l'orthogonal de e dans $\mathrm{H}_2(\widehat{X})$ pour la forme d'intersection. On pose alors pour $u \in \mathrm{H}_2(X)$

$$q_{d,H}(u) = \int_{\widehat{M}_{e,n}} \mu(e)\mu(u)^d.$$

Quand $u \in \mathrm{H}_2(X, \mathbb{Z})$ est une classe algébrique, il est facile d'interpréter les polynômes de Donaldson en termes de fibré déterminant. On peut trouver une classe $v \in K(X)$ de dimension 1, orthogonale à c, dont la classe fondamentale est u. A une telle classe est associée un fibré déterminant $\mathcal{D}_v$, caractérisé par la propriété universelle suivante: soit $\mathcal{F}$ une famille plate de faisceaux semi-stables de classe c paramétrée par une variété algébrique S, et soit $f_{\mathcal{F}} : S \to M_n$ le morphisme modulaire. Considérons les deux projections canoniques

$$\begin{array}{ccc} S \times X & \xrightarrow{p_2} & X \\ {\scriptstyle p_1}\downarrow & & \\ S & & \end{array}$$

et le produit $\mathcal{F}(v)$ de l'image inverse de v par $\mathcal{F}$ dans $K(S \times X)$; le fibré inversible sur S défini par

$$\lambda_{\mathcal{F}}(v) = \det(p_{1!}(\mathcal{F}(v)))$$

où $p_{1!}$ est le morphisme image directe en K-théorie a un sens [14], [16]. Alors

$$\lambda_{\mathcal{F}}(-v) = f_{\mathcal{F}}^*(\mathcal{D}_v).$$

Du fait que X est supposée simplement connexe, il résulte que ce fibré inversible $\mathcal{D}_v$ est indépendant du choix de v; quand n est impair, on a la formule

$$c_1(\mathcal{D}_v) = \mu(u).$$

Le fibré $\mathcal{D}_v$ peut encore être construit même quand il n'existe pas de fibré universel: le point important est que les points strictement semi-stables sont, d'après la condition (iii), obligatoirement les classes de faisceaux $F' \oplus F''$ avec

$$c(F') = c(F'') = 1 - \frac{n}{2}h^2$$

dans $K_{\text{num}}(X)$ et donc les classes $c(F')$ et $c(F'')$ sont encore orthogonales à v. La variété de modules M_n est obtenue en quotientant un ouvert R^{ss} d'un schéma de Hilbert Hilb$(H \otimes \mathcal{O}_X(-m), c)$ des quotients F de classes c, semi-stables, par l'action de $GL(H)$, où H représente un espace vectoriel de dimension $P(m)$, P désignant le polynôme de Hilbert, et m un entier suffisamment grand. Le faisceau inversible $\lambda_{\mathcal{F}}(v)$ est muni d'une action naturelle du groupe $GL(H)$; le stabilisateur des points d'orbites fermée, qui s'identifie au groupe des automorphismes du faisceau polystable correspondant à une telle orbite agit trivialement: ainsi, par le lemme de descente de Drézet, Narasimhan et Kempf [7], le faisceau se quotiente et fournit un fibré $\lambda_{M_n}(v)$ sur M_n; le dual $\mathcal{D}_v$ satisfait à la propriété universelle ci-dessus. On peut étendre la construction au cas de toutes les classes de cohomologie, même non algébriques, à condition d'introduire les opérateurs de Dirac et les homomorphismes associés en K-théorie topologique, mais nous ne servirons ici que des classes algébriques. Le polynôme de Donaldson de degré d est alors donné par la formule suivante, valable pour tout (n, H) satisfaisant aux conditions (i), (ii), (iii).

$$q_{d,H}^{alg}(u) = \int_{M_n} c_1(\mathcal{D}_v)^d.$$

Il est connu ([17], [20]) que lorsque $p_g \geq 1$, les polynômes ainsi définis coïncident avec les polynômes introduits par voie différentiable par Donaldson, à partir des équations de Yang et Mills; en particulier, ils ne dépendent pas du choix de la polarisation. Lorsque $p_g = 0$, ils dépendent en général de la polarisation.

5.2. Le cas du plan projectif

Pour tout $n \geq 2$, l'espace de modules M_n est une variété irréductible et normale de dimension $4n - 3$. Il en résulte que les conditions (i), (ii), et (iii) ci-dessus sont satisfaites. Si on prend pour u le générateur positif de $H_2(P_2)$, on obtient le nombre $q_{4n-3} = q_{4n-3}(u)$, qu'on appelle nombre de Donaldson; il peut donc se calculer en terme de fibré déterminant: le fibré déterminant $\mathcal{D}_n$ sur M_n qui permet le calcul de ce nombre est dans ce cas obtenu en prenant pour v la classe dans $K(P_2)$ du faisceau $\mathcal{O}_\ell(-1)$, où ℓ est une droite de P_2.

Considérons le morphisme

$$\gamma : M_n \to |\mathcal{O}_{\mathbf{P}_2^*}(n)|$$

qui associe à la classe d'un faisceau semi-stable F la courbe γ_F de ses droites de saut. Alors on a $\gamma^*(\mathcal{O}(1)) = \mathcal{D}_n$ [14]. Il est bien connu [2] que le morphisme γ est génériquement fini sur son image; si δ_n est le degré de $\gamma : M_n \to \gamma(M_n)$, on a donc

$$q_{4n-3} = \delta_n \deg(\gamma(M_n)).$$

En utilisant cette interprétation, on obtient $q_5 = 1$: en effet $\gamma = M_2 \to P_5$ est un isomorphisme. Pour $n = 3$, $\gamma : M_3 \to P_9$ est surjectif, et de degré 3; donc $q_9 = 3$. Pour $n = 4$, on démontre que $\delta_4 = 1$ et que $\gamma(M_4)$, hypersurface des quartiques de Lüroth dans l'espace projectif P_{14} des quartiques, est une hypersurface de degré 54 ([13], [19], [24]). Ainsi, $q_{13} = 54$. On ignore si $\delta_n = 1$ pour $n \geq 5$.

Soient $S^{n+1}(P_2)$ la puissance symétrique de P_2, $\mathcal{L}$ l'image réciproque du générateur positif de $\mathrm{Pic}(S^{n+1}(P_2)) \simeq \mathbb{Z}$ par le morphisme canonique

$$\mathrm{Hilb}^{n+1}(P_2) \to S^{n+1}(P_2).$$

On pose $\eta = c_1(\mathcal{L})$. On considère comme dans le corollaire 4.3 le fibré vectoriel $\mathcal{V}$ de rang $n+1$ sur $\mathrm{Hilb}^{n+1}(P_2)$. Le résultat suivant ramène le calcul des premiers nombres de Donaldson à un calcul de nombres d'intersection sur le schéma de Hilbert $\mathrm{Hilb}^{n+1}(P_2)$.

Théorème 5.1.
(i) *Pour $2 \leq n \leq 5$, on a*

$$q_{4n-3} = \int_{\mathrm{Hilb}^{n+1}(P_2)} \left(\frac{\eta}{2}\right)^{5-n} s_{3n-3}(\mathcal{V} \otimes \mathcal{L}).$$

(ii) *Pour $n = 6$, on a*

$$q_{21} = \frac{2}{5} \int_{\mathrm{Hilb}^{14}(P_2)} s_{14}(\mathcal{V} \otimes \mathcal{L}).$$

Dans cet énoncé, $s_j(\mathcal{V} \otimes \mathcal{L}) \in \mathrm{H}^{2j}(\mathrm{Hilb}(P_2))$ désigne la classe de Segré de degré j de $\mathcal{V} \otimes \mathcal{L}$. Pour $n = 5$ et 6 ce résultat a aussi été obtenu par Tyurin et Tikhomirov [23]; dans [9] nous présentons le même résultat, par une méthode sensiblement différente, plus géométrique. Le calcul du membre de droite sur le schéma de Hilbert a été mené a son terme par Ellingsrud et Strømme en utilisant la formule des résidus de Bott [8]. Ils obtiennent finalement $q_{17} = 2540; q_{21} = 233208$.

Remarque 5.2. Pour $n = 1$, le membre de droite a encore un sens, mais ce n'est plus un entier: on obtient en effet

$$\int_{\mathrm{Hilb}^2(P_2)} \left(\frac{\eta}{2}\right)^4 = \frac{3}{16}.$$

Démonstration. Considérons l'espace de modules $S_{\alpha,n}$ des classes d'équivalence de systèmes cohérents semi-stables de type $(c(1), 1)$. Soit w la classe de $\mathcal{O}_\ell(-2)$ dans $K(P_2)$. Si α n'est pas valeur critique, il existe un système cohérent universel $(\mathcal{V}, \mathcal{F})$, paramétré par $S_{\alpha,n}$, donc le fibré déterminant $\mathcal{D}_{\alpha,n} = \lambda_{\mathcal{F}}(-w)$ a un sens dans $\mathrm{Pic}(S_{\alpha,n})$, et, du fait que $\mathcal{F}_s$ est orthogonal à w, il résulte que ce fibré déterminant est indépendant du choix de cette famille universelle. En fait, ce fibré déterminant a encore un sens même si α est valeur critique: il suffit de constater qu'aux points strictement semi-stables, classes d'équivalence de systèmes cohérents polystables

$$\Lambda = \Lambda'' \oplus \Lambda''$$

si on écrit $\Lambda' = (\Gamma', F')$ et $\Lambda'' = (\Gamma'', F'')$ la classe $F - 2F'$ dans $K(P_2)$ est de dimension 0: ainsi, la classe w de $\mathcal{O}_\ell(-2)$ est encore orthogonale à F'. Dès lors, la construction du fibré déterminant peut se faire par descente à partir d'un fibré déterminant associé à une famille universelle paramétrée par le schéma qu'on quotiente par l'action d'un groupe linéaire $GL(H)$ pour obtenir $S_{\alpha,n}$: en effet, ce fibré est muni d'une action de $GL(H)$ et le stabilisateur de Λ, qui s'identifie encore au groupe des automorphismes de Λ, agit trivialement en un tel point. De plus, si on considère les morphismes associés à une valeur critique α

$$\begin{array}{ccc} S_{\alpha_-,n} & & S_{\alpha_+,n} \\ & p_-\searrow \quad \swarrow p_+ & \\ & S_{\alpha,n} & \end{array}$$

la propriété universelle du fibré déterminant $\mathcal{D}_{\alpha,n}$ montre que l'on a

$$p_-^*(\mathcal{D}_{\alpha,n}) = \mathcal{D}_{\alpha_-,n} \ ; \ p_+^*(\mathcal{D}_{\alpha,n}) = \mathcal{D}_{\alpha_+,n}.$$

Pour $n = 5$, le théorème 5.1 est une conséquence de la proposition suivante:

Proposition 5.3.

(i) *On suppose $n \geq 2$. L'intégrale*

$$I_{\alpha,n} = \int_{S_{\alpha,n}} c_1(\mathcal{D}_{\alpha,n})^{3n+2}$$

est indépendante de α.

(ii) *Pour $\epsilon < 1/2$, on a $I_{\epsilon,5} = q_{17}(P_2)$ et l'intégrale $I_{\alpha,n}$ est nulle pour $2 \leq n < 5$.*

(iii) *On a*

$$I_{\beta_-,n} = \int_{\mathrm{Hilb}^{n+1}(P_2)} s_{2n+2}(\mathcal{V} \otimes \mathcal{L}).$$

Démonstration. La fonction $\alpha \mapsto I_{\alpha,n}$ est évidemment constante sur tout intervalle qui évite les valeurs critiques. Les morphismes p_- et p_+ étant birationnels, on obtient

$$I_{\alpha_-,n} = I_{\alpha,n} = I_{\alpha_+,n}.$$

D'où l'assertion (i). Pour l'assertion (ii), on voit de même, par la propriété universelle du fibré $\mathcal{D}_n$ que l'on a $f^*(\mathcal{D}_n) = \mathcal{D}_{\epsilon,n}$. L'assertion (ii) résulte pour $n = 5$ du fait que f est birationnel. Pour $n < 5$, $c_1(\mathcal{D}_{\alpha,n})$ est l'image réciproque par f d'une classe de cohomologie sur M_n ; puisque $\dim M_n < \dim S_{n,\alpha} = 3n+2$, on obtient $c_1(\mathcal{D}_{\epsilon,n})^{3n+2} = 0$.

5.3. Calcul de $I_{\beta_-,n}$

On sait que l'espace de modules $S_{\beta_-,n}$ s'identifie au fibré en espaces projectifs $P(\mathcal{V})$. Sur $P(\mathcal{V}) \times P_2$, on a une suite exacte universelle

$$0 \to \mathcal{O}(1,0) \to \mathcal{F} \to \mathcal{O}(0,2) \to \mathcal{O}_\Xi(0,2) \to 0.$$

Il en résulte

$$\lambda_{\mathcal{F}}(-w) = \mathcal{O}_{P(\mathcal{V})}(1) \otimes \lambda_{\mathcal{O}_\Xi}(\mathcal{O}_\ell)$$

où le fibré inversible $\lambda_{\mathcal{O}_\Xi}(\mathcal{O}_\ell)$ est le fibré déterminant associé à la famille plate de faisceaux $\mathcal{O}_\Xi$ (*cf.* [16]); c'est exactement le fibré $\mathcal{L}$ image réciproque du générateur positif de $\mathrm{Pic}(S^{n+1}(P_2))$ par le morphisme canonique $\mathrm{Hilb}^{n+1}(P_2) \to S^{n+1}(P_2)$. Par suite, $\mathcal{D}_{\beta_-,n} = \mathcal{O}_{P(\mathcal{V}\otimes\mathcal{L})}(1)$. On a donc

$$\begin{aligned} I_{\beta_-,n} &= \int_{P(\mathcal{V}\otimes\mathcal{L})} (c_1(\mathcal{O}_{P(\mathcal{V}\otimes\mathcal{L})}(1))^{3n+2} \\ &= \int_{\mathrm{Hilb}^{n+1}(P_2)} s_{2n+2}(\mathcal{V} \otimes \mathcal{L}). \end{aligned}$$

D'où l'énoncé pour $n = 5$. □

5.4. Le cas $n = 6$

Considérons le sous-schéma de Brill-Noether de M_6 défini par $h^0(F(1)) \neq 0$. On sait que cette variété est une hypersurface irréductible et réduite (*cf.* théorème 4.7, [10]), de fibré inversible associé $\mathcal{A} = \lambda_{M_6}(-a)$, où $a = (1-h)^{-1}$ est la classe dans $K(P_2)$ de $\mathcal{O}_{P_2}(1)$; ce fibré déterminant a un sens puisque a est orthogonale à c. Notons $\mathcal{D} = \mathcal{D}_6$ le fibré détermiant de Donaldson sur M_6. Le même argument que ci-dessus fournit les égalités

$$\begin{aligned} I_{\alpha,6} &= \int_{\mathrm{Hilb}^7(P_2)} s_{14}(\mathcal{V} \otimes \mathcal{L}) \\ &= \int_{M_6} c_1(\mathcal{A})(c_1(\mathcal{D}))^{20}. \end{aligned}$$

Pour évaluer la dernière intégrale, nous allons utiliser comme dans [9] le fait que la frontière ∂M_6 est une hypersurface dont l'image par le morphisme γ est un fermé de dimension < 20. On a donc

$$\int_{\partial M_6} c_1(\mathcal{D})^{20} = 0.$$

D'autre part, on sait que le groupe de Picard de M_6 est un groupe abélien libre [5] avec pour base $\mathcal{A}$ et $\mathcal{D}$.

Lemme 5.4. *Dans* $\mathrm{Pic}(M_6)$, *noté additivement, on a* $\partial M_6 \equiv 5\mathcal{D} - 2\mathcal{A}$

Il en résulte que

$$\begin{aligned} q_{21} &= \frac{2}{5} \int_{M_6} c_1(\mathcal{A})(c_1(\mathcal{D})^{20} \\ &= \int_{\mathrm{Hilb}^7(P_2)} s_{14}(\mathcal{V} \otimes \mathcal{L}). \end{aligned}$$

Ceci achève la preuve du théorème 5.1 dans le cas $n = 6$. □

Démonstration du lemme 5.4. Ce lemme est démontré dans [9] en utilisant la formule de Riemann-Roch. La variante présentée ici utilise une fois de plus l'espace de modules des systèmes cohérents $S_{\alpha,n}$. Considérons le morphisme $f : S_{\epsilon,6} \to M_6$. Si α n'est pas valeur critique, on a évidemment $\mathrm{Pic}(S_{\alpha,6}) = \mathrm{Pic}(S_{\beta_-,6}) = \mathbb{Z}^3$. On a donc un morphisme $f^* : \mathrm{Pic}(M_6) \to \mathrm{Pic}(P(\mathcal{V}))$. Considérons la suite exacte universelle sur $P(\mathcal{V}) \times P_2$:

$$0 \to \mathcal{O}(1,-1) \to \mathcal{F} \to \mathcal{O}(0,1) \to \mathcal{O}_{\Xi}(0,1) \to 0$$

où $\mathcal{F}$ désigne ici une famille de faisceaux cohérents de classe c. Posons $\xi = \mathcal{O}_{P(\mathcal{V})}(1)$, et $\eta = \mathcal{L}$. On a dans $\mathrm{Pic}(P(\mathcal{V}))$, en notation additive:

$$\begin{aligned} f^*(\mathcal{D}) &= -\lambda_{\varphi}(\mathcal{O}_{\ell}(-1)) = \xi + \eta \\ f^*(\mathcal{A}) &= -\lambda_{\mathcal{F}}(\mathcal{O}_{P_2}(1)) = -\xi + \lambda_{\mathcal{O}_{\Xi}}(0,2) \\ &= -\xi + \det \mathcal{V} + 3\eta. \end{aligned}$$

Désignons par ℓ une droite fixée de P_2. Fixons 6 points distincts hors de cette droite; en faisant varier le dernier point sur ℓ, on obtient une inclusion

$$\ell \hookrightarrow \mathrm{Hilb}^7(P_2)$$

et on a évidement $\mathcal{V}|_{\ell} = \mathcal{O}_{\ell}^6 \oplus \mathcal{O}_{\ell}(-1)$. Le groupe de Picard de $W = P(\mathcal{V}|_{\ell})$ est $\mathbb{Z}^2$ avec pour générateurs les images réciproques de $\eta = \mathcal{L}$ et de $\xi = \mathcal{O}_{P(\mathcal{V})}(1)$, qui seront notées de la même façon. Dans $\mathrm{Pic}(W)$, on a la relation $\eta = -\det \mathcal{V}$, de sorte que l'image réciproque de $\mathcal{A}$ est $-\xi + 2\eta$. Il en résulte que le morphisme

$$\mathrm{Pic}(M_6) \to \mathrm{Pic}(W)$$

est injectif et que son image est un sous-groupe d'indice 3.

Considérons maintenant la frontière. Elle correspond dans W aux extensions qui fournissent des faisceaux singuliers, *i.e.* non localement libres. Soit $\pi : W \to \ell$ la projection canonique. L'hypersurface $f^*(\partial M_6)$ rencontre W suivant une courbe qui s'obtient en écrivant que sur W l'une des composantes du morphisme canonique $\mathcal{O}_\ell^6 \oplus \mathcal{O}_\ell(-1) = \mathcal{V}|_\ell \to \xi$ est nulle. On obtient ainsi dans Pic(W)

$$\begin{aligned} f^*(\partial M) &\equiv 7\xi + \eta \\ &\equiv 5\mathcal{D} - 2\mathcal{A} \end{aligned}$$

ce qui achève la démonstration.

Remarque 5.5. Pour n impair quelconque, le groupe de Picard est encore un groupe abélien libre, qui a pour base $(\mathcal{D}, \mathcal{B})$ où $\mathcal{B}$ désigne le fibré inversible défini par

$$\mathcal{B} := -\lambda_{M_n}(2 + 2h + (n-4)h^2).$$

Ce fibré inversible a un sens car la classe $2 + 2h + (n-4)h^2$ est orthogonale à c dans $K(P_2)$. Le même argument que ci-dessus montre que dans le groupe de Picard $\partial M_n \equiv 5\mathcal{D} - \mathcal{B}$. Ce résultat peut s'obtenir aussi à partir du résultat de Strømme [21]. Pour n pair, le fibré $\mathcal{B}$ a encore un sens, mais cette fois est divisible par 2: on a $\mathcal{B} = 2\mathcal{A}$; alors $(\mathcal{D}, \mathcal{A})$ est une base du groupe de Picard, et on a comme pour $n = 6$

$$\partial M_n \equiv 5\mathcal{D} - 2\mathcal{A}.$$

5.5. Le cas $n \leq 4$

Si α n'est pas valeur critique, il existe une famille universelle de systèmes cohérents $\Lambda_\alpha = (\mathcal{W}_\alpha, \mathcal{F}_\alpha)$ paramétrée par $S_{\alpha,n}$. Soit $u = h - h^2$ la classe du faisceau $\mathcal{O}_\ell(-1)$, où ℓ est une droite de P_2. Sur $S_{\alpha,n}$ le fibré inversible

$$\mathcal{A}_{\alpha,n} = \lambda_{\mathcal{F}_\alpha}(u) \otimes \mathcal{W}_\alpha^{\otimes -2}$$

a un sens et est indépendant du choix de cette famille universelle. Dans la suite, on oublie parfois l'indice n pour alléger les notations.

Lemme 5.6. *Sur la fibre générique P_{5-n} de $f : S_{\epsilon,n} \to M_n$ on a $\mathcal{A}_\epsilon|_{P_{5-n}} = \mathcal{O}(2)$.*

Démonstration. Chaque fois qu'on se donne une famille plate $(\mathcal{W}', \mathcal{F}')$ de systèmes cohérents α-semi-stables de type $(c(1), 1)$ paramétrée par une variété algébrique S, on dispose d'un morphisme modulaire $\varphi : S \to S_{\alpha,n}$ associé à cette famille, et on a

$$\varphi^*(\mathcal{A}_\alpha) = \lambda_{\mathcal{F}'}(u) \otimes (\mathcal{W}')^{\otimes -2}.$$

Considérons un faisceau stable F de classe $c(1)$, tel que $h^0(F) = 6 - n$. Alors l'espace projectif $P(\mathrm{H}^0(F))$ paramètre une famille plate de systèmes cohérents $(\mathcal{O}(-1), F)$, ϵ-stable, et le morphisme modulaire associé $P(\mathrm{H}^0(F)) \to S_{\epsilon,n}$ identifie l'espace projectif P_{n-5} à la fibre de f au-dessus du point de M_n défini par F. L'énoncé découle donc trivialement de la propriété universelle qu'on vient de décrire. □

On considère si α n'est pas valeur critique, et pour tous entiers (i, j) tels que $i + j = 3n + 2, i \geq 0, j \geq 0$, l'intégrale

$$J_{\alpha,n}(i,j) = \int_{S_{\alpha,n}} c_1(\mathcal{A}_{\alpha,n})^i c_1(\mathcal{D}_{\alpha,n})^j.$$

Proposition 5.7. *Supposons $2 \leq n \leq 4$, et $0 \leq i \leq 5-n$. Alors l'intégrale $\alpha \to J_{\alpha,n}(i,j)$, définie en dehors des valeurs critiques, est constante.*

Démonstration. Cet énoncé a déjà été démontré pour $i = 0$. Supposons que α soit valeur critique; désignons par Λ_- et Λ_+ les familles universelles paramétrées par $S_{\alpha_-,n}$ et $S_{\alpha_+,n}$ respectivement. Considérons le diagramme du théorème 4.4:

$$\begin{array}{ccc} & \widetilde{S}_{\alpha,n} & \\ \pi_- \swarrow & & \searrow \pi_+ \\ S_{\alpha_-,n} & & S_{\alpha_+,n} \end{array}$$

et désignons par E_α le diviseur exceptionnel dans $\widetilde{S}_{\alpha,n}$. On a alors une suite exacte de familles de systèmes cohérents paramétrés par $\widetilde{S}_{\alpha,n}$

$$0 \to \pi_+^*(\Lambda_+) \to \pi_-^*(\Lambda_-) \to \Lambda'' \to 0$$

où Λ'' est en fait une famille, paramétrée par E_α, de systèmes cohérents $(0, F'')$ où $F'' = I''(1)$ et où I'' l'idéal d'un sous-schéma fini de longueur $c_2(F'') = \frac{1}{2}(n-\alpha)$. Il en résulte, pour les images réciproques sur $\widetilde{S}_{\alpha,n}$

$$\pi_-^*(\mathcal{A}_{\alpha_-}) = \pi_-^*(\mathcal{A}_{\alpha_+}) + E_\alpha.$$

Désignons par e_α la classe fondamentale de E_α. On a alors

$$\begin{aligned} J_{\alpha_-,n}(i,j) &= \int_{S_{\alpha,n}} (c_1(\mathcal{A}_{\alpha_+}) + e_\alpha)^i c_1(\mathcal{D}_\alpha)^j \\ &= J_{\alpha_+}(i,j) + \sum_{k=1}^{i} C_i^k \int_{E_\alpha} e_\alpha^{k-1} c_1(\mathcal{A}_{\alpha_+})^{i-k} c_1(\mathcal{D}_\alpha)^j. \end{aligned}$$

Rappelons que $E_\alpha \to \Sigma_+$ est un fibré projectif localement trivial dont les fibres sont de dimension $n-2$. Il en résulte que dans la dernière somme, le terme d'indice k est nul si $k < n-1$. Alors

— Si $n = 4$, on a $5 - n < n - 1$, et par conséquent $J_{\alpha_-,4}(i, 14-i) = J_{\alpha+,4}(i, 14-i)$ pour $i \leq 1$. Ainsi, $\alpha \mapsto J_{\alpha,n}(i, 14-i)$ reste constante.

— Pour $n = 3$, et $i \leq 2$ le terme d'indice $k = 1$ est nul, ce qui donne le résultat si $i = 1$. Si $i = 2$, on obtient

$$\begin{aligned} J_{\alpha_-,3}(2,9) &= J_{\alpha_+,3} + \int_{E_\alpha} e_\alpha c_1(\mathcal{D}_\alpha)^9 \\ &= J_{\alpha_+,3}(2,9) + \int_{\Sigma_+} c_1(\mathcal{D}_\alpha)^9 \end{aligned}$$

Remarquons encore que Σ_+ est encore l'image inverse du fermé Σ des points strictement semi-stables de $S_{\alpha,n}$, et que l'on a

$$\dim \Sigma_+ - \dim \Sigma = \frac{1}{2}(n+\alpha) + 1 > 0.$$

Puisque $\mathcal{D}_\alpha$ provient de Σ, on voit que la dernière intégrale est nulle. On a donc encore $J_{\alpha_+,3}(i,j) = J_{\alpha_-,3}(i,j)$.

— Pour $n = 2$, il y a une seule valeur critique $\alpha = 2$. Alors $E_\alpha = \Sigma_+$ et $\pi_- : E_\alpha \to \Sigma_-$ est un fibré en espaces projectifs de dimension 3. On peut renverser les rôles de Σ_+ et Σ_- :

on a $\pi\,{}^*_-(\mathcal{A}_{\alpha_+}) = \pi\,{}^*_-(\mathcal{A}_{\alpha_-}) - E_\alpha$ et par conséquent

$$J_{\alpha_+,2}(i,j) = \int_{S_{\alpha,2}} (c_1(\mathcal{A}_{\alpha_-}) - e_\alpha)^i c_1(\mathcal{D}_\alpha)^j$$
$$= J_{\alpha_-,n}(i,j) + \sum_{k=1}^{i}(-1)^k C_i^k \int_{E_\alpha} e_\alpha^{k-1} c_1(\mathcal{A}_{\alpha_-})^{i-k} c_1(\mathcal{D}_\alpha)^j.$$

Si $i \leq 3$, on a $k-1 < 3$, et on voit que tous les termes de la somme ci-dessus sont nuls. Par suite $J_{\alpha_+,2}(i,j) = J_{\alpha_-,2}(i,j)$. Ceci achève la démonstration. □

Corollaire 5.8.
(i) *Pour* $2 \leq n \leq 4$ *et* $i < 5-n$, *on a* $J_{\alpha,n}(i,j) = 0$.
(ii) *Pour* $2 \leq n \leq 5$ *on a* $J_{\alpha,n}(5-n, 4n-3) = 2^{5-n} q_{4n-3}(P_2)$.

Démonstration. Remarquons que pour $i = 0$, cet énoncé se réduit à l'assertion (ii) de la proposition 5.3. On peut d'après l'énoncé ci-dessus supposer que $\alpha = \epsilon$. Puisque $\mathcal{D}_{\epsilon,n} = f^*(\mathcal{D}_n)$, on peut écrire

$$J_{\epsilon,n}(i,j) = \int_{M_n} f_*(c_1(\mathcal{A}_{\epsilon,n})^i) c_1(\mathcal{D}_n)^j$$

mais si $i < 5-n$, l'image directe $f_*(c_1(\mathcal{A}_{\epsilon,n})^i)$ est nulle. Ceci démontre l'assertion (i). Pour l'assertion (ii), on voit compte-tenu du lemme 5.6, que $f_*(c_1(\mathcal{A}_{\epsilon,n})^{5-n}) = 2^{5-n}$. D'où l'énoncé. □

5.6. Calcul de $J_{\beta_-,n}(i,j)$, pour $2 \leq n \leq 4$ et $i \leq 5-n$

On considère ici une valeur $\alpha = \beta_-$ proche de β. Considérons la suite exacte universelle

$$0 \to \mathcal{O}(1,-1) \to \mathcal{F} \to \mathcal{O}(0,1) \to \mathcal{O}_\Xi(0,1) \to 0$$

sur $P(\mathcal{V}) \times P_2$. Soit u la classe de $\mathcal{O}_\ell$ dans $K(P_2)$. On a $\lambda_\mathcal{F}(u) = -\lambda_{\mathcal{O}_\Xi}(v) = \mathcal{L}^{-1}$. La famille universelle $(\mathcal{W}_\alpha, \mathcal{F}(0,1))$ paramétrée par $S_{\beta_-,n}$ est donnée par $\mathcal{W}_\alpha = \mathcal{O}(1) \subset pr_{1*}(\mathcal{F}(0,1))$. Posons $\xi = c_1(\mathcal{O}_{P(\mathcal{V})}(1))$ et $\eta = c_1(\mathcal{L})$. On a donc,

$$c_1(\mathcal{A}_{\alpha,n}) = -(\eta + 2\xi)$$
$$= -2c_1(\mathcal{D}_{\alpha,n}) + \eta.$$

Il en résulte

$$\int_{\mathrm{Hilb}^{n+1}(P_2)} \eta^i s_{2n+2-i}(\mathcal{V}\otimes\mathcal{L}) = \sum_{k=0}^{i} 2^{k-i} C_i^k \int_{S_{\beta_-,n}} c_1(\mathcal{A}_{\alpha,n})^k c_1(\mathcal{D}_{\alpha,n})^{3n+2-k}$$
$$= \sum_{k=0}^{i} 2^{k-i} C_i^k J_{\alpha,n}(k, 3n+2-k).$$

Il résulte du corollaire 5.8 ci-dessus que cette somme est nulle si $i < 5-n$, et vaut $J_{\alpha,n}(5-n, 4n-3)$ pour $i = 5-n$. Ceci entraîne l'assertion (i) du théorème 5.1. En outre, on obtient l'énoncé suivant qui étend au cas $2 \leq n \leq 4$ le résultat obtenu pour $n = 5$ dans la proposition 5.3:

Corollaire 5.9. *Pour* $2 \leq n \leq 5$ *et* $0 \leq i < 5-n$, *on a*

$$\int_{\mathrm{Hilb}^{n+1}(P_2)} \eta^i s_{2n+2-i}(\mathcal{V}\otimes\mathcal{L}) = 0.$$

RÉFÉRENCES

[1] E. Arbarello, M. Cornalba, P. A. Griffiths et J. Harris: *Geometry of Algebraic Curves*, Springer-Verlag, (1985).

[2] W. Barth: Moduli of vector bundles on projective plane, Inventiones math. **42** (1977), 63–91.

[3] S. Donaldson: Polynomial invariants for smooth 4-manifolds,: Topology **29** (1990), 257–315.

[4] S. K. Donaldson et P. B. Kronheimer: *The Geometry of Four-Manifolds,* Oxford University Press, (1990).

[5] J. M. Drezet: Groupe de Picard des variétés de modules de faisceaux semi-stables sur $P_2(\mathbb{C})$, Ann. de l'Institut Fourier **38** (1988), 105–168.

[6] J. M. Drezet et J. Le Potier: Fibrés stables et fibrés exceptionnels sur le plan projectif, Ann. scient. Ec. Norm. Sup. 4e série **18** (1985), 193–244.

[7] J. M. Drezet et M. S. Narasimhan: Groupe de Picard des variétés de modules de fibrés semi-stables sur les courbes algébriques, Invent. math. **97** (1989, 53–94.

[8] G. Ellingsrud et S. A. Strømme: Bott's formula and enumerative geometry,: A paraître dans le Journal de l'AMS (1995).

[9] G. Ellingsrud, J. Le Potier et S. A. Strømme: Some Donaldson invariants of CP^2, In these Proceedings, (1995).

[10] Göttsche et A. Hirschowitz: Exposé à la conférence annuelle Europroj (Catania), (1993).

[11] M. He: Thèse (en préparation).

[12] K. Hulek et J. Le Potier: Sur l'espace de modules des faisceaux semi-stables de rang 2, de classes de Chern (0,3) sur P_2, Annales de l'Institut Fourier **39** (1989), 251–292.

[13] J. Le Potier: Fibrés stables sur le plan projectif et quartiques de Lüroth, Exposé au Séminaire de Géométrie algébrique de Jussieu (1989).

[14] J. Le Potier: Fibré déterminant et courbes de saut sur les surfaces algébriques, in Complex Projective Geometry, Proceedings de la Conférence de Bergen (1989), London Mathematical Society (1992), 213–240.

[15] J. Le Potier: Systèmes cohérents et structures de niveau, Astérisque **214** (1993).

[16] J. Le Potier: Faisceaux semi-stables et systèmes cohérents, Proceedings de la Conference de Durham, Cambridge University Press (1995), 179–239.

[17] J. Li: Algebraic geometric interpretation of Donaldson's polynomial invariants of algebraic surfaces, J. Differ. Geometry **37** (1993), 417–466.

[18] J. Li: Kodaira dimension of moduli space of vector bundles on surfaces, Inventiones Math. **115** (1994), 1–40.

[19] W.-P. Li et Z. Qin: Lower-degree Donalson polynomial invariants of rational surfaces, J. Algebraic Geometry **2** (1993), 413–442.

[20] J. Morgan: Comparison of the Donaldson polynomial invariant with their algebrico-geometric analogues, Topology **32**, 3 (1993), 449–488.

[21] S. A. Strømme: Ample divisors on fine moduli spaces on the projective plane, Math. Z. **187** (1984), 405–423.

[22] M. Thaddeus: Stable pairs, linear systems and the Verlinde formula, Preprint (1992).

[23] A. Tikhomirov et A. N. Tyurin: Application of geometric approximation procedure to computing the Donaldson's polynomials for CP^2,: Mathematica Gottingensis Schriftenreihe des Sonderforschungsbereichs Geometrie und Analysis **12** (1994), 1–71

[24] A. N. Tyurin: The moduli spaces of vector bundles on threefolds, surfaces and curves, Preprint Erlangen (1990).

PICARD GROUPS OF THE MODULI SPACES OF VECTOR BUNDLES OVER ALGEBRAIC SURFACES

Jun Li

Mathematics Department, University of California at Los Angeles, CA 90024,USA.
e-mail:jli@math.ucla.edu

The purpose of this note is to determine the Picard group of the moduli space of vector bundles over an arbitrary algebraic surface. Since Donaldson's pioneer work on using moduli of vector bundles to define smooth invariants of an algebraic surface, there has been a surge of interest in understanding the geometry of this moduli space. Among other things, the study of line bundles on this moduli space plays a major role in this area. One important question remain open until now is to determine the Picard group of this moduli space. This is known in some special cases, for instance for projective plane [St], ruled surfaces [Qi2, Yo] and K3 surfaces [GH]. In this note, we will settle this question by providing a general construction of line bundles that will include virtually all line bundles on this moduli space, when the second Chern class of the sheaves parameterized by this moduli space is sufficiently large. The construction is again based on Knudsen and Mumford's recipe of determinant line bundle construction. The new input from this note is that instead of using complexes on surface X we will use complexes on $X \times X$, which yields all previously known line bundles as well as new ones. After this construction, we will use our knowledge of the first two Betti numbers of this moduli space to argue that this construction contains virtually all line bundles on this moduli space. The proof relies heavily on the Grothendieck-Riemann-Roch theorem and the knowledge of the singularities of the moduli space.

Acknowledgment The author would like to thank J. Kollár and K. Yoshioka for valuable discussion.

1. Construction of line bundles on moduli spaces

We begin with a general construction of line bundles on moduli space of stable sheaves based on Knudsen and Mumford's determinant line bundle construction. One way of getting line bundles on the moduli space $\mathfrak{M}$ is by forming (perfect) complexes on $\mathfrak{M} \times X$ out of the universal sheaf $\mathcal{F}$ and then take the determinant of its push-forward on $\mathfrak{M}$. This has been worked out by many people in various settings. It is interesting to observe that we can also construct complexes on $\mathfrak{M}$ that are push-forwards of complexes on $\mathfrak{M} \times X \times X$ constructed based on $\pi_{12}^*\mathcal{F} \otimes \pi_{13}^*\mathcal{F}$, where $\pi_{12}, \pi_{13} : \mathfrak{M} \times X \times X \to \mathfrak{M} \times X$ are projections. In this way, we recover the previously known line bundles as well as a group of new line bundles. We will work out the detail of this construction in this section.

This work was partially supported by NSF grant DMS-9307892 and A. Sloan fellowship.

We begin with some word on the convention that will be used throughout this paper. We let X be a fixed smooth algebraic surface over complex numbers $\mathbb{C}$ and H a very ample line bundle on X. For any choice of $I \in \mathrm{Pic}(X)$ and $d \in \mathbb{Z}(\cong H^4(X;\mathbb{Z}))$, we form the (coarse) moduli scheme of H-stable rank 2 sheaves $\mathcal{E}$ satisfying $\det \mathcal{E} \cong I$ and $c_2(\mathcal{E}) = d$, which will be denoted by $\mathfrak{M}_H(I,d)$. (We will abbreviated it to $\mathfrak{M}(I,d)$ when the choice of H is understood.) We will denote by $\overline{\mathfrak{M}}(I,d)$ the (coarse) moduli scheme of H-semistable sheaves satisfying the same restrain, modulo certain equivalence relation. Following Gieseker [Gi1], $\mathfrak{M}(I,d)$ is quasi-projective and $\overline{\mathfrak{M}}(I,d)$ is projective. Throughout this paper, we will use Roman letters to denote vector bundles and use Calligraphic letters to denote sheaves. We denote by A^iX (resp. A_iX) the Chow cohomology (resp. homology) group of X. For any sheaf, we denote by $\tilde{c}_i$ the Chern class taking value in the Chow cohomology group and by c_i the Chern class taking value in the (ordinary) cohomology group. We denote by ch the Chern character taking value in $A^*_{\mathbb{Q}}$, the Chow cohomology group with rational coefficient. When we assign a letter, say π, to denote projection from a product space to its factor(s), we will use π_i to denote projection to the i-th copy and use π_{ij} to denote projection to the produce of i-th and j-th copies. Sometimes we will simply use, say π_X, to denote projection to the factor X.

Now we construct line bundles on $\mathfrak{M}(I,d)$. For the moment, we assume $\mathfrak{M}(I,d)$ admits a universal family, namely a sheaf $\mathcal{F}$ on $\mathfrak{M}(I,d)\times X$ flat over $\mathfrak{M}(I,d)$ whose restriction to fiber $\{w\}\times X$ is isomorphic to the sheaf represented by $w \in \mathfrak{M}(I,d)$. Note that the choice of universal family is not unique. Any two such families differ by tensoring a pullback invertible sheaf on $\mathfrak{M}(I,d)$. Let K be the Grothendieck's group of $X\times X$ and let $\widetilde{K}$ be the kernel of $\xi: K \to \mathbb{Z}$,

$$\xi(C) = \chi(C \boxtimes \pi_1^*\mathcal{E} \boxtimes \pi_2^*\mathcal{E}) \tag{1.1}$$

for some $\mathcal{E} \in \mathfrak{M}(I,d)$ and $\pi_i : X\times X \to X$ projection. (Here and later, we will use $\boxtimes$ to denote the tensor product of complexes.) We let $p_\cdot$ be the projection from $\mathfrak{M}(I,d)\times X\times X \to \mathfrak{M}(I,d)$ to its factor(s) with self-explanatory index. Then we can define a homomorphism

$$\widetilde{K} \longrightarrow \mathrm{Pic}\big(\mathfrak{M}(I,d)\big) \tag{1.2}$$

as follows: Since the universal family $\mathcal{F}$ is flat over $\mathfrak{M}(I,d)$, $\mathcal{F}$ is a perfect complex on $\mathfrak{M}(I,d)\times X$. Then for any element in $\widetilde{K}$ represented by a complex C on $X\times X$, we can form the complex $p_{23}^*C \boxtimes p_{12}^*\mathcal{F} \boxtimes p_{23}^*\mathcal{F}$ and the push forward complex

$$\big(p_1\big)_!\big(p_{23}^*C \boxtimes p_{12}^*\mathcal{F} \boxtimes p_{23}^*\mathcal{F}\big).$$

Following [KM], we can define the determinant line bundle

$$\det\Big(\big(p_1\big)_!\big(p_{23}^*C \boxtimes p_{12}^*\mathcal{F} \boxtimes p_{23}^*\mathcal{F}\big)\Big). \tag{1.3}$$

The homomorphism (1.2) is the one sending $C \in \widetilde{K}$ to the line bundle (1.3). Observe that if we replace $\mathcal{F}$ by a different universal family, say $\mathcal{F}' = \mathcal{F}\otimes\pi_1^*\mathcal{O}(L)$, $\pi_1 : \mathfrak{M}(I,d)\times X \to \mathfrak{M}(I,d)$, then

$$\det\Big(\big(p_1\big)_!\big(p_{23}^*C \boxtimes p_{12}^*\mathcal{F}' \boxtimes p_{23}^*\mathcal{F}'\big)\Big) = L^{\otimes(2\xi(C))} \otimes \det\Big(\big(p_1\big)_!\big(p_{23}^*C \boxtimes p_{12}^*\mathcal{F} \boxtimes p_{23}^*\mathcal{F}\big)\Big),$$

where $\xi(C)$ is as in (1.1). Hence (1.2) is independent of the choice of the universal family since $C \in \widetilde{K}$ and then the homomorphism (1.3) is well-defined.

We will see shortly that the homomorphism (1.2) is far from injective. In the following we will use the Grothendieck-Riemann-Roch theorem (abbreviate GRR theorem) to give an explicit description of the image of (1.2). Since $p_1 : \mathfrak{M}(I,d) \times X \times X \to \mathfrak{M}(I,d)$ is smooth,

$$\begin{aligned} \text{(1.4)} \qquad & \operatorname{ch}\Big(\big(p_1\big)_!\big(p_{23}^*C \boxtimes p_{12}^*\mathcal{F} \boxtimes p_{13}^*\mathcal{F}\big)\Big) \\ & \qquad = p_{1*}\Big(\operatorname{ch}\big(p_{23}^*C \boxtimes p_{12}^*\mathcal{F} \boxtimes p_{13}^*\mathcal{F}\big) \cdot p_{23}^* \operatorname{td}(X \times X)\Big), \end{aligned}$$

by [Fu, Example 18.3.10]. By choosing $C \in \widetilde{K}$, the summand in $A^0\big(\mathfrak{M}(I,d)\big)_{\mathbb{Q}}$ of (1.4) is trivial and the summand in $A^1\big(\mathfrak{M}(I,d)\big)_{\mathbb{Q}}$ reads

$$\begin{aligned} \text{(1.5)} \qquad & \tilde{c}_1\Big(\big(p_1\big)_!\big(p_{23}^*C \boxtimes p_{12}^*\mathcal{F} \boxtimes p_{13}^*\mathcal{F}\big)\Big) \\ & \qquad = \Big[p_{1*}\Big(\operatorname{ch}(p_{23}^*C \boxtimes p_{12}^*\mathcal{F} \boxtimes p_{13}^*\mathcal{F}) \cdot p_{23}^* \operatorname{td}(X \times X)\Big)\Big]_{[1]}. \end{aligned}$$

We will use $\big[\cdot\big]_{[i]}$ to denote its component in A^i. A simple calculation shows that it is of the form

$$\begin{aligned} &\Big[p_{1*}\Big(\big(p_{12}^*c_2(\mathcal{F}) + p_{13}^*c_2(\mathcal{F})\big) \cdot \alpha_1 + \big(p_{12}^*c_3(\mathcal{F}) + p_{13}^*c_3(\mathcal{F})\big) \cdot \alpha_2 \\ &\quad + p_{12}^*c_2(\mathcal{F}) \cdot p_{13}^*c_2(\mathcal{F}) \cdot \alpha_3 + \big(p_{12}^*c_2(\mathcal{F}) \cdot p_{13}^*c_3(\mathcal{F}) + p_{13}^*c_2(\mathcal{F}) \cdot p_{12}^*c_3(\mathcal{F})\big) \cdot \alpha_4\Big)\Big]_{[1]}, \end{aligned}$$

for some $\alpha_{\cdot} \in A^*(X \times X)_{\mathbb{Q}}$ based on type consideration and

$$\operatorname{ch}(\mathcal{F}) \equiv 2 - \tilde{c}_2(\mathcal{F}) + \frac{1}{2}\tilde{c}_3(\mathcal{F}) \mod \tilde{c}_1(\mathcal{F}).$$

Hence (1.5), which is the Chern class of (1.3), is contained in the linear span of the images

$$\text{(1.6)} \qquad p_{M*}\big(\tilde{c}_2(\mathcal{F}) \cdot p_X^*(\cdot)\big) : A^1(X) \longrightarrow A^1\big(\mathfrak{M}(I,d)\big)_{\mathbb{Q}},$$

$$\text{(1.7)} \qquad p_{M*}\big(\tilde{c}_3(\mathcal{F}) \cdot p_X^*(\cdot)\big) : A^0(X) \longrightarrow A^1\big(\mathfrak{M}(I,d)\big)_{\mathbb{Q}}$$

(p_M, p_X are projections of $\mathfrak{M}(I,d) \times X$) and

$$\text{(1.8)} \qquad p_{1*}\big(p_{12}^*\tilde{c}_2(\mathcal{F}) \cdot p_{13}^*\tilde{c}_2(\mathcal{F}) \cdot p_{23}^*(\cdot)\big) : A^1(X \times X) \longrightarrow A^1\big(\mathfrak{M}(I,d)\big)_{\mathbb{Q}}.$$

It is easy to check that the image of (1.6) is contained in the image of (1.8) when $4d > I^2$, which we assume in the sequel. Thus the Chern class of (1.3) is contained in the image of

$$\text{(1.9)} \qquad A^1\big(X \times X\big)_{\mathbb{Q}} \oplus \mathbb{Q} \longrightarrow A^1\big(\mathfrak{M}(I,d)\big)_{\mathbb{Q}}$$

that is a direct sum of (1.8) and (1.7).

We remark that since $\mathfrak{M}(I,d)$ is singular in general, the homomorphism given by the first Chern class

$$\tilde{c}_1 : \operatorname{Pic}\big(\mathfrak{M}(I,d)\big) \longrightarrow A^1\big(\mathfrak{M}(I,d)\big)_{\mathbb{Q}}$$

may have infinite kernel. However, (1.9) suggests that very likely there should be a homomorphism

$$\text{(1.10)} \qquad \Phi : \operatorname{Pic}\big(X \times X\big)^{\sigma} \oplus \mathbb{Z} \longrightarrow \operatorname{Pic}\big(\mathfrak{M}(I,d)\big)$$

compatible to (1.9) that will capture all (up to finite index) line bundles constructed in (1.3). Here $\sigma : X \times X \to X \times X$ is the map exchanging factor and $\operatorname{Pic}\big(X \times X\big)^{\sigma}$ consists of line bundles invariant under σ. We use $\operatorname{Pic}\big(X \times X\big)^{\sigma}$ instead of $\operatorname{Pic}\big(X \times X\big)$ in (1.10) since the map (1.2) so constructed is σ invariant.

We now construct this homomorphism explicitly. We first let

$$\psi : \mathrm{Pic}\Big(X \times X\Big)^{\sigma} \oplus \mathbb{Z} \longrightarrow \widetilde{K} \otimes_{\mathbb{Z}} \mathbb{Z}[\frac{1}{12}]$$

be the map sending $L \in \mathrm{Pic}\Big(X \times X\Big)^{\sigma}$ to

$$\Big(\mathcal{O}_{X\times X} - \mathcal{O}(L)\Big) + \frac{1}{2}\Big(\mathcal{O}_{X\times X} - \mathcal{O}(L)\Big)^2 - \frac{1}{3}\Big(\mathcal{O}_{X\times X} - \mathcal{O}(L)\Big)^3 + \alpha\mathbb{C}_{p_0} \in \widetilde{K} \otimes_{\mathbb{Z}} \mathbb{Q}$$

and sending $m \in \mathbb{Z}$ to

$$C_0 = m\mathcal{O}_{X\times\{x_0\}} + \alpha\mathbb{C}_{p_0} \in \widetilde{K} \otimes_{\mathbb{Z}} \mathbb{Q}\,,$$

where $p_0 \in X \times X$ is a fixed point and $\alpha \in \mathbb{Z}[\frac{1}{12}]$ is chosen so that $\psi(L)$ and $\psi(m)$ lie in $\widetilde{K} \otimes_{\mathbb{Z}} \mathbb{Z}[\frac{1}{12}]$. ($\alpha \in \mathbb{Z}[\frac{1}{12}]$ follows from L being σ-invariant.) Note that this choice gives

$$\mathrm{ch}\Big(\psi(L)\Big) = \tilde{c}_1(L) + \alpha[p_o]^{\vee}.$$

We then define $\Phi(\alpha)$, $\alpha \in \mathrm{Pic}\Big(X \times X\Big)^{\sigma} \oplus \mathbb{Z}$, to be the $\mathbb{Q}$-line bundle

$$\det\Big(\Big(p_1\Big)_!\Big(p_{23}^*\psi(L) \boxtimes p_{12}^*\mathcal{F} \boxtimes p_{13}^*\mathcal{F}\Big)\Big) \in \mathrm{Pic}\Big(\mathfrak{M}(I,d)\Big) \otimes_{\mathbb{Z}} \mathbb{Z}[\frac{1}{12}].$$

This way we have defined

$$\Phi : \mathrm{Pic}\Big(X \times X\Big)^{\sigma} \oplus \mathbb{Z} \longrightarrow \mathrm{Pic}\Big(\mathfrak{M}(I,d)\Big) \otimes_{\mathbb{Z}} \mathbb{Z}[\frac{1}{12}]. \tag{1.11}$$

Φ is a homomorphism, which can be proved similar to that on page 422 of [Li1] despite ψ is not necessarily a homomorphism. (One hint that this is true comes from the identity $\mathrm{ch}\Big(\psi(L)\Big) = \tilde{c}_1(L) + \alpha[p_o]^{\vee}$.) We shall omit the proof here since it is rather routine.

In the remainder of this section, we shall investigate how to define this homomorphism when there are no universal family on $\mathfrak{M}(I,d) \times X$ and under what condition does it extend to

$$\overline{\Phi} : \mathrm{Pic}\Big(X \times X\Big)^{\sigma} \oplus \mathbb{Z} \longrightarrow \mathrm{Pic}\Big(\overline{\mathfrak{M}}(I,d)\Big) \otimes_{\mathbb{Z}} \mathbb{Z}[\frac{1}{12}].$$

We will accomplish this by taking $\mathfrak{M}(I,d)$ as a GIT quotient of a Quot-scheme and then apply the descent theory. Following [Gi1, Ma], there is a Quot-scheme **Quot** with a linear $G = SL(N)$ action such that if we denote by $\mathbf{Quot}^s$ (resp. $\mathbf{Quot}^{ss}$) the open subset of G-stable (resp. G-semistable) points, then

$$\mathbf{Quot}^s/G = \mathfrak{M}(I,d) \quad \text{and} \quad \mathbf{Quot}^{ss}/\!/G = \overline{\mathfrak{M}}(I,d).$$

Let $\mathcal{E}$ be a universal quotient family on $\mathbf{Quot}^{ss} \times X$. Then we can define determinant line bundles on $\mathbf{Quot}^{ss}$ as we did for $\mathfrak{M}(I,d)$ by using the family $\mathcal{E}$ and complexes in $\widetilde{K}$. Let $\tilde{p}_{\cdot}$ be projections from $\mathbf{Quot}^{ss} \times X \times X$ to its factor(s) with self-explanatory index. Then for any $C \in \widetilde{K}$, we get a line bundle on $\mathbf{Quot}^{ss}$:

$$L(C) = \det\Big(\Big(\tilde{p}_1\Big)_!\Big(\tilde{p}_{23}^*C \ominus \tilde{p}_{12}^*\mathcal{E} \ominus \tilde{p}_{13}^*\mathcal{E}\Big)\Big). \tag{1.12}$$

$L(C)$ admits a canonical G-linearization since the G-action on $\mathbf{Quot}^{ss}$ lifts to a G-action on $\mathcal{E}$. By descent theory, $L(C)$ descends to $\mathfrak{M}(I,d)$ if for any $w \in \mathbf{Quot}^s$ the stabilizer $\mathrm{stab}_w \subseteq G$ of w acts trivially on the fiber of $L(C)$ over w. For the same reason, it descends to $\overline{\mathfrak{M}}(I,d)$ if similar condition holds for all $w \in \mathbf{Quot}^{ss}$ having closed orbits $G \cdot w$.

Definition 1.1. *An ample line bundle H is said to be (I,d)-generic if whenever $J \in Pic(X)$ satisfies $\Big(2c_1(J) - c_1(I)\Big) \cdot c_1(H) = 0$, then $-4c_1(J)^2 > 4d - c_1(I)^2$ unless $c_1(J)$ is torsion.*

Note that if H is (I,d)-generic, then any strictly semistable sheaf $\mathcal{F} \in \overline{\mathfrak{M}}(I,d)$ is S-equivalent to $\mathcal{F}_1 \oplus \mathcal{F}_2$ with $2c_1(\mathcal{F}_1) \equiv c_1(I)$ modulo torsions.

Proposition 1.2. *For any $C \in \widetilde{K}$, the line bundle $L(C)$ descends to $\mathfrak{M}(I,d)$. Further if H is (I,d)-generic, then it descends to $\overline{\mathfrak{M}}(I,d)$.*

Proof. For stable point $w \in \mathbf{Quot}^s$, the stabilizer stab_w is $\mathbb{Z}/N\mathbb{Z}$ acting on $\mathcal{E} \otimes k(w)$ via multiplication by an N-th root of unity, say ϵ. Here $\mathcal{E} \otimes k(w)$ is the restriction of $\mathcal{E}$ to the fiber $\{w\} \times X$. Then its induced action on $L(C)_w$ (which is the restriction to the fiber over w) is a multiplication by ϵ^m, where m is given by (1.1) that is 0 since $C \in \widetilde{K}$. Therefore by descent theory $L(C)$ descends to $\mathfrak{M}(I,d)$.

We now study descent problem on $\overline{\mathfrak{M}}(I,d)$. Let $w \in \mathbf{Quot}^{ss}$ be a strictly semistable point with closed orbit $G \cdot w$. Then $\mathcal{E} \otimes k(w)$ is a direct sum of rank one sheaves, say $\mathcal{F}_1 \oplus \mathcal{F}_2$. If we further assume H is (I,d)-generic, then $c_1(\mathcal{F}_1)$ and $c_1(\mathcal{F}_2)$ differ by a torsion element. In this case, it is straight forward to check, as did on page 426 of [Li1], that stab_w acts trivially on $L(C)_w$. Therefore, $L(C)$ descends to $\overline{\mathfrak{M}}(I,d)$ as claimed. □

Remark 1.3. *When H is not (I,d)-generic, one can determine explicitly which line bundles do descend by using the descent argument, if one knows all strictly semistable sheaves.*

Knowing that (1.12) always descends to $\mathfrak{M}(I,d)$, we can define a homomorphism

$$\Phi : \mathrm{Pic}\big(X \times X\big)^\sigma \oplus \mathbb{Z} \longrightarrow \mathrm{Pic}\big(\mathfrak{M}(I,d)\big) \otimes_{\mathbb{Z}} \mathbb{Z}[\frac{1}{12}]$$

that sends α in $\mathrm{Pic}\big(X \times X\big)^\sigma \oplus \mathbb{Z}$ to the descent of the determinant line bundle associated to the complex $\psi(\alpha)$. Similarly, if H is (I,d)-generic then the same construction yields a homomorphism

$$\overline{\Phi} : \mathrm{Pic}\big(X \times X\big)^\sigma \oplus \mathbb{Z} \longrightarrow \mathrm{Pic}\big(\overline{\mathfrak{M}}(I,d)\big) \otimes_{\mathbb{Z}} \mathbb{Z}[\frac{1}{12}]. \tag{1.13}$$

2. Injectivity of the homomorphism Φ

In this section, we will show that for sufficiently large d, the homomorphism

$$\Phi : \mathrm{Pic}\big(X \times X\big)^\sigma \oplus \mathbb{Z} \longrightarrow \mathrm{Pic}\big(\mathfrak{M}(I,d)\big) \otimes_{\mathbb{Z}} \mathbb{Z}[\frac{1}{12}]$$

has finite kernel. The method we will use is to construct a variety W and a morphism

$$g : W \to \mathfrak{M}(I,d) \tag{2.1}$$

and show that the composition

$$g^* \circ \Phi : \mathrm{Pic}\big(X \times X\big)^\sigma \oplus \mathbb{Z} \longrightarrow \mathrm{Pic}(W) \otimes_{\mathbb{Z}} \mathbb{Z}[\frac{1}{12}]$$

has finite kernel, using GRR theorem.

The variety W we shall use is the one parameterizing non-locally free sheaves that are kernels of $\mathcal{E} \to \mathbb{C}_x \oplus \mathbb{C}_y$, where $\mathcal{E}$ is fixed and $x, y \in X$ are arbitrary closed points. W will be birational to a $\mathbf{P}^1 \times \mathbf{P}^1$ bundle over $X \times X$. Let $\mathcal{E}$ be a rank two locally free sheaf on X and let $\mathbf{P}(\mathcal{E})$ be the associated projective bundle with projection

$$\pi : \mathbf{P}(\mathcal{E}) \to X$$

(we adopt the convention in [Ha, p162] that $\pi_*\big(\mathcal{O}(1)\big)=\mathcal{E}$) and let W be the blowing-up of $\mathbf{P}(\mathcal{E})\times\mathbf{P}(\mathcal{E})$ along the diagonal. We first construct a sheaf on $W\times X$ flat over W. Let q_i be the projection

$$q_i : W \longrightarrow \mathbf{P}(\mathcal{E})\times\mathbf{P}(\mathcal{E}) \xrightarrow{\mathrm{pr}_i} \mathbf{P}(\mathcal{E}),\quad i=1,2$$

and let

$$\iota_i : \mathrm{id}\times\big(\pi\circ q_i\big) : W \longrightarrow W\times X,\quad i=1,2. \tag{2.2}$$

ι_i is a closed immersion. We let

$$\beta : \pi_X^*\mathcal{E} \longrightarrow \iota_{1*}q_1^*\mathcal{O}(1)\oplus\iota_{2*}q_2^*\mathcal{O}(1)$$

be the homomorphism induced by

$$\pi_X^*\mathcal{E}\longrightarrow\pi_X^*\mathcal{E}\otimes\mathcal{O}_{\iota_i(W)}=\iota_{i*}q_i^*\pi^*\mathcal{E}\longrightarrow\iota_{i*}q_i^*\mathcal{O}(1),$$

where $\pi_X : W\times X\to X$ is the second projection. β is surjective away from $S\times X$, where S is the exceptional divisor of W. We let $\mathcal{G}$ be the kernel of β and let Ω be the cokernel of

$$\mathcal{G}\longrightarrow\pi_X^*\mathcal{E}.$$

In this way, we obtain two exact sequences:

$$0\longrightarrow\mathcal{G}\longrightarrow\pi_X^*\mathcal{E}\longrightarrow\Omega\longrightarrow 0 \tag{2.3}$$

and

$$0\longrightarrow\Omega\longrightarrow\iota_{1*}q_1^*\mathcal{O}(1)\oplus\iota_{2*}q_2^*\mathcal{O}(1)\longrightarrow\iota_{1*}q_1^*\mathcal{O}(1)_{|S\times X}\longrightarrow 0 \tag{2.4}$$

since $\mathrm{coker}(\beta)$ is isomorphic to the last term in (2.4).

We claim that $\mathcal{G}$ is a family of torsion free sheaves flat over W. Because of the exact sequence (2.3), it suffices to show that Ω is flat over W and

$$\mathrm{Tor}\big(\Omega,\mathcal{O}_{\{w\}\times X}\big)=0$$

for all closed $w\in W$. But this is clear from the exact sequence (2.4) because $\iota_i(W)\subseteq W\times X$ is smooth over W, $q_i^*\mathcal{O}(1)$ is an invertible sheaf on W and $\iota_i(W)\cap(S\times X)$ is a divisor in $\iota_i(W)$. This proves the claim.

Following the discussion in §1, we can form a homomorphism

$$\Phi_W : \mathrm{Pic}\big(X\times X\big)^\sigma\oplus\mathbb{Z}\longrightarrow\mathrm{Pic}(W)\otimes_{\mathbb{Z}}\mathbb{Z}[\frac{1}{12}] \tag{2.5}$$

using the determinant line bundle construction outlined before (1.11) using the sheaf $\mathcal{G}$ over $W\times X$ and complexes

$$\psi : \mathrm{Pic}\big(X\times X\big)^\sigma\oplus\mathbb{Z}\longrightarrow\widetilde{K}\otimes_{\mathbb{Z}}\mathbb{Z}[\frac{1}{12}].$$

We will investigate the injectivity of this homomorphism by looking at their first Chern classes, using GRR. From the formula (1.5) and the one after, for $L\in\mathrm{Pic}\big(X\times X\big)^\sigma$

$$\tilde{c}_1\big(\Phi_W(L)\big)=\Big[p_{1*}\Big(\mathrm{ch}\big(p_{23}^*\psi(L)\ominus p_{12}^*\mathcal{G}\ominus p_{13}^*\mathcal{G}\big)\cdot p_{23}^*\,\mathrm{td}(X\times X)\Big)\Big]_{[1]},$$

where $p_\cdot$ are projections from $W\times X\times X$ to its factor(s) with self-explanatory index. Since fibers of p_1 are $X\times X$, it suffices to determine

$$\begin{aligned}&\Big[\mathrm{ch}\big(p_{23}^*\psi(L)\ominus p_{12}^*\mathcal{G}\ominus p_{13}^*\mathcal{G}\big)\cdot p_{23}^*\,\mathrm{td}(X\times X)\Big]_{[5]}\\ &\quad=\Big[p_{23}^*\big(\tilde{c}_1(L)+\alpha[p_0]^\vee\big)\cdot p_{12}^*\mathrm{ch}(\mathcal{G})\cdot p_{13}^*\mathrm{ch}(\mathcal{G})\cdot p_2^*\,\mathrm{td}(X)\cdot p_3^*\,\mathrm{td}(X)\Big]_{[5]}\end{aligned} \tag{2.6}$$

modulo the kernel of p_{1*}. We first determine $\mathrm{ch}(\mathcal{G})$. By using (2.3) and (2.4) we get

$$\mathrm{ch}(\mathcal{G}) = \mathrm{ch}\Big(\pi_X^*\mathcal{E}\Big) - \mathrm{ch}\Big(\iota_{1*}q_1^*\mathcal{O}(1)\Big) - \mathrm{ch}\Big(\iota_{2*}q_2^*\mathcal{O}(1)\Big) + \mathrm{ch}\Big(\iota_{1*}q_1^*\mathcal{O}(1)_{|S\times X}\Big) \tag{2.7}$$

($\pi_\cdot$ is a projection from $W\times X$ to its factor.) Applying GRR theorem to the proper morphism $\iota_1 : W \to W\times X$, we obtain

$$\begin{aligned}\mathrm{ch}\Big(\iota_{1*}q_1^*\mathcal{O}(1)\Big) &= \iota_{1*}\Big(\mathrm{ch}(q_1^*\mathcal{O}(1))\Big)\cdot \mathrm{td}(W)\Big)\cdot \mathrm{td}(W\times X)^{-1}\\ &= \iota_{1*}q_1^*e^{\tilde{c}_1(\mathcal{O}(1))}\cdot \pi_W^*\,\mathrm{td}(W)\cdot \mathrm{td}(W\times X)^{-1}\\ &= \iota_{1*}q_1^*e^{\tilde{c}_1(\mathcal{O}(1))}\cdot \pi_X^*\,\mathrm{td}(X)^{-1}.\end{aligned}$$

Here the second identity holds because $\iota_1(W)$ is a graph of $\pi\circ q_1 : W \to X$. Similarly, we have

$$\mathrm{ch}\Big(\iota_{2*}q_2^*\mathcal{O}(1)\Big) = \iota_{2*}q_2^*e^{\tilde{c}_1(\mathcal{O}(1))}\cdot \pi_X^*\,\mathrm{td}(X)^{-1}.$$

In the following, we will choose a split $\mathcal{E}$ (i.e. $\mathcal{E}$ is a direct sum of invertible sheaves) to simplify our calculation. We assume $\det\mathcal{E} = I$. In this case, $\mathbf{P}(\mathcal{E})$ is isomorphic to $\mathbf{P}^1\times X$, $\tilde{c}_1(\mathcal{O}(1))$ is dual to a section of $\mathbf{P}(\mathcal{E})\to X$ and $\tilde{c}_1\Big(\mathcal{O}(1)\Big)^2 = 0$.

We now reorganize terms in the expansion of (2.6) in terms of (2.7) for split $\mathcal{E}$. One term in the expansion of (2.6) is

$$\begin{aligned}&\Big[p_{23}^*\Big(\tilde{c}_1(L) + \alpha[p_0]^\vee\Big)\cdot (p_{12}\circ\pi_X)^*\mathrm{ch}(\mathcal{E})\cdot (p_{13}\circ\pi_X)^*\mathrm{ch}(\mathcal{E})\Big]_{[5]}\\ &\qquad = \Big[p_{23}^*\Big(\tilde{c}_1(L) + \alpha[p_0]^\vee\Big)\cdot p_2^*\mathrm{ch}(\mathcal{E})\cdot p_3^*\mathrm{ch}(\mathcal{E})\Big]_{[5]}\end{aligned}$$

which is contained in the kernel of p_{1*}, following the vanishing of (1.1). (Recall $p_\cdot$ is designated to projections from $W\times X\times X$ to its factor, $\pi_\cdot$ is projection from $W\times X$ to its factor and $q_\cdot$ is projection from W to $\mathbf{P}(\mathcal{E})$.) The next terms are

$$\begin{aligned}&\Big[p_{23}^*\tilde{c}_1(L)\cdot p_{12}^*\Big(\iota_{i*}q_i^*e^{\tilde{c}_1(\mathcal{O}(1))}\cdot \pi_X^*\,\mathrm{td}(X)^{-1}\Big)\\ &\qquad \cdot p_{13}^*\Big(\iota_{j*}q_j^*e^{\tilde{c}_1(\mathcal{O}(1))}\cdot \pi_X^*\,\mathrm{td}(X)^{-1}\Big)\cdot p_2^*\,\mathrm{td}(X)\cdot p_3^*\,\mathrm{td}(X)\Big]_{[5]},\end{aligned}$$

where (i,j) runs through pairs $(1,1)$, $(1,2)$, $(2,1)$ and $(2,2)$. After simplification, their images under p_{1*} are

$$p_{1*}\Big(p_{23}^*\tilde{c}_1(L)\cdot p_{12}^*\big[\iota_i(W)\big]^\vee\cdot p_{13}^*\big[\iota_j(W)\big]^\vee\Big) = \begin{cases} q_{XX}^*\tilde{c}_1(L), & i\neq j;\\ q_{XX}^*\,\mathrm{pr}_1^*\Big(\tilde{c}_1(L_{|\Delta})\Big), & i,j=1;\\ q_{XX}^*\,\mathrm{pr}_2^*\Big(\tilde{c}_1(L_{|\Delta})\Big), & i,j=2,\end{cases}$$

since L is invariant under permutation σ. Here $q_{XX} : W \to X\times X$ is the projection, $\Delta\subseteq X\times X$ is the diagonal, $\tilde{c}_1(L_{|\Delta})$ is in A^1X by the isomorphism $\Delta = X$ and $\pi_\cdot : X\times X\to X$ is projection. Note that $q_{XX}^*\tilde{c}_1(L)$ appears in the expression.

We need to take care of the remainder terms. We first look at another series of terms

$$\begin{aligned}-\Big[&p_{23}^*\tilde{c}_1(L)\cdot (p_{1i}\circ\pi_X)^*\mathrm{ch}(\mathcal{E})\cdot p_{1\bar{i}}^*\Big(\iota_{1*}q_1^*e^{\tilde{c}_1(\mathcal{O}(1))}\cdot \pi_X^*\,\mathrm{td}(X)^{-1}\\ &+ \iota_{2*}q_2^*e^{\tilde{c}_1(\mathcal{O}(1))}\cdot \pi_X^*\,\mathrm{td}(X)^{-1}\Big)\cdot p_2^*\,\mathrm{td}(X)\cdot p_3^*\,\mathrm{td}(X)\Big]_{[5]},\end{aligned} \tag{2.8}$$

where $(i,\bar{i})$ is either $(2,3)$ or $(3,2)$. For simplicity, we let α_2 and $\alpha_3\in A^*(X\times X)$ be such that

$$-p_{23}^*\alpha_i = p_{23}^*\tilde{c}_1(L)\cdot p_i^*\mathrm{ch}(\mathcal{E})\cdot p_i^*\,\mathrm{td}(X).$$

For our purpose, it is sufficient to look at their components in $q_{XX}^*A^1(X\times X)$ under the decomposition

$$A^1(W)=q_{XX}^*A^1(X\times X)\oplus\mathbb{Z}[R_1]^\vee\oplus\mathbb{Z}[R_2]^\vee\oplus\mathbb{Z}[S]^\vee,$$

where $R_1,R_2\subseteq W$ are birational to $\mathbf{P}^1\times\{\mathrm{pt}\}\times X\times X$ and $\{\mathrm{pt}\}\times\mathbf{P}^1\times X\times X$ respectively, $S\subseteq W$ is the exceptional divisor of $W\to\mathbf{P}(\mathcal{E})\times\mathbf{P}(\mathcal{E})$. (Recall W is a blowing up of $\mathbf{P}^1\times\mathbf{P}^1\times X\times X$ since $\mathcal{E}$ is split.) Since $\iota_i(W)$ is a graph of $W\to X$, the component in $q_{XX}^*A^1(X\times X)$ of the image under p_{1*} of (2.8) are

$$\Big[p_{1*}\Big(p_{23}^*\alpha_2\cdot p_{13}^*\Big([\iota_1(W)]^\vee+[\iota_2(W)]^\vee\Big)\Big)\Big]_{[1]}\tag{2.9}$$

and

$$\Big[p_{1*}\Big(p_{23}^*\alpha_3\cdot p_{12}^*\Big([\iota_1(W)]^\vee+[\iota_2(W)]^\vee\Big)\Big)\Big]_{[1]}.\tag{2.10}$$

To simplify these, we proceed as follows: Consider the projection $f:W\to X\times X$ (from $W\to\mathbf{P}(\mathcal{E})\times\mathbf{P}(\mathcal{E})$) and the projection

$$P=(f,\mathrm{id}_X,\mathrm{id}_X):W\times X\times X\longrightarrow X\times X\times X\times X.$$

Then by definition of ι_1 and ι_2, we have

$$p_{12}^*\Big([\iota_j(W)]^\vee\Big)=P^*\circ\pi_{j3}^*\Big([\Delta]^\vee\Big)$$

and

$$pac^*\Big([\iota_j(W)]^\vee\Big)=P^*\circ\pi_{j4}^*\Big([\Delta]^\vee\Big),$$

where $\pi_{ij}:X\times X\times X\times X\to X\times X$ is a projection. Then the component in $A^1\Big(X\times X\Big)$ of (2.9) is the pull back (under q_{XX}^*) of

$$\Big[\pi_{12*}\circ P_*\Big([\Sigma\times X\times X]^\vee\cdot\Big(P\circ\pi_{34}\Big)^*(\alpha_2)\cdot P^*\Big(\pi_{14}^*\Big([\Delta]^\vee\Big)+\pi_{24}^*\Big([\Delta]^\vee\Big)\Big)\Big)\Big]_{[1]},$$

where Σ is a birational section of $W\to X\times X$. By projection formula, it is

$$\begin{aligned}&\Big[\pi_{12*}\Big(\Big(P_*[\Sigma\times X\times X]^\vee\Big)\cdot\pi_{34}^*(\alpha_2)\cdot\Big(\pi_{14}^*([\Delta]^\vee)+\pi_{24}^*([\Delta]^\vee)\Big)\Big)\Big]_{[1]}\\ =&\Big[\pi_{12*}\Big(\pi_{34}^*(\alpha_2)\cdot\Big(\pi_{14}^*([\Delta]^\vee)+\pi_{24}^*([\Delta]^\vee)\Big)\Big)\Big]_{[1]}\\ =&\Big[\mathrm{pr}_1^*\,\mathrm{pr}_{1*}(\alpha_2)+\mathrm{pr}_2^*\,\mathrm{pr}_{1*}(\alpha_2)\Big]_{[1]}.\end{aligned}$$

Similarly, (2.10) is

$$\Big[\mathrm{pr}_1^*\,\mathrm{pr}_{2*}(\alpha_3)+\mathrm{pr}_2^*\,\mathrm{pr}_{2*}(\alpha_3)\Big)\Big]_{[1]}.$$

Note that $\sigma^*(\alpha_2)=\sigma_3$, where σ is the permutation. Hence both are

$$\mathrm{pr}_1^*(\alpha)+\mathrm{pr}_2^*(\alpha),\quad \alpha=\mathrm{pr}_{1*}(\alpha_2).$$

The image under p_{1*} of the last set of terms in (2.6) contains factor $\mathrm{ch}\Big(\iota_{i*}q_i^*\mathcal{O}(1)_{|S\times X}\Big)$ whose components in $q_{XX}^*A^1(X\times X)$ are trivial. Put them together, we see that the component in $q_{XX}^*A^1(X\times X)$ of $\tilde{c}_1\Big(\varPhi_W(L)\Big)$ is the pull-back of

$$2\tilde{c}_1(L)+\mathrm{pr}_1^*\Big(\tilde{c}_1(L_{|\Delta})\Big)+\mathrm{pr}_2^*\Big(\tilde{c}_1(L_{|\Delta})\Big)+2\,\mathrm{pr}_1^*(\alpha)+2\,\mathrm{pr}_2^*(\alpha)\in A^1(X\times X).\tag{2.11}$$

Now we look at the component in $q_{XX}^*A^1(X\times X)$ of $\tilde{c}_1\Big(\varPhi_W(1)\Big)$. By replacing $\tilde{c}_1(L)$ in (2.6) with $\Big[X\times\{x_0\}+\{x_0\}\times X\Big]^\vee$ and going through the same line of analysis, we see that there is a $\beta \in A^1(X)$ depending only on X and I such that the component in $q_{XX}^* A^1(X\times X)$ of $\tilde c_1\big(\Phi_W(1)\big)$ is

$$q_{XX}^*\big(\mathrm{pr}_1^*\beta + \mathrm{pr}_2^*\beta\big). \tag{2.12}$$

Therefore, if $(L,n) \in \mathrm{Pic}\big(X\times X\big)^\sigma \oplus \mathbb{Z}$ makes

$$\tilde c_1\Big(\Phi\big((L,n)\big)\Big) = 0,$$

then $\tilde c_1(L)$ must be of the form $\mathrm{pr}_1^* c + \mathrm{pr}_2^* c$ for some $c \in A^1(X)$ satisfying the identity

$$F(c) + c_2(\mathcal{E})c + n\beta = 0, \tag{2.13}$$

where $F\colon A^1X \longrightarrow A^1X$ is a map depending on X and I but not $\mathcal{E}$, from (2.11) and (2.12). (The exact form of (2.13) can be derived easily from (2.11) and (2.12) but we choose not to in part to emphasize the term $c_2(\mathcal{E})\,c$ that is crucial in our later discussion.)

Now we state and prove the main proposition of this section:

Proposition 2.1. *For any choice of (I,H), there is an N such that whenever $d \geq N$, then the homomorphism*

$$\Phi : Pic\big(X\times X\big)^\sigma \oplus \mathbb{Z} \longrightarrow Pic\big(\mathfrak{M}(I,d)\big)\otimes_{\mathbb{Z}} \mathbb{Z}[\tfrac{1}{12}]$$

has finite kernel.

Proof. We let $\mathcal{E}_1$ be $\mathcal{O}\oplus\mathcal{O}(I)$ and let $\mathcal{E}_2$ be $\mathcal{O}(H)\oplus\mathcal{O}(I\otimes H^{-1})$. (When $H^2 = H\cdot I$ then we let $\mathcal{E}_2$ be $\mathcal{O}(H^{-1})\oplus\mathcal{O}(I\otimes H)$.) Let W_i be the variety constructed at the beginning of this section that is the blowing-up of $\mathbf{P}(\mathcal{E}_i)\times\mathbf{P}(\mathcal{E}_i)$ and let $\mathcal{G}_i$ be the sheaf on $W_i\times X$ constructed in (2.3). We claim that if

$$(L,n) \in \mathrm{Pic}\big(X\times X\big)^\sigma \oplus \mathbb{Z}$$

lies in the kernel of Φ_{W_i} in (2.5) for $i=1,2$, then $n=0$ and L is torsion. Indeed, by the previous argument, $\Phi_{W_i}\big((L,n)\big)=0$ implies that $\tilde c_1(L) = \mathrm{pr}_1^* c + \mathrm{pr}_2^* c$ with $c\in A^1X$ satisfying two identities that are (2.8) with $c_2(\mathcal{E})$ replaced by $c_2(\mathcal{E}_1)=0$ and by $c_2(\mathcal{E}_2)\neq 0$. Clearly, this is possible only if c is a torsion. Once we know that c is torsion, then $\Phi_W\big((L,n)\big)=0$ force $n=0$, which can be checked directly. This establishes the claim.

Now we prove the proposition. Let (L,n) be any element in the kernel of Φ. It suffices to show that $\Phi_{W_i}\big((L,n)\big)=0$ for both i. Let $\mathcal{E}$ be either $\mathcal{E}_1$ or $\mathcal{E}_2$. Then there is a constant $N\geq 0$ such that for any $d\geq N$ there is a subsheaf $\mathcal{F}\subseteq\mathcal{E}$ such that $\det\mathcal{F}=I$ and $c_2(\mathcal{F})=d$ satisfying the following property: $\mathcal{F}$ admits a deformation, say $\mathcal{F}_t$, whose general members are locally free and μ-stable. This can be proved as follows: First for large d, we can choose $\mathcal{F}\subseteq\mathcal{E}$ so that the traceless part $\mathrm{Ext}^2(\mathcal{F},\mathcal{F})^0=0$. Then we can deform $\mathcal{F}$ to locally free sheaves. When d is large enough and the support of $\mathcal{E}/\mathcal{F}$ is in general position, then the argument on page 158 of [Gi2] shows that any locally free deformation of $\mathcal{F}$ are μ-stable. This proves the existence of such deformation. Now let $0\in T$ be a smooth curve and $\mathcal{F}_T$ a deformation of $\mathcal{F}_0=\mathcal{F}$ just mentioned. Without loss of generality, we assume $\mathcal{F}_T$ is locally free over $T-0$. Let $\Lambda\subseteq X$ be the finite set over which $\mathcal{F}_0$ is not locally free.

We now construct the corresponding family of varieties W_T associated to $\mathcal{F}_T$. Let $\mathbf{P}(\mathcal{F}_T)$ be the projective bundle of $\mathcal{F}_T$ over $T\times X - \{0\}\times\Lambda$ and let W_T be the blowing-up of the diagonal of

$$\mathbf{P}(\mathcal{F}_T)\times_T\mathbf{P}(\mathcal{F}_T).$$

Similar to (2.3), we can define a sheaf $\mathcal{G}_T$ on $W_T\times X$ that is flat over W_T. Note that since $\mathcal{F}_t$ is μ-stable for $t\neq 0$, the family $\mathcal{G}_{T^*}$, $T^*=T-0$, that is the restriction of $\mathcal{G}_T$ to $W_T - W_0$

(W_0 is the fiber over $0 \in T$) is a family of H-stable sheaves with determinant I and second Chern class d. Hence, it induces a morphism

$$\mu : W_T - W_0 \longrightarrow \mathfrak{M}(I,d).$$

Also, by applying the determinant line bundle construction to the family $\mathcal{G}_T$, we can form a homomorphism

$$\Phi_{W_T} : \operatorname{Pic}\bigl(X \times X\bigr)^{\sigma} \oplus \mathbb{Z} \longrightarrow \operatorname{Pic}\bigl(W_T\bigr) \otimes_{\mathbb{Z}} \mathbb{Z}[\frac{1}{12}]$$

similar to (2.5). Since this construction commutes with base change,

$$\Phi_{W_T}\bigl((L,n)\bigr)_{|W_T - W_0} \cong \mu^* \Phi\bigl((L,n)\bigr).$$

In particular, since $\Phi\bigl((L,n)\bigr)$ is trivial, $\Phi_{W_T}\bigl((L,n)\bigr)_{|W_T - W_0}$ is trivial and consequently so is $\Phi_{W_T}\bigl((L,n)\bigr)_{|W_0}$.

Assume $\mathcal{F}^{\vee\vee} = \mathcal{E}$ is $\mathcal{E}_1$ we begin with. Then W_0 is canonically isomorphic to $W_1 - A$ for some closed $A \subseteq W_1$ of codimension two and further,

$$\Phi_{W_T}\bigl((L,n)\bigr)_{|W_0} \cong \Phi_{W_1}\bigl((L,n)\bigr)_{|W_1 - A}.$$

Hence for $(L,n) \in \ker\bigl(\Phi\bigr)$, $\Phi_{W_1}\bigl((L,n)\bigr)$ is trivial as well, since $\operatorname{codim}(A) = 2$ and W_1 is smooth. Similarly, if we choose $\mathcal{F}$ so that $\mathcal{F}^{\vee\vee} = \mathcal{E}_2$, then we must have $\Phi_{W_2}\bigl((L,n)\bigr) = 0$. This shows that $\Phi_{W_i}\bigl((L,n)\bigr)$ are trivial for $i = 1, 2$ and henceforth establishing the proposition. □

3. Surjectivity of Φ

In this section, we shall show that when $H^2(X,\mathbb{Z})$ has no torsion, then for sufficiently large d the so constructed homomorphism

$$\Phi : \operatorname{Pic}\bigl(X \times X\bigr)^{\sigma} \oplus \mathbb{Z} \longrightarrow \operatorname{Pic}\bigl(\mathfrak{M}(I,d)\bigr) \otimes_{\mathbb{Z}} \mathbb{Z}[\frac{1}{12}]$$

has finite cokernel. The tactic we will use is to first show that under certain circumstances the restriction homomorphism

$$\operatorname{Pic}\bigl(\overline{\mathfrak{M}}(I,d)\bigr) \longrightarrow \operatorname{Pic}\bigl(\mathfrak{M}(I,d)\bigr)$$

is surjective, by studying extension problem. We will then use the knowledge of the first two Betti numbers of $\mathfrak{M}(I,d)$ to determine the Picard group of $\overline{\mathfrak{M}}(I,d)$, up to finite index.

We first look at the problem of extending line bundle on $\mathfrak{M}(I,d)$ to $\overline{\mathfrak{M}}(I,d)$. This in general is rather tricky due to the singularities of $\overline{\mathfrak{M}}(I,d)$. In out case, we can give an affirmative answer to this problem because of our knowledge of singularities of $\overline{\mathfrak{M}}_H(I,d)$. To this end, we will quickly review some relevant facts about the singularities of the moduli space $\overline{\mathfrak{M}}(I,d)$. Since the discussion of $\operatorname{Pic}\bigl(\overline{\mathfrak{M}}(I,d)\bigr)$ will simplify when the polarization H is (I,d)-generic, we will work with a set of polarizations simultaneously. (Note that except when $\operatorname{Pic}(X)/\operatorname{Pic}^0(X)$ has rank one, no polarization is (I,d) generic for all d.) In the following, we will use $\mathfrak{M}_H(I,d)$ and $\overline{\mathfrak{M}}_H(I,d)$ to denote the moduli schemes of H-stable and H-semistable sheaves respectively. Also, we will speak freely of moduli scheme for $\mathbb{Q}$-ample line bundle H since $\mathfrak{M}_H(I,d)$ and $\overline{\mathfrak{M}}_H(I,d)$ only depend on the ray $\mathbb{Q}^+ \cdot c_1(H) \subseteq H^{1,1}(X,\mathbb{Q})$. We let $N_{\mathbb{Q}}^+$ be the $\mathbb{Q}$-cone in $H^{1,1}(X,\mathbb{R}) \cap H^2(X,\mathbb{Q})$ spanned by Chern classes of $\mathbb{Q}$-ample line bundles.

We say a neighborhood $\mathcal{C} \subseteq N_{\mathbb{Q}}^+$ is precompact if the closure of $\mathcal{C}$ in $H^{1,1}(X,\mathbb{R}) \cap H^2(X,\mathbb{Q})$ is still contained in $N_{\mathbb{Q}}^+$. Note that for any $c \in N_{\mathbb{Q}}^+$, a Euclidean ball of sufficiently small radius centered at c in $N_{\mathbb{Q}}^+$, after fixing an Euclidean metric on $H^{1,1}(X,\mathbb{R})$, is a precompact neighborhood of c. We will use $\mathcal{C}\ \mathbf{b}\ N_{\mathbb{Q}}^+$ to mean $\mathcal{C}$ is precompact. Also, by abuse of notation we will say a $\mathbb{Q}$-line bundle $H \in \mathcal{C}$ if $c_1(H) \in \mathcal{C}$.

We collect the relevant properties of $\overline{\mathfrak{M}}(I,d)$ in the following lemmas:

Lemma 3.1. *For any precompact $\mathcal{C}\ \mathbf{b}\ N_{\mathbb{Q}}^+$, there is an integer N depending on $(X,I,\mathcal{C})$ such that for any $d \geq N$ and $\mathbb{Q}$-ample line bundle $H \in \mathcal{C}$ (i.e. $c_1(H) \in \mathcal{C}$),*

(1) *$\overline{\mathfrak{M}}_H(I,d)$ is normal, irreducible and of dimension predicted by Riemann-Roch theorem;*

(2) *$\mathfrak{M}_H(I,d)$ is a local complete intersection scheme and the codimension of the singular locus of $\mathfrak{M}_H(I,d)$ is at least 5;*

(3) *Let $\tilde{H} \in \mathcal{C}$ be another $\mathbb{Q}$-ample line bundle with $c_1(\tilde{H}) \in \mathcal{C}$, then there are closed subsets $U_H \subseteq \mathfrak{M}_H(I,d)$ and $U_{\tilde{H}} \subseteq \mathfrak{M}_{\tilde{H}}(I,d)$ having codimension at least 5 respectively such that $\mathfrak{M}_H(I,d) - U_H$ is canonically isomorphic to $\mathfrak{M}_{\tilde{H}}(I,d) - U_{\tilde{H}}$ as schemes, by identifying two points that represent isomorphic sheaves.*

Proof. (1) and (2) are proved on page 10 of [Li3] and (3) is proved on page 458 of [Li1] by using [Qi2]. □

As was mentioned in §2, the moduli $\overline{\mathfrak{M}}_H(I,d)$ is constructed as GIT quotient of a Grothendieck's Quot-scheme, say $\mathbf{Quot}_H(I,d)$, by a reductive group G. (We take $G = PGL(m)$ this time.) The property we need from this construction is stated in the following lemma.

Lemma 3.2. *For any precompact $\mathcal{C}\ \mathbf{b}\ N_{\mathbb{Q}}^+$, there is an integer N depending on $(X,I,\mathcal{C})$ such that for any $d \geq N$ and $H \in \mathcal{C}$, we can construct a Grothendieck's Quot-scheme $\mathbf{Quot}_H(I,d)$ and a reductive group G acting linearly on $\mathbf{Quot}_H(I,d)$ of which the following holds:*

(1) *Let $\mathbf{Quot}_H(I,d)^s$ and $\mathbf{Quot}_H(I,d)^{ss}$ be the open subsets of G-stable and G-semistable (with respect to the given linearization) points in $\mathbf{Quot}_H(I,d)$. Then*

$$\mathbf{Quot}_H(I,d)^s/G = \mathfrak{M}_H(I,d), \quad \mathbf{Quot}_H(I,d)^{ss}//G = \overline{\mathfrak{M}}_H(I,d)$$

and the projection sends $w \in \mathbf{Quot}_H(I,d)^{ss}$ to $\mathcal{F}_w \in \overline{\mathfrak{M}}_H(I,d)$, where $\mathcal{F}_w$ is the corresponding quotient sheaf of w;

(2) *$\mathbf{Quot}_H(I,d)^{ss}$ is a local complete intersection scheme whose singular locus has codimension at least 5 in $\mathbf{Quot}_H(I,d)^{ss}$;*

(3) *Let $w \in \mathbf{Quot}_H(I,d)^{ss}$ be any point with closed orbit $G \cdot w \subseteq \mathbf{Quot}_H(I,d)^{ss}$, then $stab_w = \mathbb{C}^*$ if $\mathcal{F}_w \cong J_1 \oplus J_2$ with $J_1 \neq J_2$, $stab_w = PGL(2,\mathbb{C})$ if $\mathcal{F}_w \cong J \oplus J$ and $stab_w = \{1\}$ otherwise;*

(4) *Assume H is (I,d)-generic, then*

$$\operatorname{codim}\Big(\mathbf{Quot}_H(I,d)^{ss} - \mathbf{Quot}_H(I,d)^s, \mathbf{Quot}_H(I,d)^{ss}\Big) \geq 8.$$

Proof. (1) and (3) is proved by [Gi1] (see also [Ma]) and (2) is proved on page 5 of [Li2] and (4) follows from [Qi2]. □

The last is the information about the first two Betti numbers of the moduli space $\mathfrak{M}_H(I,d)$ proved recently in [Li3].

Lemma 3.3. *For any precompact $\mathcal{C}\ \mathbf{b}\ N_{\mathbb{Q}}^+$, there is an integer N depending on $(I,d,\mathcal{C})$ such that for any $d \geq N$ and $H \in \mathcal{C}$,*

$$H^1\Big(\mathfrak{M}_H(I,d);\mathbb{Q}\Big) \cong H^1\Big(X;\mathbb{Q}\Big)$$

and

$$H^2\Big(\mathfrak{M}_H(I,d);\mathbb{Q}\Big) \cong H^2\Big(X;\mathbb{Q}\Big) \oplus \wedge^2 H^1\Big(X;\mathbb{Q}\Big) \oplus H^0(X;\mathbb{Q}).$$

We also quote a lemma that concerns extending line bundles over local complete intersection singularities.

Lemma 3.4. *Let W be any quasi-projective scheme having only local complete intersection singularity and* $\operatorname{codim}\big(Sing\,W, W\big) \geq 4$. *Then for any closed subset $\Lambda \subseteq W$ of codimension at least two, the canonical map induced by inclusion*

$$Pic(W) \longrightarrow Pic(W-\Lambda)$$

is an isomorphism.

Proof. See [SGA 2] on page 132. □

Now we draw some easy consequence from these lemmas concerning the Picard group of $\mathfrak{M}_H(I,d)$. In the following, we fix an I, a $\mathcal{C}$ **b** $N_{\mathbb{Q}}^{+}$ and the integer N specified in the previous lemmas.

Lemma 3.5. *For any $H_1, H_2 \in \mathcal{C}$ and $d \geq N$, $Pic\big(\mathfrak{M}_{H_1}(I,d)\big)$ is canonically isomorphic to $Pic\big(\mathfrak{M}_{H_2}(I,d)\big)$, compatible to the birational map in Lemma 3.1.*

Proof. Let $U_{H_1} \subseteq \mathfrak{M}_{H_1}(I,d)$ and $U_{H_2} \subseteq \mathfrak{M}_{H_2}(I,d)$ be the closed subset given in Lemma 3.1. It suffices to show that the pull back homomorphism $\operatorname{Pic}\big(\mathfrak{M}_{H_i}(I,d)\big) \to \operatorname{Pic}\big(\mathfrak{M}_{H_i}(I,d) - U_{H_i}\big)$ is an isomorphism, since $\mathfrak{M}_{H_1}(I,d) - U_{H_1}$ is canonically isomorphic to $\mathfrak{M}_{H_2}(I,d) - U_{H_2}$. But this follows from Lemma 3.4 and (2) of Lemma 3.1. This proves the lemma. □

Lemma 3.6. *Suppose $H^2(X,\mathbb{Z})$ is torsion free. Then for $d \geq N$ and (I,d)-generic $H \in \mathcal{C}$, the pull back homomorphism*

$$\mu : Pic\big(\overline{\mathfrak{M}}_H(I,d)\big) \longrightarrow Pic\big(\mathfrak{M}_H(I,d)\big)$$

is an isomorphism.

Proof. It is easy to see that μ is injective. Indeed, assume L is a line bundle on $\overline{\mathfrak{M}}_H(I,d)$ admitting a non-vanishing section over $\mathfrak{M}_H(I,d)$. Since $\overline{\mathfrak{M}}_H(I,d)$ is normal and

$$\operatorname{codim}\big(\overline{\mathfrak{M}}_H(I,d) - \mathfrak{M}_H(I,d), \overline{\mathfrak{M}}_H(I,d)\big) \geq 2,$$

this section extends to a non-vanishing section over $\overline{\mathfrak{M}}_H(I,d)$. Hence L is trivial itself.

Now we show that μ is surjective. Let L be a line bundle on $\mathfrak{M}_H(I,d)$. Then its pull back π^*L on $\mathbf{Quot}_H(I,d)^s$ admits a canonical G-linearization, where $\pi : \mathbf{Quot}_H(I,d)^s \to \mathfrak{M}_H(I,d)$. By Lemma 3.4 and Lemma 3.2, π^*L extends to a line bundle $\widetilde{L}$ on $\mathbf{Quot}_H(I,d)^{ss}$ and the G-linearization of π^*L extends as well. The key observation is that $\widetilde{L}$ always descends to a line bundle on $\overline{\mathfrak{M}}_H(I,d)$, under our assumption on $H^2(X;\mathbb{Z})$. To see this, we need to check that for any $w \in \mathbf{Quot}_H(I,d)^{ss}$ with closed orbit $G\cdot w$ and non-trivial stabilizer stab_w, stab_w acts trivially on the fiber of $\widetilde{L}$ over w. Indeed, since $\widetilde{L}$ is a line bundle, this action is given by a character

$$\chi_w : \operatorname{stab}_w \longrightarrow \mathbb{C}^*.$$

By (3) of Lemma 3.2, stab_w can only take two forms, $\mathbb{C}^*$ and $PGL(2)$, unless it is trivial. Since $PGL(2)$ has no non-trivial character, we are left to show that χ_w is trivial even when $\operatorname{stab}_w = \mathbb{C}^*$. Following Drezet and Narasimhan [DN], we know that if w_1, w_2 are two points contained in the same connected component of

$$\big\{w \in \mathbf{Quot}_H(I,d)^{ss} \mid G\cdot w \text{ closed and } \operatorname{stab}_w \neq \{1\}\big\}, \tag{3.1}$$

then $\chi_{w_1} = 0$ if and only if $\chi_{w_2} = 0$. By (3) of Lemma 3.2, $G\cdot w$ closed and $\operatorname{stab}_w \neq \{1\}$ implies $\mathcal{F}_w = \mathcal{F}_1 \oplus \mathcal{F}_2$. However, since H is (I,d)-generic, $c_1(\mathcal{F}_1) - c_1(\mathcal{F}_2)$ is torsion, which is zero since $H^2(X,\mathbb{Z})$ has no torsions. Then we can deform $\mathcal{F}_1 \oplus \mathcal{F}_2$ within the subset of split quotient sheaves in $\mathbf{Quot}_H(I,d)^{ss}$, then within (3.1), to $\mathcal{F} \oplus \mathcal{F}$. Therefore w and the point $\tilde{w}$ associated to the quotient sheaf $\mathcal{F} \oplus \mathcal{F}$ lie in the same component of (3.1). Therefore $\chi_w = 0$ because $\mathrm{stab}_{\tilde{w}} = PGL(2)$ has no non-trivial characters. This proves that $\widetilde{L}$ descends to a line bundle on $\overline{\mathfrak{M}}_H(I,d)$, which is an extension of L. This proves that $\mathrm{Pic}\big(\overline{\mathfrak{M}}_H(I,d)\big) \to \mathrm{Pic}\big(\mathfrak{M}_H(I,d)\big)$ is surjective. □

As a consequence of this lemma, we see that for large d, the divisor $\Lambda \subseteq \overline{\mathfrak{M}}_H(I,d)$ consisting of non-locally free sheaves is Cartier when H is (I,d)-generic and $H^2(X,\mathbb{Z})$ is torsion free. In the following, we shall show that the condition $H^2(X;\mathbb{Z})$ has no torsion is unnecessary in this case, which is crucial in applying local Lefschetz theorem to study the Betti numbers of $\mathfrak{M}_H(I,d)$ carried out in [Li3].

Lemma 3.7. *For any precompact neighborhood $\mathcal{C}$ b $N^+_{\mathbb{Q}}$, there is a constant N such that for any $d \geq N$ and (I,d)-generic $H \in \mathcal{C}$, the divisor $\Lambda \subseteq \overline{\mathfrak{M}}_H(I,d)$ consisting of non-locally free sheaves is Cartier.*

Proof. Let k be the number of torsion elements (including 0) in $H^2(X,\mathbb{Z})$ and let $\pi : \widetilde{X} \to X$ be a k-fold (unramified) covering so that $H^2(\widetilde{X},\mathbb{Z})$ is torsion free. Let $\tilde{I} = \pi^* I$ and $\tilde{\mathcal{C}} \subseteq N^+_{\mathbb{Q}}(\widetilde{X})$ be the image in $H^2(\widetilde{X},\mathbb{Q})$ under pullback $H^2(X;\mathbb{Q}) \to H^2(\widetilde{X};\mathbb{Q})$. $\tilde{\mathcal{C}}$ b $N^+_{\mathbb{Q}}(\widetilde{X})$ is precompact. Therefore there is an N such that for any $\tilde{d} \geq N$ and $H' \in \tilde{\mathcal{C}}$, the moduli scheme $\overline{\mathfrak{M}}_{H'}(\tilde{I},kd)$ and the corresponding Quot-scheme $\mathbf{Quot}_{H'}(\tilde{I},kd)^{ss}$ satisfy the conclusion of Lemma 3.1 and 3.2. Let $\tilde{\Lambda} \subseteq \overline{\mathfrak{M}}_{H'}(\tilde{I},kd)$ be the closed subset of all non-locally free sheaves. Since $\pi : \widetilde{X} \to X$ is a finite unramified covering, the morphism

$$\overline{\pi} : \overline{\mathfrak{M}}_H(I,d) \longrightarrow \overline{\mathfrak{M}}_{\tilde{H}}(\tilde{I},kd)$$

induced by sending $\mathcal{E} \in \overline{\mathfrak{M}}_H(I,d)$ to $\pi^*(\mathcal{E})$ is an immersion, where $\tilde{H} = \pi^* H$. In particular, $\Lambda = \big(\overline{\pi}\big)^{-1}\big(\tilde{\Lambda}\big)$. We claim that $\tilde{\Lambda}$ is Cartier near $\overline{\pi}(\Lambda)$. To establish this, following the proof of Lemma 3.6, we only need to show that for any split semistable sheaf $\mathcal{E} = \mathcal{F}_1 \oplus \mathcal{F}_2$ in Λ, $\pi^*(\mathcal{F}_1) \oplus \pi^*(\mathcal{F}_2)$ can be deformed within the set of splitting semistable sheaves in $\overline{\mathfrak{M}}_{\tilde{H}}(\tilde{I},kd)$ to $\mathcal{F} \oplus \mathcal{F}$. But this is always possible because $c_1(\mathcal{F}_1) - c_1(\mathcal{F}_2)$ is torsion implies $c_1\big(\pi^*(\mathcal{F}_1)\big) = c_1\big(\pi^*(\mathcal{F}_2)\big)$. Therefore $\tilde{\Lambda}$ is Cartier near $\overline{\pi}(\Lambda)$ and hence $\Lambda = \overline{\pi}-1(\tilde{\Lambda})$ is Cartier. This completes the proof of the lemma. □

We now state and prove the main theorem of this note.

Theorem 3.8. *Suppose $H^2(X,\mathbb{Z})$ has no torsions. Then for any precompact $\mathcal{C}$ b $N^+_{\mathbb{Q}}$, there is an integer N depending on $(X,I,\mathcal{C})$ such that for any $d \geq N$ and (I,d)-generic $H \in \mathcal{C}$, the homomorphism*

$$\overline{\Phi} : Pic\big(X \times X\big)^{\sigma} \oplus \mathbb{Z} \longrightarrow Pic\big(\overline{\mathfrak{M}}(I,d)\big) \otimes_{\mathbb{Z}} \mathbb{Z}[\frac{1}{12}]$$

constructed in (1.13) has finite kernel and cokernel.

Corollary 3.9. *Suppose $H^2(X,\mathbb{Z})$ has no torsions. Then for any precompact $\mathcal{C}$ b $N^+_{\mathbb{Q}}$, there is an integer N depending on $(X,I,\mathcal{C})$ such that for any $d \geq N$ and $H \in \mathcal{C}$, the homomorphism*

$$\Phi : Pic\big(X \times X\big)^{\sigma} \oplus \mathbb{Z} \longrightarrow Pic\big(\mathfrak{M}_H(I,d)\big) \otimes_{\mathbb{Z}} \mathbb{Z}[\frac{1}{12}]$$

has finite kernel and cokernel.

Proof. By enlarging $\mathcal{C} \Subset N_{\mathbb{Q}}^{+}$ if necessary, we can assume $\mathcal{C}$ contains $\mathbb{Q}$-polarization $\tilde{H}$ that is (I,d)-generic for any $d \geq 0$. We then choose N so that all previous results holds for this N. Now for any $d \geq N$, let $\tilde{H} \in \mathcal{C}$ be (I,d)-generic. Then by Lemma 3.5, $\mathrm{Pic}\big(\mathfrak{M}_H(I,d)\big)$ is isomorphic to $\mathrm{Pic}\big(\mathfrak{M}_{\tilde{H}}(I,d)\big)$, which is isomorphic to $\mathrm{Pic}\big(\overline{\mathfrak{M}}_{\tilde{H}}(I,d)\big)$ by Lemma 3.6. On the other hand, the homomorphism Φ above certainly commutes with

$$\overline{\Phi} : \mathrm{Pic}\big(X \times X\big)^{\sigma} \oplus \mathbb{Z} \longrightarrow \mathrm{Pic}\big(\overline{\mathfrak{M}}_{\tilde{H}}(I,d)\big)$$

under these isomorphisms. Therefore, Φ has finite kernel and cokernel because $\overline{\Phi}$ does. This completes the proof of the corollary. □

Before we prove the theorem, we need two more technical lemmas. The first concerns the Hodge decomposition of $H^2\big(\mathfrak{M}_H(I,d);\mathbb{R}\big)$. Since we know $H^2\big(\mathfrak{M}_H(I,d);\mathbb{R}\big)$ explicitly, we will determine its Hodge decomposition with the aid of

$$g : W \longrightarrow \mathfrak{M}_H(I,d)$$

that is constructed in the beginning of §2 based on a μ-stable rank two locally free sheaf $\mathcal{E}$ with $\det \mathcal{E} = I$ and $c_2(\mathcal{E}) = d - 2$. Let

$$g^{\mathrm{pic}} : \mathrm{Pic}\big(\mathfrak{M}_H(I,d)\big) \longrightarrow \mathrm{Pic}\big(W\big)$$

be the induced homomorphism.

Lemma 3.10. *For any precompact $\mathcal{C} \Subset N_{\mathbb{Q}}^{+}$, there is an integer N depending on $(X,I,\mathcal{C})$ such that for any $d \geq N$, $H \in \mathcal{C}$ and $g : W \longrightarrow \mathfrak{M}_H(I,d)$ as before, the image*

$$c_1 \circ g^{pic} : Pic\big(\mathfrak{M}_H(I,d)\big) \otimes_{\mathbb{Z}} \mathbb{Q} \longrightarrow H^2(W;\mathbb{Q})$$

is isomorphic to

$$H^{1,1}(X \times X;\mathbb{R})^{\sigma} \cap H^2(X \times X;\mathbb{Q}) \oplus \mathbb{Q}\,.$$

Proof. Let Λ be the image of $c_1 \circ g^{\mathrm{pic}}$. ¿From the gauge theory and the proof of Lemma 3.3 in [Li3], the image of

$$g^* : H^2\big(\mathfrak{M}_H(I,d);\mathbb{R}\big) \longrightarrow H^2(W;\mathbb{R})$$

is spanned by the images

$$p_{W*}\big(c_2(\mathcal{G}) \cup p_X^*(\cdot)\big) : H^2(X;\mathbb{R}) \longrightarrow H^2\big(W;\mathbb{R}\big),$$

$$p_{W*}\big(c_3(\mathcal{G}) \cup p_X^*(\cdot)\big) : H^0(X;\mathbb{R}) \longrightarrow H^2\big(W;\mathbb{R}\big)$$

and the image

$$\wedge^2 p_{W*}\big(c_2(\mathcal{G}) \cup p_X^*(\cdot)\big) : \wedge^2 H^1(X;\mathbb{R}) \longrightarrow H^1(W;\mathbb{R}).$$

where $\mathcal{G}$ is the sheaf on $W \times X$ constructed in §2 and p_W, p_X are projections of $W \times X$. Based on exact sequences (2.3) and (2.4), it is straight forward to check that the direct sum of these homomorphisms

$$\mu : H^2(X;\mathbb{R}) \oplus \wedge^2 H^1(X;\mathbb{R}) \oplus H^0(X;\mathbb{R}) \longrightarrow H^2(W;\mathbb{R})$$

is injective. Since $c_i(\mathcal{G})$ is an integral class of pure Hodge type, μ preserves the integer lattice as well as the obvious Hodge structures on both sides. In particular,

$$\Lambda \subseteq \mathrm{Im}\,\big(\mu\big) \cap \big(H^{1,1}(W;\mathbb{R}) \cap H^2(W;\mathbb{Q})\big) \tag{3.2}$$

and the later is isomorphic to the direct sum of

$$H^0(X;\mathbb{Q})$$

$$H^{1,1}(X;\mathbb{R}) \cap H^2(X;\mathbb{Q})$$

and

$$\left(H^{1,0}(X;\mathbb{R}) \wedge H^{0,1}(X;\mathbb{R})\right) \cap \left(\wedge^2 H^1(X;\mathbb{Q})\right).$$

On the other hand, by our construction of the homomorphism

$$\Phi_{\mathbb{Q}} : \operatorname{Pic}\left(X \times X\right)^{\sigma} \otimes_{\mathbb{Z}} \mathbb{Q} \oplus \mathbb{Q} \longrightarrow \operatorname{Pic}\left(\mathfrak{M}_H(I,d)\right) \otimes_{\mathbb{Z}} \mathbb{Q}$$

and the formulas of their Chern classes (see (1.6), (1.7) and (1.8)), we see that the homomorphism

$$H^{1,1}(X \times X;\mathbb{R})^{\sigma} \cap H^2(X \times X;\mathbb{Q}) \oplus \mathbb{Q} \longrightarrow H^2(W,\mathbb{Q}) \tag{3.3}$$

induced by

$$g^{\text{pic}} \circ \Phi_{\mathbb{Q}} : \operatorname{Pic}\left(X \times X\right)^{\sigma} \otimes_{\mathbb{Z}} \mathbb{Q} \oplus \mathbb{Q} \longrightarrow \operatorname{Pic}(W) \otimes_{\mathbb{Z}} \mathbb{Q}$$

is injective. Clearly, the image of (3.3) is contained in Λ. Therefore, if we can show that the left hand side of (3.3) is isomorphic to the right hand side of (3.2), then for dimensional reason the equality in (3.2) must hold and then the lemma follows immediately.

We now show that

$$\begin{aligned} H^{1,1}(X \times X;\mathbb{R})^{\sigma} \cap H^2(X \times X;\mathbb{Q}) &\cong H^{1,1}(X;\mathbb{R}) \cap H^2(X;\mathbb{Q}) \\ &\oplus \left(H^{1,0}(X;\mathbb{R}) \wedge H^{0,1}(X;\mathbb{R})\right) \cap \left(\wedge^2 H^1(X;\mathbb{Q})\right). \end{aligned} \tag{3.4}$$

We will use the Kunneth decomposition

$$H^2(X \times X;\mathbb{R}) = \bigoplus_{i+j=2} H^i(X;\mathbb{R}) \otimes H^j(X;\mathbb{R}).$$

It is straightforward to check that an element

$$v = \sum_{i=1}^{n}\left(b_i(1 \otimes \alpha_i) + b_i'(\alpha_i \otimes 1)\right) + \sum_{i,j=1}^{m} a_{ij}(\beta_i \otimes \beta_j)$$

in $H^2(X \times X;\mathbb{R})$ is σ invariant if and only if

$$b_i = b_i' \quad \text{and} \quad a_{ij} = -a_{ji},$$

where $\{\alpha_i\}^n$ and $\{\beta_j\}^m$ are basis of $H^2(X;\mathbb{R})$ and $H^1(X;\mathbb{R})$ respectively. Hence

$$H^2(X \times X;\mathbb{R})^{\sigma} \cong H^2(X;\mathbb{R}) \oplus \wedge^2 H^1(X;\mathbb{R}).$$

We can derive a similar formula for $H^{1,1}(X \times X;\mathbb{R})^{\sigma}$ and $H^{1,1}(X \times X;\mathbb{Q})^{\sigma}$. Combined, we get the desired identity (3.4). This completes the proof of Lemma 3.9. □

The next lemma states that the restriction of $c_1(L)$ to $H^2\left(\mathfrak{M}_H(I,d);\mathbb{Z}\right)$, where L is a line bundle on $\overline{\mathfrak{M}}_H(I,d)$, determines $c_1(L)$ completely. This is crucial to our study since we only know the structure of $H^2\left(\mathfrak{M}_H(I,d);\mathbb{R}\right)$ rather than $H^2\left(\overline{\mathfrak{M}}_H(I,d);\mathbb{R}\right)$.

Lemma 3.11. *Let the notation be as in Theorem 3.8. Then there is an N depending on (X,I,C) such that for any $d \geq N$ and (I,d)-generic $H \in C$ the following holds: Let L be a line bundle on $\overline{\mathfrak{M}}_H(I,d)$ such that the restriction of its Chern class to $H^2\left(\mathfrak{M}_H(I,d);\mathbb{Z}\right)$ is trivial, then L is derived from a representation*

$$\rho : \pi_1\left(\overline{\mathfrak{M}}_H(I,d)\right) \longrightarrow S^1.$$

Proof. Let Z be a desingularization of $\overline{\mathfrak{M}}_H(I,d)$ with projection $\pi : Z \to \overline{\mathfrak{M}}_H(I,d)$ and let $D_1, \cdots, D_k$ be irreducible components of the exceptional divisor of π, which we assume is normal crossing. Then the kernel of the composition

$$H^2(Z;\mathbb{Z}) \xrightarrow{\pi^*} H^2\left(\overline{\mathfrak{M}}_H(I,d);\mathbb{Z}\right) \xrightarrow{\text{rest.}} H^2\left(\overline{\mathfrak{M}}_H(I,d)_{\text{reg}};\mathbb{Z}\right),$$

where $\overline{\mathfrak{M}}_H(I,d)_{\text{reg}}$ is the regular locus of $\overline{\mathfrak{M}}_H(I,d)$, is spanned by all $c_1(\mathcal{O}(D_i))$. In particular there are integers $n_1, \cdots, n_k$ such that

$$c_1\Big(\pi^* L(\textstyle\sum n_i D_i)\Big) = 0 \in H^2(Z;\mathbb{Z}),$$

since the restriction to $H^2\Big(\mathfrak{M}_H(I,d);\mathbb{Z}\Big)$ of $c_1(L)$ is trivial. Hence

$$\tilde{L} = \pi^* L\Big(\textstyle\sum n_i D_i\Big)$$

is induced by a representation $\tilde{\rho}: \pi_1(Z) \to S^1$, or induced by a local system $\tilde{\Sigma}$.

It remains to show that $\tilde{\Sigma}$ descends to a local system Σ on $\overline{\mathfrak{M}}_H(I,d)$, since then the line bundle induced by Σ is isomorphic to L on $\overline{\mathfrak{M}}_H(I,d)_{\text{reg}}$ and then by 3.1, they are isomorphic on $\overline{\mathfrak{M}}_H(I,d)$ as well. This will prove the lemma.

Instead of studying the descent of $\tilde{\Sigma}$ via projection $\pi: Z \to \overline{\mathfrak{M}}_H(I,d)$, which requires the knowledge of the singularity of $\overline{\mathfrak{M}}_H(I,d)$, we will work with the quotient $p: \mathbf{Quot}_H(I,d)^{ss} \to \overline{\mathfrak{M}}_H(I,d)$ since virtually all local information of $\overline{\mathfrak{M}}_H(I,d)$ is contained in this construction. First, since $\pi: Z \to \overline{\mathfrak{M}}_H(I,d)$ is an isomorphism over $\overline{\mathfrak{M}}_H(I,d)_{\text{reg}}$, we obtain a local system Σ' on $\overline{\mathfrak{M}}_H(I,d)_{\text{reg}}$ and its pull-back $p^*(\Sigma')$ on $p^{-1}\Big(\overline{\mathfrak{M}}_H(I,d)_{\text{reg}}\Big) \subseteq \mathbf{Quot}_H(I,d)^{ss}$. Clearly, $p^*(\Sigma')$ is G-equivalent. Let

$$R = \mathbf{Quot}_H(I,d)^{ss} - p^{-1}\Big(\overline{\mathfrak{M}}_H(I,d)_{\text{reg}}\Big).$$

By Lemma 3.1, R has codimension at least 5. Then for any closed $x \in R$ and normal slice S of $R \subseteq \mathbf{Quot}_H(I,d)^{ss}$ at x, $\pi_1(S-x) = \{1\}$, because $\mathbf{Quot}_H(I,d)^{ss}$ is a local complete intersection scheme ((2) of Lemma 3.2) and $\operatorname{codim}(R) \geq 5$, by the local Lefschetz theorem on page 155 of [GM]. Therefore $p^*(\Sigma')$ extends to a local system, say Σ'', that is G-equivalent as well, since $\mathbf{Quot}_H(I,d)^{ss}$ is normal. To show that Σ'' descends to Σ on $\overline{\mathfrak{M}}_H(I,d)$, it suffices to show that for any $w \in \mathbf{Quot}_H(I,d)^{ss}$ with closed orbit $G \cdot w$, stab_w acts trivially on the fiber of Σ'' over w. But this can be proved using the fact that $H^2(X;\mathbb{Z})$ has no torsion as we did in the proof of Lemma 3.6. This shows that Σ'' descends to a local system Σ. This completes the proof of the Lemma. □

Proof of Theorem 3.8. We let N be the integer that makes all previous results valid. Let $d \geq N$ and let $H \in \mathcal{C}$ be (I,d)-generic. By Lemma 3.8 for any line bundle L on $\overline{\mathfrak{M}}_H(I,d)$, the restriction of its Chern class $c_1(L)$ to $H^2\Big(\mathfrak{M}_H(I,d);\mathbb{Q}\Big)$ is contained in the image of

$$c_1 \circ \Phi : \operatorname{Pic}\Big(X \times X\Big)^{\sigma} \otimes_{\mathbf{Z}} \mathbb{Q} \oplus \mathbb{Q} \longrightarrow H^2\Big(\mathfrak{M}_H(I,d);\mathbb{Q}\Big).$$

Hence we can find an element $\alpha \in \operatorname{Pic}\Big(X \times X\Big)^{\sigma} \oplus \mathbb{Z}$ and an integer k such that the restriction of $c_1\Big(L^{\otimes k} \otimes L_\alpha\Big)$ to $H^2\Big(\mathfrak{M}_H(I,d);\mathbb{Z}\Big)$ is trivial, where L_α is the line bundle $\overline{\Phi}(\alpha)$ on $\overline{\mathfrak{M}}_H(I,d)$. Here k can be chosen to depend only on the induced homomorphism

$$H^2(X \times X;\mathbb{Z})^{\sigma} \oplus \mathbb{Z} \longrightarrow H^2\Big(\mathfrak{M}_H(I,d);\mathbb{Z}[\frac{1}{12}]\Big).$$

By Lemma 3.10, $L^{\otimes k} \otimes L_\alpha$ is induced by a representation $\rho : \pi_1\Big(\overline{\mathfrak{M}}_H(I,d)\Big) \longrightarrow S^1$, or an element in $H^1\Big(\overline{\mathfrak{M}}_H(I,d);\mathbb{R}\Big)$. However, $\overline{\mathfrak{M}}_H(I,d)$ is normal and $\overline{\mathfrak{M}}_H(I,d) - \mathfrak{M}_H(I,d)$ has codimension at least 2, hence

$$H^1\Big(\overline{\mathfrak{M}}_H(I,d);\mathbb{R}\Big) \longrightarrow H^1\Big(\mathfrak{M}_H(I,d);\mathbb{R}\Big)$$

is injective. On the other hand, we know

$$\dim H^1\Big(\mathfrak{M}_H(I,d);\mathbb{R}\Big) = h^1(X)$$

and the restriction of Φ

$$\Phi_0 : \operatorname{Pic}^0\big(X \times X\big)^{\sigma} \longrightarrow \operatorname{Pic}^0\big(\overline{\mathfrak{M}}_H(I,d)\big)$$

has finite kernel since

$$g^{\mathrm{pic}} \circ \Phi_0 : \operatorname{Pic}\big(X \times X\big)^{\sigma} \longrightarrow \operatorname{Pic}(W)$$

has finite kernel, by direct check. This implies $\operatorname{Im}(\Phi_0)$ has dimension $h^1(X)$. Therefore, Φ_0 is surjective. This proves that

$$\Phi : \operatorname{Pic}\big(X \times X\big)^{\sigma} \oplus \mathbb{Z} \longrightarrow \operatorname{Pic}\big(\overline{\mathfrak{M}}_H(I,d)\big) \otimes_{\mathbb{Z}} \mathbb{Z}[\frac{1}{12}]$$

has finite cokernel. □

Remark 3.12. 1. *The author inclined to believe the condition that $H^2(X;\mathbb{Z})$ is torsion free in this theorem is unnecessary, but is unable to remove it using the current technique.*

2. *It is interesting to to find a bound N in this theorem that can be determined effectively. After the effective bound of O'Grady [OG], we can find an effective bound that works for all results in this note that are independent of Lemma 3.3 whose proof relies on the result of C. Taubes that is not effective. It is interesting to get a purely algebraic proof of Lemma 3.3.*

References

[DN] Drezet, J.M. and Narasimhan, S.: Group de Picard des varietes de modules de fibres semi-stables sur les coubes algebriques, Invent. Math. **97** (1989), 53–94.

[Fu] Fulton, W.: Intersection Theory, Ergebnisse der Mathematik und ihrer Grenzgebiete,Springer-Verlag, 3 Band 2 (1984).

[GH] Göttsche, L. and Huybrechts, D.: Hodge numbers of moduli spaces of stable bundles on K3 surfaces, preprint (1994).

[Gi1] Gieseker, D.: On the moduli of vector bundles on an algebraic surface, Ann. Math. **106** (1977), 45–60.

[Gi2] Gieseker, D.: A construction of stable bundles on an algebraic surfaces, J. of Diff. Geometry **27**(1988), 137–154.

[GM] Goresky, M. and MacPherson, R.:Stratified Morse Theory Ergebnisse der Mathematik und ihrer Grenzgebiete, Springer-Verlag 3 Band 14 (1988).

[Gr] Grothendieck, A.:Techniques de construction et théorèmes d'existence en géométrie algébrique IV, les schémas de Hilbert, Séminaire Bourbaki **221** (1960/61).

[Gr] Grothendieck, A.:Cohomologie locale des faisceaux cohérents et Théorèmes de Lefschetz locaux et globaux, (SGA 2), North-Holland Publishing Company.

[Ha] Hartshorne, R.: Algebraic Geometry, Graduate Texts in Mathematics **52**, Springer-Verlag, Berlin–Heidelberg–New York.

[KM] Knudsen, F. and Mumford, D.: The projectivity of the moduli of stable curves I, preliminaries on 'det' and 'div', Math. Scand. **39** (1976), 19–55.

[Li1] Li, J.: Algebraic geometric interpretation of Donaldson's polynomial invariants, J. of Diff. Geometry **37** (1993), 417–466.

[Li2] Li, J.: Kodaira dimension of moduli space of vector bundles on surfaces, Invent. Math. **115** (1994), 1–40.

[Li3] Li, J.: The first two Betti numbers of the moduli of vector bundles over algebraic surfaces, preprint (1995).

[Ma] Maruyama, M.: Moduli of stable sheaves I, J. Math. Kyoto Univ. **17** (1977), 91–126: II, J. Math. Kyoto Univ. **18** (1978), 557–614.

[OG] O'Grady, K.: Moduli of vector bundles onprojective surfaces: some basic results, preprint.

[Qi1] Qin, Z.: Birational properties of moduli spaces of stable locally free rank-2 sheaves on algebraic surfaces, Manuscripta Math. **72** (1991), 163–180.

[Qi2] Qin, Z.: Moduli of stable sheaves on ruled surfaces and their Picard groups, J. reine angew. Math. **433** (1992), 201–219.

[St] St Strømme, S.A.: Ample divisors on fine moduli spaces on projective plane, Math. Z. **187** (1984), 405–423.

[Yo] Yoshioka, K.: The picard group of the moduli space of stable sheaves on a ruled surface, preprint (1994).

CONSTRUCTION OF MODULI SPACES OF STABLE SHEAVES VIA SIMPSON'S IDEA

Masaki Maruyama

Department of Mathematics, Graduate School of Science, Kyoto University, Sakyo-ku 606-01, Kyoto, Japan
e-mail: maruyama@kusm.kyoto-u.ac.jp

INTRODUCTION

The main body of the construction of the moduli space of semi-stable sheaves is how to make a quotient of a good part of a Quot-scheme by a semi-simple algebraic group. The construction in the framework of *Geometric Invariant Theory*, therefore, reduces to the computation of semi-stability of the good points in the Quot-scheme and the study of their orbits structure. To carry these out we had employed Gieseker's idea, that is, a morphism of the Quot-scheme to a Gieseker's space. Here we had used a *finite* resolution of a coherent sheaf by locally free sheaves and hence the underlying variety must be non-singular (see [3], [4]). To avoid this problem we can use a morphism to a product of Grassmann varieties and get moduli spaces on projective varieties which may have singularities (cf. [2]). We come then to another difficulty. In fact, if we follow this idea, then we cannot prove directly the projectivity of the moduli spaces even on a non-singular surface.

C. Simpson [9] went back to the original construction of the Quot-scheme and found that for the computation of semi-stability it is enough to exploit its embedding into a Grassmann scheme we used to prove the representability of Quot-functor. He succeeded in not only removing the assumption of smoothness but also extending the category of sheaves we handle. The role of underlying varieties is now reduced to the land where the support schemes of the sheaves can move around, and moreover the support scheme can be even non-reduced though it must satisfy Serre's condition (S_1). To introduce Simpson's notion of semi-stability we first need the following.

Definition 0.1. Let X be a projective scheme X over a field k. A coherent sheaf F on X is said to be of pure dimension d if $\dim \operatorname{Supp}(F) = d$ and if for every non-zero coherent subsheaf F' of F, we have $\dim \operatorname{Supp}(F') = d$.

It is easy to see that F is of pure dimension d if and only if F satisfies the condition (S_1) and $\operatorname{Supp}(F)$ is of pure dimension d. Thus, for every over field K of k, $F \otimes_k K$ is of pure dimension d on X_K if and only if so is F on X. It is obvious that if X is integral and $\operatorname{Supp}(F) = X$, then the pure dimensionality is nothing but the torsion freeness and that if a coherent sheaf is of pure dimension d, then so is every its coherent subsheaf. If a coherent sheaf F on X is of pure dimension d and if we fix an ample invertible sheaf $\mathcal{O}_X(1)$, then there are integers $a_0(F), a_1(F), \ldots, a_d(F)$ such that

$$\chi(F(m)) = \sum_{i=0}^{d} a_i(F) \binom{m+d-i}{d-i}.$$

Since d is not necessarily $\dim X$ and since $\mathrm{Ann}(F)$ may define a reducible or even non-reduced subscheme, we have no reasonable notion of the rank and hence we are going to use $a_0(F)$ instead of the rank.

Definition 0.2. Let $(X, \mathcal{O}_X(1))$ be a pair of a projective scheme X over a field k and an ample invertible sheaf $\mathcal{O}_X(1)$ and let F be a coherent sheaf on X of pure dimension d.

(1) The rational number $a_1(F)/a_0(F)$ is denoted by $\mu^S(F)$ and we denote the polynomial $\chi(F(m))/a_0(F)$ by $P_F^S(m)$.

(2) F is said to be μ-stable (or, μ-semi-stable) if for every coherent subsheaf E of $F \otimes_k \bar{k}$ with $0 < a_0(E) < a_0(F)$, we have $\mu^S(E) < \mu^S(F)$ ($\mu^S(E) \leq \mu^S(F)$, resp.).

(3) F is said to be stable (or, semi-stable) if for every coherent subsheaf E of $F \otimes_k \bar{k}$ with $0 < a_0(E) < a_0(F)$ and for all sufficiently large integers m, we have $P_E^S(m) < P_F^S(m)$ ($P_E^S(m) \leq P_F^S(m)$, resp.).

Note that if X is smooth and if $d = \dim X$, the above notion of (semi-)stability coincides with that for torsion free sheaves. As the case of usual semi-stable sheaves, we can prove a proposition which leads us to the notion of Seshadri equivalence.

Proposition 0.3. *Let* $(X, \mathcal{O}_X(1))$ *be as in* Definition 0.1 *and let* E *be a coherent sheaf on* X *of pure dimension* d. *Assume that* E *is semi-stable and* k *is algebraically closed.*

(1) *There exists a filtration* $0 = E_0 \subset E_1 \subset \cdots \subset E_{\alpha-1} \subset E_\alpha = E$ *by coherent* $\mathcal{O}_X$*-modules of pure dimension* d *such that* (a) E_i/E_{i-1} *is stable* $(1 \leq i \leq \alpha)$ *and* (b) $P_{E_i}^S(m) = P_E^S(m)$ $(1 \leq i \leq \alpha)$.

(2) *If* $0 = E'_0 \subset E'_1 \subset \cdots \subset E'_{\beta-1} \subset E'_\beta = E$ *is another filtration with the properties* (a) *and* (b), *then* $\alpha = \beta$ *and there is a permutation* σ *of* $\{1, \ldots, n\}$ *such that* E_i/E_{i-1} *is isomorphic to* $E'_{\sigma(i)}/E'_{\sigma(i)-1}$ $(1 \leq i \leq \alpha)$.

Definition 0.4. Let E and E' be semi-stable sheaves of pure dimension d on a projective scheme over an algebraically field.

(1) A filtration $0 = E_0 \subset E_1 \subset \cdots \subset E_{\alpha-1} \subset E_\alpha = E$ with the properties (a) and (b) of Proposition 0.3 is called a Seshadri filtration.

(2) For a Seshadri filtration $0 = E_0 \subset E_1 \subset \cdots \subset E_{\alpha-1} \subset E_\alpha = E$, we define $\mathrm{gr}(E)$ to be $\bigoplus_{i=1}^{\alpha} E_i/E_{i-1}$. Each E_i/E_{i-1} is called a component of $\mathrm{gr}(E)$.

(3) E and E' are said to be S-equivalent if $\mathrm{gr}(E) \cong \mathrm{gr}(E')$ and then we denote $E \sim_S E'$

To fix our idea let us introduce moduli functors of stable and semi-stable sheaves.

Definition 0.5. Let (Sch/S) be the category of locally noetherian S-schemes and $H(x)$ a numerical polynomial of degree d. For an object T of (Sch/S) we define two sets as follows:

$$(1)\ \Sigma_{X/S}^H(T) = \left\{ E \,\middle|\, \begin{array}{l} E \text{ is a } T\text{-flat coherent sheaf on } X \times_S T \\ \text{with the property (a)} \end{array} \right\} \Big/ \sim$$

where $\sim$ is the equivalence relation defined by (c).

$$(2)\ \bar{\Sigma}_{X/S}^H(T) = \left\{ E \,\middle|\, \begin{array}{l} E \text{ is a } T\text{-flat coherent sheaf on } X \times_S T \\ \text{with the property (b)} \end{array} \right\} \Big/ \sim$$

where $\sim$ is the equivalence relation defined by (d). The equivalence class of E is denoted by $[E]$.

(a) For every geometric point t of T, $E(t)$ is stable and its Hilbert polynomial is $H(m)$.
(b) For every geometric point t of T, $E(t)$ is semi-stable and its Hilbert polynomial is $H(m)$.
(c) $E \sim E'$ if and only if there is an invertible sheaf L on T such that $E \cong E' \otimes_{\mathcal{O}_T} L$.

(d) $E \sim E'$ if and only if for some invertible sheaf L on T, (1) $E \cong E' \otimes_{\mathcal{O}_T} L$ or (2) there exist filtrations $0 = E_0 \subset E_1 \subset \cdots \subset E_\alpha = E$ and $0 = E'_0 \subset E'_1 \subset \cdots \subset E'_\alpha = E'$ by coherent subsheaves such that $\oplus_{i=1}^{\alpha} E_i/E_{i-1}$ is flat over T, $\oplus_{i=1}^{\alpha} E_i/E_{i-1} \cong (\oplus_{i=1}^{\alpha} E'_i/E'_{i-1}) \otimes_{\mathcal{O}_T} L$ and that for every geometric point t of T, $0 = E_0(t) \subset E_1(t) \subset \cdots \subset E_\alpha(t) = E(t)$ is a Seshadri filtration of $E(t)$.

If $g : T' \to T$ is a morphism in (Sch/S), then the above properties and the equivalence relations are preserved by making the pull back $(1 \times_S g)^*(E)$. Thus both $\Sigma^H_{X/S}$ and $\bar{\Sigma}^H_{X/S}$ define contravariant functors of (Sch/S) to (Sets).

Then our main results of this article are stated as follows.

Theorem 0.6. *Let $f : X \to S$ be a projective morphism of schemes of finite type over a universally Japanese ring Ξ and $\mathcal{O}_X(1)$ be an f-ample invertible sheaf on X. Assume that there is a locally free coherent sheaf U and a closed immersion $i : X \to \mathbf{P}(U)$ over S such that $\mathcal{O}_X(1)$ is isomorphic to $i^*(\mathcal{O}_{\mathbf{P}(U)}(1))$.*

(1) $\Sigma^H_{X/S}$ has a coarse moduli scheme $M_{X/S}(H)$ in (Sch/S). *$M_{X/S}(H)$ is locally of finite type and separated over S.*

(2) $\bar{\Sigma}^H_{X/S}$ has a coarse moduli scheme $\bar{M}_{X/S}(H)$. $\bar{M}_{X/S}(H)$ is locally of finite type and separated over S and it contains the moduli scheme $M_{X/S}(H)$ of stable sheaves as an open subscheme.

(3) $\bar{M}_{X/S}(H)$ is projective over S if and only if the family of semi-stable sheaves on the fibers of X over S with Hilbert polynomial H is bounded.

Theorem 0.7. *$\bar{M}_{X/S}(H)$ is projective over S if one of the following conditions is satisfied:*

(1) Ξ is a field of characteristic zero.

(2) H is a polynomial of degree not greater than 2.

(3) $f : X \to S$ is geometrically integral, $\deg H = \dim X/S$ and the rank of sheaves is not greater than 3.

In this article we shall concentrate on showing the idea to construct a good quotient of a good part of a Quot-scheme by $PGL(N)$ and some of the detail will appear in forthcoming papers or the book"Moduli of Vector Bundles" which is now in preparation.

1. Boundedness of semi-stable sheaves

In this section we shall generalize our results of boundedness [6] to the case of Simpson's semi-stability. First we shall collect basic and remarkable property of stable or semi-stable sheaves whose proof is completely the same as in the case of torsion free sheaves.

Proposition 1.1. *Let $(X, \mathcal{O}_X(1))$ be as in* Definition 0.1 *and let E and F be coherent sheaves on X of pure dimension d.*

(1) If both E and F are μ-semi-stable and if $\mu^S(E) > \mu^S(F)$, then every homomorphism of E to F is 0.

(2) If E is μ-stable, F is μ-semi-stable and if $\mu^S(E) = \mu^S(F)$, then every non-zero homomorphism of E to F is injective.

(3) If both E and F are semi-stable and if $P^S_E(m) > P^S_F(m)$ for all sufficiently large m, then every homomorphism of E to F is 0.

(4) If E is stable, F is semi-stable and if $P^S_E(m) = P^S_F(m)$, then every non-zero homomorphism of E to F is injective and F/E is semi-stable.

(5) Assume that E is stable, then E is simple.

(6) *Assume that E is an extension of coherent sheaves of pure dimension d*

$$0 \longrightarrow E_1 \longrightarrow E \longrightarrow E_2 \longrightarrow 0$$

such that $P^S_E(m) = P^S_{E_1}(m) = P^S_{E_2}(m)$ (or, $\mu^S(E) = \mu^S(E_1) = \mu^S(E_2)$). Then E is semi-stable (μ-semi-stable, resp.) if and only if so are E_1 and E_2.

Based on these properties of semi-stable sheaves, we can see the existence and the uniqueness of Harder-Narasimhan filtrations (for the proof, see [5]).

Proposition 1.2. *Let $(X, \mathcal{O}_X(1))$ be as in as in* Definition 0.1 *and let E be a coherent $\mathcal{O}_X$-module of pure dimension d.*

(1) *There is a unique filtration $0 = E_0 \subset E_1 \subset \cdots \subset E_\alpha = E$ by coherent subsheaves such that* (a) *E_i/E_{i-1} is μ-semi-stable for $0 < i \leq \alpha$ and* (b) *$\mu^S(E_i/E_{i-1}) > \mu^S(E_{i+1}/E_i)$ for $0 < i < \alpha$.*

(2) *There is a unique filtration $0 = E'_0 \subset E'_1 \subset \cdots \subset E'_\beta = E$ by coherent subsheaves such that* (c) *E'_i/E'_{i-1} is semi-stable for $0 < i \leq \alpha$ and* (d) *$P^S_{E'_i/E'_{i-1}}(m) > P^S_{E'_{i+1}/E'_i}(m)$ for all sufficiently large integer m and for $0 < i < \beta$.*

Definition 1.3. The unique filtration of E in Proposition 1.2, (1) is called the Harder-Narasimhan filtration of E.

As in the case of usual semi-stable sheaves [5], connecting successively the points $x_i(E_i) = (a_0(E_i), a_1(E_i))$ by segments in a two dimensional real plane, we get the *Harder-Narasimhan polygon* of E and obtain the following.

Proposition 1.4. *Let E be as in* Proposition 1.2. *For every coherent subsheaf F of $E \otimes_k \bar{k}$, the point $x(F)$ is below or on the Harder-Narasimhan polygon of E.*

To consider the problem of the boundedness of semi-stable sheaves we need the notion of type of a coherent sheaf.

Definition 1.5. Let $(X, \mathcal{O}_X(1))$ be as in as in Definition 0.1 and let E be a coherent $\mathcal{O}_X$-module of pure dimension d. By setting $r = a_0(E)$, let $(\alpha) = (\alpha_1, \alpha_2, \ldots, \alpha_{r-1})$ be a sequence of rational numbers. E is said to be of type (α) if for every coherent quotient sheaf F of $E \otimes_k \bar{k}$ of pure dimension d with $0 < a_0(F) < r$, we have the inequality $\mu^S(F) \geq \mu^S(E) - \alpha_s$, where $s = a_0(F)$.

We are going to study the following families.

Definition 1.6. Let $f : X \to S$ be a projective morphism of noetherian schemes and $\mathcal{O}_{X/S}(1)$ an f-very ample invertible sheaf. Fix non-negative integers d and r, a numerical polynomial $H(x)$ of degree d with the leading coefficient $r/d!$, a sequence $(a_1, \ldots, a_d)$ of integers and a sequence $(\alpha) = (\alpha_1, \ldots, \alpha_{r-1})$ of rational numbers.

(1) Let $\mathcal{S}_{X/S}(d; r, a_1, \ldots, a_d; (\alpha))$ be the family of the classes of coherent sheaves on the fibers of X over S such that E on a geometric fiber X_s is a member of the family if E has the properties (1.6.1) and (1.6.2).

(2) Let $\mathcal{S}'_{X/S}(d; r, a_1, a_2; (\alpha))$ be the family of the classes of coherent sheaves on the fibers of X over S such that E on a geometric fiber X_s is a member of the family if E has the properties (1.6.1) and (1.6.3).

(3) Let $\mathcal{S}_{X/S}(d; H(m), (\alpha))$ be the family of the classes of coherent sheaves on the fibers of X over S such that E on a geometric fiber X_s is a member of the family if E has the properties (1.6.1) and if $\chi(E(m)) = H(m)$.

(1.6.1) E is of pure dimension d and of type (α).

(1.6.2) $a_0(E) = r$, $a_1(E) = a_1$ and $a_i(E) \geq a_i \quad (2 \leq i \leq d)$.
(1.6.3) E satisfies the condition (S_2), $a_0(E) = r$, $a_1(E) = a_1$ and $a_2(E) \geq a_2$.

We shall show first a relationship among the above three families.

Lemma 1.7. *Let the pair of* $(X, \mathcal{O}_X(1))$ *be the same as in* Definition 1.6.

(1) *If* $\mathcal{S}_{X/S}(d; r, a_1, \ldots, a_d; (\alpha))$ *is bounded for all* $(a_1, \ldots, a_d)$, *then so is* $\mathcal{S}_{X/S}(d; H(m), (\alpha))$ *for all* $H(x)$.

(2) *If* $\mathcal{S}'_{X/S}(d; r, a_1, a_2; (\alpha))$ *is bounded for all pairs* (a_1, a_2), *then so is* $\mathcal{S}_{X/S}(d; H(m), (\alpha))$ *for all* $H(x)$

Proof. (1) is obvious by the very definition of the two families. To prove (2) take a representative E on a geometric fiber X_s of a member of $\mathcal{S}_{X/S}(d; H(m), (\alpha))$. Since $H(x)$ is a numerical polynomial, we can write

$$H(m) = \sum_{i=0}^{d} a_i(H) \binom{m+d-i}{d-i}.$$

with $a_0(H) = r$ and $a_i(H)$ integers. Let Z be the closed subscheme of X_s defined by the ideal $\mathrm{Ann}(E)$ and let W be the set of points z of Z with $\dim \mathcal{O}_{Z,z} \geq 2$. Let us look at the sheaf $E' = \mathcal{H}^0_{Z/W}(E)$. Since Z is embedded in a projective space over $k(s)$, the condition (S_1) for E implies that E' is coherent by [EGA IV, Proposition 5.11.1]. Then the natural homomorphism $\rho_{Z/W} : E \to E'$ is injective and isomorphic outside a set of codimension ≥ 2 and moreover E' satisfies the condition (S_2) by [EGA IV, Corollaire 5.10.5]. Thus E' is of type (α), $a_0(E') = a_0(E) = r$, $a_1(E') = a_1(E) = a_1(H)$ and $a_2(E') \geq a_2(E) = a_2(H)$. This means that E' is a member of $\mathcal{S}'_{X/S}(d; r, a_1(H), a_2(H); (\alpha))$. The boundedness of the last family implies that there is a sequence of integers $(b) = (b_0, \ldots, b_d)$ such that E''s are (b)-sheaves and hence so are E's. □

We can reduce the boundedness of the above defined three families reduces to the case of torsion free sheaves on smooth projective schemes. The main idea here is due to C. Simpson and J. Le Potier.

Theorem 1.8. *Let the pair of* $(X, \mathcal{O}_X(1))$ *be the same as in* Definition 1.6. *Assume that* S *is over a field of characteristic 0 or that* $d \leq 2$. *Then both* $\mathcal{S}_{X/S}(d; r, a_1, \ldots, a_d; (\alpha))$ *and* $\mathcal{S}'_{X/S}(d; r, a_1, a_2; (\alpha))$ *are bounded. Hence, in these cases,* $\mathcal{S}_{X/S}(d; H(m), (\alpha))$ *is bounded, too.*

Proof. The final assertion is by Lemma 1.7. We are going to prove our theorem only for $\mathcal{S}'_{X/S}(d; r, a_1, a_2; (\alpha))$ because the proof of another family is similar and easier. For a coherent sheaf E on a geometric fiber X_s of pure dimension d, we set $\mu_{\max}(E)$ to be the $\mu^S(E_1)$ with E_1 the first filter of the Harder-Narasimhan filtration of E. Obviously if E is of type (α), then $\mu_{\max}(E)$ is bounded by an constant determined by (α). Conversely, by Proposition 1.4 we can bound the type by bounding $\mu_{\max}(E)$. Thus we may replace the family $\mathcal{S}'_{X/S}(d; r, a_1, a_2; (\alpha))$ by $\mathcal{S}'_{X/S}(d; r, a_1, a_2; C)$ which is obtained by bounding $\mu_{\max}$ from above by C instead of putting the condition on type. If we break up S into the disjoint union of suitable subschemes of S, then X is embedded in a projective space $\mathbf{P}^n_S$ over S. Hence we may assume that $X = \mathbf{P}^n_S$ and $\mathcal{O}_X(1)$ is the invertible sheaf corresponding to hyperplanes. Fix a linear subspace $Y \cong \mathbf{P}^{n-d-1}_S$ and set $\mathcal{S}_Y$ to be the subfamily of $\mathcal{S}'_{X/S}(d; r, a_1, a_2; C)$ consisting the classes of E on geometric fibers $\mathbf{P}^n_s$ such that $\mathrm{Supp}(E)$ does not meet Y_s. The projective linear group $PGL(n+1)$ so acts on $\mathcal{S}'_{X/S}(d; r, a_1, a_2; C)$ that every orbit meets the subfamily $\mathcal{S}_Y$. Thus it is enough to prove that $\mathcal{S}_Y$ is bounded. Pick a coherent sheaf E on a geometric fiber $\mathbf{P}^n_s$ which represents a member of $\mathcal{S}_Y$. Let $\pi : \mathbf{P}^n_s \setminus Y_s \to \mathbf{P}^d_s$ be the linear projection from Y_s. Since $\mathrm{Supp}(E)$ is contained in $\mathbf{P}^n_s \setminus Y_s$ and proper over $k(s)$, the direct image $F = \pi_*(E)$ is a coherent sheaf whose Hilbert polynomial is the same as that of E. The pure d-dimensionality implies that F is torsion free. We can say more.

Lemma 1.9. *If E is a coherent sheaf on an algebraic scheme Z of pure dimension d over a field k with $\mathrm{Supp}(E) = Z$, E satisfies the condition (S_2) and if $\pi : Z \to \mathbf{P}^d_k$ is a finite morphism, then $\pi_*(E)$ satisfies the condition (S_2).*

Proof. Since E satisfies the condition (S_1) and since Z is pure dimension d, E is of pure dimension d. This and the fact that π is finite imply that $\pi_*(E)$ is torsion free. Then there is an open set W of $\mathbf{P}^d_k$ such that $\pi_*(E)|_W$ satisfies the condition (S_2) and $\mathrm{codim}(\mathbf{P}^d_k \setminus W, \mathbf{P}^d_k) \geq 2$. Setting $W' = \pi^{-1}(W)$, we shall consider the commutative diagram:

$$\begin{array}{ccc} W' & \xrightarrow{\ \delta\ } & Z \\ {\scriptstyle \pi'}\downarrow & & \downarrow{\scriptstyle \pi} \\ W & \xrightarrow{\ \gamma\ } & \mathbf{P}^d_k \end{array}$$

Since $\mathrm{codim}(Z \setminus W', Z) \geq 2$, the condition (S_2) on E implies that $\delta_*\delta^*(E) = E$. Then we have simply

$$\begin{aligned} \gamma_*\gamma^*(\pi_*(E)) &= \gamma_*(\pi'_*\delta^*(E)) \\ &= \pi_*(\delta_*\delta^*(E)) \\ &= \pi_*(E). \end{aligned}$$

Thus $\pi_*(E)$ satisfies the condition (S_2). □

By this lemma F satisfies the condition (S_2). If we can prove that there is a constant C' such that for all F in the above, we have $\mu_{\max}(F) \leq C'$, then F is a member of the family $\mathcal{S}'_{\mathbf{P}^d/S}(d; r, a_1, a_2; C')$. Under our condition that S is over a filed of characteristic 0 or that $d \leq 2$, this family is bounded by virtue of [6, Main Theorem]. Then, the set of Hilbert polynomials of F's is finite. Moreover, we can find an integer m such that every F is m-regular. On the other hand, Leray's spectral sequence implies that $H^p(\mathbf{P}^d_s, F(m-p)) \cong H^p(\mathbf{P}^n_s, E(m-p))$. Thus every member of $\mathcal{S}_Y$ is m-regular. These complete our proof. We have only to prove, therefore, the following.

Lemma 1.10. *Assume that X is embedded in $\mathbf{P}^n_S$. Fix positive integers d, a_0 and C. Let $\mathcal{F}(d, a_0, C)$ be the family of classes of coherent sheaves E on the fibers of X over S such that E is in $\mathcal{F}(d, a_0, C)$ if and only if it is of pure dimension d, $0 < a_0(E) \leq a_0$ and $\mu_{\max}(E) \leq C$. There is an integer C' which is determined only by n, a_0, d and C and has the following property:*

(1.10.1) *If a coherent sheaf E on X_s is in $\mathcal{F}(d, a_0, C)$ and if $\mathbf{P}^{n-d-1}_S = Y \subset \mathbf{P}^n_S$ is a linear subspace with $Y_s \cap \mathrm{Supp}(E) = \emptyset$, then for the projection $\pi : \mathbf{P}^n_{k(s)} \setminus Y_s \to \mathbf{P}^d_{k(s)}$ and for $F = \pi_*(E)$, we have $\mu_{\max}(E) \leq C'$*

Proof. As in the argument in the above it is enough to prove our assertion for a fixed Y and E with $Y_s \cap \mathrm{Supp}(E) = \emptyset$. Let Z be the closed subscheme of $\mathbf{P}^n = \mathbf{P}^n_{k(s)}$ defined by the ideal of annihilators of E. We denote the projection of Z to $\mathbf{P}^d = \mathbf{P}^d_{k(s)}$ by π_Z. We claim

(1.10.2) Z or the structure sheaf $\mathcal{O}_Z$ is of pure dimension d.

In fact, $\mathcal{O}_Z$ is embedded in $\mathcal{H}om(E, E)$ as the multiplication by constants and hence it is enough to show that $\mathcal{H}om(E, E)$ is of pure dimension d. For suitable integers a and N, we have a surjection $\mathcal{O}_{\mathbf{P}^n}(-a)^N \to E$. Then $\mathcal{H}om(E,E)$ can be regarded as a subsheaf of $E(a)^N$. Since this is of pure dimension d, so is $\mathcal{H}om(E,E)$.

Pick a point z of Z such that $\dim \mathcal{O}_{Z,z} \geq 1$. Then the pure dimensionality of Z implies that $\operatorname{depth} \mathcal{O}_{Z,z} \geq 1$ and hence Z is Cohen-Macaulay in codimension 1. Since π_Z is finite and $\mathbf{P}^d$ is smooth, this shows

(1.10.3) There is an open set U of $\mathbf{P}^d$ such that Z is flat over U and $\mathbf{P}^d \setminus U$ is at least of codimension 2 in $\mathbf{P}^d$.

Let F_1 be the first filter of the Harder-Narasimhan filtration of F. Since $\mu(F_1) = \mu^S(F_1) = \mu_{\max}(F)$, we have to bound $\mu(F_1)$ from above. By (1.10.2) the $\mathcal{O}_{\mathbf{P}^d}$-algebra $\mathcal{A} = \pi_*(\mathcal{O}_Z)$ is torsion free. F is an $\mathcal{A}$-module and we have a bijective correspondence between the set of coherent $\mathcal{A}$-submodules of F and that of coherent subsheaves of E. The image F_1' of the canonical homomorphism $F_1 \otimes_{\mathcal{O}_{\mathbf{P}^d}} \mathcal{A} \to F$ is an $\mathcal{A}$-submodule of F and hence it is the direct image of a coherent subsheaf E_1 of E. By our hypothesis we have $\mu(F_1') = \mu^S(E_1) \leq C$. By Lemma 1.11 below there is a positive integer m determined by n, d and $a_0(E)$ such that $\mathcal{A}(m)|_U$ is generated by its global sections, that is, there is a surjection

$$\mathcal{O}_U^{\oplus \alpha} \longrightarrow \mathcal{A}(m)|_U.$$

Tensoring F_1 to this and composing the above canonical homomorphism, we have a surjection of $(F_1|_U)^{\oplus\alpha}$ to $F_1'(m)|_U$. Since F_1 is μ-semi-stable, so is $(F_1)^{\oplus\alpha}$ and hence for every non-trivial quotient coherent sheaf Q, we have $\mu(Q) \geq \mu(F_1)$. This and the fact that U contains all points of codimension 1 give rise to

$$\mu(F_1) \leq \mu(F_1'(m)) = \mu(F_1') + m \leq C + m.$$

Thus our remaining task is to prove the following.

Lemma 1.11. *There is a positive integer m determined by $a_0(E)$, n and d such that $\mathcal{A}(m)|_U$ is generated by its global sections.*

Proof. We shall first show a basic result:

(1.11.1) If E is a coherent sheaf on a projective space $\mathbf{P}^n_K$ over a field K, then for the subscheme Z defined by the ideal of annihilators of E, we have $a_0(\mathcal{O}_Z) \leq a_0(E)^2$.

In fact, we may assume that K is algebraically closed. Set $r = a_0(E)$. As in the proof of (1.10.2) $\mathcal{H}om(E,E)$ naturally contains $\mathcal{O}_Z$. Thus it is enough to show that $a_0(\mathcal{H}om(E,E)) \leq r^2$. Our proof is by induction on $d = \dim \operatorname{Supp}(E)$. First assume that $d = 0$. In this case E is generated by its global sections and $r = \dim H^0(\mathbf{P}^n, E)$. This means that there exists a surjection

$$\mathcal{O}_{\mathbf{P}^n}^{\oplus r} \longrightarrow E.$$

From this we obtain an injection $\mathcal{H}om(E,E) \to E^r$ which leads us to $a_0(\mathcal{H}om(E,E)) \leq r a_0(E) = r^2$. Assume next that $d > 0$. Take a hyperplane H which passes through none of the points in $\operatorname{Ass}(E)$. Then the multiplication by a linear form defining H provides us with the following exact sequence

$$0 \longrightarrow E(-1) \longrightarrow E \longrightarrow G \longrightarrow 0.$$

We see that $\dim \operatorname{Supp}(G) = d-1$ and $a_0(G) = a_0(E)$. This sequence gives rise to another exact sequence

$$\begin{array}{ccccccc} 0 & \longrightarrow & \mathcal{H}om(E,E(-1)) & \longrightarrow & \mathcal{H}om(E,E) & \longrightarrow & \mathcal{H}om(E,G). \\ & & \| \wr & & & & \| \wr \\ & & \mathcal{H}om(E,E)(-1) & & & & \mathcal{H}om(G,G) \end{array}$$

Since $a_0(\mathcal{H}om(E,E)) = a_0(\mathcal{H}om(E,E)/\mathcal{H}om(E,E)(-1))$, we obtain $a_0(\mathcal{H}om(E,E)) \leq a_0(\mathcal{H}om(G,G))$. On the other hand, by our induction hypothesis $a_0(\mathcal{H}om(G,G)) \leq a_0(G)^2$. Thus we come to the required inequality $a_0(\mathcal{H}om(E,E)) \leq a_0(G)^2 = r^2$.

Let us go back to the proof of Lemma 1.11. By blowing-up $\mathbf{P}^n_S$ along Y, we obtain a $\mathbf{P}^{n-d}$-bundle $p : \mathbf{P} = \mathbf{P(N)} \to \mathbf{P}^d$, where N is the vector bundle $\mathcal{O}_{\mathbf{P}^d}(1) \oplus \mathcal{O}_{\mathbf{P}^d}^{\oplus n-d}$. For a positive integer a we denote the Hilbert scheme $Hilb^a_{\mathbf{P}/\mathbf{P}^d}$ by H^a. We set

$$\mathbf{H} = \coprod_{a \leq a_0(E)^2} H^a$$

and $\tilde{Z}$ to be the universal subscheme of $\mathbf{P} \times_{\mathbf{P}^d} \mathbf{H}$. There is an open subscheme V of $\mathbf{H}$ which parametrizes the subschemes that do not meet the exceptional divisor D of the blowing-up.

$$\begin{array}{ccccc} \tilde{Z}_V & \subset & \mathbf{P} \times_{\mathbf{P}^d} \mathbf{H} & \longrightarrow & \mathbf{P} \\ {\scriptstyle p_V}\downarrow & & \downarrow{\scriptstyle p_{\mathbf{H}}} & & {\scriptstyle p}\downarrow \\ V & \subset & \mathbf{H} & \longrightarrow & \mathbf{P}^d \end{array}$$

Since the tautological line bundle $\mathcal{O}_{\mathbf{P}}(1)$ is isomorphic to $\mathcal{O}_{\mathbf{P}}(D) \otimes p^*(\mathcal{O}_{\mathbf{P}^d}(1))$, we have $\mathcal{O}_{\tilde{Z}_V}(m) \cong \mathcal{O}_{\tilde{Z}_V} \otimes p_V^*(\mathcal{O}_{\mathbf{H}}(m))$, where $\mathcal{O}_{\mathbf{H}}(1)$ is the pull-back of $\mathcal{O}_{\mathbf{P}^d}(1)$ to $\mathbf{H}$. By the projection formula we get an isomorphism $(p_V)_*(\mathcal{O}_{\tilde{Z}_V}(m)) \cong (p_V)_*(\mathcal{O}_{\tilde{Z}_V})(m)$. Since p_V is finite and flat, this direct image is compatible with base changes. There is an integer m determined by $a_0(E)$, n and d such that $R^1(p_V)_*(I_{\tilde{Z}_V}(m)) = 0$, where $I_{\tilde{Z}_V}$ is the ideal of $\tilde{Z}_V$. Then, for this m and the pull-back N_V of N to V, we have a surjection

$$S^m(N_V) \longrightarrow (p_V)_*(\mathcal{O}_{\tilde{Z}_V}(m)) \cong (p_V)_*(\mathcal{O}_{\tilde{Z}_V})(m).$$

$S^m(N_V)$ is generated by its global sections because so is N. Hence $(p_V)_*(\mathcal{O}_{\tilde{Z}_V})(m)$ is a quotient of a trivial vector bundle. On the other hand, since Z can be regarded as a subscheme of $\mathbf{P}$, (1.10.2), (1.10.3) and the universality of the Hilbert scheme provides us with a morphism f of U to V such that $Z_U = \tilde{Z} \times_{\mathbf{H}} U$ as subschemes of $\mathbf{P}_U$. Thus we have an isomorphism of $\mathcal{A}|_U(m) = (p_U)_*(\mathcal{O}_{Z_U})(m)$ to $f^*(p_V)_*(\mathcal{O}_{\tilde{Z}_V})(m)$ and hence $\mathcal{A}|_U(m)$ is generated by its global sections. □

For our later purpose let us prove the openness of Simpson's stability and semi-stability. We shall first show the openness of the pure d-dimensionality. Let E be a coherent sheaf of pure dimension d on a projective scheme X a field k. Since X is projective, we can embed X into a projective space $\mathbf{P}^n$ over k. As an $\mathcal{O}_{\mathbf{P}^n}$-module, E has a resolution by locally free coherent sheaves

$$0 \longrightarrow F_n \xrightarrow{f_n} F_{n-1} \xrightarrow{f_{n-1}} \cdots \xrightarrow{f_2} F_1 \xrightarrow{f_1} F_0 \xrightarrow{f_0} E \longrightarrow 0$$

Lemma 1.12. *Let B_i be the set of pinching points of $K_i = \ker(f_i)$ or E according to $0 \leq i \leq n$ or $i = -1$. Then the following* (1) *and* (2) *are equivalent to each other*

(1) *E is of pure dimension d.*

(2) $\dim \operatorname{Supp}(E) = d$ *and* $\dim B_j \leq n - j - 3$ *for* $n - d - 1 \leq j \leq n - 1$.

Proof. We may assume that $X = \mathbf{P}^n$. As remarked right after Definition 0.1, E is of pure dimension d if and only if

(3) $\mathrm{Supp}(E)$ is of pure dimension d and E satisfies the condition (S_1).

Assume first (3). Then outside $\mathrm{Supp}(E)$ all the K_i's are locally free and hence $B_i \subset \mathrm{Supp}(E)$. Pick a point x in $\mathrm{Supp}(E)$. If $\dim \mathcal{O}_{X,x} = n-d+t$, then depth E_x is 0 or 1 according as t is 0 or positive. Thus if $t = 0$, then $\mathrm{depth}(K_i)_x = \min\{i+1, n-d\}$ and hence $(K_i)_x$ is a free $\mathcal{O}_{X,x}$-module for $i \geq n-d-1$. If $t > 0$, then $\mathrm{depth}(K_i)_x \geq \min\{i+2, n-d+t\}$, which means that $(K_i)_x$ is a free $\mathcal{O}_{X,x}$-module if $i \geq n-d+t-2$. Thus we see that $\dim B_j \leq n-j-3$ for $n-d-1 \leq j \leq n-1$. Conversely, assume that (2) holds good. Take a point $x \in \mathrm{Supp}(E)$ such that $\dim \mathcal{O}_{X,x} = n-d+t$ for a $t > 0$. Then x is not in B_j if $n-j-3 < d+t$. Thus, for $j \geq n-d+t-2$, $(K_j)_x$ is locally free. In other words, $\mathrm{hd}(E_x) \leq n-d+t-1$ and then depth $E_x \geq 1$. This implies that x is not in $\mathrm{Ass}(E)$ and hence every irreducible component of $\mathrm{Supp}(E)$ is of dimension $\geq d$. Therefore, we get (3). □

We can now prove the openness of pure d-dimensionality.

Proposition 1.13. *Let $g : X \to S$ be a projective morphism of locally noetherian schemes and E an S-flat coherent sheaf on X. There is an open set U of S such that a point s of S if and only if $E(s)$ is of pure dimension d.*

Proof. Since our problem is local with respect to S, we may assume that S is noetherian, X is embedded in $\mathbf{P}^n_S$ and hence X itself is $\mathbf{P}^n_S$. If m is a sufficiently large integer, then $E(s)(m)$ is generated by global sections and $H^i(X_s, E(s)(m)) = 0$ for all $s \in S$ and $i > 0$. Then $g_*(E(m))$ is locally free and the natural map $g^*g_*(E(m)) \to E(m)$ is surjective. Thus there is a surjection f_0 of a locally free sheaf $F_0 = g^*g_*(E(m))(-m)$ to E. Applying the same procedure to $\ker(f_0)$ we have an exact sequence

$$F_1 \longrightarrow F_0 \longrightarrow E \longrightarrow 0.$$

Repeating this we get an exact sequence

$$0 \longrightarrow F_n \xrightarrow{f_n} F_{n-1} \xrightarrow{f_{n-1}} \cdots \xrightarrow{f_2} F_1 \xrightarrow{f_1} F_0 \xrightarrow{f_0} E \longrightarrow 0,$$

where $F_0, \ldots, F_{n-1}$ are locally free and $F_n = \ker(f_{n-1})$. Since E is flat over S and $X = \mathbf{P}^n_S$ is also flat over S, for every point $s \in S$, the restriction of the above sequence to the fiber X_s is exact. Hence $F_n(s)$ is locally free on X_s. This and the flatness of F_n over S imply that F_n is locally free. For $K_i = \ker(f_i)$ and $s \in S$, $K_i(s)_x$ is a free $\mathcal{O}_{X_s,x}$-module if and only if $(K_i)_x$ is a free $\mathcal{O}_{X,x}$-module because K_i is flat over S. Thus if B_i is the set of pinching points of K_i, then $(B_i)_s$ is the set of pinching points. Since g is proper, $S_{-1} = \{s \in S \mid \dim \mathrm{Supp}(E)_s > d\}$ and $S_j = \{s \in S \mid \dim(B_j)_s > n-j-3\}$ are closed in S. By Lemma 7.12 the open set

$$U = S \setminus \left(S_{-1} \bigcup \left(\bigcup_{j=n-d-1}^{n-1} S_j\right)\right)$$

meets our requirement. □

After replacing *torsion free* by *of pure dimension d*, P_E by P^S_E and μ_0 by μ^S, we can follow the same argument in the proof of [6, Theorem 1.7] to prove the openness of Simpson's stability and semi-stability.

Theorem 1.14. *Let $g : X \to S$ and E be the same as in* Proposition 1.13 *and $\mathcal{O}_X(1)$ a g-ample invertible sheaf.*

(1) *For every sequence* $(\alpha) = (\alpha_1, \dots, \alpha_{r-1})$, *there is an open set* $U_{(\alpha)}$ *of* S *such that for every algebraically closed field* K, $U_{(\alpha)}(K)$ *is exactly the set* $\{s \in S(K) \mid E(s) \text{ is of type } (\alpha) \text{ in the sense of Definition 1.5}\}$.

(2) *There is an open set* U^s *of* S *such that for every algebraically closed field* K, $U^s(K)$ *is exactly the set* $\{s \in S(K) \mid E(s) \text{ is stable in the sense of Simpson}\}$.

(3) *There is an open set* U^{ss} *of* S *such that for every algebraically closed field* K, $U^{ss}(K)$ *is exactly the set* $\{s \in S(K) \mid E(s) \text{ is semi-stable in the sense of Simpson}\}$.

2. Quot-schemes and semi-stable points of a Grassmann variety

Let $f : X \to S$ be a projective morphism of noetherian schemes, $\mathcal{O}_X(1)$ an f-ample invertible sheaf, F a coherent sheaf on X and P a numerical polynomial. For an S-scheme T, we set

$$\mathcal{Q}uot^P_{F/X/S}(T) = \left\{ F_T \to E \to 0 \;\middle|\; \begin{array}{l} E \text{ is flat over } T \text{ and for every geometric} \\ \text{point } t \text{ of } T,\ \chi(E(t)(m)) = P(m) \end{array} \right\} \Big/ \sim,$$

where $F_T = F \otimes_{\mathcal{O}_S} \mathcal{O}_T$ is the pull-back of F to $X \times_S T$ and where two members are equivalent under the relation $\sim$ if they are isomorphic as quotient sheaves of F_T. This defines a contravariant functor $\mathcal{Q}uot^P_{F/X/S}$ of the category (Sch/S) of locally noetherian S-schemes to that of sets. One of the main results on this functor is the following.

Theorem 2.1. $\mathcal{Q}uot^P_{F/X/S}$ *is representable by a pair* (Q, G) *of a projective* S*-scheme* $Q = \mathrm{Quot}^P_{F/X/S}$ *and a quotient coherent sheaf* G *of* F_Q. *There is an integer* m_0 *such that for every* $m \geq m_0$, $\det((f_Q)_*(G(m)))$ *is an* S*-very ample invertible sheaf on* Q. *Moreover, if we assume*

(2.1.1) X *is embedded in* $\mathbf{P}(U)$ *over* S *with* U *a locally free coherent sheaf on* S *and* $\mathcal{O}_X(1)$ *is the restriction of* $\mathcal{O}_{\mathbf{P}(U)}(1)$ *to* X

then for sufficiently large integer n, *there is an integer* m_0 *such that for every* $m \geq m_0$, $\mathrm{Quot}^P_{F/X/S}$ *is embedded in* $\mathrm{Grass}_{P(m)}(\pi_*(\mathcal{O}_{\mathbf{P}(U)}(m-n)^{\oplus N}))$ *by the Plücker invertible sheaf* $\det((f_Q)_*(G(m)))$, *where* $N = P(n)$ *and* π *is the projection of* $\mathbf{P}(U)$ *to* S.

Let S be a noetherian scheme, V a free $\mathcal{O}_S$-module of rank n and W a locally free $\mathcal{O}_S$-module of rank m. We have a canonical dual action of $G = GL(V) \cong GL(n, S)$ on V and hence on $V \otimes W = V \otimes_{\mathcal{O}_S} W$ by letting G act trivially on W. Fix a free basis $e_1, \dots, e_n$ of V. Then, for a suitable affine system of coordinates $\{x_{ij}\}$ of G over S, we have an isomorphism

$$\alpha : V_G \longrightarrow V_G$$

by sending e_j to $\sum_{i=1}^n x_{ij} e_i$ and hence another isomorphism

$$\beta : (V \otimes W)_G \longrightarrow (V \otimes W)_G.$$

Let Z be the Grassmann scheme $\mathrm{Grass}_r(V \otimes W)$ of r-dimensional quotient spaces of $V \otimes W$ and

$$\theta : (V \otimes W)_Z \longrightarrow U$$

the universal quotient bundle on Z. Composing two homomorphisms

$$\begin{array}{ccc} p_1^*((V \otimes W)_G) & \xrightarrow{p_1^*(\beta)} & p_1^*((V \otimes W)_G) \\ & & \Vert \wr \\ & & p_2^*((V \otimes W)_Z) \xrightarrow{p_2^*(\theta)} p_2^*(U) \end{array}$$

we obtain the quotient locally free sheaf $p_2^*(U)$ of $p_1^*((V\otimes W)_G)$ of rank r, where p_i is the ith projection of $G\times_S Z$. By the universality of the Grassmann scheme there exists a morphism

$$\sigma : G\times Z \longrightarrow Z$$

which makes the following diagram commutative:

$$\begin{array}{ccccc} p_1^*((V\otimes W)_G) & \xrightarrow{\sim} & \sigma^*((V\otimes W)_Z) & \xrightarrow{\sigma^*(\theta)} & \sigma^*(U) \\ {\scriptstyle p_1^*(\beta)}\downarrow\wr & & {\scriptstyle\psi}\downarrow\wr & & {\scriptstyle\varphi}\downarrow\wr \\ p_1^*((V\otimes W)_G) & \xrightarrow{\sim} & p_2^*((V\otimes W)_Z) & \xrightarrow{p_2^*(\theta)} & p_2^*(U) \end{array}$$

Now it is not hard to see that σ is an action of G on Z and ψ (or, φ) defines a G-linearization of $(V\otimes W)_Z$ (U, resp.). We have thus a G-morphism

$$\pi : Z = \mathbf{P}(\bigwedge^r U) \hookrightarrow \mathbf{P}(\bigwedge^r (V\otimes W)_Z) = \mathbf{P}(\bigwedge^r (V\otimes W))\times_S Z \longrightarrow \mathbf{P}(\bigwedge^r (V\otimes W)).$$

As is well-known, π is a closed immersion and called the Plücker embedding. The dual action of G on $V\otimes W$ induces that of $\bigwedge^r(V\otimes W)$ and hence an action on $\mathbf{P}(\bigwedge^r(V\otimes W))$. This is exactly the action we got in the above and Z is G-stable under this action. Moreover, the G-linearization of $\bigwedge^r U$ comes from that on $\mathcal{O}_{\mathbf{P}(\bigwedge^r(V\otimes W))}(1)$. Since the center $\mathbf{G}_{mS}$ acts trivially on $\mathbf{P}(\bigwedge^r(V\otimes W))$, the projective linear group scheme $\overline{G} = G/\mathbf{G}_{mS}$ acts on $\mathbf{P}(\bigwedge^r(V\otimes W))$ and $\mathcal{O}_{\mathbf{P}(\bigwedge^r(V\otimes W))}(a)$ carries a $\overline{G}$-linearization, where a is the rank of $\bigwedge^r V\otimes W$.

We shall show here one of key computations to construct the moduli spaces of stable sheaves.

Proposition 2.2. *Let k be an algebraically closed field and let S be* Spec(k). *An r-dimensional quotient k-vector space L of $V\otimes_k W$ is properly stable (or, semi-stable) with respect to the above action of $\overline{G}$ on the Grassmann variety Z and the $\overline{G}$-linearization of $(\bigwedge^r U)^{\otimes a}$ if and only if for every proper vector subspace V' of V, we have*

$$\dim L' > \frac{\dim V'}{\dim V}\dim L,$$
$$(\geq \textit{resp.})$$

where L' is the image of $V'\otimes_k W$ to L.

Proof. There is an isogeny of $SL(V)$ to $\overline{G} = PGL(V)$. By the definition of the proper stability and the semi-stability, we may replace $\overline{G}$ by $SL(V)$ and then the $GL(V)$-linearization of $\bigwedge^r U$ induces its $SL(V)$-linearization. Thus we have only to check the proper stability and the semi-stability of L with respect to the Plücker embedding of Z. Pick a 1-parameter subgroup λ of $SL(V)$ and fix a basis $e_1,\dots,e_n$ so that the dual action of λ on V is defined by

$$e_i \longmapsto t^{r_i}\otimes e_i$$

where $r_1,\dots,r_n$ satisfy the condition

(2.2.1) $\sum_{i=1}^{n} r_i = 0$, $r_1\leq r_2\leq\cdots\leq r_n$ and not all of r_i's are 0.

Take a basis $f_1,\dots,f_m$ of W and set $h_\beta = e_i\otimes f_j$ with $\beta = m(i-1)+j$. Then $h_1,\dots,h_{mn}$ is a basis of $V\otimes_k W$. If we choose the homogeneous system of coordinate of $\mathbf{P}(V\otimes_k W)$ corresponding to this basis, the action of the 1-parameter subgroup λ is given by

$$(z_1,\dots,z_{mn}) \longmapsto (t^{r_1}z_1,\dots,t^{r_1}z_m,t^{r_2}z_{m+1},\dots,t^{r_2}z_{2m},\dots,t^{r_n}z_{mn}).$$

We set $s_1 \leq s_2 \leq \cdots \leq s_{mn}$ in the following way:

$$s_\beta = r_i \text{ for } \beta = (i-1)m + j \text{ with } 0 \leq j < m.$$

Let U_β be the vector subspace of $V \otimes_k W$ generated by $\{h_1, \ldots, h_\beta\}$ and $\alpha : V \otimes_k W \to L$ the given surjection. For each integer j with $1 \leq j \leq r$, there is a unique integer μ_j such that

$$\dim \alpha(U_{\mu_j}) = j \text{ and } \dim \alpha(U_{\mu_j - 1}) = j - 1.$$

Then we have

$$0 < \mu_1 < \mu_2 < \cdots < \mu_r \leq r$$

and $\{\bar{h}_1 = \alpha(h_{\mu_1}), \ldots, \bar{h}_r = \alpha(h_{\mu_r})\}$ forms a basis of L. Using these bases we can represent the map α by the following $(r \times mn)$-matrix:

$$\begin{pmatrix} 0 & \cdots & 0 & a_{1\mu_1} & \cdots & a_{1\mu_2-1} & a_{1\mu_2} & \cdots & a_{1\mu_3-1} & \cdots & a_{1\mu_r} & \cdots & a_{1mn} \\ \vdots & & \vdots & 0 & \cdots & 0 & a_{2\mu_2} & \cdots & a_{2\mu_3-1} & \cdots & a_{2\mu_r} & \cdots & a_{2mn} \\ \vdots & & \vdots & \vdots & & \vdots & 0 & \cdots & 0 & \cdots & \vdots & & \vdots \\ \vdots & & \vdots & \vdots & & \vdots & \vdots & & \vdots & & \vdots & & \vdots \\ 0 & \cdots & 0 & 0 & \cdots & 0 & 0 & \cdots & 0 & \cdots & a_{r\mu_r} & \cdots & a_{rmn} \end{pmatrix}$$

where $a_{i\mu_i} = 1$. Let $p_{i_1 \ldots i_r}$ be the Plücker coordinate corresponding to $h_{i_1} \wedge \cdots \wedge h_{i_r}$ with $i_1 < \cdots < i_r$. If $i_j < \mu_j$ for a j, then $\{\alpha(h_{i_1}), \ldots \alpha(h_{i_j})\}$ is a subset of a vector subspace of dimension $j-1$ and hence $p_{i_1 \ldots i_r} = 0$ at the point L. On the other hand, we know that $p_{\mu_1 \ldots \mu_r}(L) \neq 0$. Our computation of Mumford's function $\mu(L, \lambda)$ [7] proceeds as follows

$$\begin{aligned} \mu(L,\lambda) &= \mu^{\wedge^r U}(L,\lambda) \\ &= -\sum_{j=1}^{r} s_{\mu_j} \\ &= -\sum_{\beta=1}^{mn} s_\beta \left(\dim \alpha(U_\beta) - \dim \alpha(U_{\beta-1})\right) \\ &= -r s_{mn} + \sum_{\beta=1}^{mn-1} (s_{\beta+1} - s_\beta) \dim \alpha(U_\beta) \\ &= -r r_n + \sum_{j=1}^{n-1} (r_{j+1} - r_j) \dim \alpha(U_{jm}). \end{aligned}$$

Lemma 2.3. *Let $r_1, \ldots r_n$ be a sequence of integers which satisfies the condition* (2.2.1). *Then there are positive rational numbers $\delta_1, \ldots, \delta_{n-1}$ such that*

$$r_j = \sum_{i=1}^{n-j} -i\delta_i + \sum_{\ell=n-j+1}^{n-1} (n-\ell)\delta_\ell$$

Proof. We have to solve the following system of linear equations

$$\begin{cases} (n-1)x_1 + (n-2)x_2 + (n-3)x_3 + \cdots + 2x_{n-2} + x_{n-1} & = & r_n \\ -x_1 + (n-2)x_2 + (n-3)x_3 + \cdots + 2x_{n-2} + x_{n-1} & = & r_{n-1} \\ \vdots & & \vdots \\ -x_1 + \cdots + (j-n)x_{n-j} + (j-1)x_{n-j+1} + \cdots + x_{n-1} & = & r_j \\ \vdots & & \vdots \\ -x_1 - 2x_2 - 3x_3 - \cdots - (n-2)x_{n-2} - (n-1)x_{n-1} & = & r_1 \end{cases}$$

Summing up from the first to the $(n-1)$-th equations, we obtain the last equation. Therefore, by Cramer's formula, our assertion reduces to the fact that the determinant of the following matrix is equal to $(r_{n-i}-r_{n-i-1})n^{n-2}$:

$$\begin{pmatrix} n-1 & n-2 & n-3 & \cdots & n-i & r_n & n-i-2 & \cdots & 2 & 1 \\ -1 & n-2 & n-3 & \cdots & n-i & r_{n-1} & n-i-2 & \cdots & 2 & 1 \\ -1 & -2 & n-3 & \cdots & n-i & r_{n-2} & n-i-2 & \cdots & 2 & 1 \\ \vdots & \vdots & \vdots & & \vdots & \vdots & \vdots & & \vdots & \vdots \\ -1 & -2 & -3 & \cdots & n-i & r_{n-i+1} & n-i-2 & \cdots & 2 & 1 \\ -1 & -2 & -3 & \cdots & -i & r_{n-i} & n-i-2 & \cdots & 2 & 1 \\ \vdots & \vdots & \vdots & & \vdots & \vdots & \vdots & & \vdots & \vdots \\ -1 & -2 & -3 & \cdots & -i & r_2 & -i-2 & \cdots & -n+2 & 1 \end{pmatrix}$$

□

Let us go back to the proof of Proposition 2.2. Since the right hand side of the equation

$$\mu(L,\lambda) = -rr_n + \sum_{j=1}^{n-1}(r_{j+1}-r_j)\dim\alpha(U_{jm})$$

is a linear function of $r_1,\ldots,r_n$, the above lemma shows us that $\mu(L,\lambda)$ is positive if and only if so is for the extreme cases

$$r_1=\cdots=r_p=p-n,\quad r_{p+1}=\cdots=r_n=p \text{ with } 1\le p\le n-1.$$

Thus $\mu(L,\lambda)>0$ for all $r_1,\ldots,r_n$ if and only if

$$rp+n\dim\alpha(U_{pm})>0$$

for $1\le p\le n-1$. If V_i is the vector subspace of V spanned by $e_1,\ldots,e_i$, then $U_{pm}=V_p\otimes_k W$ and hence the above inequality can be rewritten in the form

$$\dim\alpha(V_p\otimes_k W) > \frac{\dim V_p}{\dim V}\dim L.$$

Since every p-dimensional subspace of V is V_p for a suitable 1-parameter subgroup and a suitable basis of V, the above inequality proves our assertion for the proper stability. For semi-stability we have only to replace *positive* by *non-negative* and $>$ by $\ge$ in the final part of our argument. □

3. e-SEMI-STABLE SHEAVES

Unfortunately we have no general results on boundedness of semi-stable sheaves and hence we shall introduce a subfamily of semi-stable sheaves for a temporary expedient. Let $(X,\mathcal{O}_X(1))$ be a pair of a projective scheme over an algebraically closed field k and a very ample invertible sheaf $\mathcal{O}_X(1)$ on X. Pick a coherent sheaf E of pure dimension d. By using the complete linear system $|\mathcal{O}_X(1)|$, we can embed X into a projective space $\mathbf{P}_k^n$. There is a resolution

$$0\longrightarrow F_n\xrightarrow{f_n}F_{n-1}\xrightarrow{f_{n-1}}\cdots\xrightarrow{f_2}F_1\xrightarrow{f_1}F_0\xrightarrow{f_0}E\longrightarrow 0$$

of E by locally free $\mathcal{O}_{\mathbf{P}^n}$-modules. If we set B_j to be the set of pinching points of $K_j=\ker(f_j)$, then we know by Lemma 1.12 that $\dim B_j\le n-j-3$ for $n-d-1\le j\le n-1$. Take a hyperplane H of $\mathbf{P}_k^n$ which does not contain any point of $\mathrm{Ass}(E)\cup(\cup_{i=0}^{n-1}\mathrm{Ass}(K_i))$. Then, splitting the resolution into short exact sequences, combining them with the exact sequences to define the restriction to H and using the snake lemma, we see that the restriction of the above resolution to H is again a resolution of $E|_H$ by locally free sheaves on H. If H is sufficiently general, then $\dim \mathrm{Supp}(E|_H) = \dim \mathrm{Supp}(E) \cap H = d-1$ and $\dim B_j \cap H = \dim B_j - 1$. Since the set of pinching points B_j^H of $\ker(f_j|_H)$ is $B_j \cap H$, we see that for $n-d \leq j \leq n-2$, $\dim B_j^H \leq (n-1)-j-3$ and $\ker(f_{n-2}|_H)$ is locally free. Thus $E|_H$ is of pure dimension $d-1$ by Lemma 1.12. Repeating this procedure, we can find general hyperplanes $H_1, \dots H_{d-1}$ such that for the scheme theoretic intersection $Y = X \cap H_1 \cap \cdots \cap H_{d-1}$, E_Y is a coherent $\mathcal{O}_Y$-module of pure dimension 1. As is well-known and easy to prove, the set $\{\mu^S(F) \mid F$ is a quotient coherent sheaf of $E|_Y$ of pure dimension 1 $\}$ is bounded above.

Definition 3.1. Let $(X, \mathcal{O}_X(1))$ be a pair of a projective scheme over an algebraically closed field k and a very ample invertible sheaf $\mathcal{O}_X(1)$ on X and let E be a coherent sheaf on X of pure dimension d. Fix a positive integers e.

(1) E is said to be e-stable (or, e-semi-stable) if it is stable (semi-stable, resp.) and if for general hyperplane sections $H_1, \dots, H_{d-1}$ in $|\mathcal{O}_X(1)|$, for the subscheme $Y = H_1 \cap \cdots \cap H_{d-1}$ of X and for every coherent subsheaf F of $E|_Y$ with $0 < a_0(F) < a_0(E)$, we have

$$\mu^S(F) \leq \mu^S(E) + e.$$

(2) E is said to be strictly e-semi-stable if E is e-semi-stable and if every quotient coherent sheaf of E with $P_{E'}^S(m) = P_E^S(m)$ is e-semi-stable. Note here that the condition $P_{E'}^S(m) = P_E^S(m)$ and the semi-stability of E imply that E' is of pure dimension d.

What we have seen right before this definition is that every semi-stable sheaf is e-semi-stable for some e. As for the strict semi-stability we have a similar result.

Lemma 3.2. *Let $(X, \mathcal{O}_X(1))$ be as in* Definition 3.1 *and let*

$$0 \longrightarrow E' \longrightarrow E \longrightarrow E'' \longrightarrow 0$$

be an exact sequence of coherent $\mathcal{O}_X$-module of pure dimension d. Assume that $P_{E'}^S(m) = P_E^S(m) = P_{E''}^S(m)$.

(1) *If E' and E'' are e-semi-stable, then so is E.*

(2) *If E is e-semi-stable, then E' is e-semi-stable and E'' is $a_0(E)e$-semi-stable. Hence E is strictly $a_0(E)e$-semi-stable.*

Proof. For the semi-stability in the above assertions the results are in Proposition 1.1. If we choose general members $H_1, \dots, H_{d-1}$ of $|\mathcal{O}_X(1)|$, then for the scheme theoretic intersection $Y = H_1 \cap \cdots \cap H_{d-1}$, all the $E'|_Y$, $E|_Y$ and $E''|_Y$ are of pure dimension 1, the sequence

$$0 \longrightarrow E'|_Y \longrightarrow E|_Y \longrightarrow E''|_Y \longrightarrow 0$$

is exact and the condition in Definition 3.1 holds good for E or E', E' according as E is e-semi-stable or E' and E'' are e-semi-stable.

(1) Let F be a coherent subsheaf of $E|_Y$ with $0 < a = a_0(F) < a_0(E)$. Set $F' = F \cap E'|_Y$ and F'' to be the image of F to $E''|_Y$. Then we have

$$\begin{aligned} a_0(F) &= a_0(F') + a_0(F'') \\ a_1(F) &= a_1(F') + a_1(F'') \\ a_1(F') &\leq a_0(F')\mu^S(E') + a_0(F')e \\ a_1(F'') &\leq a_0(F'')\mu^S(E'') + a_0(F'')e \end{aligned}$$

Since $\mu^S(E') = \mu^S(E) = \mu^S(E'')$, the above relations give rise to

$$\mu^S(F) \leq \mu^S(E) + e.$$

(2) If F' be a coherent subsheaf of $E'|_Y$ with $0 < a_0(F') < a_0(E')$, then it is a subsheaf of $E|_Y$. By e-semi-stability of E we have

$$\mu^S(F') \leq \mu^S(E) + e = \mu^S(E') + e$$

as required. Next let F'' be a coherent subsheaf of E'' with $0 < a_0(F'') < a_0(E'')$. If F is the inverse image of F'' to $E|_Y$. Then we get

$$\begin{aligned} a_1(F) &= a_1(F'') + a_1(E') \\ a_1(F) &\leq (a_0(E') + a_0(F''))(\mu^S(E) + e) \end{aligned}$$

Thus we obtain

$$\begin{aligned} a_1(F'') &\leq a_0(F'')(\mu^S(E'') + e) + a_0(E')(\mu^S(E') + e) - a_1(E') \\ &= a_0(F'')(\mu^S(E'') + e) + a_0(E')e \\ &= a_0(F'')\left\{\mu^S(E'') + \frac{a_0(F'') + a_0(E')}{a_0(F'')}e\right\} \\ &\leq a_0(F'')(\mu^S(E'') + a_0(E)e) \end{aligned}$$

□

The strict e-semi-stability behaves well with respect to the S-equivalence.

Lemma 3.3. *Let $(X, \mathcal{O}_X(1))$, E', E and E'' be as in* Lemma 3.2.

(1) *If E is strictly e-semi-stable, then each component of* $\mathrm{gr}(E)$ *is e-stable.*

(2) *E is strictly e-semi-stable if and only if so are both E' and E''.*

Proof. (1) Our proof is by induction on the number α of the components of $\mathrm{gr}(E)$. If $\alpha = 1$, then we have nothing to prove. Assume that $\alpha > 1$ and take a Seshadri filtration $0 = E_0 \subset E_1 \subset \cdots \subset E_{\alpha-1} \subset E_\alpha = E$. By Lemma 3.2 we know that E_1 is e-stable. $\bar{E} = E/E_1$ is strictly e-semi-stable by the definition of strict e-semi-stability and $0 = \bar{E}_0 \subset \bar{E}_1 = E_2/E_1 \subset \cdots \subset \bar{E}_{\alpha-2} = E_{\alpha-1}/E_1 \subset \bar{E}_{\alpha-1} = \bar{E}$ is a Seshadri filtration of $\bar{E}$. Thus $\mathrm{gr}(E) = E_1 \oplus \mathrm{gr}(\bar{E})$ and our induction hypothesis tells us that each component of $\mathrm{gr}(\bar{E})$ is e-stable. We see therefore that each component of $\mathrm{gr}(E)$ is e-stable.

(2) Note first that if E is an extension of a semi-stable sheaf E'' by a semi-stable sheaf E' and if $P^S_{E'}(m) = P^S_{E''}(m)$, then E is semi-stable, $P^S_E(m) = P^S_{E'}(m) = P^S_{E''}(m)$ and $\mathrm{gr}(E) = \mathrm{gr}(E') \oplus \mathrm{gr}(E'')$. If both E' and E'' are strictly e-semi-stable, then this remark and our assertion (1) show that each component of $\mathrm{gr}(E)$ is e-stable. Let F be a coherent quotient sheaf of E with $P^S_F(m) = P^S_E(m)$. Applying the above remark to E, F and $\ker(E \to F)$, we see that $\mathrm{gr}(F)$ is a direct summand of $\mathrm{gr}(E)$ and hence each component of $\mathrm{gr}(F)$ is e-stable. Using Lemma 3.2, (1) repeatedly, we see that F is e-semi-stable. Thus E is strictly e-semi-stable. Conversely, assume that E is strictly e-semi-stable. By the very definition of the strict e-semi-stability E'' is obviously a strictly e-semi-stable sheaf. Since $\mathrm{gr}(E')$ is a direct summand of $\mathrm{gr}(E)$, each component of $\mathrm{gr}(E')$ is e-stable. Then by using the former half of this proof repeatedly, we see that E' is strictly e-semi-stable. □

Corollary 3.4. *E is strictly e-semi-stable if and only if so is* $\mathrm{gr}(E)$.

Now let us prove the openness of the strict e-semi-stability.

Proposition 3.5. *Let $f : X \to S$ be a projective morphism of locally noetherian schemes, $\mathcal{O}_X(1)$ an f-very ample invertible sheaf and F an S-flat coherent $\mathcal{O}_X$-module. Assume that the pair $(X, \mathcal{O}_X(1))$ satisfies the condition (2.1.1). Then there is an open set U of S such* *that for all algebraically closed field k, $U(k)$ is exactly the set of k-valued points s of S with $F(s)$ strictly e-semi-stable.*

Proof. Since the property that a coherent sheaf is of pure dimension d and semi-stable is open by Theorem 1.14, we may assume that for every geometric point s of S, $F(s)$ is a semi-stable sheaf of pure dimension d. By our assumption there is a locally free coherent sheaf V on S such that X is a closed S-subscheme of $\mathbf{P}(V)$ and $\mathcal{O}_X(1)$ is the restriction of $\mathcal{O}_{\mathbf{P}(V)}(1)$ to X. We may assume that V is of constant rank r. Let B be the Grassmann $Grass_{r-d+1}(V)$ of quotient bundles of V of rank $r-d+1$ and $V_B \to Q$ the universal quotient bundle. We shall consider the pull back $\tilde{F}$ of F to the scheme theoretic intersection Z of X_B and $\mathbf{P}(Q)$ in $\mathbf{P}(V_B) = \mathbf{P}(V) \times_S B$. The set B' of points t of B such that $\dim \mathrm{Supp}(\tilde{F}(t)) = 1$ and $\tilde{F}$ is flat at t is an open set of B. If I is the defining ideal of Z in X_B, then we have an exact sequence

$$I \otimes F_B \xrightarrow{u} F_B \longrightarrow \tilde{F} \longrightarrow 0$$

and we see that a point b of B is in B' if and only if $u(b)$ is injective and $\dim \mathrm{Supp}(\tilde{F}(b)) = 1$. For a over field K of $k(b)$, $u(b)$ is injective if and only if so is $u(b) \otimes_{k(b)} K$. These mean that a geometric point t of B is a geometric point of B' if and only if $u(t)$ is injective and $\dim \mathrm{Supp}(\tilde{F}(t)) = 1$. Pick a geometric point s of S. We can choose members $\sigma_1, \dots, \sigma_{d-1}$ of $V(s)$ so general that σ_1 is not zero at any point of $\mathrm{Ass}(F(s))$ and the hypersurface $\sigma_{j+1} = 0$ does not meet $\mathrm{Ass}(F(s)/(\sigma_1, \dots, \sigma_j)F(s)(-1))$. Then $\sigma_1, \dots, \sigma_{d-1}$ are linearly independent in $V(s)$ and hence gives rise to a geometric point t of B over s. By the choice of $\sigma_1, \dots, \sigma_{d-1}$ t is a geometric point of B'. Thus general points of B_s is in B'. Since $\tilde{F}$ is flat over B', Theorem 1.14 and Proposition 1.13 imply that there is an open set B_0 of B' such that a geometric point t of B' is in B_0 if and only if $\tilde{F}(t)$ is of pure dimension 1 and has the property in Definition 3.3, (1). Since B is flat over S, the image U_0 of B_0 to S is an open set and by the construction of U_0 a k-valued geometric point s of S is in $U_0(k)$ if and only if $F(s)$ is e-semi-stable.

Replacing S by U_0, we may assume that for every geometric point s of S, $F(s)$ is e-semi-stable and may assume also that the Hilbert polynomials of $F(s)$ are independent of $s \in U_0$. Take an integer i with $1 \leq i < a_0(F(s)) = a_0$. Setting $P_i = iP^S_{F(s)}(m)$ and $F_0 = F_{U_0}$, let us consider the Quot-scheme $Q_i = \mathrm{Quot}^{P_i}_{F/X/S}$ and the universal quotient sheaf F_i on $X \times_S Q_i$. Since F_i is flat over S, the above proof shows that there is a closed set Z_i of Q_i such that a k-valued geometric point z of Q_i is in $Z_i(k)$ if and only if $F_i(z)$ is not e-semi-stable. The image T_i of Z_i in S is closed because Q_i is proper over S. Now it is clear that

$$U = S \setminus \bigcup_{i=1}^{a_0-1} T_i$$

is the open set we want. □

We have also boundedness of e-semi-stable sheaves.

Proposition 3.6. *Let $f : X \to S$ be a projective morphism of noetherian schemes, $\mathcal{O}_X(1)$ an f-very ample invertible sheaf on X, $H(x)$ a numerical polynomial of degree d and let e a positive integer. Then the family of the classes of e-semi-stable sheaves with Hilbert polynomial $H(m)$ on the fibers of X over S is bounded.*

Proof. Let $\mathcal{F}$ be the family for which we have to prove the boundedness. There are integers $a_0, \dots, a_d$ such that

$$H(m) = \sum_{i=0}^{d} a_{d-i} \binom{m+i}{i}.$$

Let E be an e-semi-stable sheaf E on a geometric fiber X_s of X over S such that $\chi(E(m)) = H(m)$. For general members $D_1, \dots, D_{d-1}$ of $|\mathcal{O}_X(1)(s)|$, for $Y = D_1 \cap \cdots \cap D_{d-1}$, E_Y is of pure dimension 1 and for every coherent subsheaf F of $E|_Y$ with $0 < a_0(F) < a_0$, we have the inequality

$$\mu^S(F) \leq \mu^S(E) + e,$$

This means that $E|_Y$ has a fixed type determined by e and a_0. Since the Hilbert polynomial of $E|_Y$ is $a_0(m+1) + a_1$, Theorem 1.8 implies that the family $\{E|_Y \mid E \in \mathcal{F}\}$ is bounded. Then there is an integer m_0 such that for all integers $m \leq m_0$, $H^0(Y, E(m)|_Y) = 0$ because $E|_Y$ satisfies the condition (S_1). Set Y_i to be the scheme theoretic intersection $D_1 \cap \cdots \cap D_i$. We have an exact sequence

$$0 \longrightarrow E(m-1)|_{Y_i} \longrightarrow E(m)|_{Y_i} \longrightarrow E(m)|_{Y_{i+1}} \longrightarrow 0.$$

If we assume that for all $m \geq m_0$, $H^0(Y, E(m)|_{Y_{i+1}}) = 0$, then the above exact sequence and the fact that $E|_{Y_i}$ satisfies the condition (S_1) prove that for all $m \geq m_0$, $H^0(Y, E(m)|_{Y_i}) = 0$. Thus, by descending induction on i, we see that for every i with $0 \leq i \leq d-1$ and every $m \leq m_0$, for all $m \geq m_0$, $H^0(Y, E(m)|_{Y_i}) = 0$. This implies that $\mathcal{F}(m_0) = \{E(m_0)) \mid E \in \mathcal{F}\}$ is a (b)-family with $(b) = (0, \dots, 0)$. Thus, by a theorem of Kleiman $\mathcal{F}(m_0)$ is bounded and hence so is $\mathcal{F}$. □

4. General setting and a fundamental lemma

Let $f : X \to S$ be a morphism of projective schemes, $\mathcal{O}_X(1)$ an f-very ample invertible sheaf on X. For positive integers a_0, e and d, we set $\mathcal{F}^e(d, a_0)$ to be the family of the classes of coherent sheaves E on the fibers of X over S with the following properties:

(4.1.1) E is of pure dimension d.

(4.1.2) $0 < a_0(E) \leq a_0$.

(4.1.3) If E is on a geometric fiber X_s, $D_1, \dots, D_{d-1}$ are general members of $|\mathcal{O}_X(1)(s)|$ and if Y is the scheme theoretic intersection $D_1 \cap \cdots \cap D_{d-1}$, then $E|_Y$ is of pure dimension 1 and for every coherent subsheaf F of $E|_Y$ with $0 < a_0(F) < a_0(E)$, we have

$$\mu^S(F) \leq \mu^S(E) + e.$$

The following is due to Simpson.

Lemma 4.2. *There is an integer B such that for every member E of $\mathcal{F}^e(d, a_0)$ and for every integer $m \geq -B - \mu^S(E)$, we have*

$$\dim H^0(X_s, E(m)) \leq \frac{a_0(E)}{d!}(m + B + \mu^S(E))^d,$$

where E is on a geometric fiber X_s. Moreover, if $m \leq -B - \mu^S(E)$, then we have that $H^0(X_s, E(m)) = 0$.

Proof. Note first that for every integer t and an $E \in \mathcal{F}^e(d, a_0)$ on a fiber X_s of X over S, $a_0(E(t)) = a_0(E)$ and $a_1(E(t)) = a_1(E) + ta_0(E)$ and that for every coherent subsheaf F of $E|_Y$, $a_0(F(t)) = a_0(F)$ and $a_1(F(t)) = a_1(F) + ta_0(F)$. Hence $E(t)$ has the property (4.1.3). This means that for every integer t, $E(t)$ is also a member of $\mathcal{F}^e(d, a_0)$. Thus if we take the integer t_0 such that $0 \leq a_1(E) + t_0 a_0(E) < a_0(E)$, then $0 \leq a_1(E(t_0)) < a_0(E(t_0))$, that is,

$0 \le \mu^S(E(t_0)) < 1$ and $E(t_0)$ is in $\mathcal{F}^e(d, a_0)$. Assume that our assertions are true for every member F of $\mathcal{F}^e(d, a_0)$ with $0 \le \mu^S(F) < 1$. Then, for the above E, we have

$$\begin{aligned}\dim H^0(X_s, E(m)) &= \dim H^0(X_s, E(t_0)(m - t_0)) \\ &\le \frac{a_0(E)}{d!}(m - t_0 + B + \mu^S(E(t_0)))^d\end{aligned}$$

for all m with $m - t_0 > -B - \mu^S(E(t_0))$. Since $t_0 = \mu^S(E(t_0)) - \mu^S(E)$, the above is equivalent to the condition that for every $m > -B - \mu^S(E)$,

$$\dim H^0(X_s, E(m)) \le \frac{a_0(E)}{d!}(m + B + \mu^S(E))^d.$$

Moreover, if $m - t_0 \le -B - \mu^S(E(t_0))$ or equivalently $m \le -B - \mu^S(E)$, then $H^0(X_s, E(m)) = H^0(X_s, E(t_0)(m - t_0)) = 0$.

Therefore, we have to prove our assertion for the subfamily $\mathcal{F}^e(d, a_0)_1$ of $\mathcal{F}^e(d, a_0)$ with an additional condition $0 \le \mu^S(E) < 1$. Take an E in $\mathcal{F}^e(d, a_0)_1$ and Y for E in (4.1.3). Since $\chi(E|_Y(m)) = a_0(E)(m+1) + a_1(E)$ and since the range of $a_0(E)$ and hence that of $a_1(E)$ are between 0 and a_0, the set $\{\chi(E|_Y(m)) \mid E \in \mathcal{F}^e(d, a_0)_1\}$ is a finite set. On the other hand, the property (4.1.3) implies that the type of $E|_Y$ is bounded by rational numbers determined by a_0 and e. Then, by Theorem 1.8 $\{E|_Y \mid E \in \mathcal{F}^e(d, a_0)_1\}$ is a bounded family. Using this and the fact that $E|_Y$ satisfies the condition (S_1), we can find integers A and C such that for all $E|_Y$ and all integers m with $m \le -C$ (or, $m \ge A$), $H^0(Y, E|_Y(m)) = 0$ ($H^1(Y, E|_Y(m)) = 0$, resp.). We also have integers $\alpha_1, \ldots, \alpha_{A+C-1}$ such that for all $E|_Y$, $\dim H^0(Y, E|_Y(-C+i)) \le a_0(E)\alpha_i$. Let B_1 be the integer $\max\{\alpha_1 + C - 1, \ldots, \alpha_i + C - i, \ldots, \alpha_{A+C-1} - A + 1, C, 2a_0\}$. If $-B_1 < m \le -C$, then $\dim H^0(Y, E|_Y(m)) = 0 < a_0(E)(m + B_1)$. For every integer m with $-C + 1 \le m \le A - 1$, we have that $\dim H^0(Y, E|_Y(m)) \le a_0(E)\alpha_{m+C} = a_0(E)m + (a_0(E)\alpha_{m+C} - a_0(E)m) \le a_0(E)(m + B_1) \le a_0(E)(m + B_1 + \mu^S(E))$. Finally, if $m \ge A$, then

$$\begin{aligned}\dim H^0(Y, E|_Y(m)) &= a_0(E)(m + 1) + a_1(E) \\ &= a_0(E)\left(m + \frac{a_0(E) + a_1(E)}{a_0(E)}\right) \\ &\le a_0(E)\left(m + \frac{2a_0}{a_0(E)}\right) \\ &\le a_0(E)(m + B_1) \\ &\le a_0(E)(m + B_1 + \mu^S(E)).\end{aligned}$$

Since $B_1 \ge C$, we have that for every $m \le -B_1 - \mu^S(E)$, $\dim H^0(Y, E|_Y(m)) = 0$.

Taking $D_1, \ldots, D_{d-1}$ for E in (4.1.3), we put Y_i to be the scheme theoretic intersection $D_1 \cap \cdots \cap D_{d-i}$. We may assume that $E|_{Y_i}$ is of pure dimension i, Now suppose that there is an integer B_i such that for every $E \in \mathcal{F}^e(d, a_0)_1$ and every $m > B_i + \mu^S(E)$,

$$\dim H^0(Y_i, E|_{Y_i}(m)) \le \frac{a_0(E)}{i!}(m + B_i + \mu^S(E))^i$$

and that for every $m \le -B_i - \mu^S(E)$, $\dim H^0(Y_i, E|_{Y_i}(m)) = 0$. The multiplication by the equation of D_{d-i} provides us with an exact sequence

$$0 \longrightarrow E|_{Y_{i+1}}(m - 1) \longrightarrow E|_{Y_{i+1}}(m) \longrightarrow E|_{Y_i}(m) \longrightarrow 0.$$

If $m \le -B_i - \mu^S(E)$, then the map $H^0(Y_{i+1}, E|_{Y_{i+1}}(m - 1)) \to H^0(Y_{i+1}, E|_{Y_{i+1}}(m))$is surjective. Since $H^0(Y_{i+1}, E|_{Y_{i+1}}(-m)) = 0$ for sufficiently large m, this proves that for all

$m \le -B_i - \mu^S(E)$, $H^0(Y_{i+1}, E|_{Y_{i+1}}(m)) = 0$. Moreover, the above sequence implies inequality

$$\dim H^0(Y_{i+1}, E|_{Y_{i+1}}(n)) - \dim H^0(Y_{i+1}, E|_{Y_{i+1}}(n-1)) \le \dim H^0(Y_i, E|_{Y_i}(n)).$$

If $n > -B_i - \mu^S(E)$, then the right hand side of the above inequality is less than or equal to $\dfrac{a_0(E)}{i!}(n + B_i + \mu^S(E))^i$. Thus summing up these inequalities from $n = -B_i$ to $n = m$ and using the fact that $H^0(Y_{i+1}, E|_{Y_{i+1}}(-B_i - 1)) = 0$, we obtain

$$\begin{aligned}\dim H^0(Y_{i+1}, E|_{Y_{i+1}}(m)) &\le \sum_{n=-B_i}^{m} \frac{a_0(E)}{i!}(n + B_i + \mu^S(E))^i \\ &\le \sum_{n=-B_i}^{m} \frac{a_0(E)}{i!}(n + B_i + 1)^i \\ &\le \frac{a_0(E)}{i!} \int_{-B_i-1}^{m+1} (x + B_i + 1)^i dx \\ &= \frac{a_0(E)}{(i+1)!}(m + B_i + 2)^{i+1}\end{aligned}$$

Setting $B_{i+1} = B_i + 2$, we have that for every $E \in \mathcal{F}^e(d, a_0)_1$ and every $m > -B_{i+1}$,

$$\dim H^0(Y_{i+1}, E|_{Y_{i+1}}(m)) \le \frac{a_0(E)}{(i+1)!}(m + B_{i+1} + \mu^S(E))^{i+1}$$

and that for every $m \le -B_{i+1}$, $\dim H^0(Y_{i+1}, E|_{Y_{i+1}}(m)) = 0$. By induction on i we complete our proof. □

Let $f : X \to S$ and $\mathcal{O}_X(1)$ be the same as in Lemma 4.2. For positive integers a_0, d and integers μ, b with $b \ge \mu$, we set $\mathcal{G}^b(d, a_0, \mu)$ to be the family of the classes of coherent sheaves E on the fibers of X over S with the properties (4.1.1), (4.1.2) and the properties

(4.3.1) $\mu^S(E) \le \mu$.

(4.3.2) If E is on a geometric fiber X_s, $D_1, \dots, D_{d-1}$ are general members of $|\mathcal{O}_X(1)(s)|$ and if Y is the scheme theoretic intersection $D_1 \cap \cdots \cap D_{d-1}$, then $E|_Y$ is of pure dimension 1 and for every coherent subsheaf F of $E|_Y$ with $0 < a_0(F) < a_0(E)$, we have

$$\mu^S(F) \le b.$$

Lemma 4.4. *There is an integer B, a non-negative integer C and a polynomial $\Phi(m)$ of degree $\le d-2$ such that if E is on a geometric fiber X_s and a member of $\mathcal{G}^b(d, a_0, \mu)$, then*

(4.4.1) *for every integer m with $m \ge -B - \mu$, we have*

$$\dim H^0(X_S, E(m)) \le \frac{a_0(E)}{d!}\left(m + B + \frac{\mu + (a_0(E) - 1)b}{a_0(E)}\right)^d + \Phi(m),$$

(4.4.2) *for every integer m with $-B - b \le m \le -B - \mu$, we have* $\dim H^0(X_S, E(m)) \le C$,

(4.4.3) *for every integer m with $m \le -B - b$, we have* $\dim H^0(X_S, E(m)) = 0$,

(4.4.4) *for all real numbers $x \ge -B - \mu$, $\frac{d\Phi(x)}{dx} \ge 0$ and $\Phi(-B - \mu) \ge 0$.*

Moreover, B does not depend on μ.

Proof. Our proof is by induction on d. Pick a coherent sheaf E on a geometric fiber X_s which is a member of $\mathcal{G}^b(d, a_0, \mu)$. Assume first that $d = 1$. Let $0 = F_0 \subset F_1 \subset \cdots \subset F_{\alpha_1} \subset F_\alpha = E$ be the Harder-Narasimhan filtration of E and set $Q_i = F_i/F_{i-1}$. Since each Q_i is semi-stable, Lemma 4.2 implies that there is an integer B_1 which is independent of μ such that for every $m > -B_1 - \mu^S(Q_i)$,

$$\dim H^0(X_s, Q_i(m)) \leq a_0(Q_i)(m + B_1 + \mu^S(Q_i))$$

and for $m \leq -B_1 - \mu^S(Q_i)$, $H^0(X_s, Q_i(m)) = 0$. By the properties (4.3.1), (4.3.2) and by the definition of Harder-Narasimhan filtration, we see that $\mu^S(Q_\alpha) \leq \mu^S(E) \leq \mu$ and $\mu^S(Q_\alpha) < \cdots < \mu^S(Q_2) < \mu^S(Q_1) = \mu^S(F_1) \leq b$. Thus the above implies

$$\begin{aligned}\dim H^0(X_s, E(m)) &\leq \sum_{i=1}^{\alpha} \dim H^0(X_s, Q_i(m)) \\ &\leq \sum_{i=1}^{\alpha} a_0(Q_i)\max\{m + B_1 + \mu^S(Q_i), 0\} \\ &\leq \sum_{i=1}^{\alpha-1} a_0(Q_i)\max\{m + B_1 + b, 0\} + \\ &\quad (a_0(Q_\alpha) - 1)\max\{m + B_1 + b, 0\} + \max\{m + B_1 + \mu, 0\} \\ &= (a_0(E) - 1)\max\{m + B_1 + b, 0\} + \max\{m + B_1 + \mu, 0\}\end{aligned}$$

If $m \geq -B - \mu \geq -B - b$, then the above means

$$\dim H^0(X_s, E(m)) \leq a_0(E)\left(m + B_1 + \frac{\mu + (a_0(E) - 1)b}{a_0(E)}\right).$$

When m is between $-B_1 - b$ and $-B_1 - \mu$, we see

$$\dim H^0(X_s, E(m)) \leq (a_0(E) - 1)(b - \mu) \leq (a_0 - 1)(b - \mu) = C_1$$

Finally if $m \leq -B_1 - b$, then $\dim H^0(X_s, E(m)) = 0$. This completes the proof for $d = 1$.

Assume next that $d > 1$ and our assertion is true for $d - 1$. Let $D = D_1, \ldots, D_{d-1}$ be general members of $|\mathcal{O}_X(1)(s)|$ which have the property (4.3.2) for E. we may assume that $E|_D$ is of pure dimension $d - 1$. Then $E|_D$ is a member of $\mathcal{G}^b(d - 1, a_0, \mu)$. By our induction hypothesis we have B_{d-1}, C_{d-1} and $\Phi_{d-1}(m)$ which satisfy the conditions (4.4.1), (4.4.2), (4.4.3) and (4.4.4) for $E = E|_D$, $B = B_{d-1}$, $C = C_{d-1}$ and $\Phi(m) = \Phi_{d-1}(m)$. We may also assume that B_{d-1} is independent of μ. The exact sequence

$$0 \longrightarrow E(m-1) \longrightarrow E(m) \longrightarrow E|_D \longrightarrow 0.$$

supplies us with the inequality

$$\dim H^0(X_s, E(m)) - \dim H^0(X_s, E(m-1)) \leq \dim H^0(X_s, E|_D(m))$$

Since $H^0(X_s, E|_D(m)) = 0$ for all $m \leq -B_{d-1} - b$ and since $H^0(X_s, E(m)) = 0$ for all sufficiently small m, we see that for all $m \leq -B_{d-1} - b$, $H^0(X_s, E(m)) = 0$. Then, using this fact and (4.4.2) for $E|_D$, we have that if $-B_{d-1} - b \leq m \leq -B_{d-1} - \mu$, then $\dim H^0(X_s, E(m)) \leq C_{d-1}(b - \mu)$. The above inequalities between dimensions of cohomologies and (4.4.2) for $E|_D$ show that for all $m \geq -B_{d-1} - \mu$,

$$\begin{aligned}&\dim H^0(X_s, E(m)) - \dim H^0(X_s, E(m-1)) \\ &\qquad \leq \frac{a_0(E)}{(d-1)!}\left(m + B_{d-1} + \frac{\mu + (a_0(E) - 1)b}{a_0(E)}\right)^{d-1} + \Phi_{d-1}(m).\end{aligned}$$

Summing up this from $-B_{d-1}-b+1$ to m and using (4.4.2) for $E|_D$ again, we see that the left hand side is $\dim H^0(X_s, E(m))$ and the right hand side is bounded from above by

$$R = C_{d-1}(b-\mu) + \int_{-B_{d-1}-\mu}^{m+1} \left(\frac{a_0(E)}{(d-1)!} \left(x + B_{d-1} + \frac{\mu + (a_0(E)-1)b}{a_0(E)} \right)^{d-1} + \Phi_{d-1}(x) \right) dx$$

because $\Phi_{d-1}(x)$ is an increasing function on the interval of integration by (4.4.4) for $\Phi_{d-1}(x)$. Since

$$R = \frac{a_0(E)}{d!} \left(m + 1 + B_{d-1} + \frac{\mu + (a_0(E)-1)b}{a_0(E)} \right)^d + C_{d-1}(b-\mu) - \frac{a_0(E)}{d!} \left(\frac{(a_0(E)-1)(b-\mu)}{a_0(E)} \right)^d + \int_{-B_{d-1}-\mu}^{m+1} \Phi_{d-1}(x) dx,$$

we can find $\Phi(x)$ such that $\Phi(-B_{d-1}-1-\mu) \geq 0$ and $\frac{d\Phi(x)}{dx} = \Phi_{d-1}(x)$, by adding a suitable constant to $\int_{-B_{d-1}-\mu}^{m+1} \Phi_{d-1}(x)dx$. By (4.4.4) for $\Phi_{d-1}(x)$, $\Phi_{d-1}(x+1)$ must be non-negative if $x \geq -B_{d-1}-1-\mu$. Therefore, $B = B_{d-1}+1$, $C = C_{d-1}(b-\mu)$ and $\Phi(x)$ are integers and a polynomial we want. □

Thanks to the above lemma we can prove easily a fundamental lemma which connects semi-stable sheaves to semi-stable points of a Quot-scheme.

Lemma 4.5. *Let a_0 and d be positive integers, b be integer and let $P(m)$ be a polynomial of degree d with rational coefficients whose leading term is $m^d/d!$. Then there are integers M and L such that for all members E of $\mathcal{G}^b(d, a_0, M)$ and for all integers $m \geq L$ we have*

$$\dim H^0(X_s, E(m)) < a_0(E)P(m),$$

where E is on a fiber X_s of X over S.

Proof. Let B be the integer given in Lemma 4.4 for the integers d, a_0 and b. Let us write down the given polynomial $P(m)$ in the form

$$P(m) = \frac{m^d}{d!} + a_1 \frac{m^{d-1}}{(d-1)!} + \text{ (terms of degree } \leq n-2).$$

Take an integer M such that $M \leq b$ and

$$B + \frac{M + (a-1)b}{a} < a_1 \text{ for all integers } a \text{ with } 0 < a \leq a_0.$$

Then there is an integer $L \geq -B-M$ such that for all integers $m \geq L$ and a with $0 < a \leq a_0$,

$$\frac{1}{d!} \left(m + B + \frac{M + (a-1)b}{a} \right)^d + \frac{\Phi(m)}{a} < P(m).$$

Pick a member E of $\mathcal{G}^b(d, a_0, M)$. Then by Lemma 4.4 we know that for every integer $m > -B-M$,

$$\dim H^0(X_s, E(m)) \leq \frac{a_0(E)}{d!} \left(m + B + \frac{M + (a_0(E)-1)b}{a_0(E)} \right)^d + \Phi(m),$$

Combining these, we see that if m is an integer such that $m \geq L \geq -B - M$, then

$$\begin{aligned}\dim H^0(X_s, E(m)) &\leq \frac{a_0(E)}{d!}\left(m + B + \frac{M + (a_0(E)-1)b}{a_0(E)}\right)^d + \Phi(m) \\ &< a_0(E)P(m)\end{aligned}$$

as required. □

We shall fix, from now on, the following situation:

(4.6) Let S be a scheme of finite type over a universally Japanese ring Ξ, $f : X \to S$ a projective morphism and let $\mathcal{O}_X(1)$ be an f-very ample invertible sheaf such that X is embedded in $\mathbf{P}(U)$ with U a locally free coherent sheaf on S, $\mathcal{O}_X(1)$ is the restriction of $\mathcal{O}_{\mathbf{P}(U)}(1)$ to X and that for all points $s \in S$ and all integers $i > 0$, $H^i(X_s, \mathcal{O}_X(1)(s)) = 0$.

Since a coherent sheaf is stable (or, semi-stable) with respect to an ample invertible sheaf if and only if so is with respect to a positive power of the invertible sheaf, we will exploit the very ampleness and the vanishing of the cohomologies only to construct the moduli spaces and for the existence of the moduli spaces the ampleness of $\mathcal{O}_X(1)$ is enough.

In addition to the functors in Definition 0.5, we need functors for e-(semi)-stable sheaves.

Definition 4.7. Let (Sch/S) be the category of locally noetherian S-schemes, $H(x)$ a numerical polynomial of degree d and let e be a positive integer. For an object T of (Sch/S) we define two sets as follows:

(1) $\Sigma^{H,e}_{X/S}(T) = \left\{ E \;\middle|\; \begin{array}{l} E \text{ is a } T\text{-flat coherent sheaf on } X \times_S T \\ \text{with the property (e)} \end{array} \right\} \Big/ \sim$

where $\sim$ is the equivalence relation defined by (c) in Definition 0.5.

(2) $\bar{\Sigma}^{H,e}_{X/S}(T) = \left\{ E \;\middle|\; \begin{array}{l} E \text{ is a } T\text{-flat coherent sheaf on } X \times_S T \\ \text{with the property (f)} \end{array} \right\} \Big/ \sim$

where $\sim$ is the equivalence relation defined by (d) in Definition 0.5.

(e) For every geometric point t of T, $E(t)$ is e-stable and its Hilbert polynomial is $H(m)$.

(f) For every geometric point t of T, $E(t)$ is strictly e-semi-stable and its Hilbert polynomial is $H(m)$.

If $g : T' \to T$ is a morphism in (Sch/S), then the above properties and the equivalence relations are preserved by making the pull back $(1 \times_S g)^*(E)$. Thus both $\Sigma^{H,e}_{X/S}$ and $\bar{\Sigma}^{H,e}_{X/S}$ define contravariant functors of (Sch/S) to (Sets).

For convenience sake let us introduce families of semi-stable sheaves.

Definition 4.8. Let the pair $(X, \mathcal{O}_X(1))$ be the same as in (4.6), $H(x)$ a numerical polynomial of degree d and let e be a positive integer. $\mathfrak{S}_{X/S}(H)$ ($\bar{\mathfrak{S}}_{X/S}(H)$, $\mathfrak{S}_{X/S}(e, H)$ or $\bar{\mathfrak{S}}_{X/S}(e, H)$) denotes the family of classes of coherent sheaves on the fibers of X over S such that a coherent sheaf E on a geometric fiber is in $\mathfrak{S}_{X/S}(H)$ ($\bar{\mathfrak{S}}_{X/S}(H)$, $\mathfrak{S}_{X/S}(e, H)$ or $\bar{\mathfrak{S}}_{X/S}(e, H)$, resp.) if and only if it is stable (semi-stable, e-stable or strictly e-semi-stable, resp.) and the Hilbert polynomial of E is H.

By the argument right before Definition 3.1 and Lemma 3.2 we get directly the following.

Lemma 4.9. *For every numerical polynomial H, we have that $\mathfrak{S}_{X/S}(H) = \bigcup_e \mathfrak{S}_{X/S}(e, H)$ and $\bar{\mathfrak{S}}_{X/S}(H) = \bigcup_e \bar{\mathfrak{S}}_{X/S}(e, H)$.*

When s runs over all the geometric points of S, the set of equivalence classes of E with $[E] \in \bar{\Sigma}^{H,e}_{X/S}(\mathrm{Spec}(k(s)))$ covers the $\bar{\mathfrak{S}}_{X/S}(e, H)$.

Proposition 4.10. *Let $H(x)$ be a numerical polynomial of degree d. There exists an integer m_0 such that for every $E \in \bar{\mathfrak{S}}_{X/S}(e, H)$, the following holds:*

(4.10.1) *For every integer $m \geq m_0$ and every coherent subsheaf F of E with $0 \neq F$,*

$$\dim H^0(X_s, F(m)) \leq \frac{a_0(F)}{a_0(E)}\chi(E(m)) = \frac{a_0(F)}{a_0(E)} \dim H^0(X_s, E(m)).$$

Moreover, the equality holds if and only if $P_F^S(n) = P_E^S(n)$ as polynomials in n.

Proof. We can write $H(m)$ in the form

$$\sum_{i=0}^{d} a_{d-i}(H) \binom{m+i}{i}.$$

Put $P(m) = H(m)/a_0(H)$. Because of the boundedness of $\bar{\mathfrak{S}}_{X/S}(e, H)$ by Proposition 3.6, there is an integer m_1 such that for every $E \in \bar{\mathfrak{S}}_{X/S}(e, H)$, every integer $m \geq m_1$ and every integer $i > 0$, we have $H^i(X_s, E(m)) = 0$ and hence

$$\dim H^0(X_s, E(m))/a_0(E) = P_E^S(m) = H(m)/a_H(m) = P(m).$$

Let $\mathcal{F}$ be the family of the classes of coherent sheaves on the fibers of X over S such that F on X_s is in $\mathcal{F}$ if and only if F is a non-zero subsheaf of an $E \in \bar{\mathfrak{S}}_{X/S}(e, H)$. Put $\mu = a_1(H)/a_0(H)$ and $b = \mu + e$. Since every member F of $\mathcal{F}$ is a subsheaf of a e-semi-stable sheaf E on a geometric fiber X_s with $\mu^S(E) = \mu$ and since if $D_1, \ldots, D_{d-1}$ is sufficiently general members of $|\mathcal{O}_X(1)(s)|$, the restriction of F to $Y = D_1 \cap \cdots \cap D_{d-1}$ is a subsheaf of $E|_Y$, we see that F has the properties (4.3.1) and (4.3.2) for the above μ and b. F obviously enjoys the properties (4.1.1) and (4.1.2). Note here that if $a_0(F) = a_0(E)$, then $a_1(F)$ must be less than or equal to $a_1(E)$ because for sufficiently large integers m,

$$\chi(F(m)) = \dim H^0(X_s, F(m)) \leq \dim H^0(X_s, E(m)) = \chi(E(m)).$$

What we have seen in the above is $\mathcal{F}$ is a subfamily of $\mathcal{G}^b(d, a_0, \mu)$. Applying Lemma 4.5 to the case where $a_0 = a_0(H)$, $d = d$, $b = b$ and $P(m) = H(m)/a_0(H)$, we have integers M and L such that for every member F of $\mathcal{F}_M = \{F \in \mathcal{F} \mid \mu^S(F) \leq M\}$ and for every integer $m \geq L$, we have the inequality

$$\dim H^0(X_s, F)/a_0(F) < P(m).$$

On the other hand, the subfamily $\mathcal{F}' = \{F \in \mathcal{F} \mid \mu^S(F) > M$ and F is a subsheaf of $E \in \bar{\mathfrak{S}}_{X/S}(e, H)$ such that E/F is of pure dimension $d\}$ is bounded by [1, Lemme 2.5] and the boundedness of e-semi-stable sheaves. This means that the set $\{\chi(F(m)) \mid F \in \mathcal{F}'\}$ is finite and for sufficiently large integers m and $F \in \mathcal{F}'$, the higher cohomologies of $F(m)$ vanish. Hence there is an integer m_2 such that for every integer $m \geq m_2$ and for every $F \in \mathcal{F}'$, $F(m)$ is generated by global sections and

$$\dim H^0(X_s, F(m))/a_0(F) = \chi(F(m))/a_0(F) = P_F^S(m) \leq P_E^S(m) = P(m),$$

where F is a subsheaf of a $E \in \bar{\mathfrak{S}}_{X/S}(e, H)$ and the equality holds if and only if $P_F^S(m) = P_E^S(m)$ as polynomials in m. If F is not in $\mathcal{F}_M$ and a subsheaf of $E \in \bar{\mathfrak{S}}_{X/S}(e, H)$, then there is a subsheaf F' of E such that $F \subset F'$, $a_0(F') = a_0(F)$ and F' is in $\mathcal{F}'$. Thus for $m \geq m_2$, we have

$$\dim H^0(X_s, F(m))/a_0(F) \leq \dim H^0(X_s, F'(m))/a_0(F')$$

and the equality holds if and only if $F = F'$ because $F'(m)$ is generated by global sections. Combining all the above together, we see $m_0 = \max\{m_1, L, m_2\}$ is a desired integer. □

We shall fix a numerical polynomial $H(m)$ of degree d and write down it in the form

$$H(m) = \sum_{i=0}^{d} a_{d-i}(H) \binom{m+i}{i}.$$

with $a_i(H)$ integers. For an integer r with $1 \leq r \leq a_0(H)$, we denote $rH(m)/a_0(H)$ by $H^{(r)}(m)$. By the boundedness of $\bar{\mathfrak{S}}_{X/S}(e, H^{(r)})$ and Proposition 4.10, there is an integer $m(r,e)$ such that for all integers $m \geq m(r,e)$ and all $E \in \bar{\mathfrak{S}}_{X/S}(e, H^{(r)})$ on a geometric fiber X_s, we have

(4.11.1) $E(m)$ is generated by global sections and for all $j > 0$,

$$H^j(X_s, E(m)) = 0,$$

(4.11.2) for all coherent subsheaves F with $0 \neq F \neq E$,

$$\dim H^0(X_s, F(m))/a_0(F) \leq \dim H^0(X_s, E(m))/r.$$

and moreover the equality holds if and only if $P_F^S(x) = P_E^S(x) = H(x)/a_0(H)$ as polynomials in x.

(If $\mathfrak{S}_{X/S}(e, H^{(r)})$ is empty, then we understand $m(r,e) = 0$.)

Lemma 4.12. *If $m \geq \max_{1 \leq r \leq a_0(H)} \{m(r,e)\}$, then for all $E \in \bar{\mathfrak{S}}_{X/S}(e, H^{(r)})$ on geometric fibers X_s, we have*

(4.12.1) *$E(m)$ is generated by global sections and $H^j(X_s, E(m)) = 0$ for all $j > 0$.*

(4.12.2) *for all non-zero coherent subsheaves F of E,*

$$\dim H^0(X_s, F(m)) \leq \frac{a_0(F)}{r} \chi(E(m)) = \frac{a_0(F)}{r} \dim H^0(X_s, E(m)).$$

and moreover the equality holds if and only if $P_F^S(x) = P_E^S(x) = H(x)/a_0(H)$ as polynomials in x.

Fix an integer m with $m \geq \max_{1 \leq r \leq a_0(H)} \{m(r,e)\}$. Let $H^{(r)}[m] = H^{(r)}[m](x)$ be the polynomial $H^{(r)}(x+m)$. $H^{(r)}[m]$ is the Hilbert polynomial of members of $\bar{\mathfrak{S}}_{X/S}(e, H^{(r)})(m) = \{E(m) \mid E \in \bar{\mathfrak{S}}_{X/S}(e, H^{(r)})\}$. We are going to consider only r's for which $H^{(r)}(x)$ are numerical polynomials. Set $N_r = H^{(r)}[m](0) = H^{(r)}(m)$. Then $\dim H^0(X_s, E(m)) = N_r$ if E on a geometric fiber X_s is in $\bar{\mathfrak{S}}_{X/S}(e, H^{(r)})$.

We fix a free Ξ-module V_r of rank N_r. If Y (or, A) is a Ξ-scheme (Ξ-algebra, resp.), then $V_r(Y)$ ($V_r(A)$, resp.) denotes $V_r \otimes_\Xi \mathcal{O}_Y$ ($V_r \otimes_\Xi A$, resp.). Let us consider the Quot-scheme

$$Q_r = \mathrm{Quot}^{H^{(r)}[m]}_{V_r(X)/X/S}$$

and the universal quotient sheaf $\varphi_r : V_r(X \times_S Q_r) \to \tilde{F}_r$. Let G_r be the S-group scheme $GL(V_r(S))$. Using affine system of coordinates of G_r over s with respect to a basis $z_1, \ldots, z_{N_r}$ of V_r, we define an isomorphism θ of $V_r(G_r)$ to itself by sending z_i to $\sum_{i=1}^{n_r} z_j \otimes x_{ji}$. If $p_1 : G_r \times_S X \times_S Q_r \to G_r$ and $p_2 : G_r \times_S X \times_S Q_r \to X \times_S Q_r$ are the projections, then we can compose two homomorphisms

$$V_r(G_r \times_S X \times_S Q_r) \cong p_1^*(V_r(G_r)) \xrightarrow{p_1^*(\theta)} p_1^*(V_r(G_r)) \cong V_r(G_r \times_S X \times_S Q_r),$$

$$V_r(G_r \times_S X \times_S Q_r) \cong p_2^*(V_r(X \times_S Q_r)) \xrightarrow{p_2^*(\varphi_r)} p_2^*(\tilde{F}_r)$$

to get a surjection ψ of $V_r(G_r \times_S X \times_S Q_r)$ to $p_2^*(\tilde{F}_r)$. By the universality of $(Q_r, \varphi_r, \tilde{F}_r)$ we have a morphism $\sigma : G_r \times Q_r \to Q_r$ and an isomorphism $\rho : \sigma_X^*(\tilde{F}_r) \to p_2^*(\tilde{F}_r)$ such that the diagram

$$\begin{array}{ccc} \sigma_X^*(V_r(X \times_S Q_r)) & \xrightarrow{\sigma_X^*(\varphi_r)} & \sigma_X^*(\tilde{F}_r) \\ \| & & \rho \downarrow \wr \\ V_r(G_r \times_S X \times_S Q_r) & \xrightarrow{\psi} & p_2^*(\tilde{F}_r) \end{array}$$

is commutative. It is not hard to see that σ is an action of G_r on Q_r and ρ is a G_r-linearization on $\tilde{F}_r$ with respect to the trivial action of G_r on X. Moreover, since the action of the center of G_r induces the multiplication by units on $\tilde{F}_r$, it is canceled by the multiplication by the inverse of the units, which means that the center acts trivially on Q_r. Consequently we have a natural action of $\bar{G}_r = PGL(V_r)$ on Q_r and a G_r-linearization on $\tilde{F}_r$.

By Proposition 3.5, for each integer e' with $0 \le e' \le e$, there exists an open subscheme $R_r^{e,e'}$ of Q_r such that a geometric point y of Q_r is in $R_r^{e,e'}$ if and only if the following two conditions are satisfied:

(4.13.1) $\Gamma(\varphi_r \otimes k(y)) : V_r(k(y)) \to H^0(X_y, \tilde{F}_r(y))$ is bijective.

(4.13.2) $\tilde{F}_r(y)$ is strictly e'-semi-stable.

It is clear that $R_r^{e,e'}$ is an $\bar{G}_r$-stable open set. For every geometric point s of S and every coherent sheaf E on X_s in $\tilde{\mathfrak{S}}_{X/S}(e', H^{(r)})(m)$, there is a surjective homomorphism $\alpha : V_r(X_s) \to E$ such that $\Gamma(\alpha) : V_r(k(s)) \to H^0(X_s, E)$ is bijective by (4.12.1). By the universality of $(Q_r, \varphi_r, \tilde{F}_r)$ α corresponds to a geometric point y of Q_i. Since y is a geometric point of $R_r^{e,e'}$, we obtain a surjective map $\xi_r^{e,e'}(s)$ for every geometric point s of S:

(4.13.3)
$$\begin{aligned} \xi_r^{e,e'}(s) : R_r^{e,e'}(k(s)) \to \bar{\Sigma}_{X/S}^{H^{(r)},e'}(m)(\mathrm{Spec}(k(s))) = \\ \{[E(m)] \mid [E] \in \bar{\Sigma}_{X/S}^{H^{(r)},e'}(\mathrm{Spec}(k(s)))\}. \end{aligned}$$

On the other hand, if for two K-valued geometric points y_1, y_2 of $R_r^{e,e'}$ over the same K-valued point s of S, $\tilde{F}_r(y_1)$ is isomorphic to $\tilde{F}_r(y_2)$, then (4.13.1) provides us with an isomorphism

$$V_r(K) \xrightarrow{\sim} H^0(X_s, \tilde{F}_r(y_1)) \xrightarrow{\sim} H^0(X_s, \tilde{F}_r(y_2)) \xleftarrow{\sim} V_r(K)$$

which gives rise to a K-valued geometric point g of G_r. By the definition of the action of G_r on Q_r we see that $\sigma(K)(g, y_2) = y_1$. Conversely, if two K-valued geometric points y_1, y_2 of $R_r^{e,e'}$ are in the same $G_r(K)$-orbit, then $\tilde{F}_r(y_1) \cong \tilde{F}_r(y_2)$. Therefore, we get

(4.13.4) for every K-valued geometric point s of S, the set of isomorphism classes of strictly e'-semi-stable sheaves on X_s with Hilbert polynomial $H^{(r)}$ is in bijective correspondence with the set of $(G_r)_s(K)$-orbits of $(R_r^{e,e'})_s(K)$.

Since $V_r(X)$ is a quotient sheaf of $V_r(\mathbf{P}(U))$, Theorem 2.1 shows that for sufficiently large integer n, Q_r is embedded in $Z_r(n) = \mathrm{Grass}_{H^{(r)}[m](n)}(h_*(V_r(\mathbf{P}(U))(n)))$, where $h : \mathbf{P}(U) \to S$ is the structure morphism. Fix a sufficiently large n and set W to be the locally free coherent sheaf $h_*(\mathcal{O}_{\mathbf{P}(U)}(n))$ on S. Then $h_*(V_r(\mathbf{P}(U))(n))$ is isomorphic to $V_r \otimes W = V_r \otimes_{\Xi} W$. To construct the moduli spaces of semi-stable sheaves we may assume that W is of constant rank, for example, by replacing S by a connected component of S. On one hand, the G_r-linearization on $\tilde{F}_r$ defines that on $(f_{Q_r})_*(\tilde{F}_r(n))$. On the other hand, the action of G_r on $V_r(S)$ induces that on $h_*(V_r(\mathbf{P}(U))) \cong V_r \otimes W$, which is the same action defined right before Proposition 2.2. We know moreover that $\det((f_{Q_r})_*(\tilde{F}_r(n)))$ is the restriction of the line bundle $\mathcal{O}_{Z_r(n)}(1)$ on $Z_r(n)$ defining the Plücker embedding to Q_r. There is a positive integer

a such that $L_r = (\det((f_{Q_r})_*(\tilde{F}_r(n))))^{\otimes a}$ carries a $\bar{G}_r$-linearization (see Proposition 2.2). Combining all together, we see that

(4.13.5) the embedding of Q_r into $Z_r(n)$ is $\bar{G}_r$-equivariant and the $\bar{G}_r$-linearized invertible sheaf L_r is the restriction of the $\bar{G}_r$-linearized $\mathcal{O}_{Z_r(n)}(a)$ to Q_r.

Notation 4.14. In the above notation, if $r = a_0(H)$, then we omit r in the notation. Hence in the case of $r = a_0(H)$ we use the notation $H[m]$, N, V, Q, G, φ, F, $\bar{G}$, $R^{e,e'}$, $Z(n)$ and L instead of $H^{(r)}[m]$, N_r, V_r, Q_r, G_r, φ_r, $\tilde{F}_r$, $\bar{G}_r$, $R_r^{e,e'}$, $Z_r(n)$ and L_r, respectively.

5. Moduli spaces of stable sheaves

In this section we shall construct the coarse moduli scheme of the functor $\Sigma^H_{X/S}$ under the situation of (4.6). Let us start with the following lemma.

Lemma 5.1. *There is an integer n_0 such that for all $n \geq n_0$, $R_r^{e,e'}$ is embedded in the open subscheme of semi-stable points $Z_r(n)^{ss}(\mathcal{O}_{Z_r(n)}(a))$ of $Z_r(n)$. A geometric point y of $R_r^{e,e'}$ is in $Z_r(n)^s_0(\mathcal{O}_{Z_r(n)}(a))$ if and only if $\tilde{F}_r(y)$ is stable.*

Proof. Our proof is independent of r and hence we shall prove for $r = a_0(H)$. First there is a positive integer n_1 such that for all $n \geq n_1$, Q_r is embedded in $Z_r(n)$. Pick a geometric point y of $R^{e,e'}$ over a geometric point s of S and set $E = \tilde{F}(y)$. Take a vector subspace V' of $V(k(y))$ and let $E' = E'(V')$ be the coherent subsheaf of E generated by $\Gamma(\varphi)(V')$. Then, by (4.13.1) V' is a subspace of $H^0(X_s, E')$. The family $\mathcal{F} = \{E'(V') \mid V' \subset V(k(y)), y$ runs geometric points of $R^{e,e'}\}$ is bounded. Thus, by (4.12.2) there is a positive rational number $\epsilon < 1$ such that if $E' \in \mathcal{F}$ and $P^S_{E'}(x) \neq H[m](x)/a_0(H)$ as polynomials in x, then we have

$$\dim H^0(X_s, E') < \frac{a_0(E')}{a_0(E)} H(m)(1-\epsilon).$$

Moreover,there is a positive integer n_2 such that for all $n \geq n_2$ and all $E' \in \mathcal{F}$,

$$H^0(X_s, V' \otimes \mathcal{O}_X(n)(s)) \longrightarrow H^0(X_s, E'(n))$$

is surjective and $\dim H^0(X_s, E'(n)) = \chi(E'(n))$. On the other hand, the boundedness of $\mathcal{F}$ implies that the set of Hilbert polynomials $\{\chi(E'(a)) \mid E' \in \mathcal{F}\}$ is finite and hence we have an integer n_3 such that for every $n \geq n_3$ and every $E' \in \mathcal{F}$

$$\frac{\chi(E'(n))}{H[m](n)} > \frac{a_0(E')}{a_0(H)}(1-\epsilon)$$

Now if $n \geq n_0 = \max\{n_1, n_2, n_3\}$, then for $W = f_*(\mathcal{O}_X(n))$, the image U' of $V' \otimes W(s)$ to $U = H^0(X_y, E(n))$ is exactly $H^0(X_y, E'(n))$ and its dimension is $\chi(E'(n))$. If $P^S_{E'}(x) \neq H[m](x)/a_0(H)$, then our computation proceeds as follows:

$$\begin{aligned}\dim V' &\leq \dim H^0(X_s, E')\\ &< \frac{a_0(E')}{a_0(E)} H(m)(1-\epsilon)\\ &= \frac{a_0(E')}{a_0(E)} \dim H^0(X_s, E)(1-\epsilon)\\ &< \frac{\chi(E'(n))}{H[m](n)} \dim H^0(X_s, E)\\ &= \frac{\dim U'}{\dim U} \dim V(k(y)).\end{aligned}$$

If $\dim V' = \dim H^0(X_s, E')$ and $P^S_{E'}(x) = H[m](x)/a_0(H)$, then the equality between the first and the last terms holds. Therefore, by Proposition 2.2, y is a geometric point of $Z(n)^{ss}(\mathcal{O}_{Z(n)}(a))$ and if E is stable, then it is a geometric point of $Z(n)^s_0(\mathcal{O}_{Z(n)}(a))$. If E is not stable, then there is a coherent subsheaf E'' of E such that $P^S_{E''}(x) = P^S_E(x)$. Let V' be the subspace $H^0(X_s, E'')$ of $V(k(y)) = H^0(X_s, E)$. Then $E'(V') = E''$ by (4.12.1). Thus (4.12.2) and the above argument show that y is not a stable point. □

Fix an integer n such that $n \geq n_0$ for the n_0 in the above lemma. Let $(R^{e,e'})^s$ be the open subscheme of $R^{e,e'}$ whose geometric points consist of the points corresponding to stable sheaves. $(R^{e,e'})^s$ is a $\bar{G}$-stable open set in Q. Since Q is a $\bar{G}$-stable closed subscheme of $Z(n)$, we see easily that $Q^s(L) = Q \cap Z(n)^s_s(\mathcal{O}_{Z(n)}(a))$ (see (4.13.5)). Thus Lemma 5.1 means that $(R^{e,e'})^s$ is a $\bar{G}$-stable open subscheme of $Q^s(L)$. By [8, Theorem 4] we have a geometric quotient Y of $Q^s(L)$ by $\bar{G}$ which is quasi-projective over S. Then the image of $(R^{e,e'})^s$ to Y is open in Y and if we endow it with the structure $M^{e,e'}$ of open subscheme of Y, then the natural morphism $\pi : (R^{e,e'})^s \to M^{e,e'}$ is a geometric quotient by $\bar{G}$.

Proposition 5.2. *$M^{e,e'}$ is a coarse moduli scheme of $\Sigma^{H,e'}_{X/S}$, that is,*

(1) there is a morphism of functors $\Psi^{e,e'} : \Sigma^{H,e'}_{X/S} \to M^{e,e'}$ such that for all geometric points s of S, $\Psi^{e,e'}(\mathrm{Spec}(k(s)))$ is a bijective map of $\Sigma^{H,e'}_{X/S}(\mathrm{Spec}(k(s)))$ to $M^{e,e'}(k(s))$,

(2) if there is a morphism of functors Ψ' of $\Sigma^{H,e'}_{X/S}$ to a locally noetherian S-scheme M, then there is a unique S-morphism $h : M^{e,e'} \to M$ such that $h\Psi^{e,e'} = \Psi'$.

Proof. Let T be a locally noetherian S-scheme and E a T-flat coherent sheaf with the property (e) of Definition 4.7 for $e = e'$. By (4.12.1) $B = (f_T)_*(E(m))$ is locally free sheaf of rank N, the natural map $B(t) \to H^0(X_t, E(m)(t))$ is an isomorphism and the natural homomorphism $\beta : f^*_T(B) \to E(m)$ is surjective. Take an open covering $T = \bigcup_i U_i$ of T such that the restriction of B to each U_i is free. Fix an isomorphism of $\gamma_i : V(U_i) \to B|_{U_i}$. Then we have a surjection

$$\delta_i : V(X_{U_i}) \xrightarrow{f^*_{U_i}(\gamma_i)} f^*_{U_i}(B|_{U_i}) \xrightarrow{\beta_{U_i}} E(m)_{U_i}$$

such that for every geometric point t of U_i, the induced $\Gamma(\delta_i(t)) : V(k(t)) \to H^0(X_t, E(m)(t))$ is bijective. By the universality of $(Q, \varphi, \tilde{F})$, there is a morphism $\psi(E)_i$ of U_i to Q. The bijectivity of $\Gamma(\delta_i(t))$ implies that $\psi(E)_i$ is actually a morphism of U_i to $(R^{e,e'})^s$. The change of the isomorphism γ_i is represented by a U_i-valued point g of G and the above surjection for the new isomorphism is

$$\begin{aligned}\delta_i' &= \delta_i \cdot (gf_T)^*(\theta) = (1_X \times \psi(E)_i)^*(\varphi) \cdot (g, \psi(E)_i)_X^*(p_1^*(\theta)) \\ &= (g, \psi(E)_i)_X^*(p_2^*(\varphi) \cdot (g, \psi(E)_i)_X^*(p_1^*(\theta)) = (g, \psi(E)_i)_X^*((p_2^*(\varphi) \cdot p_1^*(\theta)) \\ &= (g, \psi(E)_i)_X^*(\sigma_X^*(\varphi)) = (\sigma_X \cdot (g, \psi(E)_i)_X)^*(\varphi),\end{aligned}$$

where the notation is as in the previous section. On the other hand, for the morphism $\psi'(E)_i$ given by the δ_i' and the universality of the Quot-scheme, we have $\delta_i' = (1_X \times \psi'(E)_i)^*(\varphi)$. These two show that $\psi'(E)_i = \sigma(g, \psi(E)_i)$. Since $M^{e,e'}$ is a geometric quotient of $(R^{e,e'})^s$ by the action of G, $\pi\psi'(E)_i = \pi\sigma(g, \psi(E)_i) = \pi\psi(E_i)$ and hence $\Psi(E)_i = \pi\psi(E)_i : U_i \to M^{e,e'}$ is independent of the choice of γ_i. Thus we see that $\Psi(E)_i = \Psi(E)_j$ on $U_i \cap U_j$. We obtain therefore a morphism $\Psi^{e,e'}(E) : T \to M^{e,e'}$. Let $\eta : T' \to T$ be a morphism of locally noetherian S-schemes. By the base change theorem $\eta^*(B) = \eta^*((f_T)_*(E(m))) \cong (f_{T'})_*((1_X \times \eta) * (E(m)))$ and hence the formation to get δ_i is compatible with base changes, that is,

$$\eta^*(\delta_i) : V(X \times_S \eta^{-1}(U_i)) \longrightarrow (1_X \times \eta)^*(E(m))_{\eta^{-1}(U_i)}$$

is the surjection obtained from $(1_X \times \eta)^*(E(m))$. This implies that

$$\psi((1_X \times \eta)^*(E))_i = \psi(E)_i\eta$$

and hence

$$\Psi^{e,e'}((1_X \times \eta)^*(E)) = \Psi^{e,e'}(E)\eta.$$

It is clear by our construction that $\Psi^{e,e'}(E \otimes_{\mathcal{O}_T} M) = \Psi^{e,e'}(E)$ for every invertible sheaf M on T. We have therefore a morphism of functors

$$\Psi^{e,e'} : \Sigma_{X/S}^{H,e'}(T) \ni [E] \longmapsto \Psi^{e,e'}(E) \in X(T).$$

For every geometric point s of S, $M_s^{e,e'}(k(s))$ is the orbit space of $R_s^{e,e'}(k(s))$ by the action of $\bar{G}_s(k(s))$. Then the bijectivity of $\Psi^{e,e'}(\mathrm{Spec}(k(s)))$ obviously follows from (4.13.4) and the above construction of $\Psi^{e,e'}$.

Since $\tilde{F}$ carries a G-linearization, the sheaf $\sigma_X^*(\tilde{F}(-m)|_{R^{e,e'}})$ is isomorphic to $(p_2)_X^*(\tilde{F}(-m)|_{R^{e,e'}})$, where $p_2 : G \times_S R^{e,e'} \to R^{e,e'}$ is the second projection and $\sigma : G \times_S R^{e,e'} \to R^{e,e'}$ is the action of G on $R^{e,e'}$. Thus $\Psi'(\tilde{F}(-m)|_{R^{e,e'}})\sigma = \Psi'(\sigma^*(\tilde{F}(-m)|_{R^{e,e'}}))$ is equal to $\Psi'(\tilde{F}(-m)|_{R^{e,e'}})p_2 = \Psi'(p_2^*(\tilde{F}(-m)|_{R^{e,e'}}))$. Since $M^{e,e'}$ is a categorical quotient of $R^{e,e'}$, there is a unique S-morphism $h : M^{e,e'} \to M$ such that $h\pi = \Psi'(\tilde{F}(-m)|_{R^{e,e'}})$. Under the notation of the previous paragraph, the functoriality of Ψ' and the construction of $\Psi^{e,e'}$ give rise to

$$\begin{aligned}h\Psi^{e,e'}(E)|_{U_i} &= h\pi\psi(E)_i = \Psi'(\tilde{F}(-m)|_{R^{e,e'}})\psi(E)_i \\ &= \Psi'(\psi(E)_i^*(\tilde{F}(-m))) = \Psi'(E_{U_i}) - \Psi'(E)|_{U_i}.\end{aligned}$$

Thus $h\Psi^{e,e'}(E) = \Psi'(E)$ for all E. □

Since both $M^{e_1,e'}$ and $M^{e_2,e'}$ are coarse moduli schemes of the same functor $\Sigma_{X/S}^{H,e'}$, we obtain a unique isomorphism $\psi_{e_1,e_2}^{e'} : M^{e_1,e'} \to M^{e_2,e'}$ such that $\psi_{e_1,e_2}^{e'}\Psi^{e_1.e'} = \Psi^{e_2.e'}$. Since $M^{e,e'}$ is an open subscheme of $M^{e,e}$, $M^{e',e'}$ can be regarded as an open subscheme of $M^{e,e}$ by means of $\psi_{e',e}^{e'}$. Taking the inductive limit of $\{M^{e,e}\}$, we obtain an S-scheme $M_{X/S}(H)$. Since each $M^{e,e}$ is quasi-projective over S, $M_{X/S}(H)$ is locally of finite type and separated over S

Theorem 5.3. *Let H be a numerical polynomial of degree d. In the situation* (4.6) $\Sigma^H_{X/S}$ *has a coarse moduli scheme $M_{X/S}(H)$ in* (Sch/S). *$M_{X/S}(H)$ is locally of finite type and separated over S.*

Proof. Take a locally noetherian S-scheme T and a coherent sheaf E on X_T with $[E] \in \Sigma^H_{M/S}$. For every positive integer e, there is an open subscheme T_e of T such that for every geometric point s of S, $T_e(k(s)) = \{t \in T(k(s)) \mid E(t) \text{ is } e\text{-stable}\}$. If $e' \leq e$, then $T_{e'} \subset T_e$ and moreover $\bigcup_e T_e = T$ because every stable sheaf is e-stable for some e by the argument right before Definition 3.3. Let $j_e : T_e \to T$ and be the open immersion and E_e be $(1_X \times j_e)^*(E)$. Since for the open immersion $j_{e',e} : T_{e'} \to T_e$, we have $E_{e'} = (1_X \times j_{e',e})^*(E_e)$, we get

$$\psi^{e'}_{e',e}\Psi^{e',e'}(E_{e'}) = \Psi^{e,e}((1_X \times j_{e',e})^*(E_e)) = j_{e',e}\Psi^{e,e}(E_e).$$

Thus we have a morphism

$$\Psi(E) = \varinjlim_e \Psi^{e,e}(E_e) : T = \varinjlim_e T_e \longrightarrow \varinjlim_e M^{e,e} = M_{X/S}(H)$$

such that $\Psi(E)|_{T_e} = \Psi^{e,e}(E_e)$. Let $\eta : T' \to T$ be a morphism in (Sch/S). By the definition of T_e and E_e, we see that $T'_e = \eta^-1(T_e)$ and $(1_X \times \eta)^*(E)_e = (1_X \times \eta_{T_e})^*(E_e)$. Then by the functoriality of $\Psi^{e,e}$ we see

$$\Psi^{e,e}((1_X \times \eta)^*(E)_e) = \Psi^{e,e}(E_e)\eta_{T_e}.$$

Taking the inductive limit of the above equality we have the functoriality

$$\Psi((1_X \times \eta)^*(E)) = \Psi^{e,e}(E)\eta$$

of Ψ. For every geometric point s of S, $\Sigma^H_{M/S}(\mathrm{Spec}(k(s))) = \bigcup_e \Sigma^{H,e}_{M/S}(\mathrm{Spec}(k(s)))$ and then we get the bijectivity of $\Psi(\mathrm{Spec}(k(s)))$.

Assume that there is an S-scheme and a morphism of functors $\Psi' : \Sigma^H_{X/S} \to M$. Let Ψ'_e be the restriction of Ψ' to the subfunctor $\Sigma^{H,e}_{X/S}$. By Proposition 5.2 there is a unique morphism h_e of $M^{e.e}$ to M such that $h_e\Psi^{e,e} = \Psi'_e$. By restricting this on $\Sigma^{H,e'}_{X/S}$, we obtain $h_e\psi^{e'}_{e',e}\Psi^{e',e'} = \Psi'_{e'} = h_{e'}\Psi^{e',e'}$. The uniqueness of $h_{e'}$ tells us that $h_e\psi^{e'}_{e',e} = h_{e'}$. Therefore, there is a unique morphism h of $M_{X/S}(H)$ to M such that $h|_{M^{e,e}} = h_e$. By a similar argument to the previous paragraph we can easily prove the equality $h\Psi = \Psi'$. □

As a corollary to the proof of Theorem 5.3 we have the following.

Corollary 5.4. *$M_{X/S}(H)$ is quasi-projective over S if and only if $\mathfrak{S}_{X/S}(H)$ is bounded. Thus if Ξ is a field of characteristic 0 or the degree of H is 1 or 2, then $M_{X/S}(H)$ is quasi-projective over S (see Theorem 1.8). If $M_{X/S}(H)$ is quasi-projective, then there is an ample invertible sheaf M on $M_{X/S}(H)$ and a positive integer c such that for every geometric point z of $M_{X/S}(H)$, $M(z)$ is canonically isomorphic to* $(\det H^0(X_z, E(z)(m+n)))^{\otimes c}$, *where $E(z)$ is the stable sheaf corresponding to z.*

6. Closed orbits of a Grassmann variety

As Simpson showed in [9], if Ξ is a field of characteristic 0, then the open subscheme R in $Q(d)$ is exactly the open set of the semi-stable points $Q(d)^{ss}(L)$. In general case this is rather hard to be proved even if we could show that $\mathfrak{S}_{X/S}(H)$ is bounded. A main point of our construction of a moduli space of semi-stable sheaves is to prove that $R^{e,e'}$ has a good quotient by the group scheme $\bar{G}$. Let $\pi : Q(d)^{ss}(L) \to Y$ be a good quotient of $Q(d)^{ss}(L)$

by $\bar{G}$. What we have to prove is that for the $\bar{G}$-stable closed set $D = Q(d)^{ss}(L) \setminus R^{e.e'}$, $\pi^{-1}\pi(D)$ is equal to D. In fact if this is true, then $\pi(R^{e,e'})$ is an open set of Y and it is the good quotient we want. We know already that closed orbits of $R^{e,e'}$ are the orbits of points corresponding to $\mathrm{gr}(E)$. Thus our problem reduces to proving that the closed orbits in $R^{e,e'}$ is closed in $Q(d)^{ss}(L)$ or equivalently in $Z(n)^{ss}(\mathcal{O}_{Z(n)}(a))$.

Throughout this section we shall fix a field k and a vector space W over k of dimension M. All vector spaces in this section are finite dimensional. For a vector space V over k and a positive integer r, $Z(V,r)$ denotes the Grassmann variety $\mathrm{Grass}_r(V \otimes_k W)$ of r-dimensional quotient spaces of $V \otimes_k W$. Take a vector space U over k of dimension r. Then, for every extension field K of k, a K-valued point of $Z(V,r)$ is represented by a surjection in $\mathrm{Hom}_K(V_K \otimes_K W_K, U_K)$ and two surjections ψ_1 and ψ_2 in $\mathrm{Hom}_K(V_K \otimes_K W_K, U_K)$ represent the same point of $Z(V,r)(K)$ if and only if there is an element g of $GL(U)(K)$ such that $\psi_2 = g\psi_1$. As a matter of fact, there is an open subscheme $Y(V,U)$ of the affine space $\mathbf{V}(\mathrm{Hom}_k(V \otimes_k W, U)^\vee)$ such that for every extension field K of k, $Y(V,U)(K)$ is exactly the set of surjections of $V_K \otimes_K W_K$ to U_K and that the natural morphism $\pi_V : Y(V,U) \to Z(V,r)$ is a principal fiber bundle with group $GL(U)$. As in section 2 $GL(V)$ acts on both $Y(V,U)$ and $Z(V,r)$ and π_V is a $GL(V)$-morphism

Definition 6.1. Let V, V' and V'' be vector spaces over k with $\dim V = \dim V' + \dim V''$, U, U' and U'' be vector spaces over k with $\dim U = \dim U' + \dim U''$ and let K be an extension field of k. Take an element ψ of $Y(V,U)(K)$ and set $\pi_V(K)(\psi) = z$.

(1) A K-valued point ψ' of $Y(V',U')$ (or, $z' = \pi_{V'}(K)(\psi')$) is said to be a subpoint of ψ (z, resp.) if there is an injective K-linear map $\alpha' : V'_K \to V_K$ and the following exact commutative diagram:

$$\begin{array}{ccccc} 0 & \longrightarrow & V'_K \otimes_K W_K & \xrightarrow{\alpha' \otimes 1} & V_K \otimes_K W_K \\ & & \downarrow{\psi'} & & \downarrow{\psi} \\ 0 & \longrightarrow & U'_K & \xrightarrow{\beta'} & U_K \end{array}$$

(2) A K-valued point ψ'' of $Y(V'',U'')$ (or, $z'' = \pi_{V''}(K)(\psi'')$) is said to be a quotient point of ψ (z, resp.) if there is a surjective K-linear map $\alpha'' : V_K \to V''_K$ and the following exact commutative diagram:

$$\begin{array}{ccccc} V_K \otimes_K W_K & \xrightarrow{\alpha'' \otimes 1} & V''_K \otimes_K W_K & \longrightarrow & 0 \\ \downarrow{\psi} & & \downarrow{\psi''} & & \\ U_K & \xrightarrow{\beta''} & U''_K & \longrightarrow & 0 \end{array}$$

(3) ψ (or, z) is said to be an extension of a K-valued point ψ'' of $Y(V'',U'')$ ($z'' = \pi_{V''}(K)(\psi'')$, resp.) by a K-valued point ψ' of $Y(V',U')$ ($z' = \pi_{V'}(K)(\psi')$, resp.) if there is an exact sequence

$$0 \to V'_K \xrightarrow{\alpha'} V_K \xrightarrow{\alpha''} V''_K \to 0$$

and there is the following exact commutative diagram:

$$\begin{array}{ccccccccc} 0 & \longrightarrow & V'_K \otimes_K W_K & \xrightarrow{\alpha' \otimes 1} & V_K \otimes_K W_K & \xrightarrow{\alpha'' \otimes 1} & V''_K \otimes_K W_K & \longrightarrow & 0 \\ & & \downarrow{\psi'} & & \downarrow{\psi} & & \downarrow{\psi''} & & \\ 0 & \longrightarrow & U'_K & \xrightarrow{\beta'} & U_K & \xrightarrow{\beta''} & U''_K & \longrightarrow & 0 \end{array}$$

Remark 6.2. Under the situation of the above definition, if ψ is an extension of ψ'' by ψ', then $g\psi$ is an extension of $g''\psi''$ by $g'\psi'$ for all $g \in GL(U)(K)$, $g' \in GL(U')(K)$ and $g'' \in GL(U'')(K)$. Indeed the following is an exact commutative diagram:

$$\begin{array}{ccccccccc} 0 & \longrightarrow & V'_K \otimes_K W_K & \xrightarrow{\alpha'\otimes 1} & V_K \otimes_K W_K & \xrightarrow{\alpha''\otimes 1} & V''_K \otimes_K W_K & \longrightarrow & 0 \\ & & \downarrow{\scriptstyle g'\psi'} & & \downarrow{\scriptstyle g\psi} & & \downarrow{\scriptstyle g''\psi''} & & \\ 0 & \longrightarrow & U'_K & \xrightarrow{g\beta' g'^{-1}} & U_K & \xrightarrow{g''\beta'' g^{-1}} & U''_K & \longrightarrow & 0 \end{array}$$

This justifies saying that z is an extension of z'' by z'. For the subpoint and the quotient point we have a similar justification.

The following plays a key role to prove our main result in this section.

Lemma 6.3. *Let V, V' and V'' be vector spaces over k with $\dim V = \dim V' + \dim V''$ and let r, r' and r'' be positive integers with $r = r' + r''$. Suppose that B' (or, B'') is a $GL(V')$-stable ($GL(V'')$-stable, resp.) closed subset of $Z(V', r')$ ($Z(V'', r'')$, resp.). Then there is a $GL(V)$-stable closed subset B of $Z(V, r)$ such that for all extension fields K of k, $B(K) = \{z \in Z(V, r)(K) \mid z$ has one of the following properties (6.3.1), (6.3.2)$\}$:*

(6.3.1) *z is an extension of a $z'' \in B''(K)$ by a $z' \in B'(K)$.*

(6.3.2) *If $\psi \in Y(V, U)(K)$ ($\psi' \in Y(V', U')(K)$ or $\psi'' \in Y(V'', U'')(K)$) represents z (z' or z'', resp.), then we have the following commutative diagram:*

$$\begin{array}{ccccccccc} 0 & \longrightarrow & V'_K \otimes_K W_K & \xrightarrow[\alpha'\otimes 1]{} & V_K \otimes_K W_K & \xrightarrow{\alpha''\otimes 1} & V''_K \otimes_K W_K & \longrightarrow & 0 \\ & & \downarrow{\scriptstyle \psi'} & & \downarrow{\scriptstyle \psi} & & \downarrow{\scriptstyle \psi''} & & \\ & & U'_K & \xrightarrow{\beta'} & U_K & \xrightarrow{\beta''} & U''_K & & \end{array}$$

where the top raw is exact and $\beta''\beta' = 0$.

Proof. Let A be the closed reduced subscheme in

$$\mathbf{P}(\mathrm{Hom}_k(U', U)^{\vee}) \times_k \mathbf{P}(\mathrm{Hom}_k(U, U'')^{\vee})$$

such that for every extension field K of k, every point of $A(K)$ is represented by a pair of linear maps $\beta' \in \mathrm{Hom}_K(U'_K, U_K)$ and $\beta'' \in \mathrm{Hom}_K(U_K, U''_K)$ with $\beta''\beta' = 0$. Set

$$Y = Y(V', U') \times_k Y(V, U) \times_k Y(V'', U'')$$

The algebraic group

$$H = GL(U') \times_k GL(U) \times_k GL(U'')$$

acts on the space $P = A \times_k Y$ as follows:

$$\begin{aligned} (H \times_k P)(K) \ni ((g', g, g''), (\beta', \beta'', \psi', \psi, \psi'')) \\ \longmapsto (g\beta' g'^{-1}, g''\beta'' g^{-1}, g'\psi', g\psi, g''\psi'') \in P(K). \end{aligned}$$

A is projective and carries a very ample invertible sheaf L with an H-linearization. Then the sheaf of graded algebras

$$\bigoplus_{a=0}^{\infty} q_*(L^{\otimes a} \otimes_{\mathcal{O}_A} \mathcal{O}_P)$$

is also H-linearized, where q is the projection of P to Y. On the other hand, since the action of H on Y is free, the quotient morphism

$$\pi : Y \longrightarrow Z = Z(V', r') \times_k Z(V, r) \times_k Z(V'', r'')$$

is a principal fiber bundle with group H. This implies that descent data on quasi-coherent sheaf on Y are nothing but H-linearizations. Therefore, there is a projective scheme P' over Z and a Cartesian diagram

$$\begin{array}{ccc} P & \xrightarrow{q} & Y \\ {\scriptstyle \pi'}\downarrow & & \downarrow{\scriptstyle \pi} \\ P' & \xrightarrow{q'} & Z \end{array}$$

Note that π' is a principal fiber bundle with group H and q' is projective.

By a routine argument of ranks of linear maps we can find an open subscheme Y' (or, Y'') of the affine space $\mathbf{V}(\mathrm{Hom}_k(V', V)^\vee)$ ($\mathbf{V}(\mathrm{Hom}_k(V, V'')^\vee)$, resp.) such that for all fields K containing k, $Y'(K)$ ($Y''(K)$, resp.) is the set of injections (surjections, resp.) of V'_K (V_K, resp.) to V_K (V''_K, resp.). There is a closed subscheme Y_0 of $Y' \times_k Y''$ such that for all extension fields K of k, $Y_0(K) = \{(\alpha', \alpha'') \in Y'(K) \times Y''(K) \mid \alpha''\alpha' = 0\}$. Let $\tilde{B}'$ and $\tilde{B}''$ be $\pi_{V'}^{-1}(B')$ and $\pi_{V''}^{-1}(B'')$, respectively. $\tilde{B}'$ (or, $\tilde{B}''$) is a $GL(V') \times_k GL(U')$-stable ($GL(V') \times_k GL(U')$-stable, resp.) closed set. There is a closed reduced subscheme F of $Y_0 \times_k P$ such that for all extension fields K of k, $F(K)$ is the set of points

$$(\alpha', \alpha'', \beta', \beta'', \psi', \psi, \psi'')$$

such that the diagram in (6.3.2) is commutative up to multiplication by non-zero constants and that $\psi' \in \tilde{B}'(K)$, $\psi'' \in \tilde{B}''(K)$. Let H' be the algebraic group $GL(V') \times_k GL(V) \times_k GL(V'')$. If we send a point

$$(g', g, g'', h', h, h'') \times (\alpha', \alpha'', \beta', \beta'', \psi', \psi, \psi'')$$

of $(H \times_k H')(K) \times (Y_0 \times_k P)(K)$ to the point

$$(h\alpha' h'^{-1}, h''\alpha'' h^{-1}, g\beta' g'^{-1}, g''\beta'' g^{-1}, g'\psi'(h' \otimes 1)^{-1}, g\psi(h \otimes 1)^{-1}, g''\psi''(h'' \otimes 1)^{-1})$$

of $(Y_0 \times_k P)(K)$, then we have an action of $H \times_k H'$ on $Y_0 \times_k P$ and we see that F is $H \times_k H'$-stable. Since $1_{Y_0} \times \pi' : Y_0 \times_k P \to Y_0 \times_k P'$ is a principal fiber bundle with group H, $F' = (1_{Y_0} \times \pi')(F)$ is closed. On the other hand, since the action of H on Y is commutative with that of H' and since $q_*(L^{\otimes a} \otimes_{\mathcal{O}_A} \mathcal{O}_P)$ carries an H'-linearization which is compatible with the H-linearization, π, π' and q' are H'-morphism and we have a very ample invertible sheaf L' on P' with an H'-linearization. Thus the closed subscheme F' with reduced structure is H'-stable. Denote $\dim V''$ by d''. Since Y_0 is a principal fiber bundle over $Z(V, d'')$ with group $GL(V') \times_k GL(V'')$ and since the projection $p : F' \to Z(V, r)$ is an H'-morphism, the same argument as in the first half of this proof gives rise to a commutative diagram

$$\begin{array}{ccccc} Y_0 & \longrightarrow & F' & \xrightarrow{p} & Z(V, r) \\ \downarrow & & \downarrow & \nearrow_{p'} & \\ Z(V, d'') & \longrightarrow & F'' & & \end{array}$$

where F' is a principal fiber bundle over F'' with group $GL(V') \times_k GL(V'')$, $GL(V)$ acts on all spaces and all the morphisms are $GL(V)$-morphisms. Since F'' is projective over $Z(V, d'')$, it is projective over k and hence $B = p'(F'')$ is a $GL(V)$-stable closed set in $Z(V, r)$. Now it is clear that B meets our requirement. □

The algebraic group $PGL(V)$ acts on $Z(V,r)$ and $\mathcal{O}_{Z(V,r)}(a)$ carries a $PGL(V)$-linearization for some positive integer a, where $\mathcal{O}_{Z(V,r)}(1)$ is the invertible sheaf defining the Plücker embedding. Proposition 2.2 provides us with a necessary and sufficient condition for a geometric point of $Z(V,r)$ to be semi-stable or stable with respect to $\mathcal{O}_{Z(V,r)}(a)$. In the sequel of this section a geometric point of $Z(V,r)$ is said to be stable (or semi-stable) if and only if the conclusion of Proposition 2.2 for the stability (semi-stability, resp.) holds good. There is an open sets $Z(V,r)^{ss}$ (or, $Z(V,r)^{s}_{0}$) of $Z(V,r)$ such that for every algebraically closed field K containing k, $Z(V,r)^{ss}(K)$ ($Z(V,r)^{s}_{0}(K)$, resp.) is the set of semi-stable (properly stable, resp.) K-valued geometric points of $Z(V,r)$. We have a result on the semi-stability of extensions, corresponding to Proposition 1.1, (6).

Proposition 6.4. *Let z', z and z'' be K-valued geometric points of $Z(V',r')$, $Z(V,r)$ and $Z(V'',r'')$, respectively. Assume that* $\dim V'/r' = \dim V/r = \dim V''/r''$.

(1) If z' is a subpoint of z and if z is semi-stable, the z' is semi-stable.

(2) Assume that z is an extension of z'' by z'. z is semi-stable if and only if so are both z' and z''.

Proof. We may assume that $K = k$. Let $\psi' \in \mathrm{Hom}_k(V'\otimes W, U')$, $\psi \in \mathrm{Hom}_k(V\otimes W, U)$ and $\psi' \in \mathrm{Hom}_k(V''\otimes W, U'')$ represent z', z and z'', respectively. Assume that z is semi-stable. Take a subspace V'_1 of V'. Since V'_1 can be regarded as a subspace of V, $\psi'(V'_1\otimes W)$ coincides with $\psi(V'_1\otimes W)$ and since z is semi-stable, we have

$$\dim \psi'(V'_1\otimes W) = \dim \psi(V_1\otimes W) \geq r\frac{\dim V'_1}{\dim V} = r'\frac{\dim V'_1}{\dim V'}.$$

Thus z' is semi-stable, which proves (1). To prove (2) let us look at the exact commutative diagram

$$\begin{array}{ccccccccc} 0 & \longrightarrow & V'\otimes_k W & \xrightarrow{\alpha'\otimes 1} & V\otimes_k W & \xrightarrow{\alpha''\otimes 1} & V''\otimes_k W & \longrightarrow & 0 \\ & & \Big\downarrow{\scriptstyle \psi'} & & \Big\downarrow{\scriptstyle \psi} & & \Big\downarrow{\scriptstyle \psi''} & & \\ 0 & \longrightarrow & U' & \xrightarrow{\beta'} & U & \xrightarrow{\beta''} & U'' & \longrightarrow & 0 \end{array}$$

Assume that z is semi-stable. We have proved that z' is semi-stable. Take a subspace V''_1 of V'' and set V_1 to be $\alpha''^{-1}(V''_1)$. Since $\dim V_1 = \dim V''_1 + \dim V'$ and ψ is semi-stable, we have

$$\frac{r}{\dim V}(\dim V''_1 + \dim V') = \frac{r}{\dim V}\dim V_1 \leq \dim \psi(V_1\otimes_k W).$$

Using the surjectivity of ψ' and the equalities $\dim V'/r' = \dim V/r = \dim V''/r''$, we obtain the desired inequality

$$\frac{r''}{\dim V''}\dim V''_1 \leq \dim \psi''(V''_1\otimes_k W).$$

Next assume that both ψ' and ψ'' are semi-stable. For a subspace V_1 of V, put $V'_1 = \alpha'^{-1}(V_1)$ and $V''_1 = \alpha''(V_1)$. We have then

$$\begin{aligned}\frac{r}{\dim V}\dim V_1 &= \frac{r'}{\dim V'}\dim V'_1 + \frac{r''}{\dim V''}\dim V''_1 \\ &\leq \dim \psi'(V'_1\otimes_k W) + \dim \psi''(V''_1\otimes_k W) \\ &\leq \dim \psi(V_1\otimes_k W)\end{aligned}$$

as required. □

To fix our idea we need a definition of isomorphisms between two points of $Z(V,r)$.

Definition 6.5. Let z_1 and z_2 be two K-valued points of $Z(V, r)$ and let ψ_1 and ψ_2 in $\mathrm{Hom}_K(V_K \otimes_K W_K, U_K)$ represent z_1 and z_2, respectively. We say that z_1 is isomorphic to z_2 if there are isomorphisms $\alpha : V_K \to V_K$ and $\beta : U_K \to U_K$ which make the following diagram commutative:

$$\begin{array}{ccc} V_K \otimes_K W_K & \xrightarrow{\alpha\otimes 1} & V_K \otimes_K W_K \\ {\scriptstyle\psi_1}\downarrow & & \downarrow{\scriptstyle\psi_1} \\ U_K & \xrightarrow{\beta} & U_K \end{array}$$

In other words, z_2 is isomorphic to z_1 if and only if z_2 is in the $GL(V)$-orbit of z_1

We can now find a typical $GL(V)$-stable closed set in $Z(V, r)$.

Theorem 6.6. *Let $V_1, \dots, V_\ell = V$ be vector spaces over k with $0 < \dim V_1 < \cdots < \dim V_\ell$, $V'_1, \dots, V'_\ell$ be vector spaces over k with $V'_1 = V_1$ and with $\dim V'_i = \dim V_i - \dim V_{i-1}$ $(i \geq 2)$ and let $r_1, \dots, r_\ell, r'_1, \dots, r'_\ell$ be positive integers with $r_i = r'_1 + \cdots + r'_i$. Assume that $\dim V'_1/r'_1 = \cdots = \dim V'_\ell/r'_\ell$. Let z_i be a point of $Z(V'_i, r'_i)^s_0(k)$. Then there exists a $GL(V)$-stable closed set $B(z_1, \dots, z_\ell)$ of $Z(V, r)^{ss}$ such that for every algebraically closed field K containing k, $B(z_1, \dots, z_\ell)(K) = \{z \in Z(V, r)(K) \mid z$ has the property (6.6.1)$\}$.*

(6.6.1) *There exists a point x_i in each $Z(V_i, r_i)(K)$ such that x_1 is isomorphic to z_1, x_i is an extension of a point isomorphic to z_i by x_{i-1} and $z = x_\ell$.*

Proof. Our proof is by induction on ℓ. If $\ell = 1$, then $B(z_1)$ is the $GL(V_1)$-orbit of z_1. The orbit is closed in $Z(V_1, r_1)^{ss}$ because z_1 is a stable point. Assume that our assertion is true for $\ell = i-1$. Then we have a $GL(V_{i-1})$-stable closed set $B'_{i-1} = B(z_1, \dots, z_{i-1})$ in $Z(V_{i-1}, r_{i-1})^{ss}$ which satisfies the conclusion of our theorem. The closure $\bar{B}'_{i-1}$ of B'_{i-1} in $Z(V_{i-1}, r_{i-1})$ is $GL(V_{i-1})$-stable closed set. The $GL(V'_i)$-orbit B''_i of z_i is closed in $Z(V'_i, r'_i)^{ss}$ and its closure $\bar{B}''_i$ in $Z(V'_i, r'_i)$ is $GL(V'_i)$-stable. Applying Lemma 6.3 to the case where $B' = \bar{B}'_{i-1}$ and $B'' = \bar{B}''_i$, we get a $GL(V_i)$-stable closed set B_i in $Z(V_i, r_i)$. If a point $x_i \in B_i(K)$ does not have the property (6.3.1), then we have a commutative diagram in (6.3.2) for $V' = V_{i-1}$, $V = V_i$ and $V'' = V'_i$. Since β'' is automatically surjective, β' cannot be injective. Thus we have

$$\dim \psi(\alpha'(V_K) \otimes W_K) = \dim \beta'\psi'(V_K \otimes W_K) < \dim U' = r_{i-1} = \frac{r_i}{\dim V_i} \dim V_{i-1},$$

which means that x_i is not semi-stable. Thus if x_i is semi-stable, then it must be an extension of z'' in $\bar{B}''_i(K)$ by a point x_{i-1} in $\bar{B}'_{i-1}(K)$. Since $\dim V_{i-1}/r_{i-1} = \dim V_i/r_i = \dim V'_i/r'_i$ by our assumption, Proposition 6.4 implies that x_i is semi-stable if and only if so are both x_{i-i} and z''. Thus if x_i is semi-stable, then it is an extension of $x_{i-1} \in B(z_1, \dots, z_{i-1})$ by a point isomorphic to z_i. Now it is clear that $B(z_1, \dots, z_i) = B_i \cap Z(V_i, r_i)^{ss}$ is the desired closed set for $\ell = i$. □

We need now the notion of a direct sum.

Definition 6.7. Let $V, V_1, \dots, V_\ell$ be vector spaces over k with $\dim V = \dim V_1 + \cdots + \dim V_\ell$ and let $r, r_1, \dots, r_\ell$ be positive integers with $r = r_1 + \cdots + r_\ell$. Fix a field K containing k. Let z (or z_i $(1 \leq i \leq \ell)$) be a K-valued point of $Z(V, r)$ ($Z(V_i, r_i)$, resp.) and let ψ (ψ_i, resp.) be a member of $\mathrm{Hom}_K(V_K \otimes_K W_K, U_K)$ ($\mathrm{Hom}_K((V_i)_K \otimes_K W_K, (U_i)_K)$, resp.) which represents z (z_i, resp.). By setting

$$\psi' = \oplus_{i=1}^{\ell} \psi_i : (\oplus_{i=1}^{\ell} V_i)_K \otimes_K W_K \longrightarrow (\oplus_{i=1}^{\ell} U_i)_K,$$

ψ is said to be a direct sum of $\psi_1, \ldots, \psi_\ell$ if there are isomorphisms $\alpha : V_K \to (\oplus_{i=1}^{\ell} V_i)_K$ and $\beta : U_K \to (\oplus_{i=1}^{\ell} V_i)_K$ which make the following diagram commutative:

$$\begin{array}{ccc} V_K \otimes_K W_K & \xrightarrow{\alpha\otimes 1} & (\oplus_{i=1}^{\ell} V_i)_K \otimes_K W_K \\ \psi \downarrow & & \downarrow \psi' \\ U_K & \xrightarrow{\beta} & (\oplus_{i=1}^{\ell} U_i)_K \end{array}$$

A direct sum of $z_1, \ldots, z_\ell$ is independent of the choice of $\psi_1, \ldots, \psi_\ell$ and hence we denote it by $z_1 \oplus \cdots \oplus z_\ell$. Direct sums of the same $z_1, \ldots, z_\ell$ are isomorphic to each other.

A direct sum $z_1 \oplus z_2$ is a special case of an extension of z_2 by z_1 and it is obvious that if we make a direct sum of $z_1, \ldots, z_\ell$ as a sequence of direct sums of two elements, then it is always isomorphic to $z_1 \oplus \cdots \oplus z_\ell$.

Lemma 6.8. *Let V, V' and V'' be vector spaces over k with $\dim V = \dim V' + \dim V''$ and let r, r' and r'' be positive integers with $r = r' + r''$. Fix a field K containing k. Let z, z' and z'' be K-valued points of $Z(V, r)$, $Z(V', r')$ and $Z(V'', r'')$, respectively and let A be a discrete valuation ring over K with residue field K and with quotient field L. Assume that z is an extension of z'' by z'. Then there exists an A-valued point $\tilde{z}$ of $Z(V, r)$ such that $\tilde{z}$ is isomorphic to z over L and $\tilde{z} \mod t$ is a direct sum of z' and z'', where t is a uniformanizing parameter of A.*

Proof. We may assume that $K = k$. Let $\psi' \in \mathrm{Hom}_k(V' \otimes W, U')$, $\psi \in \mathrm{Hom}_k(V \otimes W, U)$ and $\psi' \in \mathrm{Hom}_k(V'' \otimes W, U'')$ represent z', z and z'', respectively. We have a commutative diagram

$$\begin{array}{ccccccccc} 0 & \longrightarrow & V' \otimes_k W & \xrightarrow{\alpha'\otimes 1} & V \otimes_k W & \xrightarrow{\alpha''\otimes 1} & V'' \otimes_k W & \longrightarrow & 0 \\ & & \psi' \downarrow & & \downarrow \psi & & \downarrow \psi'' & & \\ 0 & \longrightarrow & U' & \xrightarrow{\beta'} & U & \xrightarrow{\beta''} & U'' & \longrightarrow & 0 \end{array}$$

Let $\{v_1', \ldots, v_{d'}'\}$, $\{v_1'', \ldots, v_{d''}''\}$ and $\{w_1, \ldots, w_M\}$ be bases of V', V'' and W, respectively. Take a basis $\{v_1, \ldots, v_d\}$ of V so that $v_1 = \alpha'(v_1'), \ldots, v_{d'} = \alpha'(v_{d'}')$ and $\alpha''(v_{d'+1}) = v_1'', \ldots, \alpha''(v_d) = v_{d''}''$. We shall also fix bases $\{u_1', \ldots, u_{r'}'\}$, $\{u_1'', \ldots, u_{r''}''\}$ and $\{u_1, \ldots, u_r\}$ of U', U'' and U, respectively, in the same way, that is, $u_1 = \beta'(u_1'), \ldots, u_{r'} = \beta'(u_{r'}')$ and $\beta''(u_{r'+1}) = u_1'', \ldots, \beta''(u_r) = u_{r''}''$. Our assumption that z is an extension of z'' by z' is translated into the condition that if $r' \times d'M$ matrix B and $r'' \times d''M$ matrix C represent ψ' and ψ'', respectively, then there is an $r' \times d''M$ matrix D such that

$$F = \begin{pmatrix} B & D \\ 0 & C \end{pmatrix}$$

represents ψ. Note here that z is a direct sum of z' and z'' if $D = 0$. If we denote the unit matrix of degree ℓ by I_ℓ, then we have

$$\begin{pmatrix} I_{r'} & 0 \\ 0 & t^{-1} I_{r''} \end{pmatrix} \begin{pmatrix} B & D \\ 0 & C \end{pmatrix} \begin{pmatrix} I_{d'M} & 0 \\ 0 & tI_{d''M} \end{pmatrix} = \begin{pmatrix} B & tD \\ 0 & C \end{pmatrix} = F(t).$$

$F(t)$ defines an A-module homomorphism of $(V \otimes_k W) \otimes_k A$ to $U \otimes_k A$. Since $\operatorname{rank} F(0) = \operatorname{rank} B + \operatorname{rank} C = r' + r'' = r$, $F(0) : V \otimes_k W \to U$ is surjective. Thus $F(t)$ is surjective and represents an A-valued point $\tilde{z}$ of $Z(V, r)$. By the construction of $\tilde{z}$ it is isomorphic to z over L. It is obvious that $\tilde{z} \mod t$ is a direct sum of z' and z''. $\square$

Now we come to our main result in this section.

Theorem 6.9. *Under the same situation as in* Theorem 6.6, *$GL(V)$-orbit $o(z_1, \dots, z_\ell)$ of $z_1 \oplus \cdots \oplus z_\ell$ is a unique closed orbit in $B(z_1, \dots, z_\ell)$.*

Proof. First note that $z_1 \oplus \cdots \oplus z_\ell$ is a point of $B(z_1, \dots, z_\ell)$. Our proof is by induction on ℓ. If $\ell = 1$, then $B(z_1) = o(z_1)$ and there is nothing to prove. Suppose that our assertion holds for $\ell - 1$. Let o be a closed orbit in $B(z_1, \dots, z_\ell)_K$ with K an algebraically field containing k. Since every point z of $o(K)$ is an extension of z'' isomorphic to z_ℓ by an x' in $B(z_1, \dots, z_{\ell-1})(K)$, or equivalently an extension of z_ℓ by x', there is a point $\tilde{x}$ of o such that a specialization of $\tilde{x}$ is $x' \oplus z_\ell$ by Lemma 6.5. Since $x' \oplus z_\ell$ is in $B(z_1, \dots, z_\ell)_K$ by Proposition 6.4 and o is closed, $x' \oplus z_\ell$ is a point of $o(K)$ and hence z is isomorphic to $x' \oplus z_\ell$. By induction hypothesis there is a point $\tilde{x}'$ of $B(z_1, \dots, z_{\ell-1})_K$ such that $\tilde{x}'$ is isomorphic to x' and $z_1 \oplus \cdots \oplus z_{\ell-1}$ is a specialization of $\tilde{x}'$. Since z is isomorphic to $\tilde{x}' \oplus z_\ell$ and since $z_1 \oplus \cdots \oplus z_\ell$ is a specialization of $\tilde{x}' \oplus z_\ell$, we see that z is isomorphic to $z_1 \oplus \cdots \oplus z_\ell$. □

7. Moduli spaces of semi-stable sheaves

We shall fix a pair $(X, \mathcal{O}_X(1))$ of a projective S-scheme $f : X \to S$ and an S-ample invertible sheaf $\mathcal{O}_X(1)$ on X which satisfies the condition (4.6) and shall maintain the notation of previous sections. Fix an integer m in Lemma 4.12 and consider the Quot-scheme

$$Q_r = \mathrm{Quot}^{H^{(r)}[m]}_{V_r(X)/X/S},$$

where r is a positive integer such that $r \leq a_0(H)$ and $H^{(r)}(x) = rH(x)/a_0(H)$ is a numerical polynomial and where V_r is a free Ξ-module of rank $H^{(r)}[m](0)$. Let $Q_r(d)$ be the scheme theoretic closure in Q_r of the open set of parametrizing sheaves of pure dimension d and let $\varphi_r : V_r(X \times_S Q_r(d)) \to \tilde{F}_r$ be the universal quotient sheaf on $X \times_S Q_r(d)$. Then there is open set $R_r^{e,e'}$ defined by the properties (4.13.1) and (4.13.2). For sufficiently large integer n, we have an embedding of $Q_r(d)$ into $Z_r(n)$ so that $R_r^{e,e'}$ is in the open set of semi-stable points $Z_r(n)^{ss}(\mathcal{O}_{Z_r(n)}(a))$ of $Z_r(n)$ and a geometric point y of $R_r^{e,e'}$ is in $Z_r(n)^s_0(\mathcal{O}_{Z_r(n)}(a))$ if and only if $\tilde{F}_r(y)$ is stable (see Lemma 5.1).

Lemma 7.1. *There is an integer n_0 such that for every integer $n \geq n_0$, we have the following:*

(7.1.1) *For every positive integer $r \leq a_0(H)$, $R_r^{e,e'}$ is embedded in $Z_r(n)^{ss}(\mathcal{O}_{Z_r(n)}(a)))$.*

(7.1.2) *Let y and y' be k-valued geometric points of $R_r^{e,e'}$ and $R_{r'}^{e,e'}$ over a geometric point s of S, where r' and r are positive integers such that $r' < r \leq a_0(H)$. y' is a subpoint of y as points of Grassmann varieties if and only if $\tilde{F}_r(y)$ contains a coherent subsheaf F' which is isomorphic to $\tilde{F}_{r'}(y')$.*

(7.1.3) *For F' in (7.1.2), set $F'' = \tilde{F}_r(y)/F'$. We have a surjection $H^0(X_s, F'') \otimes_k W(s) \to H^0(X_s, F''(n))$ and it defines a point y'' of $R_{r''}^{e,e'} \subset Z_{r''}(n)^{ss}(L_{r''})$, where $r'' = r - r'$. Moreover, y is an extension of y'' by y'.*

Proof. We shall take n_0 to be the maximum among n_0's in the proof of Lemma 5.1 for $r \leq a_0(H)$. (7.1.1) is Lemma 5.1. Assume that $F = \tilde{F}_r(y)$ contains a subsheaf F' which is isomorphic to $\tilde{F}_{r'}(y')$. Set $U' = H_0(X_s, \tilde{F}_{r'}(y')(n))$ and $U = H^0(X_s, \tilde{F}_r(y)(n))$. The injections

$$\alpha' : V_{r'}(k) \cong H_0(X_s, \tilde{F}_{r'}(y')) \cong H^0(X_s, F') \hookrightarrow H^0(X_s, \tilde{F}_r(y)) \cong V_r(k)$$

$$\beta' : U' = H_0(X_s, \tilde{F}_{r'}(y')(n)) \cong H^0(X_s, F'(n)) \hookrightarrow H^0(X_s, (n)) = U$$

provide us with an exact commutative diagram

$$\begin{array}{ccccc} 0 & \longrightarrow & V_{r'}(k)\otimes_k W(s) & \xrightarrow{\alpha'\otimes 1} & V_r(k)\otimes_k W(s) \\ & & \downarrow{\scriptstyle \psi'} & & \downarrow{\scriptstyle \psi} \\ 0 & \longrightarrow & U' & \xrightarrow{\beta'} & U \end{array} \tag{7.1.4}$$

Recall here that W is the locally free $\mathcal{O}_S$-module $h_*(\mathcal{O}_X(n))$ with h the structure morphism of the S=scheme $\mathbf{P}(U)$ (see (4.6)). By Lemma 5.1 ψ' and ψ represent y' and y, respectively. Thus y' is a subpoint of y. Conversely, suppose that the above exact commutative diagram is given and ψ' and ψ represent y' and y, respectively. Let F' be the coherent subsheaf of $\tilde{F}_r(y)$ generated by $V' = \alpha'(V_{r'}(k))$ which can be regarded as a k-vector subspace of $H^0(X_s, \tilde{F}_r(y))$. By our choice of m we have

$$\begin{aligned} \dim V' &= \dim U' \frac{H^{(r')}(m)}{H^{(r')}(m+n)} = \dim U' \frac{H^{(r)}(m)}{H^{(r)}(m+n)} \\ &= \dim U' \frac{\dim V_r(k)}{\dim H^0(X_s, \tilde{F}(y)(n))}. \end{aligned}$$

Thus the proof of Lemma 5.1 shows that $V' = H^0(X_s, F')$ and $P^S_{F'}(x) = H[m](x)/a_0(H)$. Since $\tilde{F}_r(y)$ is strictly e'-semi-stable, so is F' by Lemma 3.3. These and (4.12.1) provide us with the equalities

$$r'\frac{H(m)}{a_0(H)} = \dim V' = \chi(F') = a_0(F)\frac{H(m)}{a_0(H)}$$

and hence we know that $a_0(F') = r'$, that is, $\chi(F'(a)) = \chi(\tilde{F}_{r'}(a))$ as polynomials in a. On the other hand, let K' and K be the kernel of ψ' and ψ, respectively and consider the following commutative diagram of coherent sheaves on X_s

$$\begin{array}{ccccc} K'\otimes_k \mathcal{O}_{X_s}(-n) & \xrightarrow{u'\otimes 1} & V_{r'}(k)\otimes_k W(s)\otimes_k \mathcal{O}_{X_s}(-n) & \xrightarrow{1\otimes w} & V_{r'}(k)\otimes_k \mathcal{O}_{X_s} \\ \downarrow & & \downarrow{\scriptstyle \alpha'\otimes 1\otimes 1} & & \downarrow{\scriptstyle \alpha'\otimes 1} \\ K\otimes_k \mathcal{O}_{X_s}(-n) & \xrightarrow{u\otimes 1} & V_r(k)\otimes_k W(s)\otimes_k \mathcal{O}_{X_s}(-n) & \xrightarrow{1\otimes w} & V_r(k)\otimes_k \mathcal{O}_{X_s} \end{array}$$

where u' and u are natural injections and where $w : W(s)\otimes_k \mathcal{O}_{X_s}(-n) \to \mathcal{O}_{X_s}$ is defined by the map of global sections $W(s)\otimes_k \mathcal{O}_{X_s} \to \mathcal{O}_{X_s}(n)$. The construction of the Quot-scheme tells us that $\tilde{F}_r(y) = \operatorname{coker}((1\otimes w)(u\otimes 1))$ and $\tilde{F}_{r'}(y') = \operatorname{coker}((1\otimes w)(u'\otimes 1))$. Moreover, $\operatorname{im}(\alpha'\otimes 1)$ generates F' in $\tilde{F}_r(y)$. Thus we have a surjection of $\tilde{F}_{r'}(y')$ to F'. Since $\tilde{F}_{r'}(y')$ and F' have the same Hilbert polynomial, they are isomorphic. This completes the proof of (7.1.2).

Set $F'' = \tilde{F}_r(y)/F'$. Since $P^S_F(x) = P^S_{F'}(x)$ as polynomials in x and F is strictly e'-semi-stable, we have $P^S_{F''}(x) = P^S_F(x)$ and then by Lemma 3.3 F'' is strictly e'-semi-stable. The above proof shows that if we identify $V_{r'}(k)$ with $H^0(X_s, \tilde{F}_{r'}(y'))$ and $V_r(k)$ with $H^0(X_s, \tilde{F}_r(y))$, then α' and β' are the maps induced by the injection $\tilde{F}_{r'}(y') \cong F' \subset \tilde{F}_r(y)$. By (4.12.1) for $\tilde{F}_{r'}(y')$ we have an exact commutative diagram extending (7.1.4)

$$\begin{array}{ccccccccc} 0 & \longrightarrow & V_{r'}(k)\otimes_k W(s) & \xrightarrow{\alpha'\otimes 1} & V_r(k)\otimes_k W(s) & \xrightarrow{\alpha''\otimes 1} & V''\otimes_k W(s) & \longrightarrow & 0 \\ & & \downarrow{\scriptstyle \psi'} & & \downarrow{\scriptstyle \psi} & & \downarrow{\scriptstyle \psi''} & & \\ 0 & \longrightarrow & U' & \xrightarrow{\beta'} & U & \xrightarrow{\beta''} & U'' & \longrightarrow & 0 \end{array}$$

where $V'' = H^0(X_s, F'')$, $U'' = H^0(X_s, F''(n))$, ψ'' is the multiplication map and both α'' and β'' are defined by the quotient homomorphism $\tilde{F}_r(y) \to F''$. Since ψ'' is surjective, it gives rise to a k-rational point y'' of $Z_{r''}(n)$ and it comes from a point $V_{r''}(k) \otimes \mathcal{O}_{X_s} \to F''$ of $Q_{r''}$. The strict e'-semi-stability of F'' means that y'' is a geometric point of $R^{e,e'}_{r''}$. □

Corollary 7.2. *Let $E_1, \dots, E_\ell$ be e'-stable sheaves on a geometric fiber X_t such that for all i, $P^S_{E_i}(x) = H(x)/a_0(H)$ and $a_0(E_1) + \cdots + a_0(E_\ell) = a_0(H)$. There is a $\bar{G}_t$-stable closed set $B(E_1, \dots, E_\ell)$ of $(R^{e,e'})_t$ with the following properties:*

(7.2.1) *$B(E_1, \dots, E_\ell)$ is closed in $Z(n)^{ss}(L)_t$.*

(7.2.2) *For every algebraically closed field K containing $k = k(t)$,*

$$B(E_1, \dots, E_\ell)(K) = \{y \in (R^{e,e'})_t(K) \mid \mathrm{gr}(\tilde{F}(y)) \cong \oplus_{i=1}^{\ell} E_i \otimes_k K\}$$

(7.2.3) *The orbit $o(\oplus_{i=1}^{\ell} E_i)$ of a point corresponding to $\oplus_{i=1}^{\ell} E_i$ is a unique closed orbit of $B(E_1, \dots, E_\ell)(k)$.*

Proof. Let us fix an integer n such that $n \geq n_0$ for the n_0 in Lemma 7.1 and embed each $R^{e,e'}_r$ into $Z_r(n)$. Set $r'_i = a_0(E_i)$. Then, by Lemma 5.1 E_i gives rise to a k-valued geometric point z_i of $Z_{r'_i}(n)^s_0(L_{r'_i})_t$. Pick a permutation τ of $\{1, 2, \dots, \ell\}$. By Theorem 7.6 we have the $\bar{G}_t$-stable closed set $B(z_{\tau(i)}, \dots, z_{\tau(\ell)})$ in $Z(n)^{ss})_t$. If we put

$$B(E_1, \dots, E_\ell) = \bigcup_{\tau \in S_\ell} B(z_{\tau(i)}, \dots, z_{\tau(\ell)}),$$

then $B(E_1, \dots, E_\ell)$ is $\bar{G}_t$-stable closed set in $(Z(n)^{ss})_t$ and contains the unique closed orbit $o(z_1 \oplus \cdots \oplus z_\ell)$. Since

$$C = B(E_1, \dots, E_\ell) \cap ((Z(n)^{ss} \setminus R^{e,e'})_t)$$

is $\bar{G}_t$-stable closed set in $(Z(n)^{ss}(L)_t$, it must contain $o(z_1 \oplus \cdots \oplus z_\ell)$ unless it is empty. On the other hand, by Corollary 3.4 the direct sum $\oplus_{i=1}^{\ell} E_i$ corresponds to a k-valued geometric point y_0 of $(R^{e,e'})_t$ which is isomorphic to $z_1 \oplus \cdots \oplus z_\ell$ as points of the Grassmann variety $Z(n)$. Thus $o(z_1 \oplus \cdots \oplus z_\ell)$ is in $(R^{e,e'})_t$ and hence C is empty, that is, $B(E_1, \dots, E_\ell) \subset (R^{e,e'})_t$. Pick a point x in $B(z_{\tau(1)}, \dots, z_{\tau(\ell)})(K)$. Setting $r(\tau)_i = r'_{\tau(1)} + \cdots + r'_{\tau(i)}$, there is a sequence of K-valued points $x_1, \dots, x_\ell$ which has the property (7.6.1) for $Z(V_i, r_i) = Z_{r(\tau)_i}(n)$. If we take $B(z_{\tau(1)}, \dots, z_{\tau(i)})$ as in Theorem 7.6, then we see as in the above that it is a subset of $(R^{e,e'}_{r(\tau)_i})_t$. Thus each x_i is a K-valued point of $(R^{e,e'}_{r(\tau)_i})_t$. Assume that $F_{i-1} = \tilde{F}_{r(\tau)_{i-1}}(x_{i-1})$ has a filtration $0 = F'_0 \subset F'_1 \subset \cdots \subset F'_{i-1} = F_{i-1}$ such that $F'_j/F'_{j-1} \cong E_{\tau(j)} \otimes_k K$. Since x_i is an extension of $z_{\tau(i)}$ by x_{i-1} and since x_{i-1} and x_i are in $(R^{e,e'}_{r(\tau)_{i-1}})_t$ and $(R^{e,e'}_{r(\tau)_i})_t$, respectively, $F_i = \tilde{F}_{r(\tau)_i}(x_i)$ contains a coherent subsheaf F'_{i-1} which is isomorphic to F_{i-1} by Lemma 7.1. By the construction of F'_{i-1}, the point z'' defined by F_i/F'_{i-1} as in (7.1.3) is isomorphic to $z_{\tau(i)}$ over K as points of the Grassmann variety. This means that $F_i/F'_{i-1} \cong E_{\tau(i)} \otimes_k K$. Regarding $F'_1, \dots, F'_{i-1}$ as subsheaves of F_i, we have a filtration $0 = F'_0 \subset F'_1 \subset \cdots \subset F'_i = F_i$ such that $F'_j/F'_{j-1} \cong E_{\tau(j)} \otimes_k K$. Therefore, by induction on i, we get a filtration $0 = F'_0 \subset F'_1 \subset \cdots \subset F'_\ell = \tilde{F}(x)$ such that for $1 \leq j \leq \ell$, $F'_j/F'_{j-1} \cong E_{\tau(j)} \otimes_k K$. Hence $\mathrm{gr}(\tilde{F}(x)) \cong \oplus_{i=1}^{\ell} E_i \otimes_k K$. Conversely, it is not hard to see that if x is a K-valued geometric point of $R^{e,e'}$ such that $\mathrm{gr}(\tilde{F}(x)) \cong \oplus_{i=1}^{\ell} E_i \otimes_k K$, then x is in $B(E_1, \dots, E_\ell)$. This completes the proof that $B(E_1, \dots, E_\ell)$ meats our requirement. □

By [8, Theorem 4] we have a good quotient $\pi : Q(d)^{ss}(L) \to Y$. Since a good quotient separates closed orbits and the image of $\bar{G}$-stable closed set to Y is closed, the above corollary implies that $\pi^{-1}\pi(R^{e,e'}) = R^{e,e'}$ and $\bar{M}^{e,e'} = \pi(R^{e,e'})$ is an open set of Y. The open subscheme $\bar{M}^{e,e'}$ of Y is, therefore, a good quotient of $R^{e,e'}$.

Proposition 7.3. *$\bar{M}^{e,e'}$ is a coarse moduli scheme of $\bar{\Sigma}^{H,e'}_{X/S}$*

Proof. Let T be a locally noetherian S-scheme and E a T-flat coherent sheaf on $X \times_S T$ with the property (f) of Definition 4.7. As in the proof of Proposition 5.2 we have a morphism $\Psi^{e,e'}(E) : T \to \bar{M}^{e,e'}$ such that for every morphism $\eta : T' \to T$ of locally noetherian S-schemes, $\Psi^{e,e'}((1_X \times \eta)^*(E)) = \Psi^{e,e'}(E)\eta$. Since the set of geometric points of $\bar{M}^{e,e'}$ is in bijective correspondence with the set of closed orbits of $R^{e,e'}$ the construction of $\Psi^{e,e'}(E)$ and Corollary 7.2 show that if for two geometric points t_1 and t_2, $E(t_1)$ is S-equivalent to $E(t_2)$, then $\Psi^{e,e'}(E)$ maps them to the same point. We need the following well-known lemma [7, Proposition 6.1]

Lemma 7.4. *Given a commutative diagram of locally noetherian schemes:*

$$\begin{array}{ccc} P & & \\ {\scriptstyle p}\downarrow & \searrow{\scriptstyle h} & \\ T & \xleftarrow[q]{} & Y \end{array}$$

suppose that T is connected, p is flat and $H^0(P_t, \mathcal{O}_{P_t}) = k(t)$ for all points $t \in T$. If p is proper and if for a point t of T, $h(P_t)$ is set-theoretically one point, then there is a section $\theta : T \to Y$ of q such that $h = \theta p$.

Let the pair (T, E) be as above and assume that E is an extension

$$0 \longrightarrow E_1 \longrightarrow E \longrightarrow E_2 \longrightarrow 0$$

of T-flat coherent sheaves E_1 and E_2 such that for every geometric point t of T, $E_1(t)$ and $E_2(t)$ are semi-stable and $P^S_{E_1(t)} = P^S_{E_2(t)} = P^S_{E(t)}$. Take a connected affine open set $U = \mathrm{Spec}(A)$ of T and consider the affine line $\mathbf{A}^1_U = \mathrm{Spec}(A[Z])$ over U. Since $A[Z]$ is flat over A,

$$\mathrm{Ext}^1((E_2)_U, (E_1)_U) \otimes_A A[Z] \cong \mathrm{Ext}^1(\tilde{E}_2, \tilde{E}_1),$$

where $\tilde{E}_1$ and $\tilde{E}_2$ are the pull-backs of E_1 and E_2 by the projection $p_0 : X \times_T \mathbf{A}^1_U \to T$. Thus an extension class ξ determined by the given extension gives rise to an element $\xi \otimes Z$ of $\mathrm{Ext}^1(\tilde{E}_2, \tilde{E}_1)$ and it defines an extension

$$0 \longrightarrow \tilde{E}_1 \longrightarrow \tilde{E}_0 \longrightarrow \tilde{E}_2 \longrightarrow 0.$$

Let $\eta_0 : U \to \mathbf{A}^1_U$ be the zero-section. By its construction $\tilde{E}_0$ is a isomorphic to the pull-back of E on the open set $X \times_S (\mathbf{A}^1_U \setminus \eta_0(U))$. Take another affine line $\mathrm{Spec}(A[Z^{-1}])$ and the base change $\tilde{E}_\infty$ of E to the affine line. Then we can glue two affine lines to get $\mathbf{P}^1_U$ and patch $\tilde{E}_0$ and $\tilde{E}_\infty$ to obtain a U-flat coherent sheaf on $X \times_S \mathbf{P}^1_U$. Let $\eta_\infty : U \to \mathbf{P}^1_U$ be the ∞-section. Set $\bar{M} = \bar{M}^{e,e'}$. We apply the above lemma to the case where $P = \mathbf{P}^1_U$, $T = U$, $Y = \bar{M} \times_S U$, p is the projection, $h = (\Psi^{e,e'}(\tilde{E}), p)$ and q is the projection of $\bar{M} \times_S U$ to U. For a point $t \in U$, $\Psi^{e,e'}(\tilde{E})(\mathbf{P}^1_{k(t)})$ is the point corresponding to $\mathrm{gr}(E(t) \otimes \overline{k(t)})$ and hence $h(\mathbf{P}^1_{k(t)})$ is set-theoretically one point. Thus there is a section $\theta' : U \to \bar{M} \times_S U$ of q such that $h = \theta' p$. We set θ to be the composition of θ' and the projection $p_1 : \bar{M} \times_S U \to \bar{M}$.

By the functoriality of $\Psi^{e,e'}$ we obtain

$$\begin{aligned}\Psi^{e,e'}((E_1 \oplus E_2)_U) &= \Psi^{e,e'}((1_X \times \eta_0)^*(\tilde{E})) = \Psi^{e,e'}(\tilde{E})\eta_0 \\ &= p_1 h\eta_0 = p_1\theta' p\eta_0 \\ &= \theta\end{aligned}$$

and

$$\begin{aligned}\Psi^{e,e'}(E_U) &= \Psi^{e,e'}((1_X \times \eta_\infty)^*(\tilde{E})) = \Psi^{e,e'}(\tilde{E})\eta_\infty \\ &= p_1 h\eta_\infty = p_1\theta' p\eta_\infty \\ &= \theta\end{aligned}$$

Therefore, we see that $\Psi^{e,e'}(E_1 \oplus E_2) = \Psi^{e,e'}(E)$. Using this repeatedly, we find that $\Psi^{e,e'}$ is compatible with the equivalence relation in (d) of Definition 0.5 and hence $\Psi^{e,e'}$ induces a morphism $\bar{\Psi}^{e,e'} : \bar{\Sigma}^{H,e'}_{X/S} \to \bar{M}$ of functors. As is remarked right before the above lemma, for the geometric point s of S, $\bar{\Psi}^{e,e'}(\mathrm{Spec}(k(s)))$ is a bijection. A good quotient is a categorical quotient. If we use this fact, then the remaining of the proof is similar to the proof of Proposition 5.2 . □

Since $R^{e,e'}$ is a $\bar{G}$-stable open set of $R^{e,e}$, $\bar{M}^{e,e'}$ is an open subscheme of $\bar{M}^{e,e}$. $\bar{M}^{e',e'}$ and $\bar{M}^{e,e'}$ are coarse moduli schemes of the same functors and hence they are isomorphic to each other by a unique isomorphism. Thus $\bar{M}^{e',e'}$ can be regarded naturally as an open subscheme of $\bar{M}^{e,e}$. Taking the inductive limit of $\{\bar{M}^{e,e}\}$, we get an S-scheme $\bar{M}_{X/S}(H)$. Since each $\bar{M}^{e,e}$ contains $M^{e,e}$ of section 5 as such an open subscheme that $M^{e',e'} = M^{e,e} \cap \bar{M}^{e',e'}$, the moduli scheme $M_{X/S}(H)$ of stable sheaves is an open subscheme of $\bar{M}_{X/S}(H)$. Moreover, $\bar{M}_{X/S}(H)$ is locally of finite type over S because each $\bar{M}^{e,e}$ is quasi-projective.

Theorem 7.5. *Let S be a scheme of finite type over a universally Japanese ring Ξ, $f : X \to S$ a projective morphism and let $\mathcal{O}_X(1)$ be an f-very ample invertible sheaf such that X is embedded in $\mathbf{P}(U)$ with U a locally free coherent sheaf on S, $\mathcal{O}_X(1)$ is the restriction of $\mathcal{O}_{\mathbf{P}(U)}(1)$ to X. Then, $\bar{M}_{X/S}(H)$ is a coarse moduli scheme of the functor $\bar{\Sigma}^H_{X/S}$ and it contains the moduli scheme $M_{X/S}(H)$ of stable sheaves as an open subscheme. $\bar{M}_{X/S}(H)$ is locally of finite type and separated over S. Moreover, $\bar{M}_{X/S}(H)$ is projective over S if and only if $\mathfrak{S}_{X/S}(H)$ is bounded.*

Proof. The proof of the former half is as in the proof of Theorem 5.3. If $\bar{M}_{X/S}(H)$ is of finite type over S, then $\mathfrak{S}_{X/S}(H)$ is equal to $\mathfrak{S}_{X/S}(H, e)$ for some e and hence $\mathfrak{S}_{X/S}(H)$ is bounded by Proposition 3.6. Conversely, if $\mathfrak{S}_{X/S}(H)$ is bounded, then $\bar{M}_{X/S}(H) = \bar{M}^{e,e}$ for a sufficiently large e, which means that $\bar{M}_{X/S}(H)$ is of finite type over S. Then we can apply the following refinement of Langton's theorem to showing that $\bar{M}_{X/S}(H)$ is proper over S.

Theorem 7.6. *Under the situation* (4.6) *let A be a discrete valuation ring with quotient field K and residue field k. Assume that a morphism $T = \mathrm{Spec}(A) \to S$ is given. If $\bar{E}$ is a semi-stable sheaf on the generic fiber X_K of X_T, then there is a T-flat coherent sheaf $\tilde{E}$ on X_T such that $\tilde{E}_K \cong \bar{E}$ and E_k is semi-stable.*

A proof of this theorem will be given in a forthcoming article. □

Applying Theorem 1.8 and [6, Theorem 3.6], we get the following.

Corollary 7.7. *$\bar{M}_{X/S}(H)$ is projective over S if one of the following conditions is satisfied:*

(1) *Ξ is a field of characteristic zero.*

(2) *H is a polynomial of degree not greater than 2.*

(3) *$f : X \to S$ is geometrically integral, $\deg H = \dim X/S$ and the rank of sheaves is not greater than 3.*

References

[1] A. Grothendieck: Techniques de construction et théorèmes d'existence en géométrie algebrique IV: Le schémas de Hilbert, Séminaire Bourbaki, t.**13** (1960/61), n^0 221.

[2] M. Maruyama: Stable vector bundles on an algebraic surfaces, Nagoya Math. J. **58** (1975), 25–68.

[3] M. Maruyama: Moduli of stable sheaves I, J. Math. Kyoto Univ. **17** (1977), 91–126.

[4] M. Maruyama: Moduli of stable sheaves II, J. Math. Kyoto Univ. **18** (1978), 557–614.

[5] M. Maruyama: Boundedness of semi-stable sheaves of small rank, Nagoya Math. J. **78** (1980), 65–94.

[6] M. Maruyama: On boundedness of families of torsion free sheaves, J. Math. Kyoto Univ. **21** (1981), 673–701.

[7] D. Mumford: Geometric Invariant Theory, 1965, 1982, 1994, Springer-Verlag.

[8] C. S. Seshadri: Geometric reductivity over arbitrary base, Adv. Math. **26** (1977), 225–274.

[9] C. T. Simpson: Moduli of representations of the fundamental group of a smooth projective variety, to appear.

CURVES AND K3 SURFACES OF GENUS ELEVEN

Shigeru Mukai

Graduate School of Polymathematics, Nagoya University, Chikusa-ku, Nagoya, 464-01, Japan

ABSTRACT. For a general curve C of genus 11, its embeddings into K3 surfaces are unique up to isomorphisms. Such embedding $\alpha : C \longrightarrow \hat{S}$ is constructed as a non-abelian analogue of the duality between the Picard and Albanese varieties. Let S be the moduli space of stable rank two vector bundles E of canonical determinant on C with $h^0(E) \geq 7$. Then S is a smooth K3 surface and the determinant line bundle h_{det} is a polarization of genus 11. The K3 surface $\hat{S}$ which contains C is the unique 2-dimensional component of the moduli space of stable rank two sheaves of determinant h_{det} on S.

In [11], we have begun our study of the Brill-Noether locus

$$M_C(2, K, n) = \{E | h^0(E) \geq n + 2\}/\text{isom}$$

in the moduli space $M_C(2, K)$ of stable rank two vector bundles E of canonical determinant over a curve, *i.e.*, a compact Riemann surface C. In the workshop at Sanda, we discussed $M_C(2, K, 3)$ for a curve C of genus seven, for which see the forthcoming article. Here, instead, we study another interesting case:

Theorem 1. *For a general curve C of genus eleven, the Brill-Noether locus $M_C(2, K, 5)$ is a smooth K3 surface and the restriction h_{det} of the determinant line bundle of $M_C(2, K)$ is a polarization of genus eleven,* i.e., $(h^2) = 20$.

Let (X, h) be a pair of a K3 surface and a line bundle h of degree 20 and $M_X(2, h, n)$ be the moduli space of stable sheaves $\mathcal{E}$ on X with $r(\mathcal{E}) = 2$, $\det \mathcal{E} \simeq h$ and $\chi(\mathcal{E}) = 2 + n$. By [6], $M_X(2, h, n)$ is smooth and of dimension $2(11 - 2n)$ if it is not empty. Hence $M_X(2, h, n)$ is a surface only if $n = 5$ and, in fact, $\hat{X} := M_X(2, h, 5)$ is a K3 surface if it is compact. Moreover, there is a line bundle $\hat{h}$ of degree 20 on $\hat{X}$ and $(\hat{\hat{X}}, \hat{\hat{h}})$ is isomorphic to (X, h) ([7], Theorem 1.4).

Let S be $M_C(2, K, 5)$ in the theorem and $\mathcal{U}$ the Poincaré bundle on $C \times S$. We normalize $\mathcal{U}$ so that $\det \mathcal{U} \simeq K_C \boxtimes h_{det}$ (see §4) and restrict it to fibres in the other direction. Namely we consider the family of vector bundles $\mathcal{U}_p := \mathcal{U}|_{p \times S}$ on S parametrized by C.

Theorem 2. *Let C, $S = M_C(2, K, 5)$, h_{det} and $\mathcal{U}$ be as above. Then the vector bundle $\mathcal{U}_p$ on S is stable with respect to h_{det} and belongs to $M_S(2, h_{det}, 5)$ for every $p \in C$. Moreover, the classification morphism $\alpha : C \longrightarrow M_S(2, h_{det}, 5), p \mapsto [\mathcal{U}_p]$, is an embedding.*

Let $\mathcal{M}_{11}$ and $\mathcal{F}_{11}$ be the moduli spaces of curves and polarized K3 surfaces (X, h) of genus 11, respectively. Let $\mathcal{P}_{11}$ be the $\mathbf{P}^{11}$-bundle $\coprod_{(X,h) \in \mathcal{F}_{11}} |h|$ over $\mathcal{F}_{11}$ and $\phi_{11} : \mathcal{P}_{11} \cdots \rightarrow \mathcal{M}_{11}$ the rational map which associates the isomorphism class for every $C \in |h|$ (cf. [5]). By the theorem, the rational map $\psi : \mathcal{M}_{11} \cdots \rightarrow \mathcal{P}_{11}$, $[C] \mapsto (M_S(2, h_{det}, 5), \alpha(C))$, satisfies

Unité mixte de l'institute Henri Poincaré, CNRS Université Pierre et Marrie Currie, Paris.

$\phi_{11} \circ \psi = id$. Since $\mathcal{P}_{11}$ is irreducible and of the same dimension as $\mathcal{M}_{11}$, ϕ_{11} is birational and ψ is its inverse. Therefore, we have

Corollary 1. *$M_S(2, h_{det}, 5)$, with $S = M_C(2, K, 5)$, is the unique K3 surface which contains C.*

Since $\mathcal{M}_{11}$ is unirational by [4], we also have

Corollary 2. *$\mathcal{F}_{11}$ is unirational.*

Thus $M_C(2, K, 5)$ and the morphism $\alpha : C \longrightarrow M_S(2, h_{det}, 5)$ are similar to the Picard variety and Albanese map $X \longrightarrow \mathrm{Alb}\, X = (\mathrm{Pic}^0 X)^\vee$, respectively. The morphism α is the K3 hull in the following sense:

Definition An embedding $\alpha : C \longrightarrow A$ of a curve C into a variety A is a *K3 hull* if

1) there exist a line bundle L on A and its global sections $s_1, \cdots, s_{n-1}$ such that $L|_C \simeq K_C$ and C is the complete intersection $s_1 = \cdots = s_{n-1} = 0$ in A,

2) every embedding of C into a K3 surface is isomorphic to the restriction of α to the common zero locus of a codimension one subspace of $< s_1, \cdots, s_{n-1} > \subset H^0(A, L)$, and

3) there is an exact sequence

$$0 \longrightarrow H^0(N_{C/A} \otimes K_C^{-1}) \longrightarrow H^1(K_C^{-2}) \longrightarrow \bigwedge^2 H^1(\mathcal{O}_C),$$

where the first map is the coboundary map of the long exact sequence associated to the natural exact sequence

$$\begin{array}{ccccccc} [0 \longrightarrow T_C \longrightarrow T_A|_C \longrightarrow & N_{C/A} & \longrightarrow 0] \otimes K_C^{-1} \\ & \| & \\ & K_C^{\oplus(n-1)} & \end{array}$$

and the second is the dual of the Wahl map $\bigwedge^2 H^0(K_C) \longrightarrow H^0(K_C^3)$ (see [13]).

The K3 hull exists for every general curve of genus $g \geq 7$. It is the symmetric space described in [9] for $g = 7, 8$ and 9, and C itself for $g = 10$ and $g \geq 12$ (cf. [3], [8], [5]). This will be discussed elsewhere.

Let G_d^r be the locus of curves with g_d^r in the moduli space $\mathcal{M}_{11}$ and C a general member of G_6^1. C is embedded into $\mathbf{P}^5$ by $|K_C\xi^{-1}|$ and the quadric hull S of its image $C_{14} \subset \mathbf{P}^5$ is a K3 surface of degree 8, where ξ is a g_6^1 of C. There exists a family $\{\mathcal{E}_x\}_{x \in S}$ of vector bundles

$$0 \longrightarrow \mathcal{O}_S(C - A) \longrightarrow \mathcal{E}_x \longrightarrow I_x(A) \longrightarrow 0$$

of Schwarzenberger type on S parametrized by S, where A is a hyperplane section of $S \subset \mathbf{P}^5$.

Theorem 3. *For a general member C of G_6^1, the Brill-Noether locus $M_C(2, K, 5)$ is smooth and isomorphic to S by the correspondence $S \ni x \mapsto \mathcal{E}_x|_C \in M_C(2, K, 5)$.*

The quadric hull S in the theorem is the unique K3 surface which contains the hexagonal curve C. In fact, if a K3 surface contains C, then the g_6^1 on C extends to a line bundle on it.

Another interesting divisor of $\mathcal{M}_{11}$ is the locus G_9^2. For a general member C of G_9^2, $S = M_C(2, K, 5)$ is still a K3 surface and contains a line D, *i.e.*, $\deg h_{det}|_D = 1$, which parametrizes the extensions $0 \longrightarrow \zeta \longrightarrow E \longrightarrow K_C\zeta^{-1} \longrightarrow 0$ with $h^0(E) = 7$, where ζ is a g_9^2 of C. The moduli space $M_S(2, h_{det}, 5)$ is the unique quartic surface which contains the image of $\Phi_{K_C\zeta^{-1}} : C \longrightarrow \mathbf{P}^3$.

We prove the theorem in §3 after some preparations in §§1 and 2, and Theorem 1 and 2 in the final section.

Notations $\mathbf{P}_*V$ and $\mathbf{P}^*V$ are the two projective spaces associated to a vector space V. The former parametrizes one-dimensional subspaces and the latter quotient spaces. S^nV is the n-th symmetric tensor product of a vector space or vector bundle V. For an $\mathcal{O}_X$-module $\mathcal{E}$ and an $\mathcal{O}_Y$-module $\mathcal{F}$, the tensor product of their pull-backs on the product $X \times Y$ is denoted by $\mathcal{E} \boxtimes \mathcal{F}$.

1. Preliminary

Next two lemmas are useful to analyze and construct vector bundles in $M_C(2, K, n)$.

Lemma 1. *Let E be a rank two vector bundle of canonical determinant and ζ a line bundle on C. If ζ is generated by global sections, then we have*

$$\dim \operatorname{Hom}_{\mathcal{O}_C}(\zeta, E) \geq h^0(E) - \deg \zeta.$$

The proof is an easy exercise of the base-point-free pencil trick (see [10] Proposition 3.1).

Lemma 2. *Let ξ be a line bundle and consider non-trivial extensions*

$$0 \longrightarrow \xi \longrightarrow E \longrightarrow \eta \longrightarrow 0$$

of ξ by its Serre adjoint η.

1) *The extensions E with $h^0(E) = h^0(\xi) + h^0(\eta)$ are parametrized by the projective space $\mathbf{P}^* \operatorname{Coker}[S^2H^0(\eta) \longrightarrow H^0(\eta^2)]$.*

2) *Assume that the multiplication map $S^2H^0(\eta) \longrightarrow H^0(\eta^2)$ is surjective. Then $h^0(E) \leq h^0(\xi) + h^0(\eta) - 1$ for every non-trivial extension E. Moreover, the non-trivial extensions E with $h^0(E) = h^0(\xi) + h^0(\eta) - 1$ are parametrized by the quadric hull of the image of $\Phi_{|\eta|} : C \longrightarrow \mathbf{P}^*H^0(\eta)$. More precisely, for every point x of the quadric hull, there is the unique extension E such that the image of the linear map $H^0(E) \longrightarrow H^0(\eta)$ is the codimension one subspace corresponding to x.*

Proof. Let $e \in \operatorname{Ext}^1(\eta, \xi)$ be the extension class and $\delta_e : H^0(\eta) \longrightarrow H^1(\xi)$ the coboundary map. The condition that $h^0(E) = h^0(\xi) + h^0(\eta)$ is equivalent to $\delta_e = 0$, that is, e lies in the kernel of the linear map

$$\Delta : \operatorname{Ext}^1(\eta, \xi) \longrightarrow H^0(\eta)^\vee \otimes H^1(\xi), \quad e \mapsto \delta_e.$$

By the Serre duality, the linear map Δ is the dual of the multiplication map $H^0(\eta) \otimes H^0(\eta) \longrightarrow H^0(\eta^2)$. Hence $\operatorname{Ker} \Delta$ is the dual of $\operatorname{Coker}[S^2H^0(\eta) \longrightarrow H^0(\eta^2)]$, which shows (1). The first assertion of (2) is a direct consequence of (1). The condition that $h^0(E) = h^0(\xi) + h^0(\eta) - 1$ is equivalent to $\operatorname{rank} \delta_e = 1$ by our assumption. There exists a nonzero linear map $\alpha : H^0(\eta) \longrightarrow \mathbf{C}$ such that δ_e is the composite of α and its dual $\alpha^\vee : \mathbf{C} \longrightarrow H^0(\eta)^\vee \simeq H^1(\xi)$. Let x be the point of $\mathbf{P}^*H^0(\eta)$ corresponding to α. Then $I_2 = \operatorname{Ker}[S^2H^0(\eta) \longrightarrow H^0(\eta^2)]$, the degree 2 part of the homogeneous ideal of $\Phi_\eta(C)$, vanishes at x, since $S^2\alpha$ is the composite of $S^2H^0(\eta) \longrightarrow H^0(\eta^2)$ and the linear map $H^0(\eta^2) \longrightarrow \mathbf{C}$ corresponding to e by construction. Thus E with $h^0(E) = h^0(\xi) + h^0(\eta) - 1$ determines a point in the quadric hull. Conversely, a point in the quadric hull determines e with $\operatorname{rank} \delta_e = 1$. Such e is unique since Δ is injective by our assumption. □

See [11] §4 for the following criterion of smoothness:

Proposition 1. *Let E be a member of $M_C(2, K)$ with $h^0(E) = n + 2$ and put $\sigma = 3g(C) - 3 - (n+2)(n+3)/2$. Then we have*

1) $\dim_{[E]} M_C(2,K,n) \geq \sigma$, *and*

2) $M_C(2,K,n)$ *is smooth and of dimension* σ *at* $[E]$ *if and only if the multiplication map* $S^2H^0(E) \longrightarrow H^0(S^2E)$ *is injective.*

2. Hexagonal curve of genus 11

Let $S \subset \mathbf{P}^5$ be a smooth complete intersection of three quadrics. S is a K3 surface by the adjunction formula and the Lefschetz theorem. Throughout this and next sections we assume that S contains a normal elliptic curve B of degree 6 but no bisecant lines of B. Such a surface exists by the following:

Lemma 3. *Let* $B \subset \mathbf{P}^5$ *be a normal elliptic curve of degree 6. Then the intersection of three general quadrics passing through* B *is a smooth surface and does not contain a bisecant line of* $B \subset \mathbf{P}^5$.

Proof. The first assertion follows from Bertini's theorem, since $B \subset \mathbf{P}^5$ is an intersection of quadrics. All S's which contain B are parametrized by a non-empty open subset of the Grassmannian $G(3, H^0(I_B(2)))$, which is of dimension $3(9-3) = 18$. Let ℓ be a bisecant line. All S's which contain both B and ℓ are parametrized by an open subset of the Grassmannian $G(3, H^0(I_{B\cup\ell}(2)))$, which is of dimension 15. Since the bisecant lines are parametrized by a surface, we have the second assertion. □

Let $C \subset S$ be a smooth member of the complete linear system $|A+B|$, where A is a hyperplane section of $S \subset \mathbf{P}^5$. Such C exists since $|A+B|$ is free from base points. Since

$$(A^2) = 8, (A.B) = 6, (B^2) = 0 \text{ and } (A+B)^2 = 20,$$

C is of genus 11. We denote the restriction of $\mathcal{O}_S(B)$ and $\mathcal{O}_S(A)$ to C by ξ and η, respectively. By the adjunction formula, the canonical line bundle K_C of C is the tensor product of ξ and η. By the exact sequences

$$0 \longrightarrow \mathcal{O}_S(-A) \longrightarrow \mathcal{O}_S(B) \longrightarrow \xi \longrightarrow 0$$

and

$$0 \longrightarrow \mathcal{O}_S(-B) \longrightarrow \mathcal{O}_S(A) \longrightarrow \eta \longrightarrow 0,$$

and by the lemma below, the restriction maps $H^0(\mathcal{O}_S(B)) \longrightarrow H^0(\xi)$ and $H^0(\mathcal{O}_S(A)) \longrightarrow H^0(\eta)$ are isomorphisms. In particular, $|\xi|$ is a g^1_6 and its Serre adjoint η is a g^5_{14}. The embedding $C \hookrightarrow S \hookrightarrow \mathbf{P}^5$ is given by the complete linear system $|\eta|$.

Lemma 4. $H^i(\mathcal{O}_S(A-B)) = H^i(\mathcal{O}_S(B-A)) = 0$ *for every* i.

Proof. Since $(B-A.A) < 0$, we have $H^0(\mathcal{O}_S(B-A)) = 0$ and hence $H^2(\mathcal{O}_S(A-B)) = 0$ by the Serre duality. Since $(A-B)^2 = -4$, we have $\chi(\mathcal{O}_S(B-A)) = 0$. Hence it suffices to show $H^0(\mathcal{O}_S(A-B)) = 0$. Assume the contrary and let D be a member of $|A-B|$. Since $(D^2) = -4$ and $(D.A) = 2$, D is a union of two disjoint lines. Since $(D.B) = 6$, one of them meets B at at least three points. This is a contradiction. □

Tensoring $\mathcal{O}_S(A)$ with the exact sequence

$$0 \longrightarrow \mathcal{O}_S(-B) \longrightarrow \mathcal{O}_S^{\oplus 2} \longrightarrow \mathcal{O}_S(B) \longrightarrow 0,$$

we have

Lemma 5. *The multiplication map*

$$H^0(\mathcal{O}_S(B)) \otimes H^0(\mathcal{O}_S(A)) \longrightarrow H^0(\mathcal{O}_S(A+B))$$

is an isomorphism.

The multiplication map $S^2H^0(\mathcal{O}_S(A)) \longrightarrow H^0(\mathcal{O}_S(2A))$ is surjective and its kernel is of dimension 3. Since S is smooth, every quadric passing through S is smooth at $x \in S$. Therefore, we have

Lemma 6. *The multiplication map $S^2H^0(I_x(A)) \longrightarrow H^0(\mathcal{O}_S(2A))$ is injective for every $x \in S$, where I_x is the maximal ideal of $\mathcal{O}_S$ at x.*

The non-existence of bisecant lines in S implies the following:

Lemma 7. *For an effective divisor D, we have*
1) $h^0(\xi(D)) = 2$ if D is of degree ≤ 3, and
2) $h^0(\xi(D)) + h^0(\eta(-D)) \leq max\,\{8-d, d+2\}$, where d is the degree of D.

Proof. If $\deg D \leq 2$, then $h^0(\eta(-D)) = h^0(\eta) - \deg D$, since $|\eta|$ is very ample. Since $\xi\eta = K_C$, we have $h^0(\xi(D)) = h^0(\xi) = 2$ by the Riemann-Roch theorem. Assume that $\deg D = 3$ and $h^0(\xi(D)) > 2$. Then D is contained in a trisecant line ℓ of $C \subset \mathbf{P}^5$. This is a contradiction since ℓ is contained in S and $(\ell.B) = (\ell.C - A) = 2$. So we have proved (1). By the Riemann-Roch theorem, we have

$$h^0(\xi(D)) + h^0(\eta(-D)) = 2h^0(\xi(D)) + 4 - d.$$

Hence (2) immediately follows from (1) if $d = 2$ or 3. If $d \geq 4$, then $h^0(\xi(D)) \leq d - 1$ by (1). Hence we have (2). □

Now we are ready to prove the following:

Proposition 2. *Let E be a semi-stable rank two vector bundle of canonical determinant on C. Then we have*
1) $h^0(E) \leq 7$, and
2) if $h^0(E) = 7$, then E is stable and contains a line subbundle isomorphic to ξ or $\xi(p)$ for a point $p \in C$.

Proof. If $h^0(E) < 7$, then there is nothing to prove. Hence we may assume that $h^0(E) \geq 7$. By Lemma 1, E contains a subsheaf isomorphic to ξ and we have an exact sequence

$$0 \longrightarrow \xi(D) \longrightarrow E \longrightarrow \eta(-D) \longrightarrow 0$$

for an effective divisor D of degree ≤ 4. Hence we have $h^0(E) \leq h^0(\xi(D)) + h^0(\eta(-D))) \leq 7$ and $\deg D = 0$ or 1 by Lemma 7, which shows (1) and the second assertion of (2). Assume that E is not stable and let $0 \longrightarrow \alpha \longrightarrow E \longrightarrow \beta \longrightarrow 0$ be an exact sequence with $\deg\alpha = \deg\beta = 10$. Then either α or β is isomorphic to $\xi(D)$ for an effective divisor D of degree 4. Hence we have $h^0(E) \leq h^0(\xi(D)) + h^0(\eta(-D)) \leq 6$ by Lemma 7. This is a contradiction and E is stable. □

3. Proof of Theorem 3

Let A, B and $C \subset S \subset \mathbf{P}^5$ be as in the previous section. $(S, \mathcal{O}_S(A+B))$ is a polarized K3 surface of genus 11. We study $M_S(2, A+B, 5)$ and their restrictions to C, using vector bundles of Schwarzenberger type. Let

$$H^j(S, \mathcal{E}xt^i_{\mathcal{O}_S}(I_x(A), \mathcal{O}_S(B))) \Longrightarrow \mathrm{Ext}^{i+j}_{\mathcal{O}_S}(I_x(A), \mathcal{O}_S(B)),$$

be the local-global spectral sequence of extension groups, where I_x is the ideal of a point x of S. By Lemma 4, the natural map

$$(1) \qquad \mathrm{Ext}^1_{\mathcal{O}_S}(I_x(A), \mathcal{O}_S(B)) \longrightarrow H^0(S, \mathcal{E}xt^1_{\mathcal{O}_S}(I_x(A), \mathcal{O}_S(B))) \simeq \mathbf{C}$$

is an isomorphism. Let

$$(2) \qquad 0 \longrightarrow \mathcal{O}_S(B) \longrightarrow \mathcal{E}_x \longrightarrow I_x(A) \longrightarrow 0.$$

be the unique non-trivial extension of $I_x(A)$ by $\mathcal{O}_S(B)$. Since $H^1(\mathcal{O}_S(B)) = 0$, we have the exact sequence

$$0 \longrightarrow H^0(\mathcal{O}_S(B)) \longrightarrow H^0(\mathcal{E}_x) \longrightarrow H^0(I_x(A)) \longrightarrow 0.$$

The following is obvious:

Lemma 8. 1) $\dim H^0(\mathcal{E}_x) = 7$ *and* $\mathcal{E}_x$ *is generated by global sections.*
2) $H^i(\mathcal{E}_x) = 0$ *for* $i = 1, 2$.

Since the three linear maps
i) $S^2 H^0(\mathcal{O}_S(B)) \longrightarrow H^0(\mathcal{O}_S(2B))$,
ii) $H^0(\mathcal{O}_S(B)) \otimes H^0(I_x(A)) \longrightarrow H^0(I_x(A+B))$ and
iii) $S^2 H^0(I_x(A)) \longrightarrow H^0(I_x^2(2A))$.
are injective by Lemma 5 and 6, we have

Lemma 9. *The natural linear map* $S^2 H^0(\mathcal{E}_x) \longrightarrow H^0(S^2 \mathcal{E}_x)$ *is injective.*

Let E_x be the restriction of the vector bundle $\mathcal{E}_x$ to the curve C. Since $\det \mathcal{E}_x \simeq \mathcal{O}_S(C)$, we have the exact sequence

$$0 \longrightarrow \mathcal{E}_x^\vee \longrightarrow \mathcal{E}_x \longrightarrow E_x \longrightarrow 0.$$

By Lemma 8 and the Serre duality, we have

Lemma 10. *The restriction map* $H^0(\mathcal{E}_x) \longrightarrow H^0(E_x)$ *is an isomorphism.*

In particular, $\dim H^0(E_x) = 7$ and E_x is generated by global sections. Restricting the above (2) to the curve C, we have the exact sequence

$$(3) \qquad 0 \longrightarrow \xi \longrightarrow E_x \longrightarrow \eta \longrightarrow 0$$

and

$$(4) \qquad 0 \longrightarrow H^0(\xi) \longrightarrow H^0(E_x) \longrightarrow H^0(I_x(A)) \longrightarrow 0$$

if $x \notin C$. If $x \in C$, then $I_x(A) \otimes \mathcal{O}_C$ contains the sky-scraper sheaf $k(x)$ as a subsheaf and the torsion-free quotient is isomorphic to $\eta(-x)$. Hence we have the exact sequence

$$(5) \qquad 0 \longrightarrow \xi(x) \longrightarrow E_x \longrightarrow \eta(-x) \longrightarrow 0.$$

Proposition 3. E_x *is stable for every* $x \in S$.

Proof. Assume that $x \notin C$. Since $h^0(E_x) < h^0(\xi) + h^0(\eta)$ by (4), the exact sequence (3) does not split. Let ζ be a line subbundle of E_x. If the composite $\zeta \longrightarrow E_x \longrightarrow \eta$ is zero, then $\deg \zeta = \deg \xi < 10$. Otherwise ζ is isomorphic to $\eta(-D)$ for a nonzero effective divisor D. If $\deg D = 1$ and $D = (p)$ for a point $p \in C$, then the extension class e of (3) lies in the kernel of $\mathrm{Ext}^1_{\mathcal{O}_C}(\eta, \xi) \longrightarrow \mathrm{Ext}^1_{\mathcal{O}_C}(\eta(-p), \xi(p))$. Hence the point corresponding to e in the way of Lemma 2 is p. This is a contradiction. Hence we have $\deg D \geq 2$. Since $h^0(E) = 7$, we have $\deg D \geq 5$ by Lemma 7. Therefore E_x is stable.

Claim: The exact sequence (5) does not split.

Assume the contrary and let $s : E_x \longrightarrow \xi(x)$ be a splitting. Let $\mathcal{F}$ be the kernel of the composite $\mathcal{E}_x \longrightarrow E_x \longrightarrow \xi(x)$. $H^0(\mathcal{F})$ is a 5-dimensional subspace of $H^0(\mathcal{E}_x)$ and mapped onto $H^0(I_x(A))$. Since $c_1(\mathcal{F}) = 0$, the image of the evaluation homomorphism $H^0(\mathcal{F}) \otimes \mathcal{O}_S \longrightarrow \mathcal{F} \subset \mathcal{E}$ is of rank one. Hence the image is isomorphic to $I_x(A)$, which is a contradiction.

Assume that $x \in C$ and let ζ be a line subbundle of E_x. We may assume that $\zeta \simeq \eta(-x-D)$ for an effective divisor D. D is nonzero by the claim and E_x is stable by Lemma 7. □

By the proposition, $\mathcal{E}_x$ is also stable (with respect to $\mathcal{O}_S(C)$) and every endomorphism is a constant multiplication. Hence, by the exact sequence

$$0 \longrightarrow sl\ \mathcal{E}_x \longrightarrow S^2\mathcal{E}_x \longrightarrow S^2E_x \longrightarrow 0,$$

the restriction map $H^0(S^2\mathcal{E}_x) \longrightarrow H^0(S^2E_x)$ is injective. By Lemma 9 and 10, we have

Lemma 11. *The natural linear map $S^2H^0(E_x) \longrightarrow H^0(S^2E_x)$ is injective.*

Now we prove Theorem 3. Let T be a copy of S and Δ be the diagonal of $S \times T$. By the Leray spectral sequence, we have an isomorphism

$$\mathcal{E}xt^1_{\mathcal{O}_S}(I_\Delta(\sigma^*A), \mathcal{O}_{S\times T}(\sigma^*B))) \longrightarrow \tau_*\mathcal{E}xt^1_{\mathcal{O}_{S\times T}}(I_\Delta(\sigma^*A), \mathcal{O}_{S\times T}(\sigma^*B))) \simeq \mathcal{O}_T(B-A)$$

whose fibres are (1), where σ and τ are the projections of $S \times T$ onto S and T, respectively. Hence we have the universal extension

$$(6) \qquad 0 \longrightarrow \mathcal{O}_{S\times T}(\sigma^*B) \longrightarrow \mathcal{F} \longrightarrow I_\Delta(\sigma^*A + \tau^*(B-A)) \longrightarrow 0,$$

whose restriction to $S \times x$ is (2) for every $x \in T$. The restriction of $\mathcal{F}$ to $C \times T$ is a Poincaré bundle of the family $\{E_x\}_{x\in T}$. By Lemma 8, 10 and Proposition 3, we obtain the classification morphism $B_C : S \longrightarrow M_C(2, K, 5)$. B_C is injective by the following:

Lemma 12. $\dim \mathrm{Hom}(\xi, E_x) = 1$ *for every* $x \in S$.

Proof. Assume the contrary. Then E_x contains two distinct subsheaves isomorphic to ξ. Let L be the subsheaf generated by them. If L is of rank one, then L is isomorphic to $\xi(D)$ for an effective divisor D and $h^0(L) \geq 3$, which contradicts Lemma 7. If L is of rank two, then L is isomorphic to $\xi \oplus \xi$, which contradicts Lemma 11. □

Let E be a member of $M_C(2, K, 5)$. By Proposition 2, there is an exact sequence

$$0 \longrightarrow \xi \longrightarrow E \longrightarrow \eta \longrightarrow 0$$

or

$$0 \longrightarrow \xi(p) \longrightarrow E \longrightarrow \eta(-p) \longrightarrow 0$$

for a point $p \in C$. In the latter case, E is isomorphic to E_p by (1) of Lemma 2 and the claim in the proof of Proposition 3. In the former case, there exists a point $x \in S$ such that the image of $H^0(E) \longrightarrow H^0(\eta)$ is $H^0(I_x(A))$ by (2) of Lemma 2. By the stability of E, x does not belong to C. Hence E is isomorphic to E_x. Therefore, B_C is surjective. $M_C(2, K, 5)$ is smooth and of dimension 2 by Proposition 1 and Lemma 11.

4. Proof of Theorem 1 and 2

Let C be a general curve of genus 11. C does not have a g^3_{10}. Hence we have $h^0(E) \leq 6$ for every strictly semi-stable rank two vector bundle E of canonical determinant on C. Since $M_C(2, K, 5)$'s form a proper family when C varies, they are smooth K3 surfaces by Theorem 3.

Proposition 4. *There exists a vector bundle $\mathcal{U}$ on $C \times M_C(2, K, 5)$ such that $\mathcal{U}|_{C\times[E]} \simeq E$ for every $E \in M_C(2, K, 5)$ and $\det \mathcal{U} \simeq K_C \boxtimes h_{det}$. Such $\mathcal{U}$, called the normalized Poincaré bundle, is unique up to isomorphism.*

Proof. The moduli space $M_C(2, K)$ is the quotient of an open subset R of a Quot scheme by an action of $PGL(\nu)$. Let R' be the inverse image of $M_C(2, K, 5)$ by the quotient morphism $R \longrightarrow M_C(2, K)$ and $\tilde{\mathcal{U}}$ the restriction of the universal quotient bundle. By Proposition 2, $h^0(E) = 7$ for every $E \in M_C(2, K, 5)$. Hence the direct image $\pi_{R'*}\tilde{\mathcal{U}}$ is a vector bundle of rank 7. The direct image $\pi_{R'*}(\tilde{\mathcal{U}} \otimes_{\mathcal{O}_C} K_C)$ is of rank 40. Following [12], we consider the vector bundle

$$\tilde{\mathcal{U}} \otimes \pi^*_{R'}((\det \pi_{R'*}\tilde{\mathcal{U}})^{17} \otimes \det(\pi_{R'*}(\tilde{\mathcal{U}} \otimes_{\mathcal{O}_C} K_C))^{-3}).$$

The action of a central element $t \in \mathbf{G}_m \subset GL(\nu)$ on the three factors are t, $t^{7\cdot 17}$ and $t^{-40\cdot 3}$. Hence this tensor product has an action of $PGL(\nu)$ and descends to a Poincaré bundle $\mathcal{U}$ on $C \times M_C(2, K, 5)$.

Since $M_C(2, K, 5)$ is regular, there exists a line bundle L on $M_C(2, K, 5)$ such that $\det \mathcal{U} \simeq K_C \boxtimes L$. Since $\mathcal{U}^\vee \otimes_{\mathcal{O}_C} K_C \simeq \mathcal{U} \otimes_{\mathcal{O}_M} L^{-1}$, we have

$$(R^1\pi_{M*}\mathcal{U})^\vee \simeq \pi_{M*}(\mathcal{U}^\vee \otimes K_C) \simeq (\pi_{M*}\mathcal{U}) \otimes L^{-1}$$

by the (relative) Serre duality. Hence we have

$$h_{det} = (\det R^1\pi_{M*}\mathcal{U}) \otimes (\det \pi_{M*}\mathcal{U})^{-1} \simeq L^7 \otimes N^{-2}, \tag{7}$$

where we put $N = \det \pi_{M*}\mathcal{U}$. Therefore, the universal bundle $\mathcal{U} \otimes_{\mathcal{O}_M} (L^3 \otimes N^{-1})$ satisfies our requirement (cf. [1] p. 582). The normalized Poincaré bundle is unique since universal bundles differ only by $\operatorname{Pic} M(2, K, 5)$ and the Picard group has no 2-torsion. □

The determinant line bundle h_{det} is ample by [2]. We compute its degree. In the hexagonal case, $\mathcal{F}|_{C\times T}$ is a universal family. Since $\det(\mathcal{F}|_{C\times T}) \simeq K_C \boxtimes \mathcal{O}(B - A)$ by the exact sequence (6), we have $h_{det} \simeq \mathcal{O}_T(7(B - A)) \otimes (\det \tau_*(\mathcal{F}|_{C\times T}))^{-2}$ by (7). By Lemma 10 and (6), we have the exact sequence

$$0 \longrightarrow H^0(\mathcal{O}_S(B)) \otimes \mathcal{O}_T \longrightarrow \tau_*(\mathcal{F}|_{C\times T}) \longrightarrow H^0(\mathcal{O}_S(A)) \otimes \mathcal{O}_T(B - A) \longrightarrow \mathcal{O}_T(B) \longrightarrow 0$$

and $\det(\tau_*\mathcal{F}|_{C\times T}) \simeq \mathcal{O}_T(5B - 6A)$. Hence h_{det} is isomorphic to $\mathcal{O}_T(5A - 3B)$, which is a line bundle of degree 20. So we have proved Theorem 1.

For a hexagonal curve C, $\mathcal{U} = (\mathcal{F}|_{C\times T}) \otimes \mathcal{O}_T(3A - 2B)$ is the normalized Poincaré bundle. The restriction $\mathcal{U}_p$ is stable for every $p \in C$ by Proposition 3 and α is an embedding by Theorem 3. Since the properties to be shown are stable by small deformations, we have Theorem 2.

REFERENCES

[1] Atiyah, M.F. and Bott, R.: The Yang-Mills equations over Riemann surfaces, Phil. Trans. Royal Soc. London, A **308**(1982), 523-615.

[2] Beauville, A.: Fibrés de rang 2 sur une courbe, fibré determinant et fonctions thêta, Bull. Soc. math. France **116**(1988), 431-448.

[3] —— and Mérindol, J.-Y.: Sections hyperplanes des surfaces K-3, Duke Math. J., **55** (1987), 873-878.

[4] Chang, M.-C. and Ran, Z.: The unirationality of the moduli spaces of curves of genus 11, 13 (and 12), Invent. math. **76**(1984), 41-54.

[5] Mori, S. and Mukai, S.: The uniruledness of the moduli space of curves of genus 11, in *Algebraic Geometry, proceedings, Tokyo/Kyoto, 1982*, M. Raynaud and T. Shioda (eds.), Lecture Notes in Math. **1016**, Springer-Verlag, 1983, pp. 334-353.

[6] Mukai, S. : Symplectic structure of the moduli space of sheaves on an abelian or K3 surface, Invent. math. **77**(1984), 101-116.

[7] —— : On the moduli space of bundles on K3 surfaces I, in *Vector Bundles on Algebraic Varieties*, Oxford University Press, 1987, pp. 341-413.

[8] ——: Curves, K3 surfaces and Fano 3-folds of genus $\leq$ 10, in *Algebraic Geometry and Commutative Algebra in Honor of Masayoshi Nagata*, Kinokuniya, Tokyo, 1987, pp. 357-377.

[9] —— : Curves and symmetric spaces, Proc. Japan Acad. **68**(1992), 7-10.

[10] —— : Curves and Grassmannians, in *Algebraic Geometry and related topics, proceedings, Inchon, Korea, 1992*, J-H. Yang, Y. Namikawa and K. Ueno (eds.), International Press, Boston-Hong Kong, 1993, pp.19-40.

[11] —— : Vector bundles and Brill-Noether theory, in *Algebraic Geometry Special Year 1992-1993*, Math. Sci. Res. Inst. Publ. **28**, Cambridge University Press, 1995, in press.

[12] Mumford, D. and Newstead, P.E.: Periods of a moduli space of bundles on curves, Amer. J. Math. **90**(1968), 1200-1208.

[13] Wahl, J.: Introduction to Gaussian maps on algebraic curves, in *Complex Projective Geometry*, London Math. Soc. Lect. Note Ser. **179**, Cambridge University Press, 1992, pp. 304-323.

HYPER-KÄHLER STRUCTURES ON MODULI SPACES OF PARABOLIC HIGGS BUNDLES ON RIEMANN SURFACES

Hiraku Nakajima

Mathematical Institute, Tôhoku University, Aramaki, Aoba-ku, Sendai 980-77, Japan

Current Address: Department of Mathematical Sciences, University of Tokyo, Komaba, Meguro-ku, Tokyo 153, Japan

e-mail: nakajima@math.tohoku.ac.jp

On the occasion of Hiroshi Konno's wedding

1. Introduction

In the symposium, the author talked about Vafa-Witten's work [VW] on the modular invariance of the Euler numbers of moduli spaces of instantons on 4-manifolds and its relation to affine Lie algebra actions on homology groups of moduli spaces. Since these matters are explained in their original papers very comprehensively, there are no reasons to be reproduced here. Instead we shall discuss another topic in this paper. It is author's response to Yokogawa's talk in the symposium. He explained his joint work with Boden [BY] on moduli spaces of parabolic Higgs bundles on Riemann surfaces, where residues of Higgs fields at marked points are nilpotent. They are parameterized by a finite dimensional space, where each parameter (called a weight) corresponds to a particular choice of the polarization. It was observed that Betti numbers of moduli spaces for generic weights are independent of weights. In the parameter space of the weights, there are hyperplanes which correspond to singular moduli spaces. It is clear that two moduli spaces are diffeomorphic to each other if their weights lie in the same chamber of the parameter space. However, when the parameter across a hyperplane, it is not at all obvious that Betti numbers unchange. In fact, Betti numbers of moduli spaces of parabolic bundles (without Higgs fields) do change. Thus we have a remarkable contrast between moduli spaces of parabolic bundles and those of parabolic Higgs bundles.

In this paper, the author will show that moduli spaces of parabolic Higgs bundles are, in fact, diffeomorphic to each other (Theorem 4.4). Similar phenomena were observed for ALE spaces [Kr] and quiver varieties [Na]. All are explained in terms of hyper-Kähler structures on relevant spaces. A hyper-Kähler structure is a quadruple (g, I, J, K) where I, J, K are almost complex structures which satisfy the quaternion relations and are covariant constant with respect to the Riemannian metric g. The natural complex structure as the moduli space of parabolic Higgs bundles is identified with I. On the other hand, for the complex structure J, the space is identified with the moduli space of projectively flat connections with prescribed holonomy around marked points, where we introduce a certain definition of the stability. This identification between two moduli spaces is called non-abelian Hodge theory. When we have no marked points, this was first observed by Hitchin [Hi] (carried

Author supported in part by Grant-in-Aid for Scientific Research (No. 05740041), Ministry of Education, Science and Culture, Japan and also by the Inamori Foundation.

out in part by Donaldson) for a Riemann surface and by Corrlette [Co], Simpson [Si] for higher dimensional Kähler manifolds. The case when the base space is a Riemann surface with marked points was treated by Simpson [Si2]. Under this identification, the weight for the moduli space of parabolic Higgs fields corresponds to the eigenvalues of residues of projectively flat connections. The weight for projectively flat connections corresponds to eigenvalues of residues of Higgs fields (see Corollary 4.3 for more precise statement). Thus we get further flexibility of choices of parameters and can avoid singular moduli spaces to connect two moduli spaces by a path. This is the basic reason why moduli spaces are diffeomorphic to each other.

We shall not show this statement in the full generality. We restrict ourselves in the simplest case:

Assumption 1.1. The vector bundle E is of rank 2, and the flags at marked points are complete, namely $E_p = F_1(p) \supsetneq F_2(p) \supsetneq F_3(p) = 0$ for each marked point $p \in S$.

Acknowledgment. It is clear that this work does not appear unless the author heard Yokogawa's talk in the Taniguchi symposium. He is grateful to the Tanighchi Foundation for providing him with this opportunity. Warm thanks also to H. Boden, N. Hitchin and K. Yokogawa for valuable discussions.

2. Definitions

2(i). Parabolic Higgs bundles. Let X be a compact Riemann surface of genus g with n marked points $S = \{p_1, p_2, \dots, p_n\}$. We put a Kähler metric ω on X. Let us denote the canonical bundle of X by K.

First we do not assume (1.1). Let E be a C^∞-vector bundle over X. A structure of a holomorphic vector bundle over E is identified with the Dolbeault operator $\overline{\partial}_A$. We use the notation $\mathcal{E}_A$ for the holomorphic vector bundle corresponding to $\overline{\partial}_A$.

Definition 2.1. (i) A *parabolic structure* on E consists of weighted flags

$$\begin{gathered} E_p = F_1(p) \supset \cdots \supset F_{s_p}(p) \supset F_{s_p+1}(p) = 0 \\ 0 \le \alpha_1(p) < \cdots < \alpha_{s_p}(p) < 1 \end{gathered}$$

for each $p \in S$. A vector bundle with a parabolic structure is called a *parabolic bundle*.

(ii) Consider a meromorphic bundle map $\Phi\colon \mathcal{E}_A \to \mathcal{E}_A \otimes K$ which has a pole at most first order at $p \in S$. (This Φ is called a *Higgs field*.) Let us denote by $\operatorname{res}_p \Phi$ the residue at p. We say Φ *preserves* the parabolic structure if

$$\operatorname{res}_p \Phi(F_i(p)) \subset F_i(p) \qquad \text{for all } i,\ p.$$

(iii) A *parabolic Higgs bundle* is a holomorphic vector bundle $\mathcal{E}_A$ together with a parabolic structure and Φ preserving the parabolic structure.

Let $m_i(p) = \dim F_i(p) - \dim F_{i+1}(p)$. Define the parabolic degree and the slope of the parabolic bundle E by

$$\operatorname{pardeg}_\alpha E \overset{\text{def.}}{=} \deg E + \sum_{p\in S}\sum_{i=1}^{s_p} m_i(p)\alpha_i(p),$$

$$\mu_\alpha(E) \overset{\text{def.}}{=} \frac{\operatorname{pardeg}_\alpha E}{\operatorname{rank} E}.$$

A parabolic Higgs bundle $(\mathcal{E}_A, \Phi)$ is *α-stable* (*α-semistable*) if, for any proper holomorphic subbundle L of $\mathcal{E}_A$ (with induced parabolic structure) which is preserved by Φ,

$$\mu_\alpha(L) < \mu_\alpha(E) \qquad (\text{respectively } \mu_\alpha(L) \le \mu_\alpha(E)).$$

We fix a holomorphic structure on the determinant line bundle $\Lambda^{\max} E$ and consider the moduli space of α-semistable parabolic Higgs bundles with fixed determinant and trace-free Higgs fields. This moduli space was constructed by Yokogawa [Yo]. Following [BY], we denote it by $\mathcal{P}^0_\alpha$.

Now assume (1.1). Furthermore, we assume the following for simplicity.

$$\mu_\alpha(E) = 0, \qquad \alpha_2(p) = 1 - \alpha_1(p) \quad \text{for all } p \in S. \tag{2.2}$$

At each $p \in S$, the residue $\operatorname{res}_p \Phi$ is of the form

$$\operatorname{res}_p \Phi = \begin{pmatrix} \sigma(p) & 0 \\ * & -\sigma(p) \end{pmatrix}.$$

We define a natural morphism

$$\pi\colon \mathcal{P}^0_\alpha \to \mathbb{C}^n; \quad (\mathcal{E}_A, \Phi) \mapsto (\sigma(p_1), \dots, \sigma(p_n)).$$

For $\sigma \in \mathbb{C}^n$, let us denote by $\mathfrak{M}_{\mathrm{Dol}}(\alpha, \sigma)$ the inverse image $\pi^{-1}(\sigma)$. This is the moduli space of parabolic Higgs bundles with eigenvalues of residues fixed. (This notation is taken from Simpson's paper [Si3] in which he called it the Dolbeault moduli space.) When $\sigma = 0$, i.e., the residue $\operatorname{res}_p \Phi$ is nilpotent for any $p \in S$, $\mathfrak{M}_{\mathrm{Dol}}(\alpha, 0)$ turns out to be the space studied by Konno [Ko] (and denoted by $\mathcal{N}^0_\alpha$ in [BY]).

Definition 2.3. We say (α, σ) is *generic* when there are no elements $(\mathcal{E}_A, \Phi) \in \mathfrak{M}_{\mathrm{Dol}}(\alpha, \sigma)$ which are not α-stable, i.e., α-semistablity implies α-stability.

Lemma 2.4. (*cf.* [BH, 2.6]) *A parameter (α, σ) is not generic if and only if there exists an integer d and $e = (e_1, \dots, e_n) \in \{0, 1\}^n$ such that*

$$d + \sum_{i=1}^{n} e_i + (-1)^{e_i} \alpha_1(p_i) = 0 \tag{2.5}$$

$$\sum_{i=1}^{n} (-1)^{e_i} \sigma(p_i) = 0. \tag{2.6}$$

Proof. If (α, σ) is not generic, there exists an element $(\mathcal{E}_A, \Phi)$ which is a nontrivial direct sum $(\mathcal{L}_1, \Phi_1) \oplus (\mathcal{L}_2, \Phi_2)$. For each $p_i \in S$ let $e_i = \dim \mathcal{L}_1 \cap F_2(p_i) \in \{0, 1\}$. Then L_i has weight $e_i + (-1)^{e_i} \alpha_1(p_i)$ at p_i. Hence the parabolic degree is given by

$$\operatorname{pardeg}_\alpha \mathcal{L}_1 = \deg \mathcal{L}_1 + \sum_{i=1}^{n} e_i + (-1)^{e_i} \alpha_1(p_i).$$

Since $\mu_\alpha(\mathcal{L}_1) = \operatorname{pardeg}_\alpha \mathcal{L}_1 = 0$, we have the first equation in the assertion.

Since each $\mathcal{L}_i$ is a line bundle, its Higgs field Φ_i is a holomorphic 1-form. Its residue at p_i is equal to $-(-1)^{e_i} \sigma(p_i)$. Hence the residue theorem implies the second equation.

Conversely if we have two equations in the assertion, we can construct line bundles $\mathcal{L}_1$, $\mathcal{L}_2$ and Higgs fields Φ_1, Φ_2 so that $\mu_\alpha(\mathcal{L}_1) = \mu_\alpha(\mathcal{L}_2) = 0$ and $(\mathcal{L}_1, \Phi_1) \oplus (\mathcal{L}_2, \Phi_2)$ is α-semistable but not α-stable. □

Let $W = \{\alpha \mid 0 < \alpha_1(p) < 1/2 \text{ for } p \in S\}$ be the weight space and for any d and $e = (e_1, \dots, e_n)$, let $H_{d,e}$ be the hyperplane in W defined by the equation (2.5). For the residue σ, let $\Delta(\sigma) = \{(d, e) \mid (2.6) \text{ holds}\}$. Then the set

$$W \setminus \bigcup_{(d,e)\in\Delta(\sigma)} H_{d,e} \tag{2.7}$$

is the set of weights α such that (α, σ) is generic.

The following is clear.

Proposition 2.8. *If α and α' lie in the same component in (2.7), α-stability and α'-stability are equivalent and there exists an isomorphism between $\mathfrak{M}_{\mathrm{Dol}}(\alpha, \sigma)$ and $\mathfrak{M}_{\mathrm{Dol}}(\alpha', \sigma)$. In particular, if $(0, \sigma)$ is generic (which implies that (α, σ) is generic for any α), $\mathfrak{M}_{\mathrm{Dol}}(\alpha, \sigma)$ and $\mathfrak{M}_{\mathrm{Dol}}(\alpha', \sigma)$ are isomorphic for any α, α'.*

2(ii). Flat connections with prescribed holonomy around points. Let V be a C^∞ vector bundle over X, together with a parabolic structure as in the previous section.

Definition 2.9. A *filtered regular $\mathcal{D}$-module* is a projectively flat connection D on the restriction of V to $X \setminus S$ with the following conditions (i),(ii),(iii):

(i) If $D = d' + d''$ is the decomposition of the connection into operators of type $(1, 0)$ and $(0, 1)$, d'' extends across S. In other words, V together with d'' is a holomorphic vector bundle over the whole X.

(ii) If we take a local trivialization of V (respecting the holomorphic structure) around $p \in S$, D is written as

$$D = d + A\frac{dz}{z}$$

for an $\operatorname{End} V$-valued *smooth* function A. Here d is the trivial connection and z is a local coordinate vanishing to the order one around p. We call $A(p)$ the *residue* of D at p and denote it by $\operatorname{res}_p(D)$.

(iii) Furthermore, we assume that the residue preserves the parabolic structure, i.e.,

$$\operatorname{res}_p(D)(V_i(p)) \subset V_i(p) \qquad \text{for all } i,\ p \in S.$$

Remark 2.10. The residue $\operatorname{res}_p(D)$ depends on the choice of the extension V of the holomorphic vector bundle across p. For example, if we consider the extension $V(p)$, the residue is shifted by -1. Without the choice of the extension, the residue is only well-defined modulo integers.

The *parabolic degree* and *slope* of a filtered regular $\mathcal{D}$-module are just those of the underlying parabolic vector bundle. As explained in the introduction, the weights for $\mathcal{D}$-module and Higgs bundles are not the same under the identification through harmonic bundles. Hence we use the symbol β for the weight of filtered $\mathcal{D}$-modules. The corresponding slope is denoted by μ_β.

A filtered regular $\mathcal{D}$-module (V, D) is *β-stable* (*β-semistable*) if, for any proper subbundle L of V (with induced parabolic structure) which is preserved by the connection D,

$$\mu_\beta(L) < \mu_\beta(V) \qquad (\text{respectively } \mu_\beta(L) \le \mu_\beta(V)).$$

We fix a flat connection on the determinant line bundle $\Lambda^{\max} V$, which has no poles. We consider the set of β-semistable filtered regular $\mathcal{D}$-modules with fixed determinant modulo S-equivalence. As far as the author knows, there is no construction of moduli spaces in the literature. (Simpson constructed moduli spaces when there are no marked points [Si3].) Thus we just think them as sets without extra structures at this moment. Later we shall endow structures of complex manifold by an analytic method §3(ii).

Now assume (1.1). As in (2.2) we also assume

$$\mu_\beta(V) = 0, \qquad \beta_2(p) = 1 - \beta_1(p) \quad \text{for all } p \in S. \tag{2.11}$$

At each $p \in S$, the residue $\operatorname{res}_p D$ is of the form

$$\begin{pmatrix} \tau(p) & 0 \\ * & -\tau(p) \end{pmatrix}.$$

Fixing $\tau = (\tau(p_1), \ldots, \tau(p_n))$, we consider the set $\mathfrak{M}_{\mathrm{DR}}(\beta, \tau)$ whose elements are S-equivalence classes of β-semistable filtered regular $\mathcal{D}$-module D with fixed determinant and the above residues. (This notation is also taken from Simpson's paper [Si3] in which he called it the de Rham moduli space.)

We define the genericity of the parameter as Higgs bundles.

Definition 2.12. We say (β, τ) is *generic* when there are no elements $D \in \mathfrak{M}_{\mathrm{DR}}(\beta, \tau)$ which are not β-stable, i.e., β-semistablity implies β-stability.

The following can be proved exactly the same way as Lemma 2.4.

Lemma 2.13. *A parameter (β, τ) is not generic if and only if there exists an integer d and $e = (e_1, \ldots, e_n) \in \{0,1\}^n$ such that*

$$d + \sum_{i=1}^{n} e_i + (-1)^{e_i} \beta_1(p_i) = 0 \tag{2.14}$$

$$\sum_{i=1}^{n} (-1)^{e_i} \tau(p_i) = 0. \tag{2.15}$$

The chamber structure is also similar to the case of Higgs bundles. Let $W = \{\beta \mid 0 < \beta(p) < 1/2 \text{ for } p \in S\}$ be the weight space and for any d and $e = (e_1, \ldots, e_n)$, let $H_{d,e}$ be the hyperplane in W defined by the equation (2.14). For the residue τ, let $\Delta(\tau) = \{(d,e) \mid (2.15) \text{ holds}\}$. Then the set

$$W \setminus \bigcup_{(d,e) \in \Delta(\tau)} H_{d,e} \tag{2.16}$$

is the set of weights β such that (β, τ) is generic.

Then we have

Proposition 2.17. *If β and β' lie in the same component in (2.16), β-stability and β'-stability are equivalent and there exists a bijection between $\mathfrak{M}_{\mathrm{DR}}(\beta, \tau)$ and $\mathfrak{M}_{\mathrm{DR}}(\beta', \tau)$. In particular, if $(0, \tau)$ is generic (which implies that (β, τ) is generic for any β), $\mathfrak{M}_{\mathrm{DR}}(\beta, \tau)$ and $\mathfrak{M}_{\mathrm{DR}}(\beta', \tau)$ are equal for any β, β'.*

3. Analytic Constructions of Moduli Spaces

3(i). Parabolic Higgs bundles. We briefly recall Konno's analytic construction [Ko] of moduli spaces of parabolic Higgs bundles, based on the work of Biquard [Bi]. (Biquard's analysis was based on Lockhard and McOwen [LM] in turn.)

Let E be a parabolic bundle as in §2. Assume (1.1),(2.2). We introduce a hermitian metric H on E which is singular at marked points $p \in S$ as follows. For each point $p \in S$, take a local coordinate z vanishing to the order one at p. We take a local frame $\{e_1, e_2\}$ so that

$F_2(p)$ is spanned by $e_2(p)$. Then we define H so that $\{|z|^{-\alpha_j(p)}e_j\}_{1\le j\le 2}$ is a local unitary frame. In this frame, the hermitian connection for a holomorphic structure $\overline{\partial}_A$ is written as

$$d_A = d + \begin{pmatrix} \alpha_1(p) & 0 \\ 0 & \alpha_2(p) \end{pmatrix} id\theta + O(|z|^{\alpha_1(p)-\alpha_2(p)}).$$

We extend the metric H to the whole X so that it is smooth outside a neighborhood of S.

Next we introduce a certain weighted Sobolev norm on the space of C^∞ sections of $\operatorname{End} E|_{X\setminus S}$. Using the local frame $\{e_j\}$ given above, we decompose E as

$$E = G_1 \oplus G_2, \qquad \text{where } G_i \overset{\text{def.}}{=} \operatorname{span}\{e_i\}.$$

Then we have a decomposition $\operatorname{End} E = E_D \oplus E_H$, where

$$E_D \overset{\text{def.}}{=} \bigoplus_i \operatorname{End} G_i, \qquad E_H \overset{\text{def.}}{=} \bigoplus_{i\ne j} \operatorname{Hom}(G_i, G_j).$$

If u is a local section defined in a neighborhood of $p \in S$, we can write $u = u_D + u_H$ corresponding to the above decomposition. Then we put

$$\|u\|_{D^p_k} \overset{\text{def.}}{=} \sum_{i+j\le k} \left\{ \int \left| \frac{\partial^i}{\partial z^i}\frac{\partial^j}{\partial \overline{z}^j} u_D \right|^p + \left| |z|^{i+j-k} \frac{\partial^i}{\partial z^i}\frac{\partial^j}{\partial \overline{z}^j} u_H \right|^p idzd\overline{z} \right\}^{\frac{1}{p}},$$

where the local trivialization is taken with respect to $\{|z|^{-\alpha_i(p)}e_j\}$, which is unitary with respect to H. Outside a neighborhood of S, we put the usual Sobolev norm with i-th derivatives in L^p for $i \le k$. Let $D^p_k\Omega^0(\operatorname{End} E)$ be the completion of $C^\infty_0(X\setminus S, \operatorname{End} E)$ with respect to the norm $\|\ \|_{D^p_k}$. For $\operatorname{End} E$-valued forms, we define the norm in the same way and denote the corresponding Sobolev space by $D^p_k\Omega^i(\operatorname{End} E)$.

We set

$$\mathcal{G}^{\mathbb{C}} \overset{\text{def.}}{=} \{g \in D^p_2\Omega^0(\operatorname{End} E) \mid \det g_x = 1 \text{ for all } x \in X\}$$

$$\mathcal{A}_{\mathrm{Dol}} \overset{\text{def.}}{=} \{(\overline{\partial}_A, \Phi) \mid \overline{\partial}_A\Phi = 0,\ \overline{\partial}_A - \overline{\partial}_{A_0} \in D^p_1\Omega^{0,1}(\mathfrak{sl}\,E),\ \Phi - \Phi_0 \in D^p_1\Omega^{1,0}(\mathfrak{sl}\,E)\},$$

where $\mathfrak{sl}\,E$ is the bundle of trace-free endomorphisms, and $(\overline{\partial}_{A_0}, \Phi_0)$ is a fixed parabolic Higgs bundle. The complex gauge group $\mathcal{G}^{\mathbb{C}}$ acts on $\mathcal{A}_{\mathrm{Dol}}$ by pull-back.

As in [Bi, 1.5], [Ko, 2.4], we take $1 < p < 2$ sufficiently near to 1 in order to ensure that the following elliptic complex is Fredholm, where composition of operators is zero by $\overline{\partial}_A\Phi = 0$:

$$(3.1) \qquad D^p_2\Omega^0(\mathfrak{sl}\,E) \xrightarrow{\overline{\partial}_A + [\Phi\wedge\cdot]} D^p_1\Omega^{0,1}(\mathfrak{sl}\,E) \oplus D^p_1\Omega^{1,0}(\mathfrak{sl}\,E) \xrightarrow{[\Phi\wedge\cdot]+\overline{\partial}_A} D^p_0\Omega^{1,1}(\mathfrak{sl}\,E).$$

The condition can be written in terms of α, σ. This is very similar to [Ko, 2.4], but we must incorporate σ into the condition. This was unnecessary in [Ko] because spectra of residues of Higgs fields are assumed to be zero.

As observed in [Ko, 2.7], we can still define the notion of α-(semi)stability for elements in $\mathcal{A}_{\mathrm{Dol}}$. Let ${\mathcal{A}_{\mathrm{Dol}}}^{\mathrm{s}} = \{(\overline{\partial}_A, \Phi) \in \mathcal{A}_{\mathrm{Dol}} | \alpha\text{-stable}\}$.

We have the following:

Theorem 3.2. *(cf. [Ko, 1.6]) Assume (α, σ) is generic. Let $\mathfrak{M}_{\mathrm{Dol}}(\alpha,\sigma)$ be the moduli space of α-stable parabolic Higgs bundles which have fixed determinant, trace-free Higgs fields and fixed the eigenvalues of residues at marked points as in §2(i). Then ${\mathcal{A}_{\mathrm{Dol}}}^{\mathrm{s}}/\mathcal{G}^{\mathbb{C}}$ has a structure of a complex manifold which is isomorphic to $\mathfrak{M}_{\mathrm{Dol}}(\alpha,\sigma)$.*

Konno [Ko] introduced the complex structure on ${\mathcal{A}_{\mathrm{Dol}}}^{\mathrm{s}}/\mathcal{G}^{\mathbb{C}}$ by identifying it with the moduli space of Hermitian-Einstein connections. Since it is not enough for our purpose, we briefly discuss how one can define it in terms of parabolic Higgs fields.

First note that the action of $\mathcal{G}^{\mathbb{C}}$ on ${\mathcal{A}_{\mathrm{Dol}}}^{\mathrm{s}}$ is free because of the α-stability (cf. the proof of the Lemma below). Also the standard argument shows that the quotient space ${\mathcal{A}_{\mathrm{Dol}}}^{\mathrm{s}}/\mathcal{G}^{\mathbb{C}}$ is Hausdorff. In order to define the complex structure, we consider the elliptic complex (3.1).

Lemma 3.3. *If $(\overline{\partial}_A, \Phi)$ is α-stable, the first operator has trivial kernel and the second operator has trivial cokernel in the complex* (3.1).

Once this lemma is proved, the moduli space has a structure of complex manifold and the tangent space at $(\overline{\partial}_A, \Phi)$ of the moduli space is identified with the first cohomology group of the complex. One can identity it with the tangent spaces at the corresponding point of $\mathfrak{M}_{\mathrm{Dol}}(\alpha, \sigma)$ given by the complex structure in [Yo], using the standard argument in the harmonic theory. The detail will be left to the reader.

Proof of Lemma 3.3. The necessary analysis (e.g., Fredholm properties, regularity, etc.) was already developed in [LM, Bi, Ko]. So our proof is sketchy.

Suppose s lies in the kernel of the first operator. It gives a trace-free endomorphism $s\colon \mathcal{E}_A \to \mathcal{E}_A$ commuting with Φ. Here $\mathcal{E}_A$ denotes the holomorphic vector bundle corresponding to $\overline{\partial}_A$. This endomorphism s preserves the parabolic structure, i.e., $s(p)(F_i(p)) \subset F_i(p)$ by [Ko, 2.6(4a)]. The standard argument shows endomorphisms of an α-stable parabolic Higgs bundle are constant multiple of the identity. Since s is trace-free, we have $s = 0$.

If the second operator has nontrivial cokernel, there exists a nontrivial endomorphism s which lies in $\mathrm{Ker}([\Phi \wedge \cdot] + \overline{\partial}_A)^*$. Here the adjoint is defined using the hermitian metric H. Applying the Hodge star operator $*$ to s, we have a nontrivial endomorphism $*s$ which lies in $\mathrm{Ker}(\overline{\partial}_A + [\Phi \wedge \cdot])$. Thus the above argument implies $s = 0$. □

3(ii). Flat connections. Now we turn to moduli spaces of filtered regular $\mathcal{D}$-modules. Our approach is very similar to the last section. Assume (2.11). Let $p \in S$ be a marked point, and take a local coordinate z vanishing to the order one at p. We take a local frame $\{e_1, e_2\}$ as in the last section. A filtered regular $\mathcal{D}$-module D is written as

$$D = d + \begin{pmatrix} \tau(p) & 0 \\ * & -\tau(p) \end{pmatrix} \frac{dz}{z}.$$

We put a hermitian metric H so that $\{|z|^{-\beta_1 - (\tau(p) - \overline{\tau(p)})/2} e_1, z|z|^{-\beta_2 - (\tau(p) - \overline{\tau(p)})/2} e_2\}$ is a unitary frame. In this frame, D is written as

$$\text{(3.4)}\quad D = d + \begin{pmatrix} \frac{3\tau}{4} + \frac{\overline{\tau}}{4} - \frac{\beta_1}{2} & 0 \\ * & 1 - \frac{5\tau}{4} + \frac{\overline{\tau}}{4} - \frac{\beta_2}{2} \end{pmatrix} \frac{dz}{z} + \begin{pmatrix} -\frac{\tau - \overline{\tau}}{4} - \frac{\beta_1}{2} & 0 \\ * & -\frac{\tau - \overline{\tau}}{4} - \frac{\beta_2}{2} \end{pmatrix} \frac{d\overline{z}}{\overline{z}}.$$

The definition of the weighted Sobolev norms and Sobolev spaces are similar to the last section. Then we set

$$\mathcal{G}^{\mathbb{C}} \overset{\text{def.}}{=} \{g \in D_2^p\Omega^0(\operatorname{End} V) \mid \det g_x = 1 \text{ for all } x \in X\}$$

$$\mathcal{A}_{\mathrm{DR}} \overset{\text{def.}}{=} \{D \mid D \text{ is projectively flat and } D - D_0 \in D_1^p\Omega^1(\mathfrak{sl}\, V)\}.$$

The notion of β-stability can be generalized to elements in $\mathcal{A}_{\mathrm{DR}}$, and let ${\mathcal{A}_{\mathrm{DR}}}^{\mathrm{s}}$ be the set of β-stable elements.

In stead of (3.1), we consider

$$\text{(3.5)}\qquad D_2^p\Omega^0(\mathfrak{sl}\, E) \xrightarrow{D} D_1^p\Omega^1(\mathfrak{sl}\, E) \xrightarrow{D} D_0^p\Omega^2(\mathfrak{sl}\, E).$$

As in the previous section, we have

Theorem 3.6. *Assume (β, τ) is generic. Then*

(i) $\mathcal{A}_{\mathrm{DR}}{}^{\mathrm{s}}/\mathcal{G}^{\mathbb{C}}$ *has a structure of complex manifold.*

(ii) *Let $\mathfrak{M}_{\mathrm{DR}}(\beta, \tau)$ be the set of β-stable filtered regular $\mathcal{D}$-modules which have a fixed determinant and the fixed eigenvalues of residues at marked points as in §2(ii). Then there is a bijection between $\mathcal{A}_{\mathrm{DR}}{}^{\mathrm{s}}/\mathcal{G}^{\mathbb{C}}$ and $\mathfrak{M}_{\mathrm{DR}}(\beta, \tau)$.*

Hereafter we consider $\mathfrak{M}_{\mathrm{DR}}(\beta, \tau)$ as a complex manifold by the above theorem.
Next we want to discuss the complex structures of $\mathfrak{M}_{\mathrm{DR}}(\beta, \tau)$'s for different β's.

Proposition 3.7. *Suppose β and β' lie in the same component in* (2.16). *Then the bijection $\mathfrak{M}_{\mathrm{DR}}(\beta, \tau) \cong \mathfrak{M}_{\mathrm{DR}}(\beta', \tau)$ in Proposition 2.17 respects the complex structure.*

This is because the tangent space, which is the first cohomology group of (3.5), is independent of the choice of β, while the weighted Sobolev space itself *does* depend. This type of the argument is standard in the harmonic theory.

Remark 3.8. The most natural proof of this proposition is probably to construct the complex structure on $\mathfrak{M}_{\mathrm{DR}}(\beta, \tau)$ by the geometric invariant theory as in the case of the moduli space of parabolic Higgs bundles. As we mentioned before, the author cannot find the construction in the literature.

4. Harmonic Bundles and HyperKähler Structures

Now we introduce the notion of *harmonic bundles* which throw a bridge between Higgs bundles and $\mathcal{D}$-modules.

Let $(\overline{\partial}_A, \Phi)$ be a parabolic Higgs bundle. We put a hermitian metric as in §3(i). Then we have ∂_A so that $d_A = \partial_A + \overline{\partial}_A$ preserves the metric. We define the hermitian adjoint Φ^* of Φ. Then $(\overline{\partial}_A, \theta)$ is called *harmonic* if

$$R_A^{\perp} + [\Phi, \Phi^*] = 0,$$

where $R_A^{\perp}$ is the trace-free part of the curvature of d_A.

Next suppose D is a filtered regular $\mathcal{D}$-module. We put a hermitian metric as in §3(ii). Then the bundle of trace-free endomorphisms decomposes as

$$\mathfrak{sl}\, E = \mathfrak{su}\, E \oplus i\, \mathfrak{su}\, E,$$

where $\mathfrak{su}\, E$ is the bundle of trace-free skew adjoint endomorphisms. We can decompose D into

$$D = d_A + \Psi \colon \Omega^0(\mathfrak{su}\, E) \to \Omega^1(\mathfrak{su}\, E) \oplus \Omega^1(i\, \mathfrak{su}\, E).$$

Further, we have the decomposition into operators of types $(1,0)$ and $(0,1)$:

$$d_A = \partial_A + \overline{\partial}_A, \quad \Psi = \Phi + \Phi^*.$$

We say D is *harmonic* if

$$\overline{\partial}_A \Phi = 0.$$

Thus if D is a harmonic $\mathcal{D}$-module, $(\overline{\partial}_A, \Phi)$ is a harmonic Higgs bundle, and vice versa.

The local behaviors around a marked point $p \in S$ correspond as follows: First let $(\overline{\partial}_A, \Phi) \in \mathcal{A}_{\mathrm{Dol}}{}^{\mathrm{s}}$. In the local frame $\{|z|^{-\alpha_i(p)} e_i\}$, we have the following behaviors:

$$d_A = d + \begin{pmatrix} \alpha_1(p) & 0 \\ 0 & \alpha_2(p) \end{pmatrix} id\theta + E_1, \quad \text{where } E_1 \in D_1^p\Omega^1(\mathfrak{su}\, E),$$

$$\Phi = \begin{pmatrix} \sigma(p) & 0 \\ 0 & -\sigma(p) \end{pmatrix} \frac{dz}{z} + E_2, \quad \text{where } E_2 \in D_1^p\Omega^{1,0}(\mathfrak{sl}\, E).$$

Next let $D \in \mathcal{A}_{\mathrm{DR}}{}^{\mathrm{s}}$. If $D = d_A + \Phi + \Phi^*$, the local behaviors in the frame

$$\{|z|^{-\beta_1 - (\tau(p) - \overline{\tau(p)})/2} e_1, z|z|^{-\beta_2 - (\tau(p) - \overline{\tau(p)})/2} e_2\}$$

is given by

$$d_A = d + \begin{pmatrix} \frac{\tau(p) + \overline{\tau}(p)}{2} & 0 \\ 0 & 1 - \frac{\tau(p) + \overline{\tau}(p)}{2} \end{pmatrix} id\theta + E_3 \quad \text{where } E_3 \in D_1^p\Omega^1(\mathfrak{su}\, E),$$

$$\Phi = \begin{pmatrix} \frac{\tau(p) - \beta_1(p)}{2} & 0 \\ 0 & \frac{1 - \tau(p) - \beta_2(p)}{2} \end{pmatrix} \frac{dz}{z} + E_4 \quad \text{where } E_4 \in D_1^p\Omega^{0,1}(\mathfrak{sl}\, E).$$

This follows from (3.4).

Hence parameters should correspond by

$$\alpha_1(p) = \frac{\tau(p) + \overline{\tau}(p)}{2}, \qquad \sigma(p) = \frac{\tau(p) - \beta_1(p)}{2}. \tag{4.1}$$

(This was given in the table in the introduction in [Si2].) Note that the extensions of the underlying vector bundles across S differ for Higgs bundles and filtered $\mathcal{D}$-modules. Hence this correspondence should be understood modulo integers by Remark 2.10.

The moduli space of harmonic parabolic Higgs bundles, or equivalently harmonic filtered regular $\mathcal{D}$-modules, has a hyper-Kähler structure. As explained in [Ko, 3.9], this is because the moduli space is a hyper-Kähler quotient. Since it is important for our purpose, we shall explain. Let

$$\mathcal{G} \stackrel{\text{def.}}{=} \{g \in D_2^p\Omega^0(\operatorname{End} E) \mid g_x \in \mathrm{SU}(E_x) \text{ for all } x \in X\}$$

$$\mathcal{A} \stackrel{\text{def.}}{=} \{(d_A, \Phi) \mid d_A - d_{A_0} \in D_1^p\Omega^1(\mathfrak{su}\, E),\ \Phi - \Phi_0 \in D_1^p\Omega^{1,0}(\mathfrak{sl}\, E)\},$$

where (A_0, Φ_0) is a fixed parabolic Higgs bundle. The space $\mathcal{A}$ is an infinite dimensional affine space. Its tangent space is

$$D_1^p\Omega^1(\mathfrak{su}\, E) \oplus D_1^p\Omega^{1,0}(\mathfrak{sl}\, E) \cong D_1^p\Omega^{0,1}(\mathfrak{sl}\, E) \oplus D_1^p\Omega^{1,0}(\mathfrak{sl}\, E),$$

which has the natural Riemannian metric given by the L^2-inner product and three almost complex structures given by

$$I(\xi, \phi) = (i\xi, i\phi), \quad J(\xi, \phi) = (i\phi^*, -i\xi^*), \quad K(\xi, \phi) = (-\phi^*, \xi^*).$$

These give a hyper-Kähler structure on $\mathcal{A}$. The natural action of the gauge group $\mathcal{G}$ on $\mathcal{A}$ preserves the hyper-Kähler structure and the hyper-Kähler moment map (in the sense of [HKLR]) is given by

$$\mu_I(d_A, \Phi) = R_A^{\perp} + [\Phi, \Phi^*], \qquad (\mu_J + i\mu_K)(d_A, \Phi) = -2\overline{\partial}_A\Phi.$$

Hence the moduli space of harmonic parabolic Higgs bundles

$$\mu_I^{-1}(0) \cap \mu_J^{-1}(0) \cap \mu_K^{-1}(0)/\mathcal{G}$$

is a hyper-Kähler quotient, and has a hyper-Kähler structure induced from $\mathcal{A}$.

Theorem 4.2. (i) *There exists a natural diffeomorphism* $\mu_I^{-1}(0) \cap \mu_J^{-1}(0) \cap \mu_K^{-1}(0)/\mathcal{G} \to \mathfrak{M}_{\mathrm{Dol}}(\alpha, \sigma)$. *The complex structure on* $\mathfrak{M}_{\mathrm{Dol}}(\alpha, \sigma)$ *corresponds to the complex structure* I.

(ii) *There exists a natural diffeomorphism* $\mu_I^{-1}(0) \cap \mu_J^{-1}(0) \cap \mu_K^{-1}(0)/\mathcal{G} \to \mathfrak{M}_{\mathrm{DR}}(\beta, \tau)$. *The complex structure on* $\mathfrak{M}_{\mathrm{DR}}(\beta, \tau)$ *corresponds to the complex structure* J.

The existence of bijections between moduli spaces was given in [Si2, Theorem 6, 7]. In fact, his statements are not strong enough, since he only show that the complex gauge transformation, which transforms an α-stable parabolic Higgs bundle to a harmonic one, is bounded and respects residues of Higgs fields. This is because he did not introduce weighted Sobolev spaces. Once we know $R_A^{\perp} + [\Phi, \Phi^*] \in D_0^p \Omega^2(\mathfrak{su}\, E)$, the complex gauge transformation is in D_2^p by the a-priori estimate for the weighted Sobolev space. Another approach is to follow Konno's argument [Ko], where he solved the partial differential equation in the weighted Sobolev space directly.

The statement concerning the complex structure of the moduli spaces are easy to prove, and we leave it to the reader as an exercise.

Combining the statements (i),(ii), we obtain the following non-abelian Hodge theorem:

Corollary 4.3. *There exists a diffeomorphism between* $\mathfrak{M}_{\mathrm{Dol}}(\alpha, \sigma)$ *and* $\mathfrak{M}_{\mathrm{DR}}(\beta, \tau)$ *where parameters correspond by* (4.1).

Combining this with Propositions 2.17, 3.7, we obtain

Theorem 4.4. *The differentiable structures on* $\mathfrak{M}_{\mathrm{Dol}}(\alpha, \sigma)$ *and* $\mathfrak{M}_{\mathrm{DR}}(\beta, \tau)$ *are independent of the choice of generic parameters* (α, σ), (β, τ).

References

[Bi] O. Biquad, *Fibrés paraboliques stables et connecxions singulières plates*, Bull. Soc. Math. France **119** (1991), 231–257.

[BH] H. Boden and Y. Hu, *Variations of moduli of parabolic bundles*, Math. Ann. **301** (1995), 539–559.

[BY] H. Boden and K. Yokogawa, *Moduli spaces of parabolic Higgs bundles and parabolic K(D)-pairs over smooth curves: I*, preprint.

[Co] K. Corlette, *Flat G-bundles with canonical metrics*, Jour. of Diff. Geom. **28** (1988), 361–382.

[Hi] N.J. Hitchin, *The self-duality equations on a Riemann surface*, Proc. London Math. **55** (1987), 59–126.

[HKLR] N.J. Hitchin, A. Karlhede, U. Lindström and M. Roček, *Hyperkähler metrics and supersymmetry*, Comm. Math. Phys. **108** (1987) 535–589.

[Ko] H. Konno, *Construction of the moduli space of stable parabolic Higgs bundles on a Riemann surface*, J. Math. Soc. Japan **45** (1993), 253–276.

[Kr] P.B. Kronheimer, *The construction of ALE spaces as a hyper-Kähler quotients*, J. Differential Geom. **29** (1989) 665–683.

[LM] R.B. Lockhart and R.C. McOwen, *Elliptic differenetial operators on noncompact manifolds*, AnnScuola. Norm. Sup. Pisa Cl. Sci. **12** (1985), 205–239.

[Na] H. Nakajima, *Instantons on ALE spaces, quiver varieties, and Kac-Moody algebras*, Duke Math. **76** (1994), 365–416.

[Si] C.T. Simpson, *Constructing variations of Hodge structure using Yang-Mills theory and applications to uniformization*, Jour. of the A.M.S. **1** (1988), 867–918.

[Si2] ———, *Harmonic bundles on noncompact curves*, Jour. of the A.M.S. **3** (1990) 713–770.

[Si3] ———, *Moduli of representations of the fundamental group of a smooth projective variety*, preprint.

[VW] C. Vafa and E. Witten, *A strong coupling test of S-duality*, Nucl. Phys. B **431** (1994), hep-th/9408074.

[Yo] K. Yokogawa, *Compactification of moduli of parabolic sheaves and moduli of parabolic Higgs sheaves*, J. Math. Kyoto Univ. **33** (1993), 451–504.

NOTES ON SYMPLECTIC 4-MANIFOLDS WITH $b_2^+ = 1$

Hiroshi Ohta

Graduate School of Polymathematics, Nagoya University, Chikusa-ku, Nagoya 464, Japan

Kaoru Ono

Department of Mathematics, Ochanomizu University, Otsuka, Tokyo 112, Japan

1. Statement of Results

In this short note, we shall explore some implications of constraints for symplectic structures via Seiberg-Witten theory due to Taubes [T1,2]. In Kähler geometry, there are characterizations of complex projective spaces (e.g. Hirzebruch-Kodaira-Yau, Kobayashi-Ochiai, Siu-Yau, Mori). We review a theorem of Kobayashi and Ochiai [K-O]. Let M be a compact Kähler manifold of complex dimension n with positive first Chern class. If $c_1(M)$ is divisible by a positive integer l in $H^2(M;\mathbf{Z})$, then we have $l \leq n+1$ and the equality holds if and only if M is biholomorphic to the complex projective space $\mathbf{C}P^n$. Positivity of the first Chern class is crucial for characterization of $\mathbf{C}P^n$. For instance, Mumford constructed a fake complex projective plane which has the same Betti numbers as $\mathbf{C}P^2$, but has an infinite fundamental group [M]. On the other hand, Fujita [F] showed that Kähler manifolds with positive first Chern class and having the same integral cohomology ring are complex projective spaces provided the dimension $n \leq 5$. Our first result concerns numerical restrictions for symplectic four manifolds with "positive" first Chern class.

Theorem A *Let (X,ω) be a closed symplectic 4-manifold with integral cohomology class $[\omega] \in H^2(X;\mathbf{Z})$. Suppose $c_1(X) = \lambda\omega$ for some $\lambda > 0$. Then we have*
(1) $\lambda \leq 3$ *and* $c_1^2(X) \leq 9$.
(2) $b_2^+(X) = 1$, $b_1(X) = 0$ *and* $\operatorname{Todd}(X) = 1$.
(3) $\pi_1(X)$ *has no subgroup with finite index.*

If one can show X is simply connected, the homeomorphism types of X above are classified as follows:
(1) When $\lambda = 1$, $X = \mathbf{C}P^2 \# k\overline{\mathbf{C}P^2}$ with $0 \leq k \leq 8$.
(2) When $\lambda = 2$, $X = \mathbf{C}P^1 \times \mathbf{C}P^1$.
(3) When $\lambda = 3$, $X = \mathbf{C}P^2$.
The authors do not know whether there exists a symplectic 4-manifold with non trivial fundamental group, satisfying the properties in Theorem A.

Theorem B *Let X be a closed smooth 4-manifold. Suppose that X satisfies the following two conditions:*
(1) $2e(X) + 3\operatorname{sign}(X) \geq 0$, *where $e(X)$ denotes the Euler number of X and* $\operatorname{sign}(X)$ *is the signature of X.*
(2) *X admits a metric with positive scalar curvature.*

Then X can not admit a symplectic structure ω such that

$$\int_X c_1(K_X) \wedge \omega > 0.$$

Here K_X denotes the canonical bundle with respect to an almost complex structure compatible with ω.

For example, $\mathbf{C}P^1 \times \mathbf{C}P^1$ and $\mathbf{C}P^2 \# k\overline{\mathbf{C}P^2}$ with $0 \leq k \leq 9$ satisfy the conditions in Theorem B. Taubes proved the case $X = \mathbf{C}P^2$ in [T2].

The inequality in Theorem B does not hold, in general, for k points blow-up of a complex projective plane. Note that symplectic blow-up is equivalent to a symplectic embedding of a round ball. Since the complex projective plane with the standard symplectic structure admits full packing by 16 balls of the same size, we have a symplectic structure on 16 points blow-up of the complex projective plane which violates the inequality.

Let H be a line in $\mathbf{C}P^2$ avoiding blowing up points and $\tilde{H}$ the inverse image of H in the blown-up manifold. Denote by E_i the exceptional divisors. Then the Poincare dual of the first Chern class is $3\tilde{H} - \sum E_i$ and the Poincare dual of the symplectic form is $\tilde{H} - \sum a_i E_i$, where a_i is the symplectic area of E_i. Because of the existence of full packing by 16 balls, we can choose a_i arbitrary close to $1/4$. This violates the inequality.

As a corollary to Theorem B, we have

Corollary *No rational surface can be diffeomorphic to a minimal surface of general type.*

This corollary was proved by Friedman and Qin [F-Q], using Donaldson theory. Corollary above implies an affirmative answer to Hirzebruch conjecture solved by Qin [Q] and Kotschick's result that the Barlow surface is not diffeomorphic to a rational surface $\mathbf{C}P^2 \# 8\overline{\mathbf{C}P^2}$ [K].

2. Proof of Theorem A

Since

$$\int_X c_1(X) \wedge \omega = \lambda \int_X \omega^2 > 0,$$

we have $b_2^+(X) = 1$ by Taubes' theorem in [T2]. Thus $\lambda^2\omega^2 = c_1^2(X) = 2e(X) + 3\text{sign }(X) \leq 9$. As the cohomology class $[\omega]$ is integral, we have $\lambda \leq 3$. The possibility of $b_1(X)$ satisfying $1 \leq c_1^2(X) \leq 9$ is $b_1(X) = 0, 1, 2$. But since X is a symplectic 4-manifold, $b_1(X) + b_2^+(X)$ must be odd. So $b_1(X) = 0$ or 2.

Claim $b_1(X) = 0$.

Suppose that $b_1(X) = 2$. Then $b_2^-(X) = 0$. We have a d-fold covering $\tilde{X}$ of X for any $d \geq 2$. Pulling back the symplectic form ω by the projection $\pi : \tilde{X} \to X$, we have a symplectic form $\tilde{\omega} = \pi^*\omega$ on $\tilde{X}$. In this case $(\tilde{X}, \tilde{\omega})$ satisfies $c_1(\tilde{X}) = \lambda\tilde{\omega}$, hence we have $b_2^+(\tilde{X}) = 1$ as well. Comparing with both signatures, we have

$$1 - b_2^-(\tilde{X}) = d(1 - b_2^-(X)) = d.$$

Therefore $d = 1$. This is a contradiction. Hence Todd $(X) = 1$ follows from $b_1(X) = 0$ and $b_2^+(X) = 1$. The proof of the assertion (3) in Theorem A is similar to that of the claim above. So we omit it.

3. Proof of Theorem B and Corollary

Let ω be a symplectic form. First of all, since X admits a metric with positive scalar curvature, we note that $b_2^+(X) = 1$ by Taubes' theorem [T2]. We choose and fix a Riemanian metric g_0 for which the two form ω is self-dual. Following the argument of the proof of Theorem 3 in [T2], also [T1], the perturbed Seiberg-Witten equation for $c_1(X)$ has a unique solution up to gauge transformations. The invariant for $c_1(X)$ is ± 1. On the other side, X has a metric g_1 with positive scalar curvature by assumption. Hence the Seiberg-Witten invariant for $c_1(X)$ is zero ([W]). Since $c_1^2(X) = 2e(X) + 3\text{sign }(X) \geq 0$, $c_1(X)$ is in the closure of positive cone in $H^2(X)$. Therefore $c_1(X)$ can not be represented by an anti-self-dual harmonic two form for any metrics. Hence the Seiberg-Witten invariant for $c_1(X)$ is zero not only for the metric g_1 but also for g_0. ("Wall crossing" phenomena with respect to perturbation of metrics does not happen.) This disagreement is arisen from "wall crossing" with respect to perturbation of equations. Exactly the same argument as Taubes' [T2] shows there exists some $0 \leq s \leq 1$ so that

$$(1-s)\int_X c_1(X) \wedge \omega = s\int_X \omega^2 \geq 0.$$

It follows that there does not exist a symplectic form ω such that

$$\int_X c_1(K_X) \wedge \omega > 0.$$

Proof of Corollary. By the classification theory, any rational surface X is diffeomorphic to $\mathbf{C}P^1 \times \mathbf{C}P^1$ or $\mathbf{C}P^2 \# k\overline{\mathbf{C}P^2}$ for some $k \geq 0$. Assume X is diffeomorphic to a minimal surface of general type. Then since $c_1^2(X) > 0$ for any minimal surface of general type, it is enough to consider the cases $\mathbf{C}P^1 \times \mathbf{C}P^1$ or $\mathbf{C}P^2 \# k\overline{\mathbf{C}P^2}$ with $0 \leq k \leq 8$. By Hitchin's result [H], they all have a metric with positive scalar curvature. Hence by Theorem B, there is no symplectic form ω such that

$$\int_X c_1(K_X) \wedge \omega > 0.$$

However if X is a surface of general type, we have a holomorphic section of $K_X^{\otimes m}$ for sufficiently large $m \geq 1$. Thus

$$m\int_X c_1(K_X) \wedge \omega > 0.$$

This is a contradiction.

4. Remarks

A. Hattori noticed that our Theorem A holds under a weaker condition. Let X be a symplectic manifold with $c_1(X)^2 > 0$ and

$$\int_X c_1(X) \wedge \omega > 0.$$

Then the same conclusion as in Theorem A holds.

In the introduction, we mentioned fake projective planes. For these manifolds, we can show the following:

Proposition *Let X be a fake projective plane. Namely, X is a complex surface of general type with the same Betti numbers as the complex projective plane. Then for any symplectic form ω on X, we have*

$$\int_X c_1(K_X) \wedge \omega > 0.$$

Proof. Because the Euler number and signature of X is 3 and 1 respectively, we have $c_1(X)^2 = 3c_2(X)$. Note that $c_2(X)[X] = 3$, hence $c_1(X)$ is not zero either. Then X must be a quotient of complex ball in $\mathbf{C}^2$ by the theorem of Yau [Y]. Hence the fundamental group of X is realized as a subgroup in $SU(2,1)$. On the other hand, such a group has proper subgroups of finite index [R, 6.11 Theorem]. Taking a covering space of X with respect to a proper subgroup of finite index, we get a closed four manifold $\widetilde{X}$ with $b_2^+ > 1$. Since the pull back of ω to $\widetilde{X}$ is a symplectic form, which we denote by $\tilde{\omega}$, Taubes' constraint gives

$$\int_{\widetilde{X}} c_1(K_{\widetilde{X}}) \wedge \tilde{\omega} \geq 0.$$

This implies that

$$\int_X c_1(K_X) \wedge \omega \geq 0.$$

Because $c_1(X)$ and $[\omega]$ are non-zero and $b_2(X) = 1$, the left hand side of the above inequality is not zero. Hence we get the desired inequality.

References

[F-Q] R. Friedman and Z. B. Qin: On complex surfaces diffeomorphic to rational surfaces, Invent. Math., to appear.

[F] T. Fujita: On topological characterization of complex projective spaces and affine linear spaces, Proc. Japan Acad., **56** Ser. A (1980), 231–234.

[H] N. J. Hitchin: On the curvature of rational surfaces, In Differential Geometry, Proc. Sympos. Pure Math., **27** (1975), 65–80.

[K-O] S. Kobayashi and T. Ochiai: Characterizations of complex projective spaces and hyperquadrics, J. Math. Kyoto Univ., **13** (1973), 31–47.

[K] D. Kotschick: On manifolds homeomorphic to $\mathbf{C}P^2\#8\overline{\mathbf{C}P^2}$, Invent. Math., **95** (1989), 591–600.

[M] D. Mumford: An algebraic surface with K ample, $K^2 = 9$, $p_g = q = 0$, Am. J. Math., **101** (1979), 233–244.

[Q] Z.B.Qin: Complex structures on certain differentiable 4-manifolds, Topology **32** (1993), 551–566.

[R] M. S. Raghunathan: *Discrete subgroups in Lie groups*, Springer Verlag (1972).

[T1] C.H.Taubes: The Seiberg-Witten invariants and symplectic forms, Math. Research Letters, **1** (1994), 809–822.

[T2] C.H.Taubes: More constrains on symplectic manifolds from Seiberg-Witten equations, Math. Research Letters **2** (1995), 9–14.

[W] E.Witten: Monopoles and four-manifolds, Math. Research Letters, **1** (1994), 769–796.

[Y] S-T. Yau: Calabi's conjecture and some new results in algebraic geometry, Proc. Nat. Acad. Sci. U.S.A., **74** (1977), 1798–1799.

ON THE LOCAL GEOMETRY OF MODULI SPACES OF LOCALLY FREE SHEAVES

Ziv Ran

Department of Mathematics, University of California, Riverside, Riverside, California CA92521-0001 USA
e-mail: ziv@math.ucr.edu

In [R1] the author gave a canonical construction of the universal deformation of a compact complex manifold. Here we shall give an analogous construction for the universal deformation of a simple locally free sheaf (over a fixed locally-ringed base space $(X, \mathcal{O}_X)$; this construction differs somewhat from that of [R1], featuring a systematic use of the Jacobi complexes of a module. The resulting construction of universal or Poincaré bundle is applicable in other problems as well, such as deformations of manifold, where it yields a simplification of the method employed in [R1] to construct the structure sheaf of the total space of the universal deformation.

As an application of our method, we shall prove the closedness of *trace forms*—these are $H^2(\mathcal{O}_X)$–valued 2–forms on the moduli space of locally free sheaves, E, on X induced by the trace or Killing form on $\mathcal{E}\mathrm{nd}(E)$—

$$\tau : H^1(\mathcal{E}\mathrm{nd}(E)) \otimes H^1(\mathcal{E}\mathrm{nd}(E)) \to H^2(\mathcal{E}\mathrm{nd}(E)) \xrightarrow{\mathrm{tr}} H^2(\mathcal{O}_X).$$

For $(X, \mathcal{O}_X)$ a compact smooth complex surface with trivial canonical bundle, τ is in fact, by Serre duality, a non–degenerate scalar valued 2–form, hence, being closed, yields a symplectic structure on the moduli space, a result first proven by Mukai [M]; see also Kobayashi's book [K] for an analytic approach. Similarly, for $(X, \mathcal{O}_X)$ a compact Riemann surface with its sheaf of locally constant functions, in which case a locally free sheaf E is just a local system, τ again yields a symplectic structure, first discovered in this case by Hitchin [H].

Acknowledgment I am grateful to Professors Maruyama and Mukai for their invitation to attend the symposium and to Professor Fujiki for suggesting the problem of closedness of trace forms. I also wish to thank Professors V. Ginzburg and L. Jeffrey for subsequest communications.

1. Calculus

1.1. Enveloping Algebras and Differential Operators

Let $\mathfrak{g}$ be a Lie algebra, say over a field $\mathbb{C}$ of irrelevant characteristic (i.e. 0 or >> 0), and E a faithful $\mathfrak{g}$–module. As is well known, the universal enveloping algebra $\mathcal{U} = \mathcal{U}(\mathfrak{g})$ may be identified as vector space, though not as algebra, with $\oplus_{i=0}^{\infty} S^i(\mathfrak{g})$, and this identification takes $\mathcal{U}^m \subset \mathcal{U}$, the elements of order $\leq m$, to $\oplus_{i=0}^{m} S^i(\mathfrak{g})$. The algebra structure on $\mathcal{U}(\mathfrak{g})$ may be described as follows. Let $\mathcal{S}^{\cdot} = \mathcal{S}^{\cdot}(E) = \oplus_{i=0}^{\infty} S^i(E)$ be the symmetric algebra on

E and note that elements of $\mathfrak{g}$ acting on E extend to graded (i.e. homogeneous of degree zero) derivations on $\mathcal{S}^{\cdot}$; in fact it is easy to see that the Lie algebra of all graded derivations coincides with $\mathfrak{gl}(E) = \mathrm{End}(E)$. Therefore if we denote by $\mathcal{D}$ (resp. $\mathcal{D}^m$) the algebra (resp. $\mathfrak{g}$–module) of all (resp. all degree $\leq m$) graded $\mathbb{C}$–linear differential operators on $\mathcal{S}^{\cdot}$, we obtain an injective filtration preserving algebra and $\mathfrak{g}$–module homomorphism

$$\rho : \mathcal{U}(\mathfrak{g}) \to \mathcal{D}.$$

Moreover it is easy to see that for $\mathfrak{g} = \mathfrak{gl}(E)$, ρ is in fact an isomorphism.

We want now to identify $\mathrm{im}(\rho)$ in case $\mathfrak{g} = \mathfrak{sl}(E)$, the traceless endomorphisms (assuming E is finite–dimensional). To this end, note that a graded endomorphism $\varphi^{\cdot}$ of $\mathcal{S}^{\cdot}$ admits a graded trace $\mathrm{tr}^{\cdot}(\varphi^{\cdot})$, which itself is a graded endomorphism of $\mathcal{S}^{\cdot}$ given by

$$\mathrm{tr}^i(\varphi^{\cdot}) = \text{image of } (\varphi^{i+1} \in S^{i+1}(E) \otimes S^{i+1}(E^*) \subset S^iE \otimes S^i(E^*) \otimes E \otimes E^*) \\ \text{in } S^i(E) \otimes S^i(E^*) \text{ under id} \otimes \text{id} \otimes \text{tr}$$

In other words, $\mathrm{tr}^{\cdot}$ is the natural extension of $\mathrm{tr} : E \otimes E^* \to \mathbb{C}$ as a derivation of degree -1 on the algebra of all graded endomorphisms of $\mathcal{S}^{\cdot}$. Clearly $\mathrm{tr}^{\cdot}(\varphi^{\cdot}) = 0$ iff for all $e_1, \ldots, e_i \in E$, $e_1^*, \ldots, e_i^* \in E^*$, acting on $\mathcal{S}^{\cdot}$ in the standard manner (interior multiplication), the trace (in the usual sense) of $e_1^* \ldots e_i^* \circ \varphi^{i+1} \circ e_1 \ldots e_i$ is zero. From this it is easy to see that $\rho(\mathcal{U}(\mathfrak{sl}(E)))$ coincides with the sub–algebra $\mathcal{SD} \subset \mathcal{D}$ of traceless operators.

The above considerations extend immediately to the case where E is a locally free sheaf over a $\mathbb{C}$–ringed space $(X, \mathcal{O}_X)$ and $\mathfrak{g}$ is a Lie sub–algebra of $\mathfrak{gl}(E) = \mathcal{E}\mathrm{nd}_{\mathcal{O}_X}(E)$. Also, as far as $\mathcal{D}^m$ is concerned, $char(\mathbb{C})$ need only be $> m$

1.2. Functors on S–Modules

In [R1] we showed how an Artin local $\mathbb{C}$–algebra S may be reconstructed from a certain 'order–symbolic' or OS structure on the space of $\mathbb{C}$–valued differential operators on S. Our purpose here is to note an analogue of this for S–modules.

Now fix a local $\mathbb{C}$–algebra S with maximal ideal $\mathfrak{m}$ and residue field $S/\mathfrak{m} = \mathbb{C}$ and put

$$B_0^i = D^i(S, \mathbb{C}) = \mathrm{Hom}(S_i, \mathbb{C}) = S_i^*, \qquad S_i = S/\mathfrak{m}^{i+1},$$

and

$$B^i = D_+^i(S, \mathbb{C}) = (\mathfrak{m}/\mathfrak{m}^{i+1})^*.$$

For an S–module E, put

$$B^i(E) = B_0^i \otimes_S E,$$

where B_0^i is viewed as S–bi–module and E as (symmetric) S–bi–module. At least when $E \otimes S_i$ is S_i–free, $B^i(E)$ may be identified with the right S–module of differential operators $D^i(E^\vee, \mathbb{C})$, $E^\vee = \mathrm{Hom}_S(E, S)$.

We have a symbol map

$$\sigma^i : B_0^i \to B^i \otimes_{\mathbb{C}} B_0^{i-1}$$

which factors through $F_i(B^i \otimes_{\mathbb{C}} B^{i-1})$, where F_i is the filtration induced by the order filtration on B^i, and this gives rise to a symbol map

$$\sigma_E^i : B^i(E) \to B^i \otimes_{\mathbb{C}} B^{i-1}(E),$$

which again factors through $F_i(B^i \otimes_{\mathbb{C}} B^{i-1}(E))$. These σ_E^i, $i \leq m$, together with the obvious maps $B^0(E) \to B^1(E) \to \ldots \to B^m(E)$ are referred to as a 'modular order–symbolic' (MOS) structure on $B^m(E)$. Note that $B^m(E)$ itself is a right S_m–module, called the m–th transpose of E.

"Dually", suppose we are given an MOS structure $G^{\cdot}$, G^i a right S_i–module. We then define an S_m–module $C^m(G^{\cdot})$, called the module of quasi–scalar homomorphisms $B_0^m \to G^{\cdot}$, inductively as follows.

$$C^0(G) = G^0$$

$$C^i(G) = \text{all right } S\text{–linear maps } \varphi^i : B_0^i \to G^i$$

such that for some $\varphi^{i-1} : B_0^{i-1} \to G^{i-1}$ the following diagrams commute.

$$\begin{array}{ccc} B_0^{i-1} & \xrightarrow{\varphi^{i-1}} & G^{i-1} \\ \downarrow & & \downarrow \\ B_0^i & \xrightarrow{\varphi^i} & G^i \end{array} \qquad\qquad \begin{array}{ccc} B_0^i & \xrightarrow{\varphi^i} & G^i \\ \sigma^i \downarrow & & \downarrow \sigma_G^i \\ B^i \otimes B_0^{i-1} & \xrightarrow{\mathrm{id}\otimes\varphi^{i-1}} & B^i \otimes G^{i-1} \end{array}$$

Note that we have natural maps

$$E \to C^m(B^m(E))$$

$$B^m(C^m(G^{\cdot})) \to G^{\cdot}.$$

At least when E is S_m–free (resp. $G^{\cdot}$ is 'co–free', i.e. a sum of copies of B_0^m with the standard MOS structure), these are isomorphisms.

1.3. Exterior Derivative

Our purpose here is to give a convenient interpretation of the Cartan formula for exterior derivative. Let M be a manifold germ with tangent sheaf T, and consider the sheaf $B^1(\bigwedge^i T)$, which may be identified as the sheaf of $\mathcal{O}_M$–valued first–order differential operators on the dual $\bigwedge^i T^{\vee} = \Omega^i$, and fits in an exact sequence

$$0 \to \bigwedge^i T \to B^1(\bigwedge^i T) \to T \otimes \bigwedge^i T \to 0.$$

Over the sub–sheaf $\bigwedge^{i+1} T \subset T \otimes \bigwedge^i T$, this sequence admits a canonical splitting φ_i : $\bigwedge^{i+1} T \to B^1(\bigwedge^i T)$ given by

$$(*)\qquad \begin{aligned} \varphi_i(v_0 \wedge \ldots \wedge v_i)(\omega) = &\sum (-1)^j v_j \omega(v_0, \ldots, \hat{v}_j, \ldots, v_i) + \\ &\sum_{j<k} (-1)^{j+k} \omega([v_j, v_k], v_0, \ldots, \hat{v}_j, \ldots, \hat{v}_k, \ldots, v_i) \end{aligned}$$

for $\omega \in \Omega^i$. The Cartan formula says in particular that φ_i is $\mathcal{O}_M$–linear. Now given an i–form, ω, viewed as a map $\bigwedge^i T \to \mathcal{O}_M$, it extends to a map $B^1(\bigwedge^i T) \to B^1(\mathcal{O}_M) = \mathcal{O}_M \oplus T$. Projecting on the $\mathcal{O}_M$ factor, we get a map

$$\tilde{\omega} : B^1(\bigwedge^i T) \to \mathcal{O}_M.$$

Cartan's formula then says that the exterior derivative $d\omega$ coincides with $\tilde{\omega} \circ \varphi_i$.

Note that $\mathrm{im}(\varphi_1)$ coincides with the kernel of the natural map

$${}^t d_0 : B^1(T) \to T^2 = D_+^2(\mathcal{O}, \mathcal{O})$$

$${}^t d_0(a) = a \circ d_0$$

where $a : \Omega^1 \to \mathcal{O}$ is a first–order differential operator and $d_0 : \mathcal{O} \to \Omega^1$ is exterior derivative.

Indeed we have an exact sequence

$$\begin{array}{ccccccccc}
 & & & & & & \bigwedge^2 T & & \\
 & & & & & \overset{\varphi_1}{\swarrow} & \downarrow & & \\
0 & \longrightarrow & T & \longrightarrow & B^1(T) & \longrightarrow & T\otimes T & \longrightarrow & 0 \\
 & & \| & & {}^t d_0 \downarrow & & \downarrow & & \\
0 & \longrightarrow & T & \longrightarrow & T^2 & \longrightarrow & S^2 T & \longrightarrow & 0
\end{array}$$

Similarly, using the inclusion $B^1(\bigwedge^i T) \subset B^1(T \otimes \bigwedge^{i-1} T) = B^1(T) \otimes_{\mathcal{O}} \bigwedge^{i-1} T$, we may identify $\mathrm{im}(\varphi_i)$ as the kernel of the natural map $B^1(\bigwedge^i T) \xrightarrow{{}^t d_0 \otimes \mathrm{id}} T^2 \otimes \bigwedge^{i-1} T = B^2(\bigwedge^{i-1} T)/\Omega^{i-1} =: B^2_+(\bigwedge^{i-1} T)$, which in fact coincides with the composite $B^1(\bigwedge^i T) \xrightarrow{{}^t d_{i-1}} B^2(\bigwedge^{i-1} T) \to B^2_+(\bigwedge^{i-1} T)$. Note that the above considerations hold in all characteristics > 3.

2. Jacobi Complexes and Universal Deformations

Fix a base space X which for convenience we assume to be a paracompact locally-ringed space over a field $\mathbb{C}$ of sufficiently large or zero characteristic, and a simple (nonconstant-endomorphism-free) locally free sheaf E on X. Let $\mathfrak{g} = \mathfrak{sl}(E)$, the Lie algebra of traceless endomorphisms of E, so that $H^0(\mathfrak{g}) = 0$. As in [R1], we have Jacobi complexes $J^{\cdot}_m(\mathfrak{g})$ which may be described as follows. Let $X{<}m{>}$ be the m-fold very symmetric product of X, i.e. the space of subsets of X of cardinality $\in [1, m]$, with the topology induced by the natural map from the cartesian product $X^m \to X{<}m{>}$. For $i \le m < \mathrm{char}(\mathbb{C})$ let $\lambda^i(\mathfrak{g})$ be the image of the exterior alternating product of $\mathfrak{g}$, supported on $X{<}i{>} \subset X{<}m{>}$. Then the bracket on $\mathfrak{g}$ gives rise to a map $\lambda^i(\mathfrak{g}) \to \lambda^{i-1}(\mathfrak{g})$, and these fit together to form the complex $J^{\cdot}_m(\mathfrak{g})$—

$$\lambda^m(\mathfrak{g}) \to \lambda^{m-1}(\mathfrak{g}) \to \ldots \to \lambda^1(\mathfrak{g}) = \mathfrak{g},$$

in which we put $\lambda^i(\mathfrak{g})$ in degree $-i$. Similarly, the action of $\mathfrak{g}$ on E gives rise to a complex $J^{\cdot}_m(\mathfrak{g}, E)$ on $X{<}m{>} \times X$—

$$\lambda^m(\mathfrak{g}) \boxtimes E \to \ldots \to \mathfrak{g} \boxtimes E \to E,$$

in degrees $\in [-m, 0]$, where the last term E is supported on the diagonal in $X{<}1{>} \times X = X \times X$. The natural maps

$$J^{\cdot}_i(\mathfrak{g}) \to J^{\cdot}_m(\mathfrak{g}), \qquad i \le m,$$

$$J^{\cdot}_m(\mathfrak{g}) \to F_m(J^{\cdot}_{m-1}(\mathfrak{g}) \boxtimes J^{\cdot}_{m-1}(\mathfrak{g}))$$

give rise to an OS structure on $V_m = \mathbb{H}^0(J^{\cdot}_m(\mathfrak{g}))$ which, as in [R1], yields a $\mathbb{C}$-algebra structure on $R_m = \mathbb{C} \oplus V_m^*$, which turns out to be the ring of the universal m-th order deformation of E (see below). To get the universal or Poincaré bundle on $X \times \mathrm{Spec} R_m$, consider

$$G^m = \mathbb{R}^0 p_{2*}(J^{\cdot}_m(\mathfrak{g}, E)).$$

This forms a sheaf of $\mathcal{O}_X \otimes R_m$-modules, and the natural map

$$J^{\cdot}_m(\mathfrak{g}, E) \to F_m(J^{\cdot}_m(\mathfrak{g}) \boxtimes J^{\cdot}_{m-1}(\mathfrak{g}, E))$$

endows G^m with a MOS structure compatible with the OS structure on V^m, whence as in § 1.2 a sheaf

$$\mathcal{P}_m(E) := C^m(G^m).$$

As G^m is locally isomorphic to $(V_m \oplus \mathbb{C}) \otimes \mathcal{O}_X$, clearly $\mathcal{P}_m(E)$ is a locally free $R_m \otimes \mathcal{O}_X$ module, and it turns out to be the Poincaré bundle for E. (The last construction may be applied in other situations, such as deformations of manifolds, where it yields a simplification of the construction in [R1] of the structure sheaf of the universal deformation of a manifold X.)

Notwithstanding the simple construction of $\mathcal{P}_m(E)$ and R_m, the proof of their universality involves the somewhat complicated construction of Kodaira–Spencer bi–complexes associated to a given deformation of E. Given such a deformation, i.e. a locally free sheaf $\mathcal{E}$ on $X \times \mathrm{Spec}(S)$, S Artin local, with $\mathcal{E} \otimes k(0) \xrightarrow{\sim} E$, we construct bi–complexes $W_m^{\cdot,\cdot}$, and $K_m^{\cdot,\cdot}$, which fit in exact (up to quasi–isomorphism) sequences (in which $B^m = D_+^m(S, \mathbb{C})$)—

$$0 \to J_m^{\cdot}(\mathfrak{g}) \to W_m^{\cdot,\cdot} \to B^m[1-m, m] \to 0 \qquad \text{on } X{<}m{>}, \tag{2.1}$$

$$0 \to J_m^{\cdot}(\mathfrak{g}, E) \to K_m^{\cdot,\cdot} \to B_S^m(\mathcal{E})[1-m, m] \to 0 \qquad \text{on } X{<}m{>} \times X, \tag{2.2}$$

whose associated coboundary maps

$$\alpha_m : B^m \to \mathbb{H}^0(J_m^{\cdot}(\mathfrak{g})) = V_m,$$
$$\beta_m : B_S^m(\mathcal{E}) \to \mathbb{R}^0 p_{2*}(J_m^{\cdot}(\mathfrak{g}, E)) = G_m$$

respect the respective OS structures, hence give rise to a ring homomorphism $t_{\alpha_m} : R_m \to S_m = S/\mathfrak{m}^m$ and to an S_m–linear map $t_{\beta_m} : \mathcal{E}_m := \mathcal{E} \otimes S_m \to \mathcal{P}_m(E) \otimes_{R_m} S_m$, which turns out to be an isomorphism.

We sketch the construction of these bi–complexes. Let $\mathcal{S}^{\cdot}(E)$ be the symmetric algebra on E and similarly for $\mathcal{E}$ and $\mathcal{E}_m$ (as $\mathcal{O}_X \otimes S$–modules). A graded, $\mathcal{O}_X$–linear map $\varphi : \mathcal{S}^{\cdot}(\mathcal{E}_i) \to \mathcal{S}^{\cdot}(\mathcal{E}_j)$ is said to be a differential operator of degree $\leq k$ if for all $e^* \in E^* \subset \mathcal{E}_i^*$ and $\mathfrak{m}_{e^*}$, the ideal in $\mathcal{S}^{\cdot}(\mathcal{E}_i)$ generated by $\ker(e^*) \subset \mathcal{E}_i$ and $\mathfrak{m}$, we have $\varphi(\mathfrak{m}_{e^*}^l) \subset \mathfrak{m}_{e^*}^{l-k}$; similarly for $\psi : S_i \to S_j$. φ is said to be traceless if it can be locally written as $\sum \lambda_r \psi_r$, with $\lambda_r : \mathcal{S}^{\cdot}(E) \to \mathcal{S}^{\cdot}(E)$ traceless, $\psi_r : S_i \to S_j$. Define

$$\mathcal{A}^{i,j} = \mathcal{A}^{i,j}(\mathcal{E}) = \left\{ (\varphi, \psi) \;\middle|\; \begin{array}{l} \varphi : \mathcal{S}^{\cdot}(\mathcal{E}_i) \to \mathcal{S}^{\cdot}(\mathcal{E}_j), \psi : S_i \to S_j, \mathcal{O}_X\text{-linear graded} \\ \text{traceless (resp. } \mathbb{C}\text{-linear) diff. op. of order} \leq i-j \text{ such} \\ \text{that } \varphi|_{S_i} = \psi, \text{and } \psi(1) = 0 \end{array} \right\}.$$

Also define

$$\mathcal{A}_0^{i,j} = \mathcal{A}^{i,j} \oplus \mathbb{C}.$$

These are (S_j, S_i) bi–modules and admit filtrations $F.$, and $G.$ by total order (resp. horizontal order); in particular $G_0(\mathcal{A}^{i,j})$ consists of the S_j–linear or 'relative' operators on $\mathcal{E}_j$ and $G_0(\mathcal{A}^{i,j})/\mathfrak{m}G_0 A^{i,j} = \mathcal{U}_{i-j}(\mathfrak{g})$. Put $\bar{\mathcal{A}}^{i,j} = \mathcal{A}^{i,j}/\mathfrak{m}G_0(\mathcal{A}^{i,j})$. As in [R1] define

$$W_m^{i,-j} = \lambda^{j-i-1}(\mathfrak{g}) \boxtimes F_m(\sigma^{i+1}(\bar{\mathcal{A}}^{m,j-1})), \qquad (i,j) \neq (m-2, m)$$

with $\sigma^{\cdot}$ = exterior symmetric product (over $\mathbb{C}$), and

$$W_m^{m-2,-m} = \ker(\mathcal{A}^{m,m-1} \boxtimes F_m(\sigma^{m-1}(\bar{\mathcal{A}}^{m,m-1})) \to \sigma^m(B^{m,m-1}) \to B^m).$$

Then let $W_{m,0}^{\cdot,\cdot}$ be the analogous complex with $\mathcal{A}^{i,j}$ replaced by $\mathcal{A}_0^{i,j}$ and put

$$K_m^{i,-j} = W_{m,0}^{i,-j} \boxtimes_{S_m} \mathcal{E}_m, \qquad (i,j) \neq (0,0)$$
$$K_m^{0,0} = E.$$

As in [R1], $W_m^{\cdot,\cdot}$ fits in an exact sequence as in (2.1), and using the fact that $\mathcal{E}_m$ is S_m–flat, $K_m^{\cdot,\cdot}$ fits in (2.2).

For instance, for $m = 1$ (2.2) has the form

$$\begin{array}{ccccccccc}
0 & \longrightarrow & \mathfrak{g} \boxtimes E & \longrightarrow & A_0^{1,0} \boxtimes_{S_1} \mathcal{E} & \longrightarrow & B^1(\mathcal{E}) & \longrightarrow & 0 \\
 & & \downarrow & & \downarrow & & & & \\
 & & E & = & E & & & & \\
 & & \| & & \| & & & & \\
 & & J_1^{\cdot}(\mathfrak{g}, E) & & K_1^{\cdot,\cdot} & & & &
\end{array}$$

3. Closed 2–forms

Let M be a manifold germ parameterizing a deformation of a locally free sheaf E on X with $H^0(\mathfrak{sl}(E)) = 0$. For a point $0 \in M$ we have the Kodaira–Spencer map

$$\alpha_1 : T_0 M \to H^1(\mathfrak{g}), \qquad \mathfrak{g} = \mathfrak{sl}(E)$$

and its higher order analogues. For simplicity we assume E is unobstructed, which means that the spectral sequence

$$E_1^{p,q} = \mathbb{H}^q(J_m^p(\mathfrak{g})) \Rightarrow \mathbb{H}^i(J_m^{\cdot}(\mathfrak{g}))$$

degenerates at E_1 for $p + q = 0$ for all m; in particular the map $S^2 H^1(\mathfrak{g}) \to H^2(\mathfrak{g})$ induced by bracket vanishes. Actually, only 3rd order deformations will come into play, so $m = 3$ would suffice. Similarly, we need only assume char$(\mathbb{C}) > 3$ .Now we have a trace or Killing form

$$\mathrm{tr} : \mathfrak{g} \boxtimes \mathfrak{g} \to \mathcal{O}_X$$
$$\mathrm{tr}(A, B) = \mathrm{tr}(AB).$$

As is well known, this is a non–degenerate symmetric bilinear form and induces an alternating form

$$\tau : \bigwedge^2 H^1(\mathfrak{g}) \to H^2(\mathcal{O}_X)$$

hence a map

$$\tau_{M,0} : \bigwedge^2 T_{M,0} \to H^2(\mathcal{O}_X)$$

which we view as yielding a 2–form τ_M on M with values in the fixed vector space $H^2(\mathcal{O}_X)$. When X is a complex manifold with $n = \dim X = 2$ and $K_X = \mathcal{O}_X$, the isomorphism $\mathfrak{g} \to \mathfrak{g}^*$ induced by tr yields an isomorphism $H^1(\mathfrak{g}) \to H^1(\mathfrak{g}^*) = H^1(\mathfrak{g})^*$, hence the 2–form τ is non–degenerate on the moduli space. More generally, any $\eta \in H^{n,n-2}(X)$ yields via cup–product with η, a (scalar–valued) 2–form on M.

Theorem 3.1. *τ_M is a closed 2–form on M.*

Proof. It suffices to prove this for the universal second–order deformation. tr yields a map

$$\mathrm{tr} : \sigma^2 \mathfrak{g} \to \mathcal{O}_X$$

which on cohomology gives rise to

$$\tau : H^2(\sigma^2 \mathfrak{g}) = \bigwedge^2 H^1(\mathfrak{g}) \to H^2(\mathcal{O}_X).$$

Now we may identify $B^1(\wedge^2 T_M)_0$ as $\mathbb{H}^0(J_1^{\cdot}(\mathfrak{g}, \sigma^2 \mathfrak{g}))$, i.e. as $\mathbb{H}^0$ of the following complex on $X<2> \times X$:

$$\mathfrak{g} \boxtimes \sigma^2 \mathfrak{g} \to \sigma^2 \mathfrak{g}$$

$$A \times (B, C) \mapsto ([A, B], C) + (C, [A, B]) + (B, [A, C]) + ([A, C], B)$$

Now applying tr we get

$$2\,\mathrm{tr}(ABC - BAC + BAC - BCA) = 2\,\mathrm{tr}(A(BC) - (BC)A) = 0,$$

hence tr is a map of $\mathfrak{g}$–modules, where $\mathfrak{g}$ acts trivially on $\mathcal{O}_X$, hence tr extends to a map of complexes

$$J_1^{\cdot}(\mathfrak{g}, \sigma^2\mathfrak{g}) \to J_1^{\cdot}(\mathfrak{g}, \mathcal{O}).$$

Now the differential $\mathfrak{g} \boxtimes \sigma^2\mathfrak{g} \to \sigma^2\mathfrak{g}$ clearly vanishes on $\sigma^3\mathfrak{g}$, so that $\sigma^3\mathfrak{g}[2]$ forms a sub–complex of $J_1(\mathfrak{g}, \sigma^2\mathfrak{g})$ and we have a diagram

$$\begin{array}{ccc} \sigma^3(\mathfrak{g}) & & \\ \downarrow & & \\ \mathfrak{g} \boxtimes \sigma^2\mathfrak{g} & \longrightarrow & \sigma^2(\mathfrak{g}) \\ \downarrow & & \downarrow \\ \lambda^2(\mathfrak{g}) \boxtimes \mathfrak{g} & \longrightarrow & \mathfrak{g} \boxtimes \mathfrak{g} \end{array}$$

where the left vertical arrows compose to zero and $\mathbb{H}^0$ of the bottom complex is $T^2_{R_2} \otimes_{R_2} T^1$; in fact the bottom complex splits as a direct sum of $J_3(\mathfrak{g})/J_1^{\cdot}(\mathfrak{g})$ and a complex $C^{\cdot} : \sigma^{2,1}(\mathfrak{g}) \to \sigma^2(\mathfrak{g})$, $\sigma^{2,1}$ being the mixed tensor power and $J_{\cdot}^1(\mathfrak{g}, \sigma^2(\mathfrak{g})) = \sigma^3(\mathfrak{g})[2] \oplus C^{\cdot}$. As discussed in §1.3, $d\tau : \bigwedge^3 H^1(\mathfrak{g}) \to H^2(\mathcal{O}_X)$ coincides with the composite

$$\bigwedge^3 H^1(\mathfrak{g}) = H^3(\sigma^3\mathfrak{g}) \to \mathbb{H}^2(J_1(\mathfrak{g}, \sigma^2\mathfrak{g}) \to$$
$$\mathbb{H}^2(\mathfrak{g}, \mathcal{O}) = H^2(\mathcal{O}) \oplus H^1(\mathfrak{g}) \otimes H^2(\mathcal{O}) \to H^2(\mathcal{O})$$

But in view of the diagram

$$\begin{array}{ccccc} \sigma^3\mathfrak{g}[2] & \longrightarrow & J_1(\mathfrak{g}, \sigma^2\mathfrak{g}) & & \\ \downarrow & & \downarrow & & \\ \mathfrak{g} \boxtimes \mathcal{O}[2] & \longrightarrow & J_1^{\cdot}(\mathfrak{g}, \mathcal{O}) & \longrightarrow & \mathcal{O}[1] \end{array}$$

and the fact that the bottom arrows compose to zero, $d\tau$ clearly vanishes. □

Example 1 If X admits a symplectic form ω, and is moreover Kähler, it is shown by Mukai [M] for $n = 2$, and Kobayashi [K] in general that the 2–form $\tau \wedge \omega^{n/2-1} \wedge \bar{\omega}^{n/2}$ is non–degenerate (and closed), yielding a symplectic structure on the moduli space.

Example 2 For X a compact Riemann surface (with its sheaf of locally constant functions), a volume form ω similarly gives rise to a symplectic structure on the moduli space of local systems, first discovered by Hitchin [H]. More generally we get closed 2-forms on moduli of local systems on any (reasonable) space.

References

[H] Hitchin, N.J.:

[K] Kobayashi, S.: 'Differential Geometry of Complex Vector Bundles'. Iwanami and Princeton University Press, 1987.

[M] Mukai, S.: 'Symplectic Structure on the Moduli Space of Sheaves on an Abelian or K3 Surface'. Invent. Math. **77** (1984), 101-116.

[R1] Ran, Z.: 'Canonical Infinitesimal Deformations'. preprint.

[R2] Ran, Z.: 'Infinitesimal Deformations of Vector Bundles and Their Cohomology Groups'. (in preparation).

SMALL FIELDS IN CONSTRUCTIVE ALGEBRAIC GEOMETRY

Frank-Olaf Schreyer
Mathematik, Universität Bayreuth, D-95447 Bayreuth, Germany
e-mail: schreyer@btm8×2.mat.uni-bayreuth.de

0. Introduction

Let $\mathbf{M}$ be a parameter space for some objects in algebraic geometry, eg. $\mathbf{M}$ could be a Hilbert scheme or a moduli space. Suppose we want to know whether $\mathbf{M}$ is non-empty and whether an open part of $\mathbf{M}$ corresponds to smooth objects.

A frequently used method for this is to construct explicitly a point $p \in \mathbf{M}$. For example the class room proof for the existence of smooth projective curves of every genus consist in observing that the equation

$$y^2 = \prod_{i=1}^{2g+1} (x - a_i)$$

defines such a curve, if $a_i \neq a_j$ for $i \neq j$. Typically for such examples is that the actual point constructed has as its field of definition a rational function field over $\mathbb{Q}$. In the example above the field of definition is $\mathbb{Q}(a_1, \dots, a_{2g+1})$.

Implicitly in this approach to the non-emptiness of $\mathbf{M}$, we use the fact that $\mathbf{M}$ is uni-rational, or as in the example above, that $\mathbf{M}$ contains a uni-rational subvariety. However if $\mathbf{M}$ has general type we cannot expect to find $\mathbb{Q}$ rational points at all.

On the other hand for a finite field $\mathbb{F}_q$ almost every variety has a lot of points. Roughly if we can describe $\mathbf{M}$ as a subvariety of codimension c in a rational variety $\mathbb{G}$ the chance for a point $p \in \mathbb{G}(\mathbb{F}_q)$ to lie in $\mathbf{M}(\mathbb{F}_q)$ is $1 : q^c$. Thus for small q and c and sufficiently fast computer and programs we can hope to find a point $p \in \mathbf{M}(\mathbb{F}_q)$ within a reasonable amount of time. Furthermore, if $p \in \mathbf{M}(\mathbb{F}_q)$ is a point in which $\mathbf{M}$ has expected codimension, one can deduce the existence of a point in $\mathbf{M}$ defined over some number field L, ie. one obtains that $\mathbf{M}(\mathbb{C})$ is non-empty.

In this paper we establish the existence of certain smooth rational surfaces in $\mathbb{P}^4$ by this method.

1. Motivation and Main Result

By a result of Ellingsrud and Peskine [EP] it is known that the Hilbert scheme of surfaces in $\mathbb{P}^4$ has only finitely many components parametrizing smooth rational surfaces. In the moment the best known bound for the degree of rational surfaces is $d \leq 82$, cf. [BF], [C]. On the other hand 19 families of smooth rational surface have been constructed, cf. [DES], [P], which all have degree ≤ 11. In this paper we establish the existence of three further families.

Main Result *The Hilbert scheme of smooth surfaces in $\mathbb{P}^4_{\mathbb{C}}$ of degree $d = 11$, sectional genus $\pi = 10$ and $p_g = q = 0$ has at least four components. One component parametrizes Enriques surfaces blown up in 6 points embedded by a linear system of type*

$$(a) \qquad H = H_{min} - 2E_1 - \textstyle\sum_{i=2}^{6} E_i$$

where H_{min} denotes a line bundle on the minimal model with self-intersection $H_{min}^2 = 20$ and $E_1, \dots, E_6$ denote the exceptional curves. The other three components parametrize rational surfaces $\mathbb{P}^2$ blown up in 15 points embedded by the linear system

$$(b) \qquad H = 18L - \textstyle\sum_{i=1}^{5} 6E_i - \sum_{j=6}^{10} 5E_j - 2E_{11} - \sum_{k=12}^{15} E_k,$$

$$(c) \qquad H = 15L - \textstyle\sum_{i=1}^{5} 5E_i - \sum_{j=6}^{9} 4E_j - \sum_{k=10}^{11} 3E_k - 2E_{12} - \sum_{l=13}^{15} E_l,$$

and

$$(d) \qquad H = 16L - 6E_1 - \textstyle\sum_{i=2}^{7} 5E_i - \sum_{j=8}^{10} 4E_j - \sum_{k=11}^{12} 2E_l - \sum_{l=13}^{15} E_k,$$

respectively. Here L denotes the class of a line in $\mathbb{P}^2$. The corresponding components of the Hilbert scheme are generically smooth of dimension $36 = 24 + 12$.

Remark Notice that for a surface with these invariants we have $h^1(X, \mathcal{O}(H)) = 2$. Thus we expect that surfaces with $h^0(X, \mathcal{O}(H)) = 5$ form a subscheme of codimension $10 = h^0(\mathcal{O}(H)) * h^1(\mathcal{O}(H))$ in a moduli space of polarized surfaces (X,H). The basic difficulty for the construction of these surfaces is, that we have no simple description of this subscheme. The actual construction of these surfaces avoids the approach via linear systems. We proceed in several steps starting with the methods of [DES].

Step 1 (Beilinson's spectral sequence) We want to construct two vector bundles $\mathcal{F}$ and $\mathcal{G}$ on $\mathbb{P}^4$ with $rank\mathcal{G} - rank\mathcal{F} = 1$ and a morphism $\phi \in Hom(\mathcal{F}, \mathcal{G})$ such that $cokern(\phi) \cong I_X(4)$ is the ideal sheaf of the desired surface X up to a twist.

Assume that the dimensions of the the cohomology groups $H^i(\mathbb{P}^4, I_X(m))$ of the ideal sheaf I_X have the following values:

$i \uparrow$

	2			
		1	5	5

$\rightarrow m$

By Riemann-Roch these are the smallest possible values for a surface with the given invariants. Thus surfaces with these values form an open part of their Hilbert scheme. Applying Beilinson's spectral sequence to $I_X(4)$ yields the E_2-diagram

	$2\Omega(3)$			
			$\mathcal{G}$	

Thus we may take $\mathcal{F} = 2\Omega^3(3)$, $\mathcal{G}$ the homology of $0 \to \Omega^2(2) \to 5\Omega^1(1) \to 5\mathcal{O} \to 0$, and ϕ the d_2-differential of the spectral sequence.

Step 2 (Syzygies of $\mathcal{F}$ and $\mathcal{G}$) The description of $\mathcal{G}$ as homology of a complex is two clumsy for the following. The description as a sheafified syzygy module is more convenient: Set

$$M := \bigoplus_m H^1(\mathbb{P}^4, I_X(4+m)).$$

Then M is an $S = \bigoplus_m H^0(\mathbb{P}^4, \mathcal{O}(m)) = k[x_0, \ldots, x_4]$-module with Hilbert function $(1, 5, 5, \ldots)$. Assume that M is a cyclic module. Then M has a presentation

$$0 \leftarrow M \leftarrow S(2) \leftarrow 10S$$

and we can recover $\mathcal{G}$ as the sheafified syzygy module of M

$$0 \leftarrow \mathcal{O}(2) \leftarrow 10\mathcal{O} \leftarrow \mathcal{G} \leftarrow 0.$$

Thus a parameter space for vector bundles $\mathcal{G}$ is the Grassmannian

$$\mathbb{G} = Grass(10, H^0(\mathbb{P}^4, \mathcal{O}(2))).$$

A general point $p \in \mathbb{G}$ gives a module M with Hilbert function $(1, 5, 5, 0, \ldots)$ and syzygies

$$0 \leftarrow M \leftarrow S(2) \leftarrow 10S \leftarrow 15S(-1) \oplus 5S(-2) \leftarrow 26S(-3) \leftarrow 20S(-4) \leftarrow 5S(-5) \leftarrow 0.$$

The corresponding vector bundle $\mathcal{G}$ has a presentation

$$0 \leftarrow \mathcal{G} \leftarrow 15\mathcal{O}(-1) \oplus 5\mathcal{O}(-2) \leftarrow 26\mathcal{O}(-3).$$

However, since $\mathcal{F} = 2\Omega^3(3)$ has the presentation

$$0 \leftarrow 2\Omega^3(3) \leftarrow 10\mathcal{O}(-1) \leftarrow 2\mathcal{O}(-2),$$

and $Hom(\mathcal{O}(-2), \mathcal{O}(-3)) = 0$ the corresponding group $Hom(\mathcal{F}, \mathcal{G}) = 0$. What we need are points $p \in \mathbb{G}$ such that the corresponding modules M have additional 3rd syzygies. Let

$$\mathbf{M} = \{p \in \mathbb{G} \mid \dim_k Tor_3^S(M, k)_2 \geq 2\}.$$

For a general point $p \in \mathbf{M}$ we expect a module M with syzygies

$$\begin{aligned} 0 \leftarrow M \leftarrow S(2) \leftarrow 10S \leftarrow 15S(-1) \oplus 7S(-2) \\ \leftarrow 2S(-2) \oplus 26S(-3) \leftarrow 20S(-4) \leftarrow 5S(-5) \leftarrow 0 \end{aligned}$$

Let $\mathbf{M}^\circ = \{p \in \mathbb{G} \mid M_1 = 0 \quad \text{and} \quad \dim_k Tor_3^S(M, k)_2 = 2\} \subset \mathbf{M}$ denote the open subscheme of points $p \in \mathbf{M}$ whose corresponding module has expected syzygies. For $p \in \mathbf{M}^\circ$ there is up to automorphisms of $\mathcal{F} = 2\Omega^3(3)$ at most one inclusion $\mathcal{F} \hookrightarrow \mathcal{G}$. Thus we obtain:

Proposition 1. *An open subset of $\mathbf{M}^\circ$ is isomorphic to an open part of the Hilbert scheme of smooth surfaces in $\mathbb{P}^4$ of degree $d = 11$, sectional genus $\pi = 10$ and $p_g = q = 0$. $\mathbf{M}^\circ \subset \mathbb{G}$ has in every point codimension at most* 14.

Proof. The condition that $cokern(\mathcal{F} \to \mathcal{G}) \cong I_X(4)$ is the twisted ideal sheaf of a smooth surface is an open condition for points in $\mathbf{M}^\circ$. Conversely the condition that the H^1-module M of a smooth surface corresponds to a point $p \in \mathbf{M}^\circ$ is open in the Hilbert scheme. This proves the statement for the underlying reduced structures of $\mathbf{M}^\circ$ and the Hilbert scheme.

For the second statement suppose we have a local subscheme $SpecR \subset \mathbb{G}$, R a local ring, whose closed point lies in $\mathbf{M}^\circ$. Then since the Hilbert function of M cannot become smaller, the corresponding family M_R of modules over $SpecR$ is automatically flat and its syzygies lift to a complex

$$0 \leftarrow M_R \leftarrow S_R(2) \leftarrow 10S_R \leftarrow 15S_R(-1) \oplus 7S_R(-2)$$
$$\leftarrow 2S_R(-2) \oplus 26S_R(-3) \leftarrow 20S_R(-4) \leftarrow 5S_R(-5) \leftarrow 0$$

of $S_R = R[x_0, \dots, x_4]$-modules. The scheme structure of $\mathbf{M}^\circ$ is defined by the condition $SpecR \subset \mathbf{M}^\circ$ iff the map

$$7S_R(-2) \leftarrow 2S_R(-2)$$

is identically zero. Since this map is given by 14 elements of R we obtain that $\mathbf{M}^\circ \subset \mathbb{G}$ has codimension at most 14 everywhere. Moreover, with this structure an open part of $\mathbf{M}^\circ$ is scheme theoretically isomorphic to an open part of the Hilbert scheme.

Step 3 (Search for good points in $\mathbf{M}(\mathbb{F}_q)$) For a field with q elements we expect to find points $p \in \mathbf{M}(\mathbb{F}_q) \subset \mathbb{G}(\mathbb{F}_q)$ at a rate of $(1 : q^{14})$ by proposition 1. Using the computer algebra system MACAULAY [BS], the speed to compute 2nd linear syzygies of an ideal $m = (q_1, \dots, q_{10}) \subset \mathbb{F}_q[x_0, \dots, x_4]$ generated by 10 quadrics is about $2\frac{examples}{seconds}$ on a moderate work station. Thus if we take the field $\mathbb{F}_3$ with 3 elements we expect to find a point in $\mathbf{M}(\mathbb{F}_3)$ within 2 months. Fortunately we can improve our description of $\mathbf{M}$ to obtain smaller codimension. Consider

$$\Omega^2(4) \hookrightarrow 10\mathcal{O}(2).$$

A section $s \in H^0(\mathbb{P}^4, \Omega^2(4))$ determines 10 quadrics with at least one 2nd linear syzygy. This defines a rational map

$$\rho\colon \mathbb{P}^{44} \cong \mathbb{P}(H^0(\mathbb{P}^4, \Omega^2(4))^*) \longrightarrow \mathbb{G},$$

whose image contains the good part of $\mathbf{M}$. Since $\rho^{-1}(\mathbf{M}^\circ) \to \mathbf{M}^\circ$ is a $\mathbb{P}^1$-bundle, $\rho^{-1}(\mathbf{M}^\circ) \subset \mathbb{P}^{44}$ should have codimension $7 = 44 - 50 + 14 - 1$. Using this we can hope to find points in $\mathbf{M}(\mathbb{F}_3)$ within hours by taking points at random in $\mathbb{P}^{44}(\mathbb{F}_3)$. The following points were found by this method:

$$\begin{aligned} p_a = (\ & x_1^2 + x_3x_0, \quad x_1x_2 + (x_1 + x_2 - x_4)x_0, \quad x_1x_3 + (-x_1 - x_4 + x_0)x_0, \\ & x_1x_4 + (-x_1 + x_2)x_0, \quad x_2^2 + (x_1 + x_3 - x_0)x_0, \quad x_2x_3 + (x_1 + x_2 - x_0)x_0, \\ & x_2x_4 + (-x_1 - x_2 - x_3 - x_4 - x_0)x_0, \quad x_3^2 + (x_2 - x_3 + x_4 - x_0)x_0, \\ & x_3x_4 + (-x_1 - x_2 - x_3 - x_4 + x_0)x_0, \quad x_4^2 + (-x_2 + x_3 + x_4)x_0 \ ), \end{aligned}$$

$$\begin{aligned} p_b = (\ & x_1^2 + (x_1 - x_3 + x_4 - x_0)x_0, \quad x_1x_2 + (x_1 - x_2 - x_3 - x_4 - x_0)x_0, \\ & x_1x_3 + (-x_3 + x_0)x_0, \quad x_1x_4 + (x_3 - x_0)x_0, \quad x_2^2 + (x_2 + x_3 - x_0)x_0, \\ & x_2x_3 + (x_1 + x_2 - x_3 + x_4 + x_0)x_0, \quad x_2x_4 + (x_1 + x_2 + x_3 + x_4)x_0, \\ & x_3^2 + (x_2 - x_4 + x_0)x_0, \quad x_3x_4 + (-x_2 + x_4)x_0, \quad x_4^2 + (x_1 - x_2 - x_3 - x_4 - x_0)x_0 \ ), \end{aligned}$$

$$\begin{aligned} p_c = (\ & x_1^2 + (x_2 + x_3 + x_4 - x_0)x_0, \quad x_1x_2 + (x_1 + x_2 + x_0)x_0, \\ & x_1x_3 + (-x_1 - x_2 + x_3 + x_4 + x_0)x_0, \quad x_1x_4 - x_2x_0, \quad x_2^2 + (x_1 - x_2 - x_3 + x_4)x_0, \\ & x_2x_3 + (-x_1 + x_2)x_0, \quad x_2x_4 + (-x_1 - x_2 + x_0)x_0, \\ & x_3^2 + (-x_1 + x_2 + x_3 - x_4 + x_0)x_0, \quad x_3x_4 + (-x_2 - x_3 + x_0)x_0, \\ & x_4^2 + (x_1 + x_2 + x_0)x_0 \ ), \end{aligned}$$

$$p_d = (\ x_1^2 + (-x_1 + x_2 + x_3 - x_4)x_0,\ x_1x_2 + (x_1 - x_3 + x_4)x_0,\ x_1x_3 + (-x_1 + x_4 + x_0)x_0,$$
$$x_1x_4 + (-x_1 + x_3 + x_4 - x_0)x_0,\quad x_2^2 + (x_1 - x_2 - x_4 - x_0)x_0,$$
$$x_2x_3 + (x_1 - x_2 + x_3 + x_4 - x_0)x_0,\quad x_2x_4 + (x_1 + x_2 - x_3 - x_4 - x_0)x_0,$$
$$x_3^2 + (x_3 + x_4 - x_0)x_0,\quad x_3x_4 + (-x_3 - x_4 + x_0)x_0,\quad x_4^2 + (x_1 + x_3 - x_4 - x_0)x_0\ \ ).$$

A computation in MACAULAY shows, that each of these points gives a smooth surface with desired invariants defined over $\mathbb{F}_3$. These surfaces differ by their configuration of 6-secant lines. p_a, p_b, p_c define a surface with 5, 6 or 7 6-secants respectively. For the surface given by p_d there is a plane, which intersects the surface in a plane quartic plus 3 points, counting for 3 6-secant lines, and further 4 6-secants lines.

Note, that only a small fraction of point $p \in \mathbf{M}^\circ(\mathbb{F}_3)$ lead to smooth surfaces. Out of ca. 350,000 points in $\mathbb{P}^{44}(\mathbb{F}_3)$ tested, ca. 970 would lead to an injection $2\Omega^3(3) \hookrightarrow \mathcal{G}$ out of which only 107 gave surfaces with at most hypersurface singularities, and only 13 points actually gave smooth surfaces, 4 with 5 6-secants, 5 with 6 6-secants, 3 with 7 6-secants and 1 of type (d). This search took about 60 hours of cpu.

Step 4 (Lift to characteristic zero) Given a point $p \in \mathbf{M}^\circ(\mathbb{F}_q)$ we try to find a number field L, a prime $\wp \subset L$ with residue field $\mathcal{O}_{L,\wp}/\wp\mathcal{O}_{L,\wp} \supset \mathbb{F}_q$ such that p lifts to an $\mathcal{O}_{L,\wp}$-valued point of $\mathbf{M}^\circ$. The following simple criterion applies:

Proposition 2. *If $p \in \mathbf{M}^\circ(\mathbb{F}_q)$ is a point, in which $\mathbf{M}(\mathbb{F}_q) \subset \mathbb{G}(\mathbb{F}_q)$ has expected codimension* 14, *then there exists a number field L, a prime $\wp \subset L$ with residue field $\mathcal{O}_{L,\wp}/\wp\mathcal{O}_{L,\wp} \supset \mathbb{F}_q$ such that p lifts to an $\mathcal{O}_{L,\wp}$-valued point of* $\mathbf{M}$. *Moreover, if the surface $X_q/\mathbb{F}_q$ corresponding to p is smooth, then surface X/L corresponding to the generic point of $Spec\, L \subset Spec\, \mathcal{O}_{L,\wp} \to \mathbf{M}$ is also smooth.*

Proof. For notational convenience we assume that $\mathbb{F}_q$ is a prime field. If this is not the case, $\mathbb{Z} \subset \mathbb{Q}$ has to be replaced by the ring of integers in a number field which has $\mathbb{F}_q$ as a residue class field.

Since $\mathbf{M}(\mathbb{F}_q)$ has pure codimension 14 in p, there are hypersurfaces $H_1, \ldots, H_{14} \subset \mathbb{G}$ such that p is an isolated point of $\mathbf{M}(\mathbb{F}_q) \cap H_1 \cap \ldots \cap H_{14}$. We may assume that $H_1, \ldots, H_{14}$ are defined over $\mathbb{Z}$. Since $\mathbf{M}$ (the model of the Hilbert scheme) is also defined over $\mathbb{Z}$, we can consider $\mathbf{M} \cap H_1 \cap \ldots \cap H_{14}$ over $Spec\mathbb{Z}$ and an irreducible component $\mathcal{L}$, which contains p. $\mathcal{L}$ is 1-dimensional. The residue class field of the generic point of $\mathcal{L}$ is a number field L finitely generated over $\mathbb{Q}$, because $\mathbf{M}/\mathbb{Z}$ is projective. Let $\mathcal{O}_L$ denote the ring of integers of L, and let $\wp \subset \mathcal{O}_L$ be a prime ideal which lies over $p \in \mathcal{L}$. Then $Spec\, \mathcal{O}_{L,\wp} \to \mathcal{L} \subset \mathbf{M}$ is an $\mathcal{O}_{L,\wp}$-valued point which lifts p. The residue class field $\mathcal{O}_{L,\wp}/\wp\mathcal{O}_{L,\wp}$ is a finite extension of $\mathbb{F}_q$.

Performing the construction of the surfaces over $\mathcal{O}_{L,\wp}$ gives a flat family

$$\begin{array}{ccccc} X_\wp & \hookrightarrow & \mathcal{X} & \supset & X_L \\ \downarrow & & \downarrow & & \downarrow \\ \wp & \in & Spec\, \mathcal{O}_{L,\wp} & \supset & Spec\, L. \end{array}$$

Since smoothness is an open property, the general fiber X_L is smooth with the special fiber $X_\wp$.

For the points $p_a, \ldots, p_d \in \mathbf{M}(\mathbb{F}_3)$ above one can compute the tangent spaces $T_{\mathbf{M}(\mathbb{F}_3),p_i} \subset T_{\mathbb{G}(\mathbb{F}_3),p_i}$. The result is, that all have codimension 14. By proposition 2 we deduce the existence of four smooth surfaces defined over some number fields $L \subset \mathbb{C}$. Moreover the corresponding points in $\mathbf{M}(\mathbb{C})$ are smooth. By proposition 1 the Hilbert scheme is smooth of dimension $36 = 50 - 14$ in these points.

Step 5 (Adjunction process) We recall a result of Sommese and Van de Ven [SV] for surface X over $\mathbb{C}$. Let H a very ample divisor on X and K_X its canonical divisor. Then unless X is ruled by lines, ie. ruled by curves L with $L.H = 1$, the adjoint linear system $\mid H + K_X \mid$ is base point free. Moreover the morphism

$$\varphi = \varphi_{H+K} : X \to \mathbb{P}^r$$

is birational onto its smooth image unless X is a conic bundle, or X is a Del Pezzo surfaces and $H = -K_X$, or (X, H) is one of a few exceptions. φ blows down all exceptional curves $E \subset X$ with $E^2 = E.K = -1$ and $E.H = 1$.

Let X_1 be the image of φ, and $H_1 = \varphi_*(H + K_X)$. Repeating the process with (X_1, H_1) we obtain a sequence

$$X \to X_1 \to X_2 \to \ldots \to X_k.$$

The curves blown down by the kth map correspond to exceptional curves E with $E^2 = E.K = -1$ and $E.H = k$. For a surface of Kodaira dimension ≥ 0 one obtains finally the minimal model. If the Kodaira dimension $= -\infty$ one ends up with a ruled surfaces, a conic bundle, a Del Pezzo surface, $\mathbb{P}^2$, or with one of the few exceptions of Sommese and Van de Ven. Notice that the adjoint surfaces are defined over the field of definition of the surface.

For surface X over a finite prime field one can compute the adjoint surfaces explicitly with MACAULAY. For the surface defined by $p_a, \ldots, p_d$ over $\mathbb{F}_3$ the computation is small enough to be carried through. Let $X_\pi^d \subset \mathbb{P}^r$ indicate a surface of degree d, sectional genus π in $\mathbb{P}^r$, and let $\xrightarrow{k}$ denote the blow down of k exceptional curves simultaneously. The computation gives:

$$(a)\quad \begin{matrix} \mathbb{P}^4 & & \mathbb{P}^9 & & \mathbb{P}^{10} \\ \cup & & \cup & & \cup \\ X_{10}^{11} & \xrightarrow{5} & X_{11}^{19} & \xrightarrow{1} & X_{11}^{20} \end{matrix}$$

$$(b)\quad \begin{matrix} \mathbb{P}^4 & & \mathbb{P}^9 & & \mathbb{P}^{10} & & \mathbb{P}^9 & & \mathbb{P}^7 & & \mathbb{P}^4 \\ \cup & & \cup & & \cup & & \cup & & \cup & & \cup \\ X_{10}^{11} & \xrightarrow{4} & X_{11}^{19} & \xrightarrow{1} & X_{10}^{19} & \xrightarrow{0} & X_8^{16} & \xrightarrow{0} & X_5^{11} & \xrightarrow{5} & X_1^4 \end{matrix}$$

$$(c)\quad \begin{matrix} \mathbb{P}^4 & & \mathbb{P}^9 & & \mathbb{P}^{10} & & \mathbb{P}^8 & & \mathbb{P}^4 \\ \cup & & \cup & & \cup & & \cup & & \cup \\ X_{10}^{11} & \xrightarrow{3} & X_{11}^{19} & \xrightarrow{1} & X_9^{18} & \xrightarrow{2} & X_5^{12} & \xrightarrow{4} & X_1^4 \end{matrix}$$

$$(d)\quad \begin{matrix} \mathbb{P}^4 & & \mathbb{P}^9 & & \mathbb{P}^{10} & & \mathbb{P}^8 & & \mathbb{P}^5 \\ \cup & & \cup & & \cup & & \cup & & \cup \\ X_{10}^{11} & \xrightarrow{3} & X_{11}^{19} & \xrightarrow{2} & X_9^{18} & \xrightarrow{0} & X_6^{13} & \xrightarrow{3} & X_2^6 \end{matrix}$$

By the double point formula for surfaces in $\mathbb{P}^4$,

$$d^2 - 5d - 10(\pi - 1) - 2K^2 + 12\chi = 0$$

one has $K^2 = -6$ on the original surfaces in $\mathbb{P}^4$.

In case (a) the 2nd adjoint surface satisfies $H_2.K_2 = K_2^2 = 0$. By the Hodge index theorem this implies that H_2 and K_2 are numerical dependent. Hence K_2 is numerically trivial. Since $p_g = q = 0$ this implies, that we have an Enriques surface blown up in 6 points in case (a).

Case (b) and (c) both end with the complete intersection of two quadrics in $\mathbb{P}^4$, ie. a Del Pezzo surface. In case (d) the final surface is a conic bundle: the intersection of $\mathbb{P}^1 \times \mathbb{P}^2 \subset \mathbb{P}^5$ with a quadric hypersurface.

In all four cases the computation shows, that all higher adjoint surfaces are arithmetically Cohen-Macaulay, and that their ideals are generated by quadrics.

We argue now that the adjunction process of the surfaces over the number fields L of step 4 has the same numerical behavior: Let $f\colon \mathcal{X} \to Spec\,\mathcal{O}_{L,\wp}$ be one of these families, and let $K = K_{\mathcal{X}/\mathcal{O}_{L,\wp}}$ denote the relative canonical divisor. Then

$$H^i(\mathcal{X}, \mathcal{O}(m(H+K))) = 0 \quad \text{for} \quad i > 0 \quad \text{and} \quad m \geq 0.$$

Indeed $\Gamma(Spec\,\mathcal{O}_{L,\wp}, R^i f_*\mathcal{O}(m(H+K))) = H^i(\mathcal{X}, \mathcal{O}(m(H+K)))$ and $R^i f_*\mathcal{O}(m(H+K)) \otimes \mathcal{O}_{L,\wp}/\wp \cong H^i(X_\wp, \mathcal{O}(m(H+K))) = 0$ for $i \geq 1$ follows from the base change theorem, [H] III Thm 12.11, because the image of $X_\wp$ under the adjunction map is arithmetically Cohen-Macaulay. Moreover $H^0(\mathcal{X}, \mathcal{O}(m(H+K)))$ is a free $\mathcal{O}_{L,\wp}$-module and $Sym_m H^0(\mathcal{X}, \mathcal{O}(H+K)) \to H^0(\mathcal{X}, \mathcal{O}(m(H+K)))$ surjective for $m \geq 0$. Hence

$$\mathcal{X}_1 = Proj_{\mathcal{O}_{L,\wp}} \oplus_m H^0(\mathcal{X}, \mathcal{O}(m(H+K))) \subset \mathbb{P}(H^0(\mathcal{X}, \mathcal{O}(H+K))),$$

the family of adjoint surfaces, is again flat over $\mathcal{O}_{L,\wp}$. In particular the special and the generic fiber have the same Hilbert polynomial.

The same argument applies to $f_1\colon \mathcal{X}_1 \to Spec\,\mathcal{O}_{L,\wp}$. Hence all the higher adjoint surfaces form flat families. Concerning the exceptional divisors we note

$$K_i^2 = d_i + d_{i+1} - 4\pi_i + 4.$$

Hence the number of exceptional curves of the ith adjunction map

$$K_i^2 - K_{i-1}^2 = d_{i+1} - d_{i-1} + 4(\pi_{i-1} - \pi_i)$$

coincide in generic and special fibers.

Finally over the complex numbers a Del Pezzo surface of degree 5 is always isomorphic to $\mathbb{P}^2_{\mathbb{C}}$ blown up in five points embedded by the linear system

$$3L - \sum_{i=1}^{5} E_i.$$

The intersection of a quadric with $\mathbb{P}^1_{\mathbb{C}} \times \mathbb{P}^2_{\mathbb{C}} \subset \mathbb{P}^5_{\mathbb{C}}$ is isomorphic to $\mathbb{P}^2_{\mathbb{C}}$ blown up in 7 points embedded by the linear system

$$4L - 2E_1 - \sum_{i=2}^{7} E_i,$$

if we choose the minimal model over $\mathbb{C}$ appropriate.

Hence the surfaces (a),(b),(c) and (d) have the type as claimed. Note however, that the plane description of these surfaces is in general not defined over L the field of definition of these surfaces. This completes the verification of the main result.

Remarks 1. A family of Enriques surfaces as in the main result was already constructed in [DES]. However those Enriques surfaces have an H^1-module M with $\dim_{\mathbb{C}} Tor_3^S(M, \mathbb{C})_2 = 5$. Presumably they lie in the boundary of the family above.

2. I found the fact, that very few points in $\mathbf{M}(\mathbb{F}_3)$ lead to smooth surfaces, while many more would give singular surfaces, somewhat surprising. What is the explanation for this phenomenon?

3. Complete documentation of the computer computation and MACAULAY scripts can be obtained upon request.

REFERENCES

[BS] D. Bayer and M. Stillman: *MACAULAY: A system for computations in algebraic geometry and commutative algebra.* Source and object code available for Unix and Macintosh computers. Contact the authors, or download from **Zariski.havard.edu** via anonymous ftp.

[Bei] A. Beilinso: Coherent sheaves on $\mathbb{P}^N$ and problems in linear algebra, Functional Anal. Appl. **12** (1978), 214–216.

[BF] R. Braun and G. Fløystad: A bound for the degree of smooth surfaces in $\mathbb{P}_4$ not of general type, Compositio **93** (1994), 211–229.

[C] M. Coo: An improved bound for the degree of smooth surfaces in $\mathbb{P}_4$ not of general type, preprint (1994),1–4.

[DES] W. Decker, L. Ein and F.-O. Schreyer: Construction of surfaces in $\mathbb{P}_4$, J. Alg. Geom. **2** (1993), 185–237.

[EP] G. Ellingsrud and C. Peskine: Sur les surfaces lisse de $\mathbb{P}_4$, Invent. Math. **95** (1989), 1–12.

[H] R. Hartshorne: *Algebraic Geometry*, Springer Berlin, Heidelberg, New York, Tokio 1977.

[P] S. Popsecu: On smooth surfaces of degree ≥ 11 in $\mathbb{P}^4$, Dissertation, Saarbrücken, (1992).

[SV] A.J. Sommese and A. Van de Ven: On the adjunction mapping, Math. Ann. **278** (1987), 593–603.

HOMOTOPY OVER THE COMPLEX NUMBERS AND GENERALIZED DE RHAM COHOMOLOGY

Carlos Simpson

Laboratoire de Topologie et Géométrie, URA 1408, Université Paul Sabatier, 31062 Toulouse cedex, France
e-mail: simpson@cicx.cict.fr

1. Introduction

In "Pursuing stacks" [Gr2] A. Grothendieck sketched a vision of a theory of homotopy relative to $Spec(\mathbf{Z})$. The objects were to be "homotopy types" in the category of schemes. We take up this theme in the more restrained setting of schemes over $Spec(\mathbf{C})$ (or over any field k of characteristic zero). The first task is to elaborate a notion of homotopy type in the category $\mathcal{X}$ of schemes over k. This has already been done by Illusie and Jardine, as is mentioned in [Gr2]. The basic idea is to replace the notion of homotopy type in $\mathcal{X}$ by the functor from $\mathcal{X}$ to homotopy types that it would represent: concretely, a presheaf of topological spaces over $\mathcal{X}$. The appropriate sheaf condition takes into account the homotopy-theoretic nature of the problem, and is what Jardine called "fibrant" (because of the way it fits into the closed model category framework). We use the terminology "homotopy sheaf" for a fibrant presheaf of spaces. The right notion of equivalence in the category of presheaves of spaces is Illusie's weak equivalence, that is the condition of giving an isomorphism on the homotopy group sheaves (the sheaves associated to the homotopy presheaves which are defined in the obvious way). Finally, note that to have a good notion of morphism between two presheaves of spaces, the source needs to be constructed in a "free" way—this is the cofibrant condition we explain below.

Not every presheaf of spaces should qualify as a genuine "homotopy type" over $Spec(k)$. We would like to impose some algebraicity or finite-type conditions, which should be compatible with truncations of homotopy groups and with fiber products. In particular, the conditions must be expressed just as conditions on the homotopy group sheaves (which are sheaves over objects $S \in \mathcal{X}$, that is to say sheaves over $\mathcal{X}/S$). Looking at the homotopy exact sequence for a fiber product, these conditions should be preserved by kernels, cokernels and extensions. We present below several conditions like this. The first, a condition for a sheaf of abelian groups (thus a condition we would want to use for the π_i with $i \geq 2$), is the notion of *vector sheaf*. This is what was called *U-coherent sheaf* by Hirschowitz [Hi]. Next we need some conditions for sheaves of nonabelian groups. We present a sequence of such conditions numbered P1, P2, P3 (not used very often), P4, P5 (the main condition), P6, and we add the numbering P7 for the vector sheaf condition. Alternate terminology for P5 is *presentable*, and for P6 is *affine presentable*. The condition P2 says that G has a presentation with surjections $X \rightarrow G$ and $R \rightarrow X \times_G X$ where R and X are schemes of finite type over the base S. The conditions P3, P4 and P5 concern certain lifting properties for the morphisms in the presentation, designed essentially to guarantee that the multiplication

$G \times_S G \to G$ lifts to a map $\mu : X \times_S X \to X$ with $\mu(x,e) = \mu(e,x) = x$ for $e : S \to X$ a lift of the identity.

We arrive at our notion of good fibrant presheaf of spaces T: we say that T is *very presentable* if $\pi_0(T)$ is a P3 sheaf on $\mathcal{X}$, and if the sheaves $\pi_i(T|_{\mathcal{X}/S}, t)$ are P6 (affine presentable) for $i = 1$ and P7 (vector sheaves) for $i \geq 2$. The class of very presentable fibrant presheaves is closed under fiber product (and obviously closed under truncation since it concerns only the homotopy groups).

This notion of homotopy type over $Spec(k)$ is an attempt to do what Grothendieck was suggesting in the long discussion about "special unipotent schemes" in the middle of [Gr2].

An important remark is that in the case of base $S = Spec(k)$, then a group sheaf G is presentable (P5) if and only if it is represented by a group scheme of finite type over $Spec(k)$; it is affine presentable (P6) if and only if it is represented by an affine group scheme of finite type; and it is a vector scheme (P7) if and only if it is a vector space. Consequently, supposing $k = \overline{k}$, the *connected* very presentable presheaves of spaces (i.e. those having $\pi_0(T) = Spec(k)$) are just those with π_1 represented by an affine group scheme of finite type, and with π_i vector spaces for $i \geq 2$. If k is not algebraically closed then it is possible that T has no basepoint over $Spec(k)$, and we get a Galois-twisted version of the case described in the previous phrase.

The utilisation we have in mind for our very presentable homotopy types over $\mathcal{X}$ is as the forms for a sort of "shape theory". We measure the topology of an arbitrary presheaf of spaces W by the functor

$$T \mapsto \underline{Hom}(W, T).$$

We have two basic cases in mind. The first is when W is a constant presheaf associated to a finite CW complex. The fact that the category of presentable (or very presentable) presheaves is closed under fiber product, implies in this case that if T is presentable (resp. very presentable) then $\underline{Hom}(W, T)$ is presentable (resp. very presentable). The "shape" of W is then a functor from presentable spaces to presentable spaces. One can see (although we don't state this here) that this functor is representable if W is simply connected. The homotopy group sheaves of the presentable space representing this functor are the sheaves represented by the tensor products $\pi_i(W) \otimes_{\mathbf{Z}} \mathbf{C}$. This gives an interpretation for the notion of tensoring a space with $\mathbf{C}$.

The second case is related to algebraic de Rham cohomology of smooth projective varieties. Suppose X is a smooth projective variety. Put

$$X_{DR}(Y) := Hom(Y^{\mathrm{red}}, X).$$

This is a sheaf on $\mathcal{X}$. We have

$$\pi_0 \underline{Hom}(X_{DR}, K(\mathcal{O}, n)) = H^n_{DR}(X, \mathbf{C})$$

where $\mathcal{O}$ denotes the structure sheaf of rings on $\mathcal{X}$ and the right hand side does Grothendieck's algebraic de Rham cohomology of X. In view of this, we can say that the *de Rham shape of* X is the functor of fibrant very presentable presheaves of spaces

$$T \mapsto \underline{Hom}(X_{DR}, T).$$

It is difficult to show that $\underline{Hom}(X_{DR}, T)$ is very presentable. Note that we no longer speak of presentable T but only very presentable ones, because algebraic de Rham cohomology behaves badly if the coefficients are not vector spaces (degree ≥ 2) or affine algebraic groups (degree 1).

Suppose $k = \mathbf{C}$. Let X_B be the constant presheaf whose values are the topological space underlying X^{an}. The functor $T \mapsto \underline{Hom}(X_B, T)$ is the *Betti shape* of X (this falls into the first case discussed above). We have a notion of associated sheaf on the analytic site $\mathcal{X}^{an}$ of complex analytic spaces with the etale topology, and we sketch a proof that

$$\underline{Hom}(X_B, T)^{an} \cong \underline{Hom}(X_{DR}, T)^{an}.$$

This generalizes the already known isomorphisms between Betti and de Rham cohomology. In the case of connected T, this follows relatively easily from the Postnikov tower for T. The main problem in the proof is the case where T is not connected, for example with π_0 a nontrivial scheme of finite type.

This paper is an announcement of results with some sketches of proofs. The full details will appear elsewhere.

I would like to thank J. Tapia, L. Breen, and H. Miller for pointing me towards a better understanding of the homotopy theory of presheaves of spaces; and C. Walter and A. Hirschowitz for discussions about the notion of U-coherent sheaf.

2. Presheaves of topological spaces and homotopy-sheaves

Let k be a field of characteristic zero, and let $\mathcal{X}$ be the site of noetherian schemes over $Spec(k)$ with the etale topology. Recall that for $X \in \mathcal{X}$, a *sieve over* X is a full subcategory $\mathcal{B} \in \mathcal{X}/X$ with the property that if $(Y, f) \in \mathcal{B}$ and $(Z, g) \to (Y, f)$ is a morphism in $\mathcal{X}/X$ then $(Z, g) \in \mathcal{B}$, and such that $\mathcal{B}$ contains a surjective family of etale morphisms $\{U_\alpha \to X\}$.

We have a notion of presheaf of sets over $\mathcal{X}$, or more generally of objects in any category $\mathcal{D}$, over $\mathcal{X}$ (this is just a contravariant functor $\mathcal{X} \to \mathcal{D}$).

If F is a presheaf of sets over $\mathcal{X}$, then $\Gamma(\mathcal{X}, F)$ is the same as $Hom(*_{\mathcal{X}}, F)$, where $*_{\mathcal{X}}$ denotes the presheaf associating to each object of $\mathcal{X}$ the set containing one element. The final object $Spec(k) \in \mathcal{X}$ gives a natural isomorphism $\Gamma(\mathcal{X}, F) \cong F(Spec(k))$. This remark is also relevant to any category of the form $\mathcal{X}/X$, where the final object is X. We will sometimes confuse $F(X)$ and $\Gamma(\mathcal{X}/X, F|_{\mathcal{X}/X})$.

A presheaf F of sets is a *sheaf* if, for every $X \in \mathcal{X}$ and every sieve $\mathcal{B}$ over X, the restriction morphism

$$F(X) \cong \Gamma(\mathcal{X}/X, F|_{\mathcal{X}/X}) \to \Gamma(\mathcal{B}, F|_{\mathcal{B}})$$

is an isomorphism.

Presheaves of spaces

A *presheaf of topological spaces over* $\mathcal{X}$ is (following the above definitions) a contravariant functor $T : \mathcal{X} \to Top$ to the category of topological spaces. Note that a presheaf of sets can be interpreted as a presheaf of topological spaces by giving the sets the discrete topology.

Suppose R and T are presheaves of topological spaces over $\mathcal{X}$. The set of morphisms $Hom(R, T)$ has a natural topology (combining compact-open and Tychonoff).

We say that a morphism $f : R \to T$ of presheaves of topological spaces is *locally a weak equivalence* if for each $X \in \mathcal{X}$, the morphism $R(X) \to T(X)$ is a weak equivalence, that is it induces isomorphisms $\pi_0(R(X)) \cong \pi_0(T(X))$ and

$$\pi_i(R(X), r) \xrightarrow{\cong} \pi_i(T(X), f(r))$$

for all i and all $r \in R(X)$. We say that a morphism $R \to T$ of presheaves of topological spaces is *locally a fibration* if for each $X \in \mathcal{X}$, the morphism $R(X) \to T(X)$ is a fibration.

Suppose $R \subset T$ is an inclusion of presheaves of topological spaces (this just means a morphism that is injective object-by-object). We say that the pair (T, R) is *obtained by* *freely adding an n-cell over $X \in \mathcal{X}$* if there is a functorial isomorphism

$$T(Y) \cong (R(Y) \cup \coprod_{Y \to X} B^n)/\sim$$

where B^n denotes the n-ball, and $\sim$ denotes the equivalence relation obtained from a gluing morphism $g : S^{n-1} = \partial B^n \to R(X)$: we say that a point $y \in S^{n-1} \subset B^n_{f:Y\to X}$ is equivalent to the restriction $g(y)|_Y$ in $R(Y)$ (and then close up to make this an equivalence relation, which just has the effect of possibly identifying certain points on different S^{n-1}'s). In particular, note that $R(Y)$ is a closed subset of $T(Y)$ with the induced topology, and $(T(Y), R(Y))$ satisfies excision.

The restriction morphisms $T(Y) \to T(Z)$ corresponding to $Z \to Y$ are obtained from restriction morphisms of the presheaf

$$Y \mapsto R(Y) \cup \coprod_{Y\to X} B^n;$$

these restrictions come from the restrictions of R and from composition of $Y \to X$ with $Z \to Y$ to give $Z \to X$—the restriction then is the identity

$$B^n_{Y\to X} \to B^n_{Z\to X}.$$

One thing to note is that if X has any nontrivial automorphisms, then $T(X)$ will be bigger than $R(X)$ union the cell in question (which is the cell corresponding to $1 : X \to X$).

If $R \subset T$ is an inclusion of presheaves of topological spaces we say that the pair (T, R) is *obtained by free cell additions* if there is a sequence of subspaces $R = T_0 \subset \ldots \subset T_\alpha \subset \ldots \subset T_\infty = T$ indexed by an ordinal $\alpha \in A$, such that $0 \in A$ is the first element, $\infty \in A$ is the last element, and such that if α has no predecessor then T_α is the union of the preceding T_β; such that $T_\alpha \subset T$ is a closed subset with the induced topology, for any α; and finally such that $T_{\alpha+1}$ is obtained from T_α by freely adding an n-cell over an object $X \in \mathcal{X}$.

A morphism $R \to T$ of presheaves of topological spaces is a *cofibration* if it is an inclusion such that (T, R) is obtained by free cell additions.

Suppose $P, R \subset T$ are presheaves of topological spaces and $f \in Hom(R, P)$. Let $Hom_f((T, R), P)$ denote the inverse image of f in $Hom(T, P)$. Use the same notation for topological spaces. Suppose T is obtained from R by freely adding an n-cell over X with gluing map $g : S^{n-1} \to R(X)$. Then the natural morphism

$$Hom_f((T, R), P) \to Hom_{f_X \circ g}((B^n, S^{n-1}), P(X))$$

is an isomorphism.

Proposition 2.1. *Suppose $R \to T$ is cofibration. Suppose $h : P \to Q$ is locally a weak equivalence. Suppose $f : R \to P$ is given. Then the morphism*

$$Hom_f((T, R), P) \to Hom_{hf}((T, R), Q)$$

is a weak equivalence. □

We say that two morphisms $P \to Q$ are *homotopic* if they correspond to points in $Hom(P, Q)$ joined by an arc. A *homotopy* is such an arc; this can be interpreted as a morphism $P \times [0, 1] \to Q$ inducing the two given morphisms over the endpoints.

Corollary 2.2. *Suppose $f : R \to T$ is a cofibration and locally a weak equivalence. Then the space of retractions $T \to R$ (that is, the space $Hom_1((T, R), R)$) is weakly contractible. If r is any retraction, then rf and fr are homotopic to the identities.* □

We say that a presheaf of topological spaces T is *cofibrant* if the natural morphism $\emptyset \to T$ is a cofibration, where $\emptyset$ is the presheaf assigning to each $X \in \mathcal{X}$ the empty set.

Corollary 2.3. *Suppose T is cofibrant, and $f : R \to T$ is locally a weak equivalence. Then there exists a homotopy inverse, that is a morphism $g : T \to R$ such that fg and gf are homotopic to the identity.* □

Replacement of morphisms by cofibrations

It is useful to have a canonical replacement of a presheaf of spaces by a cofibrant presheaf of spaces. We define this directly by modifying the classical construction, the realization of the singular simplicial complex.

Let $[0,1]^b$ denote the b-cube. Let Δ^a denote the simplex of dimension a. Suppose T is a presheaf of spaces. For $X \in \mathcal{X}$ put

$$M(T)(x) = \left\{ \coprod_{\alpha,\beta.} \Delta^a \times [0,1]^b \right\} / \sim$$

where the product is taken over composable sequences

$$Y_0 \overset{\beta_0}{\leftarrow} Y_1 \overset{\beta_1}{\leftarrow} Y_2 \overset{\beta_2}{\leftarrow} \dots \overset{\beta_b}{\leftarrow} Y_{b+1} = X$$

and singular simplices $\alpha : \Delta^a \to T(Y_0)$. The equivalence relation $\sim$ identifies points on the boundary of $\Delta^a \times [0,1]^b$, with points on other copies in the disjoint union, as follows. Points on boundary copies of $\Delta^{a-1} \times [0,1]^b$ are identified with points in the appropriate copies of $\Delta^{a-1} \times [0,1]^b$ in the product, according to the usual ordering mechanism. On the other hand, a point (z,t) with $t_i = 0$ corresponds to the point $(z, t_1, \dots, \widehat{t_i}, \dots, t_b)$ in the copy of $\Delta^a \times [0,1]^{b-1}$ which corresponds to $\beta_0, \dots, \beta_{i-1}\beta_i, \dots, \beta_b$. A point (z,t) with $t_i = 1$ corresponds to the point $(z, t_{i+1}, \dots, t_b)$ in the copy of $\Delta^a \times [0,1]^{b-i}$ corresponding to

$$(\beta_0 \cdots \beta_{i-1})^* \circ \alpha; \beta_i, \dots, \beta_b.$$

If $\varphi : X' \to X$ is a morphism then $\varphi^* : M(T)(X) \to M(T)(X')$ is defined by

$$\varphi^*(z,t)_{\alpha,\beta} = (z,t)_{\alpha,\beta'}$$

where

$$\beta' = (\beta_1, \dots, \beta_{b-1}, \beta_b\varphi).$$

It is clear that $M(T)$ is cofibrant. The cells are added in order of increasing dimension. It is easy to see that each stage consists of free cell additions where the cells $\{(z,t)_{\alpha,\beta}\}$ are added over the objects Y_{b-1}. In order to check this, let $M_n(T)(X)$ denote the subspace obtained from cells of dimension $\leq n$. Then

$$M_n(T)(X) = \left\{ M_{n-1}(T)(X) \cup \coprod_{\alpha,\beta_0\dots\beta_{b-1};a+b=n} \; \coprod_{\beta_b : X \to Y_b} \Delta^a \times [0,1]^b \right\} / \sim_n .$$

The equivalence relation $\sim_n$ identifies points of the cell $((\alpha, \beta_0, \dots, \beta_{b-1}); \beta_b)$ with points of $M_{n-1}(T)(X)$ via the composition map

$$\partial(\Delta^a \times [0,1]^b) \to M_{n-1}(T)(Y_b) \overset{\beta_b^*}{\to} M_{n-1}(T)(X).$$

To see this, note that in fact we have a commutative square

$$\begin{array}{ccc} \Delta^a \times [0,1]^b & = & \Delta^a \times [0,1]^b \\ \downarrow & & \downarrow \\ M(T)(Y_b) & \xrightarrow{\beta_b^*} & M(T)(X) \end{array}$$

where the first vertical arrow corresponds to $((\alpha, \beta_0, \ldots, \beta_{b-1}); 1_{Y_b})$, and the second corresponds to $((\alpha, \beta_0, \ldots, \beta_{b-1}); \beta_b)$. The cell gluing maps are the restrictions of these vertical arrows to the boundary of the cell, noting that the boundaries map into $M_{n-1}(T)$. Since the equivalence relation $\sim_n$ can be expressed as it is, the cell additions going from $M_{n-1}(T)$ to $M_n(T)$ conform to the model given for free cell additions.

The map

$$(z,t)_{\alpha,\beta} \mapsto (\beta_0 \cdots \beta_{b+1})^*(\alpha(z))$$

gives a map $M(T) \to T$. We claim that this is a local weak equivalence. Note that it factors through the map

$$(z,t)_{\alpha,\beta} \mapsto (z)_{(\beta_0 \cdots \beta_{b+1})^* \circ \alpha}$$

$$\psi : M(T) \to Sing(T).$$

We show that ψ is a local weak equivalence. Note that for each X we have a morphism $\xi : Sing(T)(X) \to M(T)(X)$ sending the cell α to the cell $(\alpha, 1_X)$. It is clear that $\psi\xi = 1$. We show that $\xi\psi$ is homotopic to the identity. Define this homotopy F_s (for $s \in [0,1]$) as follows:

$$F_s(z, t_1, \ldots, t_b)_{\alpha,\beta_0,\ldots,\beta_b} := (z, t_1, \ldots, t_b, s)_{\alpha,\beta_0,\ldots,\beta_b,1_X}.$$

It preserves the equivalence relations defining the gluing of cells. Note that (from these equivalence relations) F_0 is the identity, whereas

$$F_1(z, t_1, \ldots, t_b)_{\alpha,\beta_0,\ldots,\beta_b} = (z)_{(\beta_0 \cdots \beta_b)^* \circ \alpha} = \xi\psi(z, t_1, \ldots, t_b)_{\alpha,\beta_0,\ldots,\beta_b}.$$

We have now shown that $M(T) \to Sing(T)$ is a local weak equivalence. Since $Sing(T) \to T$ is a local weak equivalence, we obtain that $M(T) \to T$ is a local weak equivalence.

We obtain the following theorem in its absolute case.

Theorem 2.4. *If $f : R \to T$ is a morphism of presheaves of spaces on $\mathcal{X}$ then there is a factorisation*

$$R \to M(R \xrightarrow{f} T) \to T$$

where $R \to M(R \xrightarrow{f} T)$ is cofibrant and $M(R \xrightarrow{f} T) \to T$ is a local weak equivalence. Furthermore, this factorization is functorial in $f : R \to T$.

Proof: The above construction treats the case $R = \emptyset$; the general case is then easy to obtain. □

Illusie weak equivalence, homotopy-sheaves and fibrant morphisms

We now bring into play the Grothendieck topology on $\mathcal{X}$. Suppose T is a presheaf of topological spaces. We define $\pi_0^{\rm pre}(T)$ to be the presheaf of sets $Y \mapsto \pi_0(T(Y))$. We define $\pi_0(T)$ to be the sheaf associated to this presheaf. If $X \in \mathcal{X}$ and $t \in T(X)$ we define $\pi_i^{\rm pre}(T_{/X}, t)$ to be the presheaf of groups on $\mathcal{X}/X$ given by $Y \mapsto \pi_i(T(Y), t|_Y)$. Define $\pi_i(T_{/X}, t)$ to be the sheaf on $\mathcal{X}/X$ associated to this presheaf.

We recall the definition introduced by Illusie [Il]. A morphism $f : R \to T$ is an *Illusie weak equivalence* if the induced morphism $\pi_0(R) \to \pi_0(T)$ is an isomorphism, and if for any object $X \in \mathcal{X}$ and for any $y \in R(X)$ the morphism $\pi_i(R_{/X}, r) \to \pi_i(T_{/X}, f(r))$ is an isomorphism.

The condition of being locally a weak equivalence may be expressed in a similar way but using the homotopy presheaves rather than the associated sheaves. In particular, if f is locally a weak equivalence, then f is an Illusie weak equivalence, but the converse is not necessarily true—one could have morphisms of homotopy presheaves which are not isomorphisms but which induce isomorphisms between the associated sheaves.

We say that a morphism $f : P \to Q$ is *globally fibrant* if, for any cofibration $R \to T$ which is an Illusie weak equivalence, and for any $h : R \to P$, the morphism

$$Hom_h((T,R),P) \to Hom_{fh}((T,R),Q)$$

is a weak equivalence of topological spaces.

We say that f is *fibrant* if it is globally and locally fibrant. We say that P is a *homotopy sheaf* (or *fibrant*) if the morphism $P \to *_{\mathcal{X}}$ is globally fibrant (or equivalently, fibrant, since it is automatically locally fibrant).

We have chosen, for each $X \in \mathcal{X}$ and each sieve $\mathcal{B} \subset \mathcal{X}/X$, a cofibrant presheaf $U_{\mathcal{B}} = M(*_{\mathcal{B}})$ over $\mathcal{B}$ such that the morphism $U_{\mathcal{B}} \to *_{\mathcal{B}}$ is locally a weak equivalence. Note that for any presheaf of topological spaces T there is a natural composition of morphisms

$$T(X) \cong \Gamma(\mathcal{X}/X, T|_{\mathcal{X}/X}) \to \Gamma(\mathcal{B}, T_{\mathcal{B}}) \to Hom_{\mathcal{B}}(U_{\mathcal{B}}, T|_{\mathcal{B}}).$$

We have the following criterion.

Theorem 2.5. *A morphism $f : P \to Q$ of presheaves of topological spaces is globally fibrant if and only if for each X and each sieve $\mathcal{B} \subset \mathcal{X}/X$, the diagram*

$$\begin{array}{ccc} P(X) & \to & Q(X) \\ \downarrow & & \downarrow \\ Hom_{\mathcal{B}}(U_{\mathcal{B}}, P|_{\mathcal{B}}) & \to & Hom_{\mathcal{B}}(U_{\mathcal{B}}, Q|_{\mathcal{B}}) \end{array}$$

is weak-homotopy cartesian (that is, it induces a weak equivalence between $P(X)$ and the homotopy fiber product). □

One can see with this characterization that the notion of homotopy sheaf is a generalization of the notion of sheaf. As an example, if $X = U_1 \cup U_2 \cup U_3$ is an open covering and $\mathcal{B}$ is the sieve of objects over X which factor through one of the U_i, then $Hom_{\mathcal{B}}(U_{\mathcal{B}}, T|_{\mathcal{B}})$ is homotopic to the space of $(a_i, \sigma_{ij}, \xi_{ijk})_{i,j,k\in\{1,2,3\}}$ where $a_i \in T(U_i)$, σ_{ij} is a path from $a_i|_{U_i\cap U_j}$ to $a_j|_{U_i\cap U_j}$ in $T(U_i \cap U_j)$, and where ξ_{ijk} is a triangle in $T(U_i \cap U_j \cap U_k)$ bounded by the restrictions of the paths. The homotopy sheaf condition requires that this space be homotopic to $T(X)$.

Jardine has proved the existence of an associated homotopy sheaf in general [Ja]. His theorem gives the following (in our context which differs in a few technical ways).

Theorem 2.6. *Suppose $R \to T$ is a morphism of presheaves of spaces over $\mathcal{X}$. There is a factorization*

$$R \to R' \to T$$

such that $R \to R'$ is an Illusie weak equivalence, and $R' \to T$ is fibrant. □

In the case of the morphism $R \to *$, one can proceed more concretely as follows. Put

$$sR(X) := \lim_{\to, \mathcal{B}\subset\mathcal{X}/X} Hom_{\mathcal{B}}(U_{\mathcal{B}}, R|_{\mathcal{B}})$$

with the homotopy limit realized by a mapping telescope, for example. If R is n-truncated (see below) then

$$R' = s^{n+2}R$$

is the associated homotopy sheaf. The exponent $n+2$ generalizes the well-known fact that to get the sheaf associated to a presheaf of sets, one has to do the obvious operation two times.

Corollary 2.7. *If R is a presheaf of spaces then there is an Illusie weak equivalence $R \to R'$ with R' fibrant (that is, R' is a homotopy sheaf).*

Proof: Take $T = *$ in the above theorem. □

Corollary 2.8. *If $f : R \to T$ is an Illusie weak equivalence with R and T cofibrant homotopy sheaves, then f is a homotopy equivalence (i.e. has a quasi-inverse).*

Proof: After replacing f by a cofibration, the existence of the quasi-inverse follows from the definition of T being fibrant. □

Theorem 2.9. *The category of presheaves of topological spaces, with the fibrations, cofibrations, and Illusie weak equivalences as defined above, is a closed model category.* □

Remark: Aside from the fact that he works with simplicial objects instead of spaces, Jardine's closed model category is slightly different from ours. He admits any injection of simplicial presheaves as a cofibration, and as a result has an additional condition (between the local Kan condition and the global sheaf condition) for fibrations. This is an example of Quillen's philosophy that several different closed model structures can be talking about the "same" underlying homotopy theory.

Stacks

We say that T is *n-truncated* if the homotopy groups vanish in degrees strictly greater than n. The 1-truncated topological spaces are the "same thing" as groupoids, by the Poincaré groupoid Π_1 and geometric realization $|\cdot|$ constructions. In the same way, the 1-truncated presheaves of spaces over $\mathcal{X}$ are the same as the categories fibered in groupoids over $\mathcal{X}$. We do not give a precise statement but simply note that if T is a presheaf of spaces then the functor

$$\mathcal{X} \to Cat, \quad X \mapsto \Pi_1(T(X))$$

can be integrated to give a split fibered category, and given a fibered category we can first replace it by a split one and then take the geometric realization fiber by fiber, giving a 1-truncated presheaf of spaces. These constructions are quasi-inverses with homotopies or equivalences satisfying appropriate compatibility conditions.

A 1-truncated presheaf of spaces T over $\mathcal{X}$ is globally fibrant if and only if the corresponding fibered category is a stack of groupoids over $\mathcal{X}$ (i.e. satisfies the descent condition given in [D-M]).

If $\mathcal{C} \to \mathcal{X}$ is a stack of groupoids and T a presheaf of spaces associated to $\mathcal{C}$ by the correspondence, then $T(X)$ is a union of connected components corresponding to the isomorphism classes of objects of $\mathcal{C}_X$, and the component of $T(X)$ corresponding to $Y \in \mathcal{C}_X$ is an Eilenberg-MacLane space $K(Aut_{\mathcal{C}_X}(Y), 1)$ (that is, the fundamental group is the group of automorphisms $Aut_{\mathcal{C}_X}(Y)$ of Y and the other homotopy groups are trivial).

The point of view of 1-truncated homotopy sheaves gives a geometric way to think about stacks of groupoids. Furthermore, the notion of n-truncated homotopy sheaf will be "the same" as the notion of n-stack of n-groupoids when an appropriate definition of the latter is found, so for the moment the topological point of view gives the easiest way to make this generalization. Note, however, that topological spaces are irremediably associated to *groupoids* rather than more general *categories*—by using topological spaces we lose any information about the direction of arrows. The eventual notion of n-stack of n-categories will thus be more general than the notion of n-truncated homotopy sheaf.

Local systems

Suppose T is a presheaf of topological spaces. A *local system* over T is a fibrant morphism $L \to T$ which, over each object, is a covering space. A *local system of groups (resp. abelian groups, rings, etc.)* is a local system L together with the appropriate structural morphisms

$L \times_T L \to L$, identity $T \to L$ and inverse $L \to L$ making each covering space $L(X)$ into a local system of groups (resp. abelian groups etc.) over the topological space $T(X)$.

Eilenberg-MacLane presheaves

Suppose T is a presheaf of spaces on $\mathcal{X}$. If $\mathcal{G}$ is a local system of groups over T and if $n \geq 1$ is an integer (with $n = 1$ unless $\mathcal{G}$ is abelian), then we can construct a fibrant morphism $\varphi : K(\mathcal{G}/T, n) \to T$ with a section $0 : T \to K(\mathcal{G}/T, n)$. The fiber of φ over $t \in T(Y)$ is a presheaf of spaces $K(\mathcal{G}/T, n) \times_T Y$ on $\mathcal{X}/Y$ such that $\pi_0(K(\mathcal{G}/T, n) \times_T Y) = *$, such that $\pi_i(K(\mathcal{G}/T, n) \times_T Y, 0) = \{1\}$ for $i \neq n$, and together with an isomorphism $\varphi : \pi_n(K(\mathcal{G}/T, n) \times_T Y, 0) \cong \mathcal{G}|_Y$.

This is constructed by first constructing a presheaf version (with homotopy presheaves as given above) and then taking the associated sheaf relative to T.

A fibration $R \to T$ is of *twisted Eilenberg-MacLane type* if $R \times_T R \to R$ is isomorphic to $K(G/R, n)$ for some local system G and some n (which we call the degree). If $n \geq 2$ then G is a local system of groups on T given by the n-th homology of R/T. The fibration is classified by a section $T \to K(G/T, n+1)$, or more precisely it is the pullback via this section of the standard fibration

$$* = K(G \times_T B/B, n) \to B := K(G/T, n+1).$$

A presheaf of spaces can be decomposed into a Postnikov tower in the same way as in the usual topological case. More precisely, we have functors $\tau_{\leq n}$ truncating the homotopy groups above n (that is, killing the π_i for $i > n$). We get morphisms $\tau_{\leq n} T \to \tau_{\leq n-1} T$ which are fibrations of twisted Eilenberg-MacLane type. The homotopy local systems π_i are all pulled back from local systems on $\tau_{\leq 1} T$ which itself corresponds to a stack.

Morphism spaces

The fact that fiber products exist in $\mathcal{X}$ yields the following nice property.

Lemma 2.10. *Suppose $T_0 \subset T_1$ is cofibrant. Then for any $X \in \mathcal{X}$, $T_0|_{\mathcal{X}/X} \subset T_1|_{\mathcal{X}/X}$ is cofibrant.*

Proof: If T_1 is obtained from T_0 by adding an n-cell over Y then $T_1|_{\mathcal{X}/X}$ is obtained from $T_0|_{\mathcal{X}/X}$ by adding an n-cell over $X \times Y \in \mathcal{X}/X$. The general case follows from the definition of cofibrant morphism. □

If R and T are presheaves of spaces with R cofibrant, we define the presheaf

$$\underline{Hom}(R, T)(X) := Hom(R|_{\mathcal{X}/X}, T|_{\mathcal{X}/X}).$$

This is functorial in X. If T is fibrant then $\underline{Hom}(R, T)$ is fibrant.

Suppose R is cofibrant and $T \to R$ is a fibrant morphism. Then we define $\Gamma(R, T)$ to be the space of sections $R \to T$. We define the sheaf

$$\underline{\Gamma}(R, T)(X) := \Gamma(R|_{\mathcal{X}/X}, T|_{\mathcal{X}/X}).$$

If R is not cofibrant then let $M(R) \to R$ denote the functorial replacement of R by a cofibrant object. Put

$$\underline{Hom}(R, T) := \underline{Hom}(M(R), T)$$

and

$$\underline{\Gamma}(R, T) := \underline{\Gamma}(M(R), T \times_R M(R)).$$

If T is a presheaf of spaces on $\mathcal{X}$ then we obtain a complex of presheaves $C.(T, \mathbf{Z})$ which associates to $X \in \mathcal{X}$ the singular complex of $T(X)$.

Suppose $i \geq 1$, T is a presheaf of spaces and $\mathcal{F}$ is a local system of groups over T (abelian for $i \geq 2$). Then we define the sheaf

$$\underline{H}^i(T, \mathcal{F}) := \pi_0(\underline{\Gamma}(T, K(\mathcal{F}/T, i)).$$

Lemma 2.11. *If T is a presheaf of spaces on $\mathcal{X}$ let $C.(T, \mathbf{Z})$ be the complex of presheaves which associates to $X \in \mathcal{X}$ the singular complex of $T(X)$. If $\mathcal{F}$ is a sheaf of abelian groups on $\mathcal{X}$ then we $\underline{H}^i(T, \mathcal{F})$ is isomorphic to the sheaf $\underline{Ext}^i(C.(T, \mathbf{Z}), \mathcal{F})$. If $i = 1$ and $\mathcal{F}$ is sheaf of groups then $\underline{H}^1(T, \mathcal{F})$ is the sheaf of isomorphism classes of principal $\mathcal{F}$-bundles over T.*

There is also a relative version. Suppose $T \rightarrow R$ is a morphism. Suppose $W \rightarrow R$ is a fibrant morphism. Then we obtain

$$\underline{Hom}(T/R, W/R) \rightarrow R,$$

a fibration whose fiber over $r \in R(X)$ is the $Hom(T_r, W_r)$ in $\mathcal{X}/X$. If $W \rightarrow T \rightarrow R$ then we obtain

$$\underline{\Gamma}(T/R, W) \rightarrow R,$$

a fibration whose fiber over $r \in R(X)$ is $\Gamma(T_r, W_r)$. We have the composition formula

$$\underline{\Gamma}(T, W) = \underline{\Gamma}(R, \underline{\Gamma}(T/R, W)).$$

Suppose $\mathcal{F}$ is a local system of groups on T, abelian for $i \geq 2$. Then we can define

$$\underline{H}^i(T/R, \mathcal{F}) := \pi_{0/R}(\Gamma(T/R, K(\mathcal{F}/T, i)))$$

using a relative version of the set of components $\pi_{0/R}$ with the obvious definition. The composition formula gives rise to a Leray spectral sequence

$$\underline{H}^i(R, \underline{H}^j(T/R, \mathcal{F})) \Rightarrow \underline{H}^{i+j}(T, \mathcal{F}).$$

The "flexible sheaves" point of view

A *flexible presheaf* on $\mathcal{X}$ is the data of a topological space T_X for each $X \in \mathcal{X}$, and for each composable sequence of morphisms

$$\phi = (X_0 \stackrel{\phi_1}{\leftarrow} X_1 \ldots \stackrel{\phi_n}{\leftarrow} X_n)$$

a continuous function

$$(t_1, \ldots, t_{n-1} \in [0,1]^{n-1} \mapsto T(\phi_1, \ldots, \phi_n; t_1, \ldots, t_{n-1}) \in Hom(X_0, X_n)$$

satisfying the following conditions:
–if $t_i = 0$ then $T(\phi; t) = T(\phi_1, \ldots, \phi_i\phi_{i+1}, \ldots, \phi_n; t_1, \ldots, \widehat{t_i}, \ldots, t_{n-1})$;
–if $t_i = 1$ then $T(\phi; t) = T(\phi_{i+1}, \ldots, \phi_n; t_{i+1}, \ldots, t_{n-1}) \circ T(\phi_1, \ldots, \phi_i; t_1, \ldots, t_{i-1})$.

A *flexible morphism* from R to T is a flexible presheaf F on $\mathcal{X} \times \mathcal{I}$—where $\mathcal{I}$ is the category with objects 0 and 1 and one non-identity morphism $0 \leftarrow 1$—such that $F|_{\mathcal{X} \times \{0\}} = R$ and $F|_{\mathcal{X} \times \{1\}} = T$. Composition of flexible morphisms is only defined up to homotopy which is well defined up to homotopy etc. The space $Hom^{\mathrm{flex}}(R, T)$ of flexible morphisms from R to T has a natural topology.

A flexible presheaf T is a *flexible sheaf* if, for all $X \in \mathcal{X}$ and all sieves $\mathcal{B} \subset \mathcal{X}/X$, the morphism

$$T_X \rightarrow Hom^{\mathrm{flex}}(*_{\mathcal{B}}, T|_{\mathcal{B}})$$

is a weak equivalence (compare with the criterion 2.5).

We could work throughout this paper with flexible presheaves and flexible sheaves, instead of working with the fibrant and cofibrant morphisms of presheaves of spaces. These two points of view are equivalent:

0. Presheaves of spaces can be considered as flexible presheaves in an obvious way;

1. Any flexible presheaf is equivalent (in the flexible sense) to a presheaf of spaces;
2. If R is a cofibrant presheaf of spaces and T is a presheaf of spaces then $Hom(R,T) \rightarrow Hom^{\text{flex}}(R,T)$ is a weak equivalence; and
3. If T is a presheaf of spaces then T is a homotopy sheaf (fibrant) if and only if it is a flexible sheaf.

In view of (1) we could, even in the flexible theory, restrict ourselves to objects which are actual presheaves. We then have the choice of two procedures for getting the "right" space of morphisms from R to T—one way is to look at the space of flexible morphisms, and the other way is to replace R by a cofibrant presheaf $M(R)$ and then look at the usual morphisms of presheaves. Both ways lead to the same result.

We have chosen the cofibrant-fibrant version for our exposition, since the statements seem easier to understand in an elementary way.

3. Vector sheaves

We have a sheaf of rings $\mathcal{O}$ on $\mathcal{X}$, defined by $\mathcal{O}(X) := \Gamma(X, \mathcal{O}_X)$. Note that it is represented by the affine line.

Suppose $S \in \mathcal{X}$ is a scheme over k, and let $\mathcal{X}/S$ denote the category of schemes over S. We will define a notion of *vector sheaf* on $\mathcal{X}/S$. This was defined by Hirschowitz under the name "U-coherent sheaf" [Hi]. We feel that the terminology "vector sheaf" is more suggestive. Many of the results below are due to Hirschowitz (in particular, the observation that duality is involutive).

A *vector scheme over* S is a sheaf V of abelian groups on $\mathcal{X}/S$ which is a sheaf of $\mathcal{O}$-modules and such that there exists an etale covering $\{S_\alpha \rightarrow S\}$ such that each $V|_{S_\alpha}$ is representable by a scheme F_α which is affine of finite type over S_α. In this case the $\mathcal{O}$-module structure is unique and the category of vector schemes, with morphisms being those which are compatible with the $\mathcal{O}$-module structure, is a full subcategory of the category of sheaves of abelian groups on $\mathcal{X}$ (cf Lemma 3.2 below).

Suppose V is a vector scheme on $\mathcal{X}/S$, and suppose S is affine. Then there is an exact sequence

$$0 \rightarrow V \rightarrow \mathcal{O}^m \rightarrow \mathcal{O}^n$$

of abelian sheaves on $\mathcal{X}$ [Hi].

We come now to the main definition of this section. A *vector sheaf on* S is a sheaf of abelian groups F on $\mathcal{X}/S$ such that there exists an etale covering $\{S_\alpha \rightarrow S\}$ such that for each α there exists an exact sequence

$$U_\alpha \rightarrow V_\alpha \rightarrow F|_{S_\alpha} \rightarrow 0$$

of sheaves of abelian groups, with U_α and V_α vector schemes over S_α.

Denote by $\mathcal{V}(S)$ the category of vector sheaves over S.

If T is a presheaf of spaces then we obtain the notion of *vector local system over* T. This is a local system L over T such that for any $S \in \mathcal{X}$ and $t \in T(S)$, the pullback $t^*(L)$ is a vector sheaf on S. Since all of our constructions and definitions are compatible with pullback, we concentrate on the case of a base scheme S.

If $X \rightarrow S$ is an element of $\mathcal{X}/S$, we denote by $F|_X$ the restriction of F to the category $\mathcal{X}/X$. It is a vector sheaf over X (this is easy to see from the definitions). If F is a vector sheaf and $f \in F(Y)$ and $a : X \rightarrow Y$ is a morphism, we denote the restriction of f to X by $a^*(f)$ or just $f|_X$.

Remark: If F is a vector sheaf, and S is an affine variety, then the cohomology groups $H^i(S, F)$ vanish for $i > 0$. If

$$F_1 \to F_2 \to F_3$$

is an exact sequence of vector sheaves (that is, an exact sequence in the category of abelian sheaves on $\mathcal{X}$, where the elements are vector sheaves) then for any X over S which is itself an affine scheme, the sequence

$$F_1(X) \to F_2(X) \to F_3(X)$$

is exact.

Remark: One can show that a vector sheaf V over an affine S has a resolution by vector schemes, over S rather than over an etale covering of S [Hi].

Lemma 3.1. *Suppose F is a vector sheaf over S. Then for any $X \in \mathcal{X}/S$ and Y a scheme of finite type over k, we have*

$$F(X \times_{Spec(k)} Y) = F(X) \otimes_k \mathcal{O}(Y).$$

The isomorphism is given by the pullback $F(X) \to F(X \times_k Y)$ and the scalar multiplication by the pullback of functions on Y. □

We use this lemma as follows. Suppose F is a vector sheaf over S. Then for any $X \in \mathcal{X}/S$ we have $F(X \times \mathbf{A}^1) = F(X) \otimes_k k[t]$. The isomorphism is given by the pullback $F(X) \to F(X \times \mathbf{A}^1)$ and the scalar multiplication by the pullback of the coordinate function t on $\mathbf{A}^1$.

Lemma 3.2. *A vector sheaf has a unique structure of $\mathcal{O}$-module, and any morphism of vector sheaves is automatically compatible with the $\mathcal{O}$-module structure. Thus the category of vector sheaves, with morphisms equal to those morphisms of abelian sheaves compatible with the $\mathcal{O}$-module structure, is a full subcategory of the category of sheaves of abelian groups on $\mathcal{X}/S$.*

Proof: A structure of $\mathcal{O}$-module is given by an automorphism

$$m : F(X \times \mathbf{A}^1) \to F(X \times \mathbf{A}^1),$$

scalar multiplication by the coordinate t of the second variable. The distributive property implies that for $n \in \mathbf{Z} \subset \mathbf{A}^1$ the induced morphism $m|_{X \times \{n\}}$ is multiplication by n (i.e. addition n times). Using the above lemma we see that this determines the map m (essentially because of Zariski density of $\mathbf{Z} \subset \mathbf{A}^1$). □

Theorem 3.3. *The subcategory of vector sheaves in the category of sheaves of abelian groups on $\mathcal{X}$ is closed under the operations of kernel, cokernel, and extension. Similarly for the subcategory of vector local systems in the category of local systems of abelian groups over a presheaf of spaces T.* □

The proof for extensions again uses a Zariski density argument. We note that these proofs strongly use the condition that k is of characteristic zero—in characteristic p the image of $\mathbf{Z}$ is no longer Zariski dense in $\mathbf{A}^1$, and indeed the lemma and the theorem are no longer true in positive characteristic. For example, the endomorphism $x \mapsto x^p$ of the additive group scheme, isn't compatible with the $\mathcal{O}$-module structure. See [Br] for what happens with extensions.

Morphism sheaves and duality

If F and G are sheaves of abelian groups, we denote by $\underline{Hom}(F,G)$ the internal Hom, that is the sheaf of homomorphisms of sheaves of abelian groups from F to G. The value $\underline{Hom}(F,G)(X)$ is the space of morphisms of sheaves of abelian groups from $F|_{\mathcal{X}/X}$ to $G|_{\mathcal{X}/X}$ (this is already a sheaf).

If T is a presheaf of spaces and F and G are local systems of abelian groups over T then we obtain a local system $\underline{Hom}(F,G)$ of abelian groups over T. This construction is compatible with pullback.

Suppose F, G are vector sheaves (or vector local systems). We have defined $\underline{Hom}(F,G)$ which is for now a sheaf of abelian groups. Put

$$F^* := \underline{Hom}(F, \mathcal{O}).$$

If $\phi : F \to G$ is a morphism of vector schemes, then we obtain a morphism $\phi^t : G^* \to F^*$, and the construction $\phi \mapsto \phi^t$ preserves composition (reversing the order, of course).

Lemma 3.4. (Hirschowitz [Hi]) *Suppose*

$$0 \to U \to V \to W \to F \to 0$$

is an exact sequence with U, V and W vector schemes. Then taking the dual gives an exact sequence

$$0 \to F^* \to W^* \to V^* \to U^* \to 0.$$

The dual of a vector sheaf is again a vector sheaf. If F is a vector sheaf, then $F^{**} = F$ (via the natural morphism). Duality is thus an exact involution of the category of vector sheaves. The same thus hold for vector local systems. □

If $F \to G$ is a surjection of vector sheaves, and if A is a vector scheme, then the morphism

$$\underline{Hom}(A,F) \to \underline{Hom}(A,G)$$

is surjective. If X is affine then

$$\underline{Hom}(A,F)(X) \to \underline{Hom}(A,G)(X)$$

is surjective.

A *coherent sheaf* is a sheaf which (locally) has a presentation of the form

$$\mathcal{O}^n \to \mathcal{O}^m \to F \to 0.$$

In particular, note that it is a vector sheaf. This coincides with the usual definition: if S is affine and $X \to S$ is a morphism, then $F(X) = F(S) \otimes_{\mathcal{O}(S)} \mathcal{O}(X)$ (this is because the same is true for $\mathcal{O}$, and the presentation remains exact on the right after tensoring). We can in fact assume that a presentation as above exists globally over any affine base.

Duality interchanges coherent sheaves and vector schemes: the dual of a coherent sheaf is a vector scheme and vice-versa.

The vector schemes are projective objects in $\mathcal{V}$, and the coherent sheaves are injective objects. There exist enough projectives and injectives (assuming that S is affine). Note that this property could break down for vector local systems.

If A is a vector scheme, then the functor $V \mapsto \underline{Hom}(A,V)$ from $\mathcal{V}$ to the category of abelian sheaves, is exact. If F is a coherent sheaf, then the functor $V \mapsto \underline{Hom}(V,F)$ is exact.

Taking the dual of the three step resolution by vector schemes shows that every vector sheaf F admits a resolution

$$0 \to F \to U \to V \to W \to 0,$$

with U, V and W coherent sheaves (in particular, injective). Again, this is not necessarily the case for vector local systems so it would have to be applied individually to the pullback vector sheaves.

Internal Hom and tensor products

Lemma 3.5. *If F and G are vector sheaves, then $\underline{Hom}(F,G)$ is a vector sheaf. (Idem for vector local systems.)*

Proof: This follows from the result for duality and the above considerations. □

We now define the *tensor product* $F \otimes^{\mathcal{V}} G$ of two vector sheaves to be

$$F \otimes_{\mathcal{O}} G := (\underline{Hom}(F, G^*))^*.$$

Beware that this is not just the tensor product of sheaves of $\mathcal{O}$-modules (although this will be the case if F and G are coherent sheaves). We can also define the *cotensor product*

$$F \otimes^{\mathcal{O}} G := \underline{Hom}(F^*, G).$$

Again, beware here that this is not equal to the tensor product. The difference is seen in noting that the tensor product is right exact as usual, whereas the cotensor product is left exact (in either variable). Duality permutes the tensor and cotensor products:

$$(F \otimes_{\mathcal{O}} G)^* = F^* \otimes^{\mathcal{O}} G^*$$

and

$$(F \otimes^{\mathcal{O}} G)^* = F^* \otimes_{\mathcal{O}} G^*.$$

Define recursively

$$V_1 \otimes \ldots \otimes V_n := V_1 \otimes (V_2 \otimes \ldots \otimes V_n)$$

for either one of the tensor products.

By *multilinear form* $V_1 \times \ldots \times V_n \to W$ we mean simply a multilinear morphism of sheaves of groups. In the same way as above for the linear morphisms, we obtain a vector sheaf $Mult(V_1 \times \ldots \times V_n, W)$ of multilinear forms (denoted $Bil(\ )$ when $n = 2$).

Theorem 3.6. *Suppose V_i are vector sheaves, $i = 1, \ldots, n$. There is a multilinear form*

$$\mu : V_1 \times \ldots \times V_n \to V_1 \otimes_{\mathcal{O}} \ldots \otimes_{\mathcal{O}} V_n$$

which is universal in the sense that if

$$\phi : V_1 \times \ldots \times V_n \to W$$

is a multilinear form then there is a unique morphism

$$\psi : V_1 \otimes_{\mathcal{O}} \ldots \otimes_{\mathcal{O}} V_n \to W$$

such that $\phi = \psi \circ \mu$. □

Corollary 3.7. *The tensor and cotensor products have natural commutativity and associativity isomorphisms satisfying the usual constraints.* □

We can define symmetric and exterior powers, either with respect to the tensor product or with respect to the cotensor product. Let S_n denote the symmetric group on n objects. Let $V^{\otimes_{\mathcal{O}} n}$ (resp. $V^{\otimes^{\mathcal{O}} n}$) denote the tensor product (resp. cotensor product) of n copies of a vector sheaf V. Then S_n acts on $V^{\otimes_{\mathcal{O}} n}$ (resp. $V^{\otimes^{\mathcal{O}} n}$) because of the commutativity and associativity. This representation is completely reducible (this can be see object-by-object). The components are vector sheaves; this can be seen by noting that the fixed part is the kernel of a morphism of vector sheaves (the direct sum of $1 - \gamma$ for $\gamma \in S_n$); to get the other components, apply that to tensor products with the irreducible representations of S_n. The trivial component of $V^{\otimes_{\mathcal{O}} n}$ (resp. $V^{\otimes^{\mathcal{O}} n}$) is denoted by $Sym^n_{\mathcal{O}}(V)$, the symmetric power (resp. $Sym^{\mathcal{O}}_n(V)$, the symmetric copower). The component corresponding to the sign representation is denoted by $\bigwedge^n_{\mathcal{O}}(V)$, the exterior power (resp. $\bigwedge^{\mathcal{O}}_n(V)$, the exterior copower).

We have the expression $U \otimes^{\mathcal{O}} V = Bil(U^* \times V^*, \mathcal{O})$. The inequality mentioned above implies in particular that there are bilinear functions on $U^* \times V^*$ which are not sums of tensors $u \otimes v$. This is a big difference from the case of schemes (for example if U and V are coherent so that U^* and V^* are vector schemes, then the bilinear functions *are* sums of tensor products).

Cohomology of Eilenberg-MacLane presheaves

The results in this section are analogous to Breen's results in characteristic p [Br] (and in fact, the case $S = Spec(k)$ is just the characteristic 0 version of his result). The additional complications of Breen's results in characteristic p show why we have restricted to the case of characteristic zero.

Proposition 3.8. *Suppose V is a vector sheaf on S. Then the following complexes are exact:*

$$\ldots \bigwedge_{\mathcal{O}}^{j} V \otimes_{\mathcal{O}} Sym^k_{\mathcal{O}} V \to \bigwedge_{\mathcal{O}}^{j-1} V \otimes_{\mathcal{O}} Sym^{k+1}_{\mathcal{O}} V \to \ldots$$

and

$$\ldots \bigwedge_{\mathcal{O}}^{j} V \otimes_{\mathcal{O}} Sym^k_{\mathcal{O}} V \to \bigwedge_{\mathcal{O}}^{j+1} V \otimes_{\mathcal{O}} Sym^{k-1}_{\mathcal{O}} V \to \ldots.$$

Similarly for the cotensor product.

Proof: Let $C = (V \to V)$ be the complex with V in degrees 0 and 1, and with the identity as differential. Then the complex $C^n = C \otimes_{\mathcal{O}} \cdots \otimes_{\mathcal{O}} C$ is acyclic. On the other hand, the symmetric group S_n acts on C^n with signs for transpositions of elements of odd degree. The subcomplex $Sym^n_{\mathcal{O}}(C)$ of elements fixed by this action is a direct summand. This fixed subcomplex is exactly one of the complexes considered above. To get the other complex, apply the same construction to the shifted complex $C[1]$. For the cotensor product just take the dual (or repeat the same proof). □

Corollary 3.9. *Suppose V is a vector scheme over S, and suppose $\mathcal{F}$ is a coherent sheaf on S. For n odd we have*

$$H^i(K(V/S,n)/S, \mathcal{F}) = \mathcal{F} \otimes_{\mathcal{O}} \bigwedge_{\mathcal{O}}^{\frac{i}{n}}(V^*),$$

and for n even we have

$$H^i(K(V/S,n), \mathcal{F}) = \mathcal{F} \otimes_{\mathcal{O}} Sym^{\frac{i}{n}}_{\mathcal{O}}(V^*).$$

The answer is zero if $\frac{i}{n}$ is not an integer. The multiplicative structures in the case $\mathcal{F} = \mathcal{O}$ are the obvious ones.

Proof: Proceed by induction on n; we suppose the theorem is true for $n \leq m-1$ and we prove it for $n = m$. Use the fibration sequence

$$K(V/S, m-1) \to * \xrightarrow{f} K(V/S, m).$$

We then show, by the usual argument using the Leray spectral sequence, that $K(V/S, m)$ has the desired cohomology H^i. The above proposition gives exactness of the complexes at each stage in the spectral sequence, except at the end where we get the new cohomology. Note that tensor product $\mathcal{F} \otimes_{\mathcal{O}} \mathcal{G} = \underline{Hom}(\mathcal{F}^*, \mathcal{G})$ is exact for coherent sheaves. □

Corollary 3.10. *Suppose V is a vector sheaf over S. There is a complex $C_*^{\text{vect}}(K(V/S,n)/S)$ of vector sheaves over S such that for any vector sheaf U on S,*

$$\underline{H}^i(K(V/S,n)/S, U) = \underline{\mathbf{Ext}}^i(C_*^{\text{vect}}(K(V/S,n)/S), U)$$

is the hyper-ext. In particular, $\underline{H}^i(K(V/S,n)/S, U)$ are vector sheaves over S.

Proof: We get the complex in question when V is a vector scheme, from the above results. If V is a vector sheaf then resolve it by vector schemes and patch together the complexes. □

See also the generalization 5.7 below.

Corollary 3.11. *Suppose U and V are vector sheaves over S. Let $\underline{Ext}^i_{Gp}(U,V)$ denote the $\underline{Ext}$ sheaves in the category of sheaves of groups over the site $\mathcal{X}/S$. Then the natural maps*

$$\underline{Ext}^i_{\mathcal{O}}(U,V) \to \underline{Ext}^i_{Gp}(U,V)$$

are isomorphisms. In particular if U, V are vector sheaves over S then $\underline{Ext}^i_{Gp}(U,V) = 0$ for $i \geq 3$.

Proof: For any n we have that $C_*(K(U/S,n), \mathbf{Z}) \sim U[n]$ in degrees $< 2n$. For n large it follows that $\underline{Ext}^i_{Gp}(U,V) = \pi_0 \underline{Hom}(K(U/S,n), K(V/S,n+i))$. On the other hand, we can show (using the above calculations) that this is equal to $\underline{Ext}^i_{\mathcal{O}}(U,V)$. Since we have injective or projective resolutions of length ≤ 2 in the category of vector sheaves, we get the last statement. □

This theorem generalizes the results from earlier in this section, for $\underline{Ext}^0 = \underline{Hom}$ and $\underline{Ext}^1$ (which parametrizes extensions).

4. Presentability conditions for sheaves

In the previous section, we have seen a very convenient type of presentability notion for sheaves of abelian groups: the notion of being a vector sheaf. For sheaves in general, for sheaves of nonabelian groups, or even for sheaves of abelian groups which might arise in a nonabelian context, the right presentability condition is harder to come by.

We will define several conditions, numbered $P1$, $P2$, $P3$, $P4$, $P5$, $P6$ and $P7$.

Fix a base scheme $S \in \mathcal{X}$. In what follows, we work in the category of sheaves on $\mathcal{X}/S$. Thus a sheaf is supposed to be on $\mathcal{X}/S$ unless otherwise specified.

P1. We say that $\mathcal{F}$ is *P1* if there is a surjective morphism of sheaves $X \to \mathcal{F}$ where X is represented by a scheme of finite type over S. We may assume that X is affine.

P2. We say that $\mathcal{F}$ is *P2* if there are surjective morphisms of sheaves $X \to \mathcal{F}$ and $Y \to X \times_{\mathcal{F}} X$ where X and Y are represented by schemes of finite type over S. We may assume that X and Y are affine.

Lemma 4.1. *If G is a sheaf of groups which is P1, and G acts on a sheaf F which is P2, then the quotient sheaf F/G is again P2.* □

Boundedness conditions

We consider the following boundedness conditions for a sheaves on $\mathcal{X}$.

B1. We say that a sheaf $\mathcal{F}$ is *B1* if, for any k-algebra B, we have that

$$\lim_{\rightarrow} \mathcal{F}(Spec(B')) \rightarrow \mathcal{F}(Spec(B))$$

is an isomorphism, where the limit is taken over the subalgebras $B' \subset B$ which are of finite type over k. This is equivalent to the local finite type condition of Artin [Ar].

B2. We say that a sheaf $\mathcal{F}$ is *B2* if, for any complete local ring A, we have that the morphism

$$\mathcal{F}(Spec(A)) \rightarrow \lim_{\leftarrow} \mathcal{F}(Spec(A/\mathbf{m}^i)$$

is an isomorphism.

We can make the same conditions on presheaves of spaces, replacing the limits by homotopy limits and replacing equality by weak equivalence of topological spaces.

Lemma 4.2. *1. If $\mathcal{F}$ and $\mathcal{G}$ are B1 (resp. B2) and f, g are two morphisms from $\mathcal{F}$ to $\mathcal{G}$ then the equalizer is again B1 (resp. B2).*
2. Suppose $\mathcal{F} \rightarrow \mathcal{G}$ is a surjective morphism of sheaves. If $\mathcal{F}$ and $\mathcal{F} \times_{\mathcal{G}} \mathcal{F}$ are B1 then $\mathcal{G}$ is B1.
3. Suppose $\mathcal{F}$ is a sheaf which is P2. Then $\mathcal{F}$ is B1. If the ground field is uncountable, then $\mathcal{F}$ is B2. □

(I imagine that property 3 should hold even if k is countable but I don't know the proof.)

Lifting properties

We say that a morphism of sheaves $a : \mathcal{F} \rightarrow \mathcal{G}$, is *vertical* (or *S-vertical*, if the base needs to be specified), if it satisfies the following lifting properties for all n:
$Lift_n(Y; Y_i)$: Suppose Y is a scheme with n closed subschemes $Y_i \subset Y$, with retractions $r_i : Y \rightarrow Y_i$—commuting pairwise ($r_i r_j = r_j r_i$)—such that for $j \leq i$, r_i retracts Y_i to $Y_j \cap Y_i$. Suppose given a morphism $Y \rightarrow \mathcal{G}$, and liftings $\lambda_i : Y_i \rightarrow \mathcal{F}$ such that $\lambda_i|_{Y_i \cap Y_j} = \lambda_j|_{Y_i \cap Y_j}$. Then locally in the etale topology over Y near at least one Y_i, there exists a lifting λ which agrees with the given liftings λ_i on Y_i.

The notion of vertical morphism in some sense replaces the notion of smooth morphism when we work relative to a base scheme S. This notion serves two purposes: (1) it allows us to define the tangent sheaf to certain sheaves; and (2) if G is a group sheaf over S and $X \rightarrow G$ is a vertical surjection, we can lift the multiplication to a map $\mu : X \times_S X \rightarrow X$ with $\mu(x, e) = \mu(e, x) = x$ (where $e : S \rightarrow X$ is a lift of the identity).

The notion of a vertical morphism satisfies several nice properties. Principal among these are the following:
—If $a : \mathcal{F} \rightarrow \mathcal{G}$ and $b : \mathcal{G} \rightarrow \mathcal{H}$ are two morphisms which are vertical, then ba is vertical;
—If $a : \mathcal{F} \rightarrow \mathcal{G}$ and $b : \mathcal{G} \rightarrow \mathcal{H}$ are two morphisms such that a and ba are vertical, and a is surjective, then b is vertical;
—Any surjective morphism of sheaves of groups is vertical.
—Etale surjections between schemes are vertical;
—Any injective morphism $\mathcal{F} \hookrightarrow S$ is vertical.

It is sometimes useful to know that if $f : \mathcal{F} \rightarrow \mathcal{G}$ is a morphism of sheaves which are P2, then f is vertical if and only if $Lift_n(Y; Y_i)$ holds for all systems $(Y; Y_i)$ with Y (and hence Y_i) artinian.

Remark: The property of being vertical is stable under base change of S: suppose $p : S' \rightarrow S$ is a morphism of schemes. If $f : \mathcal{F} \rightarrow \mathcal{G}$ is vertical then $p^*(f) : p^*(\mathcal{F}) \rightarrow p^*(\mathcal{G})$ is vertical. This follows from the form of the lifting properties.

More presentability properties

We can now complete our list of presentability properties.
P3. We say that a sheaf $\mathcal{F}$ is *P3* if there exists a surjective morphism

$$X \to \mathcal{F}$$

and a surjective $X \times X$-vertical morphism

$$R \to X \times_{\mathcal{F}} X$$

with X and R represented by affine schemes of finite type over $Spec(k)$.
P4. We say that a sheaf $\mathcal{F}$ on $\mathcal{X}/S$ is *P4* if there exist surjective S-vertical morphisms

$$X \to \mathcal{F}$$

and

$$R \to X \times_{\mathcal{F}} X$$

with X and R represented by affine schemes of finite type over S.
P5. We say that $\mathcal{F}$ is *P5* if it is P4 and if, in addition, the structural morphism $\mathcal{F} \to S$ is vertical. If $\mathcal{F}$ is a group sheaf we then say that it is a *presentable group sheaf*.
P6. We say that $\mathcal{F}$ is *P6* if it is P5 and if, in addition, for any artinian S-scheme S', $\mathcal{F}(S')$ is an affine scheme over $Spec(k)$ (the fact that it is a scheme over $Spec(k)$ will be mentioned below). If $\mathcal{F}$ is a group sheaf we then say that it is an *affine presentable group sheaf*.
P7. We say that a sheaf of groups $\mathcal{F}$ on $\mathcal{X}/S$ is *P7* if it is a vector sheaf (as a sheaf of groups).

Corollary 4.3. *If $\mathcal{G}$ is a P4 group sheaf then it is also P5.*

Indeed, the third property of vertical morphisms mentioned above, applied to the surjection of sheaves of groups, $\mathcal{F} \to S$, gives that this structural morphism is vertical. □

Kernels and extensions

Theorem 4.4. *Let $A = 2, 3, 4, 5, 6$ or 7, and denote by PA the property P2,P3, P4, P5, P6 or P7 respectively.*
1. If $f, g : \mathcal{G} \to \mathcal{H}$ are two morphisms, and if $\mathcal{G}$ and $\mathcal{H}$ are PA, then the equalizer $\mathcal{F}$ is PA.
2. If $\mathcal{F} \to \mathcal{H}$ and $\mathcal{G} \to \mathcal{H}$ are two morphisms between PA sheaves, then the fiber product $\mathcal{F} \times_{\mathcal{H}} \mathcal{G}$ is PA.
3. Suppose $A = 2$ or $A \geq 5$. If

$$1 \to \mathcal{F} \to \mathcal{E} \to \mathcal{G} \to 1$$

is an extension of group sheaves and if any two of the elements are PA, then the third one is PA too (for $A = 7$ we have to assume that $\mathcal{E}$ is abelian). □

As a first example, note that any sheaf of groups which is representable by a scheme of finite type G over S, is presentable. This is because we can take $X = G$ and R equal to the diagonal G in the definition of property P4; and property P5 is then Corollary 4.3. Applying the above theorem, we find that the category of presentable group sheaves contains the category generated by representable group sheaves under the operations of extensions, kernels, and division by normal subgroups. In particular, it is much bigger than the category of representable group sheaves. I believe that the category of presentable group sheaves is strictly larger than that generated by representable group sheaves under these operations. For example the group sheaves $Aut(V)$ for a vector sheaf V, which are presentable as shown below, are probably not generated from representable group sheaves by kernels, extensions and quotients (although I don't have a counterexample). In an intuitive sense the two categories are about the same.

The properties PA satisfies an etale descent condition ($A = 2, \ldots, 7$). Suppose $\mathcal{F}$ is a sheaf over S, and suppose $S' \to S$ is a surjective etale morphism such that $\mathcal{F}|_{S'}$ is PA over S'. Then $\mathcal{F}$ is PA over S.

Remark: The notion of presentable group sheaf, and all of the lemmas we have proved since then, work equally well over a field k of positive characteristic (this is in contrast with the part about vector schemes, where virtually none of the arguments work in positive characteristic). I don't know if the notion of presentable group sheaf is the "right" one in positive characteristic. Some of the arguments we now give will not work as well in positive characteristic.

Tangent sheaves

Theorem 4.5. *Suppose $p : \mathcal{F} \to S$ is a P4 sheaf on S. Suppose $u : S \to \mathcal{F}$ is a section. Then the tangent sheaf $T(\mathcal{F}/S)_u$ over S, defined by*

$$T(\mathcal{F}/S)_u(b : S' \to S) := \{\eta : S' \times Spec(k[\epsilon]/(\epsilon^2)) \to \mathcal{F} \quad : \quad p\eta = bp_1 \ \ and \ \ \eta|_{S'} = ub\},$$

has a natural structure of sheaf of abelian groups making it a vector sheaf over S. These fit together into a vector local system $T(\mathcal{F})$ over $\mathcal{F}$ whose fiber over $u : S' \to \mathcal{F}$ is $T(\mathcal{F} \times_S S'/S')_u$.

We explain how to add two tangent vectors. Let $X \to \mathcal{F}$ and $R \to X \times_{\mathcal{F}} X$ be the presentation of $\mathcal{F}$. Suppose

$$S' \times Spec(k[\epsilon_i]/(\epsilon_i^2)) \to \mathcal{F}$$

are two tangent vectors ($i = 1, 2$). Then we can lift these to morphisms

$$\varphi_i : S' \times Spec(k[\epsilon_i]/(\epsilon_i^2)) \to X.$$

Furthermore, the morphism

$$(\varphi_1|_{S'}, \varphi_2|_{S'}) : S' \to X \times_{\mathcal{F}} X$$

lifts to a morphism $\psi = S' \to R$. By verticality of the first projection $p_1 : R \to X$ (which follows from the verticality of the morphisms in the presentation of $\mathcal{F}$), we can lift the morphism φ_1 to a morphism

$$\psi_1 : S' \times Spec(k[\epsilon_i]/(\epsilon_i^2)) \to R$$

agreeing with ψ over S'. Then $\varphi_1' := p_2\psi_1$ is a morphism like φ_1 but with $\varphi_1'|_{S'} = \varphi_2|_{S'}$. Now, since X is a scheme, we can complete the pair (φ_1', φ_2) to a morphism

$$S' \times Spec(k[\epsilon_1, \epsilon_2]/(\epsilon_1^2, \epsilon_2^2, \epsilon_1\epsilon_2)) \to X.$$

The restriction φ_3 of this morphism to the diagonal gives the sum of the two tangent vectors. Associativity is checked in a similar way, and the inverse is defined by $\epsilon \mapsto -\epsilon$, to get the group structure.

The vertical map $X \to \mathcal{F}$ gives a surjection $T(X)_{\tilde{u}} \to T(\mathcal{F})_u$ for any lift $\tilde{u}$ of the section u. This, together with a similar consideration for R, allows us to prove that $T(\mathcal{F})_u$ is a vector sheaf. □

If $f : \mathcal{F} \to \mathcal{G}$ is a morphism of P4 sheaves over S then we obtain a morphism of local systems on $\mathcal{F}$,

$$df : T(\mathcal{F}/S) \to f^*(T(\mathcal{G}/S)).$$

We can define the relative tangent sheaf $T(\mathcal{F}/\mathcal{G})$ to be the kernel of this morphism.

If f is vertical then df is surjective (*Question:* is this a criterion for verticality?).

Remark: It is in the proof of Theorem 4.5 that we first use the notion of verticality of the morphisms in the presentation of $\mathcal{F}$. Here is an example of a P2 sheaf over $Spec(k)$ where the tangent sheaf as defined above isn't a sheaf of groups. Let Y be a smooth surface and let $X \to Y$ be the blow-up of a point. Let $X' \subset X$ be an open set intersecting but not containing the exceptional divisor. Let $\mathcal{F}$ be the sheaf-theoretic image of $X' \to Y$. Then $\mathcal{F}$ is P2 with surjection $X' \to \mathcal{F}$ and relation scheme $R = X' \times_Y X'$. However, the morphism $X' \to \mathcal{F}$ is not vertical. The tangent vectors at the origin (corresponding to the blown-up point downstairs) are those elements of the tangent space to Y at the blown-up point which correspond to points in the exceptional divisor—because of the choice of X', we are now missing some tangent vectors (and in particular the tangent space is not a group in any natural way)!

Remark: If $\mathcal{G}$ is a presentable group sheaf over S, then $Lie(\mathcal{G}) := T(\mathcal{G})_1$ is a vector sheaf and there is a bilinear form (Lie bracket)

$$[\cdot,\cdot] : Lie(\mathcal{G}) \times Lie(\mathcal{G}) \to Lie(\mathcal{G})$$

which, over artinian base schemes, reduces to the usual Lie bracket. One can construct this in the same way as a purely algebro-geometric construction of the Lie bracket on any algebraic group.

In fact, if $\mathcal{G}$ is any sheaf of groups then $Lie(\mathcal{G}) = T(\mathcal{G})_1$ has an abelian group structure, since we can use the group structure of $\mathcal{G}$ to add tangent vectors. However, without any other conditions on $\mathcal{G}$ this will not necessarily be a vector sheaf. A couple of questions come to mind.

Question: If $\mathcal{G}$ is a P2 group sheaf, then is $Lie(\mathcal{G})$ a vector sheaf? One can show that it is an inductive limit of vector sheaves, but the category of vector sheaves is not closed under inductive limits.

Question: If $\mathcal{G}$ is a P2 group sheaf, is $\mathcal{G}$ automatically P5? If so, the theory sketched here could be considerably simplified.

Automorphisms of vector sheaves

One of the main examples of presentable group sheaves is given by the following theorem.

Theorem 4.6. *If V is a vector sheaf over S, then V is presentable. Also the group sheaf $Aut(V)$ is a presentable group sheaf.*

The presentation of V as a quotient of vector schemes gives the desired presentation. For the second part, $\underline{Hom}(V,V)$ is a vector sheaf, hence presentable, and $\underline{Aut}(V)$ is the equalizer of two maps

$$\underline{Hom}(V,V) \times \underline{Hom}(V,V) \to \underline{Hom}(V,V) \times \underline{Hom}(V,V).$$

□

Direct images by finite morphisms

Suppose $f : S' \to S$ is a finite morphism. If G is a presentable group sheaf on S' then we define the direct image $f_*(G)$ on S by the formula

$$f_*(G)(Y) := G(Y \times_S S').$$

One can show that it is a presentable group sheaf on S'.

The case $S = Spec(k)$

We now analyse the definitions of the previous section in the case where the base scheme is $Spec(k)$

Lemma 4.7. *Suppose $f : X \to Spec(k)$ is a scheme of finite type. Then X is smooth over $Spec(k)$ if and only if f is vertical.* □

Theorem 4.8. *If G is a presentable group sheaf on $\mathcal{X}/Spec(k)$ then it is represented by a smooth separated scheme of finite type over k.*

Proof: One shows, using the above lemma, that G is a smooth algebraic space. Then using the group structure one finds that it is separated. Finally, by looking carefully at the presentation one obtains an open subset which is a scheme, and again by the group structure G is a scheme. □

Application: Suppose $S' \to Spec(k)$ is an artinian scheme of finite type. Let $\pi : S' \to Spec(k)$ denote the structural morphism. If G is a presentable group sheaf over $Spec(k)$ the pullback $G|_{S'}$ is presentable and the direct image $\pi_*(G|_{S'})$ is presentable over $Spec(k)$. By the above result, $\pi_*(G|_{S'})$ is represented by a group scheme of finite type which we denote $G_{S'}$ over $Spec(k)$. We have

$$G(S') = G_{S'}(Spec(k)).$$

The Lie algebra of the group scheme $G_{S'}$ is $Lie(\mathcal{G})(S')$.

If $X \to G$ is a vertical surjection then we obtain a scheme of finite type $X_{S'} = \pi_*(X|_{S'})$ with a smooth morphism $X_{S'} \to G_{S'}$.

Local study of presentable group sheaves

In this section we return to the case of general base scheme S.

This is the second main place where we use the condition of verticality of the morphisms in the presentations.

Theorem 4.9. *Suppose G is a presentable group over a base S. Suppose $X \to G$ is a vertical surjection with X an affine scheme over S as in Property P4. Let $TX \to S$ be the tangent vector scheme at the identity section e. Suppose $s \in S$. Then there is an etale neighborhood*

$$e(s) \in U \xrightarrow{p} X$$

and an etale S-morphism $qU \to TX$, such that $q = p$ over the identity section (mapping to the zero section of TX).

Proof: The verticality of $X \to G$ allows us to lift the multiplication of G to a map

$$\mu : X \times_S X \to X, \ \ \mu(x, e) = \mu(e, x)x.$$

Let $\mathbf{2} : X \to X$ denote the map $x \mapsto \mu(x, x)$. The existence of this homothety implies that, formally around the identity section, X is isomorphic to its tangent vector scheme TX. Using Artin approximation we obtain the same for an etale neighborhood. □

Theory of the connected component

We need to develop a suitable theory of the connected component of a presentable group sheaf G.

Theorem 4.10. *If G is a presentable group sheaf over S, then there is a unique subsheaf of groups $G^0 \subset G$ such that G^0 is presentable and such that for any artinian S-scheme S', we have $G^0(S')$ equal to the connected component of $G(S')$ (when these are considered as* *algebraic groups over the ground field of S'—cf the application at the end of the section on the situation over Spec(k)).*

Proof: Let $U \to G$ be the morphism given by Theorem 4.9 (here U is etale over the presentation scheme X). Possibly after replacing U by a subset, one can show that the fibers $U(S')$ are connected (using the fact that U is etale over a vector scheme). Then the product $U \times U \to G$ gives a morphism which surjects onto a subgroup-scheme G^0 (the proof of this uses the fact that if W is a Zariski open subset of a connected group scheme then $W \cdot W$ is the whole group scheme). □

We say that a presentable group sheaf G is *finite* if $G^0 = \{1\}$. If G is any presentable group sheaf, then the connected component G^0 is a normal subgroup sheaf, and the quotient $C := G/G^0$ is again presentable. Over artinian S', this quotient is just the group of connected components, in particular the connected component is trivial. Thus C is finite.

Lemma 4.11. *If G is a finite presentable group sheaf, then there is an integer N such that for any henselian local S-scheme S' (with algebraically closed residue field), the number of elements in $G(S')$ is less than or equal to N.* □

We can now state a structure theorem.

Theorem 4.12. *If G is a presentable group sheaf over S then there is an exact sequence*

$$1 \to E \to G \to G' \to 1$$

where G' is a presentable subgroup sheaf of $\underline{Aut}(V)$ for V a vector sheaf, and E is presentable with abelian connected component E^0. If G is connected then E is the center of G.

Proof: Let $V = Lie(G)$. The adjoint action gives a morphism $G \to \underline{Aut}(V)$, whose image G' is presentable (Theorem 4.4 parts 1 and 3). Let E be the kernel. It centralizes G^0 (as can be seen by looking over all artinian bases) so we get the remaining statements. □

5. Presentable presheaves of topological spaces

We introduce the notation $\varpi_1(T, t_1, t_2)$ for the paths in T from t_1 to t_2 up to homotopy. Thus

$$\varpi_1(T, t_1, t_2) = \pi_0(P^{t_1,t_2}T).$$

If T is a homotopy sheaf and if $Y \in \mathcal{X}$ with $t \in T(Y)$ then we have the homotopy group sheaves $\pi_i(T/Y, t)$ (although for π_0 there is no need for a base point—$\pi_0(T)$ is a sheaf on $\mathcal{X}$). We make the following definition.

—We say that T is *presentable* if it satisfies the following conditions:

(1) The sheaf $\pi_0(T)$ is P1 over k.
(2) For any scheme Y and any two sections $\eta, \eta' : Y \to T$ the sheaf $\varpi_1(T|_{Z/Y}, \eta, \eta')$ is P4 over Y.
(3) For any scheme Y and section $\eta : Y \to T$, the higher homotopy group sheaves $\pi_i((T|_{Z/Y}), \eta)$, for $i \geq 1$, are presentable group sheaves (P5) over Y.

Note that condition 3 when $i = 1$ is equivalent to condition 2 when $\eta = \eta'$.

Caution: If T is 0-truncated, that is a sheaf of sets, and happens to have a group structure, then this notion is not the same as the notion that T be a presentable group sheaf. The presentability in T as defined here refers to the higher homotopy groups.

We call the presentable homotopy sheaves T, the *presentable spaces over* $\mathcal{X}$ or just *presentable spaces.*

A *morphism* of presentable spaces is just a morphism of presheaves. When we speak of morphisms we usually want the domain presheaf to be cofibrant.

We remark that property 1 implies the seemingly stronger statement that there is a section $f : Z \to T$ over a scheme Z of finite type over k, such that the associated morphism $Z \to \pi_0(T)$ is surjective.

Similarly, the property P1 in condition 2 implies the existence of a section $X \to P^{\eta,\eta'}(T|_{Z/Y})$ giving a surjection from X to $\varpi_1(T|_{Z/Y}, \eta, \eta')$.

The second condition reduces, in the case $\eta = \eta'$, to the statement that for any scheme Y and section $\eta : Y \to T$, the fundamental group sheaf $\pi_1((T|_{Z/Y}), \eta)$ is a P4 group sheaf over Y.

We can give an alternative characterization, from which it follows that any truncation $\tau_{\leq n}T$ of a presentable space is again presentable.

Theorem 5.1. *Suppose $T \in Top/\mathcal{X}$. Then T is presentable if and only if the sheaf π_0 is P3, and for any $Y \in \mathcal{X}$ and $t \in T(Y)$, the sheaves $\pi_i(T|_{\mathcal{X}/Y}, t)$ are presentable (P5) over Y.* □

We make the following more restrictive definition. Say that a presentable group sheaf G on $\mathcal{X}/S$ is *affine* if, for any artinian S-scheme S', the group scheme $G(S')$ over $Spec(k)$ is affine. A truncated homotopy sheaf T is *very presentable* if T is presentable and if for any $\eta \in T_Y$ we have that $\pi_1(T/Y, \eta)$ is affine, and $\pi_i(T/Y, \eta)$ are vector sheaves for $i \geq 2$.

Even more generally, we can say that T is $(A_0, A_1, \ldots)$-presentable if $\pi_0(T)$ is PA_0 and if for any $\eta \in T_Y$ we have that $\pi_i(T/Y, \eta)$ is PA_i. The condition of presentability is $(3, 5, 5, \ldots)$-presentability, where as very presentability is $(3, 6, 7, 7, \ldots)$-presentability. In this notation we could also admit $A_i = 0$ to mean no condition at all.

Lemma 5.2. *If $R \to T$ is an Illusie weak equivalence between spaces over $\mathcal{X}$ then R is presentable (resp. very presentable, $(A_0, A_1, \ldots)$-presentable) if and only if T is presentable (resp. very presentable).* □

Remark: The category of presentable (resp. very presentable) spaces is closed under weak equivalences and fiber products but not under cofiber products (push-outs); thus it is not a closed model category.

The idea behind the definition of very presentable is that we want to require the higher homotopy groups to be unipotent. Note that if we don't require π_1 to be affine, or π_i to be unipotent ($i \geq 2$), then the comparison between algebraic and analytic de Rham cohomology presented below is no longer true, even over the base $S = Spec(k)$ where all of the groups are representable. This is the reason for making the definition of very presentable.

Question: If G is an abelian affine presentable group sheaf on $\mathcal{X}/S$ such that for any artinian $S' \to S$ the group scheme $G(S')$ over k is a direct sum of additive groups, then is G a vector sheaf?

If we knew this, we could replace the condition of being a vector sheaf by the condition that the $G(S')$ are unipotent (hence additive) for $G = \pi_i$, $i \geq 2$; this would then be along the same lines as the affineness condition for π_1. As it is, we need to require the condition of π_i being vector schemes ($i \geq 2$) for many of the arguments below.

Proposition 5.3. *Suppose R, S and T are spaces over $\mathcal{X}$, with locally fibrant morphisms $R \to T$ and $S \to T$. Suppose $Z \in \mathcal{X}$ and $(r, s) \in R \times_T S(Z)$. Let $t \in T(Z)$ be the common* *image of r and s. Then we have the following long exact sequence of homotopy group sheaves on $\mathcal{X}/Z$:*

$$\ldots \to \pi_i(R\times_T S|_{\mathcal{X}/Z},(r,s)) \to \pi_i(R|_{\mathcal{X}/Z},r)\times\pi_i(S|_{\mathcal{X}/Z},s)\to$$
$$\pi_i(T|_{\mathcal{X}/Z},t)\to\pi_{i-1}(R\times_T S|_{\mathcal{X}/Z},(r,s))\to\ldots,$$

terminating with the sequence

$$\pi_2(R|_{\mathcal{X}/Z},r)\times\pi_2(S|_{\mathcal{X}/Z},s)\to\pi_2(T|_{\mathcal{X}/Z},t)\to\pi_1(R\times_T S|_{\mathcal{X}/Z},(r,s))\to$$
$$\pi_1(R|_{\mathcal{X}/Z},r)\times\pi_1(S|_{\mathcal{X}/Z},s)\rightrightarrows\pi_1(T|_{\mathcal{X}/Z},t)$$

(the last part meaning that the image is equal to the equalizer of the two arrows). Furthermore, we have a similar sequence for the path spaces. Suppose (r_1,s_1) and (r_2,s_2) are two points, with images t_1 and t_2. We have the exact sequence

$$\pi_2(R|_{\mathcal{X}/Z},r_1)\times\pi_2(S|_{\mathcal{X}/Z},s_1)\to$$
$$\pi_2(T|_{\mathcal{X}/Z},t_1)\overset{acts\ on}{\to}\varpi_1(R\times_T S|_{\mathcal{X}/Z},(r_1,s_1),(r_2,s_2))\ \textit{with quotient the equalizer of}$$
$$\varpi_1(R|_{\mathcal{X}/Z},r_1,r_2)\times\varpi_1(S|_{\mathcal{X}/Z},s_1,s_2)\rightrightarrows\varpi_1(T|_{\mathcal{X}/Z},t_1,t_2).$$

□

Theorem 5.4. *If R, S and T are presentable (resp. very presentable) spaces over $\mathcal{X}$ with morphisms $R\to T$ and $S\to T$; then the fiber product $R\times_T S$ is presentable (resp. very presentable).*

Proof: We first show that $R\times_T S$ is P1. In general, if $A\to B$ induces a surjection on π_0 and if $C\to B$ is a morphism, then $A\times_B C\to C$ induces a surjection on π_0. We apply this principle several times. Choose a morphism $Z\to T$ such that $Z\to\pi_0(T)$ is surjective. Then

$$\pi_0((R\times_T S)\times_T Z)\to\pi_0(R\times_T S)$$

is surjective. But

$$(R\times_T S)\times_T Z=(R\times_T Z)\times_Z(S\times_T Z).$$

If $X\to R$ is surjective on π_0, then

$$X\times_T Z\to R\times_T Z$$

is a surjection on π_0. Finally,

$$X\times_T Z=P^{\eta,\eta'}T$$

where $\eta,\eta':X\times Z\to T$ are the compositions of the two projections with the morphisms $X\to T$ and $Z\to T$. By the assumption that the π_0 of the path spaces are P4, there is a scheme of finite type W and a morphism

$$W\to X\times_T Z$$

inducing a surjection on π_0. We obtain $W\to R\times_T Z$ inducing a surjection on π_0. Similarly there is $V\to S\times_T Z$ inducing a surjection on π_0. Finally,

$$W\times_Z V=W\times_{R\times_T Z}(R\times_T Z)\times_Z(S\times_T Z)\times_{S\times_T Z}V$$

so applying our general principle two more times, we find that the morphism

$$W\times_Z V\to R\times_T S$$

induces a surjection on π_0. This shows that $R\times_T S$ is P1.

Now the conditions that the π_0 of the path spaces be P4, and that the homotopy group sheaves be presentable (resp. very presentable, $i\geq 2$), follow from the exact sequences of the previous lemma, and Theorem 4.4. □

Exercise: For which values of $(A_0, A_1, \dots)$ does Theorem 5.4 hold for $(A_0, A_1, \dots)$-presentable spaces? Place these conditions in Corollary 5.6 below.

Proposition 5.5. *Suppose $R \to T$ is a locally fibrant morphism of spaces, such that T is presentable (resp. very presentable) and for all $Y \in \mathcal{X}$ and $t \in T(Y)$, the fiber $R_t := R \times_T Y$ is presentable (resp. very presentable). Suppose also that for any fiber R_t over Y, $\pi_0(R_t)$ is P4. Then R is presentable (resp. very presentable).*

Proof: This follows easily by looking at the long exact homotopy sequences. □

Presentable shapes

Recall the definition of $\underline{Hom}(R,T)$ from the first section: it is the presheaf $X \mapsto Hom(R'_X, T|_{\mathcal{X}/X})$ where R'_X is a functorial replacement of $R|_{\mathcal{X}/X}$ by a cofibrant presheaf.

Corollary 5.6. *Suppose W is a finite CW complex, and let $W_{\mathcal{X}}$ denote the constant homotopy sheaf with values W (that is, the homotopy sheaf associated to the constant presheaf). If $T \in Top/\mathcal{X}$ is presentable (resp. very presentable) then $\underline{Hom}(W_{\mathcal{X}}, T)$ is presentable (resp. very presentable).*

Proof: We first show this for $W = S^n$, the n-sphere. Do this by induction on n It is clear for $n = 0$ because then W consists of two points and $\underline{Hom}(W_{\mathcal{X}}, T) = T \times T$. For any n, write S^n as the union of two copies of B^n joined along S^{n-1}. We get

$$\underline{Hom}(S^n_{\mathcal{X}}, T) = T \times_{\underline{Hom}(S^{n-1}_{\mathcal{X}}, T)} T,$$

since $\underline{Hom}(B^n_{\mathcal{X}}, T) = T$. By Theorem 5.4, $\underline{Hom}(S^n_{\mathcal{X}}, T)$ is presentable (resp. very presentable). This shows the corollary for the spheres.

We now treat the case of general W, by induction on the number of cells. We may thus write $W = W' \cup B^n$ with the cell B^n attached over an attaching map $S^{n-1} \to W'$, and where we know the result for W'. Then

$$\underline{Hom}(W_{\mathcal{X}}, T) = \underline{Hom}(W'_{\mathcal{X}}, T) \times_{\underline{Hom}(S^{n-1}_{\mathcal{X}}, T)} T.$$

Again by Theorem 5.4, we obtain the result for W. □

Let $Pres/\mathcal{X}$ denote the category of presentable spaces. We define the *presentable shape of* W to be the functor

$$Shape(W) : T \mapsto \underline{Hom}(W_{\mathcal{X}}, T)$$

from $Pres/\mathcal{X}$ to $Pres/\mathcal{X}$.

One can show (using the calculations of Corollary 3.9 over $S = Spec(k)$) that if W is connected and simply connected then this functor is homotopy-representable by an object $Hull(W) \in Pres/\mathcal{X}$. On the other hand, in most cases where W is not simply connected, the presentable shape is not representable. We could try to interpret the hull of W as the inverse limit of $Shape(W)$, but this is not a standard kind of inverse limit. It is a question for further study, just what information is contained in $Shape(W)$.

The calculations of Corollary 3.9 generalize in the following way.

Theorem 5.7. *Suppose T is a $(0,0,7,7,\dots)$-presentable cofibrant and fibrant presheaf. Let $R := \tau_{\leq 1} T$. There is a complex $C^{\rm vect}_*(T/R)$ of vector local systems over R such that for any vector local system U on T (which necessarily comes from R),*

$$\underline{H}^i(T/R, U) = \underline{\mathbf{Ext}}^i(C^{\rm vect}_*(T/R), U)$$

is the hyper-ext in the category of vector local systems over R. In particular, $\underline{H}^i(T/R, U)$ are vector local systems over R.

Proof: This comes from Corollary 3.10 and the Leray spectral sequence for the Postnikov tower of T. □

Analytic sheaves

When we speak of analytic sheaves, it is understood that the base field is $k = \mathbf{C}$. Let $\mathcal{X}^{\mathrm{an}}$ denote the site of complex analytic spaces with the etale topology. There is again a natural sheaf of rings which we denote $\mathcal{O}^{\mathrm{an}}$ over $\mathcal{X}^{\mathrm{an}}$. Suppose $\mathcal{F}$ is a sheaf (resp. homotopy sheaf) on $\mathcal{X}$, and $Y \in \mathcal{X}^{\mathrm{an}}$. Put

$$\mathcal{F}^{\mathrm{an,pre}}(Y) := \mathcal{F}(Spec(\mathcal{O}^{\mathrm{an}}(Y))).$$

Then let $\mathcal{F}^{\mathrm{an}}$ be the sheafification of $\mathcal{F}^{\mathrm{an,pre}}$. We call this the *analytic sheaf (resp. homotopy sheaf) associated to* $\mathcal{F}$. Note for example that $\mathcal{O}^{\mathrm{an}}$ (in the sense originally defined) is indeed the analytic sheaf associated to $\mathcal{O}$. More generally, if $\mathcal{F}$ is a sheaf on $\mathcal{X}$ represented by a scheme of finite type Z, then $\mathcal{F}^{\mathrm{an}}$ is represented by the associated analytic space Z^{an}.

6. De Rham cohomology

Suppose T is a homotopy sheaf over $\mathcal{X}$. Then

$$Y \mapsto T(Y^{\mathrm{red}})$$

is a presheaf of spaces which we denote by T_{DR}. It is a homotopy sheaf (globally fibrant) because the restriction of the etale topology on Y is the etale topology of Y^{red}.

We will use this construction in particular in the case of a sheaf of sets represented by an object X; we obtain another sheaf of sets X_{DR}. This *de Rham construction* is functorial in an obvious way.

Suppose $Z \to S$ is a morphism. We define

$$Z_{DR/S} := Z_{DR} \times_{S_{DR}} S.$$

Note, for example, that $(X \times S)_{DR/S} = X_{DR} \times S$.

The following theorems show that the de Rham construction gives a sheaf-theoretical way to interpret de Rham cohomology.

Theorem 6.1. *Suppose $p : Z \to S$ is smooth with connected geometric fibers. Suppose $\mathcal{F}$ is a sheaf of sets on $\mathcal{X}$ which is P2. Then any morphism $\varphi : Z_{DR/S} \times S \to \mathcal{F}$ is relatively constant, that is factors through a morphism $\psi : S \to \mathcal{F}$.*

Proof: By Lemma 4.2, $P2 \Rightarrow B1 + B2$. Pick a point $z_0 \in Z$ lying over $s_0 \in S$; replacing S by an etale neighborhood of s_0 if necessary, we may then choose a section $\sigma : S \to Z$ passing through z_0. Composing with φ gives a morphism $\psi : S \to \mathcal{F}$. We claim that there is an etale neighborhood U of z_0 such that $\varphi|_U = (\psi \circ p)|_U$. This will show that φ is relatively constant on an etale open covering of Z, and since Z has connected fibers this gives the theorem. To prove the claim, let $\hat{Z} := Z \times_{Z_{DR}} \{z_0\}$ be the completion of Z at z_0, and let $Y := Spec(\hat{\mathcal{O}}_{Z,z_0})$ be the spectrum of the completion of the local ring at z_0. We have a factorization

$$\hat{Z} \xrightarrow{p} S \xrightarrow{p} Z_{DR/S}$$

of the inclusion morphism $\hat{Z} \to Z_{DR/S}$, thus $\varphi|_{\hat{Z}} = (\psi \circ p)|_{\hat{Z}}$. But now the property B2 for $\mathcal{F}$ implies that $\varphi|_Y = (\psi \circ p)|_Y$. Property B1 and Artin approximation give an etale neighborhood U of z_0 with a $\varphi|_U = (\psi \circ p)|_U$. □

Theorem 6.2. *If X is a smooth quasiprojective variety over k, then*

$$\pi_i(\underline{Hom}(X_{DR}, K(\mathbf{G}_a, n)), 0) = \mathbf{H}^{n-i}(X^{\mathrm{zar}}; (\Omega^{\cdot}_X, d))$$

is the algebraic de Rham cohomology (or zero if $i > n$). Here 0 is the constant morphism at the basepoint section. The equation means that the sheaf on the left is represented by the k-vector space on the right.

Proof: We have

$$\pi_i(\underline{Hom}(X_{DR}, K(\mathbf{G}_a, n)), 0) = \underline{H}^i(X_{DR}, \mathcal{O}).$$

We will define a complex of local systems denoted $(p_*\Omega^{\cdot}_X, d)$ on X_{DR}. Let Ω^i_X denote the coherent sheaf on $\mathcal{X}/X$ of differential i-forms on X. Let $p : X \to X_{DR}$ denote the canonical morphism. We claim that the exterior derivative is well defined on the direct image $p_*\Omega^{\cdot}$ making this into a complex resolving $\mathcal{O}$. If $Y \to X_{DR}$ is a morphism then it corresponds to a morphism $f : Y^{red} \to X$. Let $G \subset Y \times X$ denote the graph of f (noting that Y^{red} is a closed subscheme of Y). Then $X \times_{X_{DR}} Y$ is the formal completion $\hat{G}$ of $X \times Y$ along G. Replacing Y by an open subset if necessary, we can assume that the morphism f extends to a morphism $f' : Y \to X$. Furthermore, we may assume that the tangent bundle of X admits a frame in a neighborhood of the image of Y. Flowing formally along the vector fields of the framing gives a trivialization $\hat{G} \cong Y \times \hat{Z}$ where $\hat{Z}$ is the completion of an affine space (of dimension $dim(X)$) at the origin. We get

$$(p_*\Omega^i_X)(Y) = \Omega^i_{\hat{G}/Y}(\hat{G}) \cong \mathcal{O}(Y)\hat{\otimes}_k\Omega^i_{\hat{Z}}$$

where the tensor product is completed in an appropriate way. The differential d is provided by the relative differential of $\Omega^i_{\hat{G}/Y}(\hat{G})$. A formal version of the Poincaré lemma shows that the resulting complex is a resolution of $\mathcal{O}(Y)$. This gives the claim.

We can calculate using this resolution. We get

$$\underline{H}^i(X_{DR}, \mathcal{O}) = \underline{\mathbf{H}}^i(X_{DR}, (p_*\Omega^{\cdot}_X, d))$$

(where the underlined boldface H designates the sheaf of hypercohomology groups). We just have to show that the hypercohomology sheaf on the right is the sheaf represented by the classical (Zariski) hypercohomology of X with coefficients in the usual de Rham complex on the Zariski site. We show this for any smooth X. Note that for affine sufficiently small Y and for any coherent sheaf V on X,

$$\underline{H}^i(X_{DR}, p_*(V))(Y) = H^i(X \times Y, p_1^*V)$$

because the higher direct images are trivial. Take an open covering $U \to X$ with U a disjoint union of affine open subsets of X. Note that U_{DR} is cohomologically trivial for the component sheaves $p_*\Omega^i_X$ (by the above applied to U instead of X). Putting $U^n = U \times_X \ldots \times_X U$, we get a double complex

$$(p_*\Omega^i_X)(U^n \times Y)$$

with differentials d and the usual δ corresponding to the simplicial structure of the U^n, which calculates

$$\underline{\mathbf{H}}^i(X_{DR}, (p_*\Omega^{\cdot}_X, d))(Y).$$

But this double complex is equal to the double complex which calculates the usual algebraic de Rham cohomology $\mathbf{H}^i(X^{\mathrm{zar}}, (\Omega^{\cdot}_X, d))$, tensored over k by $\mathcal{O}(Y)$. The cohomology of the total complex is therefore the value on Y of the sheaf represented by the vector space $\mathbf{H}^i(X^{\mathrm{zar}}, (\Omega^{\cdot}_X, d))$. □

We need the following lemma in preparation for the next theorem.

Lemma 6.3. *If $\mathcal{G}$ is a sheaf of groups and $\mathcal{F}$ a sheaf, then principal $\mathcal{G}$-bundles over $\mathcal{F}$ are classified by morphisms $\mathcal{F} \to K(\mathcal{G}, 1)$. In fact, the moduli stack of principal bundles corresponds to the presheaf of spaces $\underline{Hom}(\mathcal{F}, K(\mathcal{G}, 1))$.* □

Theorem 6.4. *If X is a smooth quasiprojective variety and G is an affine group scheme over k, then*

$$\underline{Hom}(X_{DR}, K(G, 1))$$

is the 1-truncated homotopy sheaf associated to the moduli stack of principal G-bundles with integrable connection over X; thus $\pi_0 \underline{Hom}(X_{DR}, K(G, 1))$ is the sheaf of isomorphism classes of principal G-bundles with integrable connection, and if $\eta, \eta' : X_{DR} \to K(G, 1)$ are morphisms corresponding to principal bundles P, P' with flat connections, then $\pi_0 P^{\eta,\eta'} \underline{Hom}(X_{DR}, K(G, 1))$ is a sheaf represented by the scheme of flat sections of $Hom_G(P, P')$.

Proof: By the lemma, the space $\underline{Hom}(X_{DR}, K(G, 1))$ is the moduli stack of principal G-bundles $\mathcal{P}$ over X_{DR} (that is, sheaves $\mathcal{P}$ with free action of G such that the quotient is X_{DR}). Now a principal G-bundle over X_{DR} is the same thing as a principal G-bundle P on X with a relative trivialization of P on X/X_{DR}. A relative trivialization means a compatible system of trivializations τ_Y of $p_1^*(P)$ on $X \times_{X_{DR}} Y$ over Y, for every $Y \to X_{DR}$. On the one hand, given such a system of trivializations and applying it to the first infinitesimal neighborhood of the diagonal in $X \times X$, we obtain a connection ∇ on P. By looking at the compatibility condition and the neighborhood of the total diagonal in $X \times X \times X$, we get integrability of this connection. On the other hand, suppose ∇ is an integrable connection on P. Formal integration gives the desired trivialization τ_Y on $X \times_{X_{DR}} Y$ in view of the fact that this space is isomorphic to $Y \times \hat{Z}$ as in the previous proof (for Y suitably small).

The space of paths between two morphisms $X_{DR} \to K(G, 1)$ is 0-truncated, equal to the set of isomorphisms between the principal bundles $\mathcal{P}$ and $\mathcal{P}'$. By following the above argument this is the same as the space of isomorphisms between P and P' which are compatible with the connections. □

Suppose that $G = G\ell(V)$ for an n-dimensional vector space V. A morphism $\rho : X_{DR} \to K(G, 1)$ corresponds to a vector bundle $\mathcal{V}$ of rank n on X with integrable connection ∇. Let $T \to K(G, 1)$ be the fibration with a section, and fiber $K(V, m)$, which corresponds to the standard action of G on V. Let $F \to X_{DR}$ be the pullback of this fibration by ρ. This is a fibration with section σ, with fiber $K(V, m)$.

Lemma 6.5. *In the above situation,*

$$\pi_i(Sect(X_{DR}, F), \sigma) = \mathbf{H}^{m-i}(X; (\mathcal{V} \otimes \Omega^{\cdot}_X, \nabla))$$

is the algebraic de Rham cohomology of X with coefficients in the bundle $\mathcal{V}$ with connection ∇.

Proof: Similar to the proof of Theorem 6.2. □

In view of these results, if T is a very presentable homotopy sheaf we define the *generalized de Rham cohomology of X with coefficients in T* to be the homotopy sheaf $\underline{Hom}(X_{DR}, T)$.

Singular varieties

The theory sketched above and below should work also for singular projective varieties (with exactly the same definition of X_{DR})—note that if $\hat{Y}$ is the formal completion of a smooth space along X then $\hat{Y}_{DR} = X_{DR}$. I haven't checked the details of this. See [Ha] for the algebraic de Rham theory in the singular case. On the other hand, if X is a quasiprojective variety then the theory should work for the higher dimensional cohomology, but it breaks down at nonabelian H^1 where we get classes corresponding to systems of differential equations with irregular singularities.

7. Formal and algebraic de Rham cohomology

Theorem 7.1. *Suppose T is a very presentable homotopy sheaf on $\mathcal{X}$, and suppose X is a smooth projective variety. Then*

$$\underline{Hom}(X_{DR}, T) := \underline{\Gamma}(X_{DR}, X_{DR} \times T)$$

satisfies the boundedness properties B1 and B2.

It is not difficult to check the property B1, essentially because everything is given by cocycles.

We will give an outline of the proof of property B2. Suppose that A is a complete local k-algebra (completion of an algebra of finite type) and let $S = Spec(A)$. Let $\hat{S}$ denote the formal completion of S at the maximal ideal, that is it is the functor defined by $\hat{S}(Y)$ equal to the set of elements of $S(Y)$ such that Y^{red} maps to the closed point defined by the maximal ideal of A. We have to prove that the natural restriction morphism

$$GFGA(X_{DR} \times S, T) : Hom(X_{DR} \times S, T) \to Hom(X_{DR} \times \hat{S} \to T)$$

is a homotopy equivalence. Looking at a Postnikov tower and using Theorem 6.1, it suffices to treat the the spaces of sections $\Gamma(X_{DR} \times S, T)$ where $T = K(G/X_{DR} \times S, n)$ and G is either an affine presentable group sheaf pulled back from S ($n = 1$) or G is a local system which is a twist of the pullback of a vector sheaf from S ($n \geq 2$).

Theorem 7.2. *Suppose V is a vector sheaf. Suppose $\xi \in H^1(X_{DR} \times S, p_2^*(Aut(V)))$, and let K be the local system over $X_{DR} \times S$ obtained by twisting $p_2^*(V)$ by ξ. Then for any i the restriction morphism*

$$H^i(X_{DR} \times S, K) \to H^i(X_{DR} \times \hat{S}, K|_{X_{DR} \times \hat{S}})$$

is an isomorphism.

The proof is based on induction on the dimension of the support, and on the fact that if V is a vector sheaf then the global sections correspond to a coherent sheaf $\mathcal{F}$; then apply Grothendieck's GFGA to $\mathcal{F}$. □

This treats the case $n \geq 2$ in the Postnikov reduction above.

The main case is when $T = K(G, 1)$. We state this separately. If R is a topological space over $\mathcal{X}$ and G a local system of groups over R, let $H^1_{\text{cat}}(R, G)$ denote category of principal G-bundles over R.

Theorem 7.3. *Suppose S is complete and G is an affine presentable group sheaf over S. Then*

$$GFGA(X_{DR} \times S, G) : H^1_{\text{cat}}(X_{DR} \times S, G) \to H^1_{\text{cat}}(X_{DR} \times \hat{S}, G)$$

is an equivalence of categories.

Preliminary remarks

Here are a few statements (without the proofs) which are important parts of the argument.

Suppose G is an affine presentable group scheme over S, and suppose $Z \to S$ is a formally smooth S-scheme which is locally the formal completion of a smooth S-scheme of finite type along a closed smooth subscheme. If P is a principal G-bundle on $Z \times_S \hat{S}$, then there is a fixed etale covering $U \to Z$ such that $P|_{U \times_S \hat{S}} \cong U \times_S G \times_S \hat{S}$.

Suppose $f : G \to H$ is a morphism of presentable group sheaves over S, and suppose that the induced morphism $f(S) : G(S) \to H(S)$ is an isomorphism. Then for any formally smooth $Y \to S$, we have that $f(Y)$ is an isomorphism. If Z is the formal completion of a smooth S-scheme along an S-smooth subscheme, the induced morphisms

$$H^1_{\text{cat}}(Z_{DR/S}, G) \to H^1_{\text{cat}}(Z_{DR/S}, H)$$

and

$$H^1_{\text{cat}}(Z_{DR/S}|_{\hat{S}}, G) \to H^1_{\text{cat}}(Z_{DR/S}|_{\hat{S}}, H)$$

are equivalences of categories.

Suppose that G is a presentable group sheaf over S, and suppose that $S' \subset S$ is a closed subscheme. Let $U = S - S'$ be the complement, and suppose that $G|_U = \{1\}$. Then there is a closed subscheme $j : S'' \hookrightarrow S$ with the same reduced support as S, and a presentable group sheaf H on S'', together with a morphism $G \to j_*(H)$ inducing an isomorphism on sections over S. If G is affine (resp. abelian, affine abelian) then H can be chosen to be affine (resp. abelian, affine abelian). Furthermore, if S'' has dimension n we can then take a finite morphism $f : S'' \to Spec(k[[x_1, \ldots, x_n]])$ and consider the presentable group sheaf $f_*(H)$. In order to prove the theorem for G it suffices to prove it for $f_*(H)$. In this way we reduce to the case $S = Spec(k[[x_1, \ldots, x_n]])$ where n is the dimension of the support of G.

Finally, suppose $S = Spec(k[[x_1, \ldots, x_n]])$ and let η denote the generic point of S. For certain presentable group sheaves H including direct images of $\mathcal{O}^a \times \mathbf{G}_m^b$, if $G_\eta \to H_\eta$ is a morphism of generic fibers then there is a morphism $g : G' \to G$ with G' affine and presentable, such that g induces an isomorphism on generic fibers and such that the morphism $G'_\eta \to H_\eta$ extends to a morphism $G' \to H$.

Outline of the proof of 7.3:

There are basically two cases to treat: the case G affine abelian, and the case G a subgroup of $\underline{Aut}(V)$ for a vector sheaf V.

We first discuss the affine abelian case. Fix the base scheme S (complete as usual) and a smooth projective X. We need to show that if G is an abelian affine presentable group sheaf, then the morphism

$$GFGA^i(G) : H^i(X_{DR} \times S, G) \to H^i(X_{DR} \times \hat{S}, G)$$

is injective for $i = 2$ and bijective for $i \le 1$.

We reduce first to the case $S = Spec(k[[x_1, \ldots, x_n]])$, and note that the generic fiber G_η, which is affine and abelian, is a subgroup scheme of a direct image (for a field extension of $k(\eta)$) of a group scheme of the form $H' = \mathcal{O}^a \times \mathbf{G}_m^b$. Let S' be the integral closure of S in this field extension and let H be the direct image from S' to S of H'. For $\lambda \in \mathcal{O}(S)$ let $H'(\lambda) \subset H'$ be the union of the identity section and the subscheme defined by λ. Let $H(\lambda) \subset H$ be the direct image of $H'(\lambda)$.

By arguments with exact sequences and the 5-lemma, by induction on the dimension of the support, and using the extension results referred to above, we can reduce to the case $G = H/H(\lambda)$. This case can then be treated rather easily by hand.

Remark: The GFGA theorem is not true if we don't include the affine hypothesis; it breaks down in the abelian case, and we have used the fact that all fibers of G are affine in the above argument. One can get a counterexample by looking at a group scheme G over $S = Spec(k[[t]])$ such that the special fiber is $\mathbf{G}_m$ and the general fiber is an elliptic curve.

We need a slight improvement in the affine abelian case. Suppose G is an affine abelian presentable group sheaf. A group-object L over $X_{DR} \times T$ is said to be a *finite twisting of* G if there is a finite etale covering $X' \to X$ such that the pullback of L to $X'_{DR} \times T$ is isomorphic to the constant sheaf G. The GFGA isomorphism for finite twists of an affine abelian presentable group sheaf, follows from the untwisted case by a Hochschild-Serre spectral sequence.

Finally we treat the case of subgroups of $\underline{Aut}(V)$. Suppose $S = Spec(A)$ with A complete as usual, and suppose V is a vector sheaf over S. Then $M := V(S)$ is an A-module of finite type. Let $\mathcal{F}$ be the coherent sheaf on S associated to the A-module M. We have a morphism $j : \mathcal{F} \to V$ inducing $\mathcal{F}(S) \cong V(S)$. Let $\underline{Aut}(\mathcal{F}, V)$ be the group of automorphisms of the pair $(\mathcal{F}, V)$ preserving the morphism j. It is presentable (indeed, it is the equalizer of two maps $\underline{Aut}(\mathcal{F}) \times \underline{Aut}(V) \to \underline{Hom}(\mathcal{F}, V)$). We have a diagram

$$\underline{Aut}(\mathcal{F}) \stackrel{\alpha}{\leftarrow} \underline{Aut}(\mathcal{F}, V) \stackrel{\beta}{\rightarrow} \underline{Aut}(V).$$

Furthermore, the morphism β induces an isomorphism on sections over S, because an automorphism of V defined over S induces an unique automorphism of M and hence of $\mathcal{F}$.

Suppose $G \subset \underline{Aut}(V)$ is a presentable subgroup. To prove that $GFGA(X_{DR} \times S/S, G)$ is an isomorphism, we proceed by induction on the dimension of the support of G: suppose this dimension is n and suppose that we know the isomorphism (and in fact, suppose that we know Theorem 7.3) for presentable groups supported in dimension $\leq n-1$. As stated previously we can suppose that $S = Spec(k[[x_1, \ldots, x_n]])$. Let K be the fraction field of $A = k[[x_1, \ldots, x_n]]$. Then the restriction of G to $Spec(K)$ is an algebraic subgroup G_K of $G\ell(V_K)$. In particular, there is a line $L_K \subset V_K^{\otimes o(a,b)}$ (this notation means the tensor power a times of the dual, tensored with the power b times of V), such that G_K is the subgroup of $G\ell(V_K)$ fixing L. Note that $\mathcal{F}_K = V_K$. Let $\mathcal{L} \subset \mathcal{F}^{\otimes o(a,b)}$ be the subsheaf of sections which lie in L_K over $Spec(K)$. Note that it is a coherent subsheaf.

Using the induction on the dimension of the support, the relevant cohomology exact sequences, and the type of remarks given in the preceding subsection, we can reduce to the case of the group $H = \underline{Aut}(\mathcal{F}, \mathcal{L})$. The proof that $GFGA(X_{DR} \times S/S, H)$ is an isomorphism is then straightforward based on Grothendieck's GFGA theorem for coherent sheaves. We obtain the isomorphism in question for subgroups of $\underline{Aut}(V)$.

This completes the verification in the two cases. The argument for putting these cases together is the following. Suppose G is a presentable group sheaf. Its Lie algebra sheaf V is then a vector sheaf, and we have a morphism $G \to \underline{Aut}(V)$. Let G' denote the image. Let $C := G/G^o$ be the finite group sheaf of components. The tautological morphism yields a morphism

$$G \to G' \times C.$$

Let $G'' \subset G' \times C$ denote the image of this morphism. Let C' be a constant finite group equal to $C(S)$, then we get a morphism

$$G' \times C' \to G' \times C$$

inducing an isomorphism on sections over S. Pulling this back we get

$$G^3 \to G''$$

inducing an isomorphism on sections over S. Now C' is a finite group so it is a subgroup of the automorphisms of a vector space, so G^3 is a subgroup of the automorphisms of a vector sheaf. By the second case above, we know the theorem for G^3 and by the second preliminary remark we know GFGA for G''.

Let E denote the kernel of $G \to G''$. In this case E is abelian and the action of G on E factors through a finite group sheaf.

We now show the GFGA isomorphism for G. Note that there is a natural morphism

$$p : H^1(X_{DR} \times T, G) \to H^1(X_{DR} \times T, G'')$$

and if $\eta'' \in H^1(X_{DR} \times T, G'')$ then the fiber of p over η'' is a principal homogeneous space over $H^1(X_{DR} \times T, L)$ which is nonempty if and only if a certain obstruction in $H^2(X_{DR} \times T, L)$ vanishes; here L is the local system on $X_{DR} \times T$ obtained by twisting E by the class η''. Since the action of G'' on E factors through the finite group sheaf, there is a finite covering $X' \to X$ such that $L|_{X'} \cong E$, in other words L is a finite twisting of E.

The connecting morphism giving the obstruction is compatible with the GFGA morphism, and since we know that the GFGA morphism is injective on $H^2(X_{DR} \times T, L)$, the condition of vanishing of the obstruction is preserved by the GFGA morphism on $H^1(X_{DR} \times T, G'')$. On the other hand, the GFGA isomorphism for $H^1(X_{DR} \times T, L)$ implies that the GFGA morphism on $H^1(X_{DR} \times T, G)$ induces an isomorphism on the fiber over η''. Since this is true for any η'', we obtain the GFGA isomorphism for G. □

8. Betti and de Rham cohomology on the analytic site

Suppose now that $k = \mathbf{C}$.

Let X_B denote the homotopy sheaf associated to the presheaf of spaces obtained by setting $X_B^{\mathrm{pre}}(S)$ equal to the usual topological space underlying X^{an}. Let X_B^{an} denote the same on the analytic site.

Let X_{DR}^{an} denote the analytic sheaf associated to X_{DR} (this is the same as taking the usual definition of X_{DR} but in the analytic site).

Theorem 8.1. *Suppose T is a presentable homotopy sheaf over $\mathcal{X}$. The morphism of homotopy sheaves on the analytic site $\mathcal{X}^{\mathrm{an}}$*

$$\underline{Hom}(X_B, T)^{\mathrm{an}} \to \underline{Hom}(X_B^{\mathrm{an}}, T^{\mathrm{an}})$$

is a homotopy equivalence.

Proof: Prove this by induction on cell-attachments defining K. At each stage, the new $\underline{Hom}$ is a fiber product of ones already calculated. We reduce in this way to proving that if R, S and T are presentable flexible sheaves with morphisms $R \to T$ and $S \to T$, then

$$(R \times_T S)^{\mathrm{an}} \to R^{\mathrm{an}} \times_{T^{\mathrm{an}}} S^{\mathrm{an}}$$

is a homotopy equivalence. This follows from the exact sequences of Proposition 5.3. □

For an element S in the analytic site, $X_B^{\mathrm{an}}(S)$ is the space of continuous maps from S to the topological space underlying X. On the other hand (modulo the question), $X_{DR}^{\mathrm{an}}(S)$ is the space of analytic maps from S^{red} to X. Such an analytic map gives a continuous map, so we obtain a natural morphism

$$DRB : X_{DR}^{\mathrm{an}} \to X_B^{\mathrm{an}}.$$

Theorem 8.2. *Suppose T is a very presentable flexible sheaf, and suppose X is a smooth projective variety over $k = \mathbf{C}$. The morphism $X^{an}_{DR} \to X^{an}_B$ induces a homotopy equivalence*

$$\underline{Hom}(X^{an}_B, T^{an}) \cong \underline{Hom}(X^{an}_{DR}, T^{an}).$$

Proof: Note that the analogue of Theorem 6.1 holds in the analytic case. We may use the usual Postnikov reduction to reduce to the following statements—let DRB^* denote the pullback via the morphism DRB. Suppose U is a Stein space. First, if G is an affine presentable group sheaf on U (pullback of an affine presentable group sheaf on an algebraic scheme S by a map $U \to S$) then DRB^* induces an equivalence between the categories of principal G-bundles on X^{an}_{DR} and X^{an}_B. And second, if V is a vector sheaf on U and L is a twist of V over X^{an}_B, then DRB^* induces an isomorphism on cohomology with coefficients in L.

We may choose a covering of $X^{an} \times U$ consisting of open sets of the form $W_\alpha \times U_\alpha$ with W_α and all intersections contractible. Consider the statement about principal bundles. The category of principal bundles on $X^{an}_B \times U$ is equivalent to the category of principal G-bundles on $X \times U$ whose transition functions are constant in the X-direction (and with morphisms the sections constant in the X-direction). On the other hand, the principal bundles on $X^{an}_{DR} \times U$ are the pairs (P, ∇) where P is a principal G-bundle on $X^{an} \times U$ and ∇ is an integrable connection (relative to U)—which means that ∇ is, with respect to local trivializations, a section of $Lie(G) \otimes_{\mathcal{O}} \Omega^1_{X \times U/U}$. To show the equivalence of categories, it suffices to show that if P is a trivial principal G-bundle on $W_\alpha \times U_\alpha$ with connection ∇ then there is a trivialization of P such that ∇ corresponds to the zero section, and this trivialization is unique up to conjugation by a section of G which is constant in the W_α-direction (i.e. factoring through U_α).

This trivialization is obtained by the generalized Baker-Campbell-Hausdorff formula for the integration of an ordinary differential equation. To put this into effect, we note that we can choose a vertical surjection $Y \to G$ and an analytic neighborhood of the identity section $e \subset Y' \subset Y$ such that Y' is isomorphic to a neighborhood of the zero section in $T(Y)_e$. We may, using the good lift of the multiplication map given by verticality, exponentiate from a neighborhood of the zero section in $T(Y)_e$ into Y' (we can assume that the above identification is equal to this exponential). Then, noting that $T(Y)_e \to Lie(G)$ is a surjection of vector sheaves, we can apply the generalized B-C-H formula to get a section $W_\alpha \times U_\alpha \to Y'$. Note that in this formula we have to choose a parsing because we lift the operations of G and $Lie(G)$ to operations in Y and $T(Y)_e$ but which no longer satisfy the same associativity. The projection of our section into G gives the desired trivialization. For uniqueness, suppose $f : W_\alpha \times U_\alpha \to Y$ is a section, and let g denote the section $(f|_{P \times U_\alpha}) \circ p_2$ which is constant in the W_α-direction (here $P \in W_\alpha$ is a point). We have to show that (f, g) lifts into a section $W_\alpha \times U_\alpha \to R$ (at least locally around P) where $R \to Y \times_G Y$ is the relation scheme. For this, note that the lift exists over completions of the local rings, and then apply the analytic version of Artin approximation.

We turn now to the case of higher cohomology. Let L be a local system obtained as a twist of a vector sheaf V on U. We may assume that the twisting cocycle has coefficients which are constant in the X-direction. Let X^a (resp. U^a) denote the usual complex analytic spaces (rather than sheaves on our analytic site). We obtain a sheaf $\underline{L}$ over the topological space $X^a \times U^a$ whose sections are locally those sections of V which are constant in the X^a-direction. The Betti cohomology in question is just the Çech cohomology with coefficients in $\underline{L}$. Define similarly the sheaf of holomorphic functions which are constant in the X^a-direction, $\underline{\mathcal{O}}$. The Poincaré lemma says that the usual de Rham complex of sheaves $\Omega^{\cdot}_{X^a \times U^a/U^a}$ on $X^a \times U^a$ resolves $\underline{\mathcal{O}}$. The components of this complex are locally free sheaves of $\mathcal{O}_{X^a \times U^a}$-modules although the complex is only $\underline{\mathcal{O}}$-linear. The tensor product

$$\underline{L} \otimes_{\underline{\mathcal{O}}} \Omega^{\cdot}_{X^a \times U^a / U^a}$$

resolves $\underline{L}$. Thus we can calculate the Betti cohomology with this complex. On the other hand, a proof similar to the proof of Theorem 6.2 shows that this complex calculates the de Rham cohomology. □

9. The Betti-de Rham relationship in general

Lemma 9.1. *Suppose A and B are presheaves of spaces on $\mathcal{X}$ which are B1 and B2. Suppose*

$$\varphi : A^{\mathrm{an}} \to B^{\mathrm{an}}$$

*is a fibrant morphism of associated analytic presheaves, such that for any artinian S' of finite type over Spec(**C**), $\varphi(S')$ is a weak equivalence. Then φ is an Illusie weak equivalence.*

Proof: Suppose $K \subset L$ is an inclusion of finite CW complexes (we can take $L = B^n$ and $K = S^{n-1}$). Suppose U is a Stein analytic space with ring of functions $\mathcal{O}_U$. Suppose we have morphisms $K \to A^{\mathrm{an}}(U)$ and $L \to B^{\mathrm{an}}(U)$ forming a commutative diagram. We have to show that for any $u \in U$ there is an open neighborhood $u \in V \subset U$ and a lifting $L \to A^{\mathrm{an}}(V)$. For this, note that

$$A^{\mathrm{an}}(U) = A(Spec(\mathcal{O}_U)), \quad B^{\mathrm{an}}(U) = B(Spec(\mathcal{O}(U))).$$

By the condition B2 and the hypothesis that the liftings exist over artinian schemes, we obtain a lifting $L \to A(Spec(\hat{\mathcal{O}}_{U,u}))$ over the complete local ring. By the condition B1, there is an algebra $\mathcal{O}_U \subset C \subset \hat{\mathcal{O}}_{U,u}$ with C of finite type over $\mathcal{O}_U$, such that there is a lifting $L \to A(Spec(C))$. Note that in fact $C = \mathcal{O}_U \otimes_D C'$ for a subalgebra $D \subset \mathcal{O}_U$ of finite type over **C** and an extension $D \subset C'$ of finite type. The inclusion $D \subset \mathcal{O}_U$ corresponds to an analytic morphism $U \to Spec(D)$. Over the completion of the local ring we get a lifting into $Spec(C')$ so by Artin approximation in the analytic setting, there is an open neighborhood $u \in V \subset U$ and an analytic morphism $V \to Spec(C')$ lifting the morphism $V \subset U \to Spec(D)$. We obtain, in particular, $Spec(\mathcal{O}_V) \to Spec(C)$ and hence we get a lifting $L \to A(Spec(\mathcal{O}_V)) = A^{\mathrm{an}}(V)$ as desired. □

Theorem 9.2. *Suppose T is a very presentable homotopy sheaf, and suppose X is a smooth projective variety over $k = \mathbf{C}$. The morphism of homotopy sheaves on the analytic site*

$$\underline{Hom}(X_{DR}, T)^{\mathrm{an}} \to \underline{Hom}(X_{DR}^{\mathrm{an}}, T^{\mathrm{an}})$$

is a homotopy equivalence.

Proof: From the previous results the right hand side is homotopic to $\underline{Hom}(X_B, T)^{\mathrm{an}}$. The result over Spec(**C**) follows from Postnikov induction, from Theorems 6.1, 6.2, 6.4 and 6.5, and from the usual GAGA comparisons between de Rham and Betti cohomology. If S' is artinian of finite type over **C** then $\underline{Hom}(S', T)$ is again very presentable (the homotopy group sheaves are direct images of those for T and direct image preserves the conditions in question). Note that for any presheaf of spaces $\mathcal{F}$ we have

$$\underline{Hom}(\mathcal{F}, T)^{\mathrm{an}}(S') = \underline{Hom}(\mathcal{F}, T)(S') = \underline{Hom}(\mathcal{F}, \underline{Hom}(S', T))(Spec(k)).$$

We get the result in question over artinian schemes of finite type. The result in general follows from the preceding lemma, from Theorem 7.1 saying that $\underline{Hom}(X_{DR}, T)$ is B1 and B2, and from Corollary 5.6 and Lemma 4.2 which together imply that $\underline{Hom}(X_B, T)$ is B1 and B2. □

Corollary 9.3. *Suppose T is a very presentable flexible sheaf, and suppose X is a smooth projective variety over $k = \mathbf{C}$. There is a natural homotopy equivalence*

$$\underline{Hom}(X_B, T)^{\mathrm{an}} \cong \underline{Hom}(X_{DR}, T)^{\mathrm{an}}.$$

□

Presentability of generalized de Rham cohomology

Theorem 9.4. *If T is a very presentable flexible sheaf and X is smooth and projective over k then $\underline{Hom}(X_{DR}, T)$ is presentable.* □

The proof of this is a bit long, but somewhat similar to the proof of the GFGA result Theorem 7.1. The main case, where $T = K(G/S, 1)$, goes through an analysis of the cases G affine abelian, and G a subgroup of $\underline{Aut}(V)$. It is worthy to note that the proof uses quite strongly the condition P1 already known for $H^1(X_B \times S, G)$ and the GAGA comparisons established above.

Theorem 9.4 allows us to define the *de Rham shape* of X to be the functor

$$Shape(X_{DR}) : T \mapsto \underline{Hom}(X_{DR}, T)$$

from the category of very presentable spaces to itself.

References

[Ar] M. Artin. Algebraic approximation of structures over complete local rings, *Publ. Math. I.H.E.S.* **36** (1969), 23–58.

[Br] L. Breen. Extensions du groupe additif, *Publ. Math. I.H.E.S.* **48** (1978), 39–126.

[D-M] P. Deligne, D. Mumford. The irreducibility of the space of curves of a given genus, *Publ. Math. I.H.E.S.* **36** (1969), 75–110.

[Gr1] A. Grothendieck. On the de Rham cohomology of algebraic varieties. *Publ. Math. I.H.E.S.* **29** (1966), 95–103.

[Gr2] A. Grothendieck. *Pursuing Stacks*, unpublished manuscript.

[Ha] R. Hartshorne. On the de Rham cohomology of algebraic varieties. *Publ. Math. I.H.E.S.* **45** (1975), 5–99.

[Hi] A. Hirschowitz. Cohérence et dualité sur le gros site de Zariski, *Algebraic curves and projective geometry* (Ballico, Ciliberto eds.), *Lecture Notes in Math.* **1389** (1989), 91–102.

[Il] L. Illusie. *Le complexe cotangent I et II, Lecture Notes in Math.* **239, 283**, Springer-Verlag, Berlin (1971,1972).

[Ja] J.F. Jardine. Simplicial presheaves, *J. Pure and Appl. Algebra* **47** (1987), 35–87.

[Th] Thomason. Algebraic K-theory and etale cohomology, *Ann. Sci. E.N.S.* **18** (1985), 437–552.

REMARKS ON ZERO-CYCLES OF SELF-PRODUCTS OF VARIETIES

Claire Voisin

URA 1169 du CNRS, Université d'Orsay, Mathématique, Bâtiment 425, 91405 Orsay cedex, France

1. Introduction

1.1 This paper proposes a few results and rises many questions on algebraic cycles on self-products X^k of a variety X. If X is a curve, it is well known that $CH_0(X^{(k)}) = CH_0(X^{(k+1)})$ for $k \geq g(X)$ where $X^{(k)}$ denotes the kth symmetric product of X, and the isomorphism is given by the map μ_{x_0}: $X^{(k)} \to X^{(k+1)}$, $\mu_{x_0}(z) = z + x_0$, for any point $x_0 \in X$. Correspondingly, one has: $\forall k \geq g(X)$, $\forall l \in \mathbb{N}$, $H^0(\Omega^l_{X^{(k+1)}}) = H^0(\Omega^l_{X^{(k)}})$, which is the effect on holomorphic forms of the previous equality, using Mumford-Roitman theorem ([16], [22]). This last fact generalizes to higher dimensional varieties as follows: Assume $H^0(\Omega^l_X) = 0$ for l even, $l \neq 0$; then for $k \geq \sum_{i=1}^{\dim X} h^{i,0}$ and any l, $H^0(\Omega^l_{X^k})^{\mathfrak{S}_k} = H^0(\Omega^l_{X^{k+1}})^{\mathfrak{S}_{k+1}}$, where $()^{\mathfrak{S}_\cdot}$ means the invariant part under the action of the symmetric group $\mathfrak{S}_\cdot$. According to Bloch-Beilinson conjectures [11], this should imply that $CH_0(X^k)^{\mathfrak{S}_k} = CH_0(X^{k+1})^{\mathfrak{S}_{k+1}}$ for $k \geq \sum_{i=1}^{\dim X} h^{i,0}$. I have no general result on this but I will construct in section 3 families of threefolds with no $H^{2,0}, H^{1,0}$ and which satisfy: $CH_0(X^k)^{\mathfrak{S}_k} = CH_0(X^{k+1})^{\mathfrak{S}_{k+1}}$ for $k \geq h^{3,0}(X)$.

1.2 Now, let us consider surfaces: If we have the equality $CH_0(S^k)^{\mathfrak{S}_k} = CH_0(S^{k+1})^{\mathfrak{S}_{k+1}}$ for some k, then, again by Mumford-Roitman theorem, this implies that the symmetric holomorphic forms of degree $2(k+1)$ on S^{k+1} vanish. Since the space of such forms is equal to $S^{k+1}H^{2,0}(S)$ it follows that $H^{2,0}(S) = 0$. Now Bloch's conjecture for surfaces asserts that if $H^{2,0}(S) = 0, CH_0(S)_0 \cong Alb(S)$, where $CH_0(S)_0$ denotes the group of zero cycles of degree zero, and the isomorphism is given by the Albanese map. So one can hope to split Bloch's conjecture into steps:

1) Show that $H^{2,0}(S) = 0 \Rightarrow CH_0(S^k)^{\mathfrak{S}_k} = CH_0(S^{k+1})^{\mathfrak{S}_{k+1}}$ for some $k \geq 0$.

2) Show that $CH_0(S^k)^{\mathfrak{S}_k} = CH_0(S^{k+1})^{\mathfrak{S}_{k+1}}$ for some $k \geq 0$ implies $CH_0(S)_0 \cong Alb(S)$.

We will show in section 2 the following first non trivial case of 2):

Theorem.(See 2.2) *Suppose $CH_0(S^2)^{\mathfrak{S}_2} = CH_0(S)$; then $CH_0(S)_0 = Alb(S)$.*

The remainder of this section is devoted to some speculations on how the equality $CH_0(X^k)^{\mathfrak{S}_k}$ $= CH_0(X^{k+1})^{\mathfrak{S}_{k+1}}$ for a threefold (we will construct such threefolds with $k = h^{3,0}$ arbitrary, $h^{2,0} = h^{1,0} = 0$ in section 3) could lead to a proof of the representability of $CH_1(X)_{alg}$.

1.3 In the third section, we will turn to the study of skew zero-cycles on self-products of a surface S, that is zero-cycles $z \in CH_0(S^k)$ such that $\sigma_*(z) = \epsilon(\sigma)z$, $\forall \sigma \in \mathfrak{S}_k$.

Suppose S is a smooth projective surface with $h^{1,0} = 0$; then the space of skew holomorphic forms on S^k, that is the set $\{\omega \in \bigoplus_l H^0(\Omega^l_{S^k})/\sigma^*(\omega) = \epsilon(\sigma)\omega, \forall\sigma \in \mathfrak{S}_k\}$ is equal to $\bigwedge^k H^0(\Omega^2_S) \oplus \bigwedge^{k-1} H^0(\Omega^2_S)$. It follows that for $k \geq h^{2,0}(S)$, one has: the map $\widetilde{pr^*_{1...k}} : H^0(\Omega^{\cdot}_{S^k})^{skew} \to H^0(\Omega^{\cdot}_{S^{k+1}})^{skew}$, obtained as the composite of $pr^*_{1...k}$ and antisymmetrization, is surjective. Again according to Bloch-Beilinson conjectures, one expects the injectivity of the map $pr_{1...k*} : CH_0(S^{k+1})^{skew} \to CH_0(S^k)^{skew}$ for $k \geq h^{2,0}(S)$. This is clearly equivalent to the following:
(*) for $z_1, \ldots, z_k$, zero-cycles of degree zero on S, for $k \geq h^{2,0}(S)+1$, one has $\sum_{\sigma\in\mathfrak{S}_k} \epsilon(\sigma)\sigma^*(z_1 * \ldots * z_k) = 0$ in $CH_0(S^k)$. This section provides several constructions of surfaces satisfying (*). This property is especially beautiful in the case of $K3$-surfaces, where it simply means: Let z, z' be zero-cycles of degree zero on S; then $z * z' = z' * z$ in $CH_0(S \times S)$. We prove this for Kummer surfaces, and for a certain ten dimensional family of $K3$ surfaces with generic Picard number equal to 10 (3.3, 3.4). We use this to show the existence of threefolds with arbitrary $h^{3,0} \neq 0$ (hence CH_0 non representable) satisfying the equality $CH_0(X^k)^{\mathfrak{S}_k} = CH_0(X^{k+1})^{\mathfrak{S}_{k+1}}$ for $k \geq h^{3,0}(X)$ which contrasts very much with Theorem 2.2 for $k = h^{3,0} = 1$. We study also higher dimensional skew cycles on self-products of Kummer surfaces, in order to illustrate the conjecture 3.8, which generalizes the property mentioned above. In section 4, we describe the Kuga-Satake construction and try to show that it should be related to this conjecture.

I would like to thank the Tata Institute, where the main part of this work was done, for its hospitality, and the "Indofrench Centre for the Promotion of advanced Research" for its financial support during my stay at Bombay. I thank the Taniguchi foundation and Professor Maruyama for inviting me to this symposium and allowing me to present these ideas.

2. Symmetric zero-cycles

2.1 Let S be a smooth projective variety over $\mathbb{C}$ and let $x_0 \in S$. The following properties are equivalent:

2.1.1 *$CH_0(S^k)^{\mathfrak{S}_k}$ is supported on $\bigcup_{i=1}^k x_0^i(S^{k-1})$* , where $x_0^i : S^{k-1} \to S^k$ is the inclusion with x_0 on the ith factor.

2.1.2 *The map $pr_{1...k-1} : S^k \to S^{k-1}$ induces a bijective map $(pr_{1...k-1})_* : CH_0(S^k)^{\mathfrak{S}_k} \to CH_0(S^{k-1})^{\mathfrak{S}_{k-1}}$.*

2.1.3 *For $z \in CH_0(S)_0$, one has $z^{*k} = 0$ in $CH_0(S^k)$, where for $z = \sum_i n_i x_i$, $z' = \sum_j m_j y_j$, $z * z' = \sum_{i,j} n_i m_j (x_i, y_j)$.*

To see this, one notes first that it suffices to prove these equivalences for Chow groups with rational coefficients, using the divisibility of $CH_0{}^0$ and Roitman's theorem on torsion [5]. Next one has the following fact:

2.1.4 *For any k, one has a direct sum decomposition:*

$$CH_0(S^k)^{\mathfrak{S}_k}{}_{\mathbb{Q}} = (\bigoplus_{l=0}^{k} x_0{}^{*l} * CH_0{}^0(S)^{*(k-l)})^{\mathfrak{S}_k}{}_{\mathbb{Q}}.$$

Proof of 2.1.4. The surjectivity of the natural map μ from the right hand side to the left hand side of this equality comes from the fact that $CH_0(S^k)^{\mathfrak{S}_k}{}_{\mathbb{Q}}$ is generated by 0-cycles z^{*k}, $z \in CH_0(S)$: any such z can be written as $z' + (d^0(z)x_0$, where $z' \in CH_0{}^0(S)$, so

$z^{*k} = (z' + (d^0(z)x_0)^{*k}$, and the development of the right hand side in powers of x_0 gives the surjectivity of μ.

To see the injectivity, assume:

(*) $0 = \sum_{l=0}^{k} z_l$, where $z_l \in (x_0{}^{*l} * CH_0{}^0(S)^{*(k-l)})^{\mathfrak{S}_k}{}_{\mathbb{Q}}$ are not all zero, and let k_0 be the greatest integer such that $z_{k_0} \neq 0$.

Then applying the projection $pr_{1\ldots k-k_0} =: p$ on the first factor S^{k-k_0} of S^k, one gets:

(**) $0 = p_*(\sum_{l=0}^{k_0} z_l)$.

But each z_l can be written by definition as $z_l = (\sum_{d^0(I)=k-l} j_{I*})(z'_l)$, where $I \subset \{1, \ldots, k\}$, $j_I : S^{k-l} \longrightarrow S^k$ is the inclusion with x_0 on the j-th factor for $j \notin I$, and $z'_l \in (CH_0{}^0(S)^{*(k-l)})^{\mathfrak{S}_{k-l}} \subset CH_0(S^{k-l})$. It follows that $p_*(z_l) = 0$ for $l < k_0$, and $p_*(z_{k_0}) = z'_{k_0}$. So (**) gives $z'_{k_0} = 0$, hence $z_{k_0} = 0$, a contradiction.

2.1.4 implies the equivalence of 2.1.1 and 2.1.3, because it shows that the quotient

$$CH_0(S^k)^{\mathfrak{S}_k}{}_{\mathbb{Q}} / (\sum_{i=1}^{k} x^i_{0*}(CH_0(S^{k-1})))^{\mathfrak{S}_k}{}_{\mathbb{Q}}$$

is isomorphic to $CH_0{}^0(S)^{*k})^{\mathfrak{S}_k}{}_{\mathbb{Q}}$.

2.1.2 $\Rightarrow$ 2.1.3 follows from the fact that $(CH_0{}^0(S)^{*k})^{\mathfrak{S}_k}{}_{\mathbb{Q}}$ is contained in $\mathrm{Ker}((pr_{1\ldots k-1})_*)$.

Finally the proof of 2.1.4 gives the implication 2.1.3 $\Rightarrow$ 2.1.2 as follows: under the assumption 2.1.3, any $z \in CH_0(S^k)^{\mathfrak{S}_k}{}_{\mathbb{Q}}$ is written as $z = \sum_{l=1}^{k_0} z_l$, with $1 \leq k_0 \leq k$ and $z_{k_0} \neq 0$. Then we have shown: $pr_{1\ldots k-k_0 *}(z) \neq 0$; so a fortiori $pr_{1\ldots k-1 *}(z) \neq 0$ and $pr_{1\ldots k-1*}$ is injective. Its surjectivity is always true, so 2.1.3 $\Rightarrow$ 2.1.2 is proved.

Now we will prove:

2.2 Theorem. *Let S be a surface satisfying one of these properties for $k = 2$; then $CH_0(S)_0 = Alb(S)$.*

Proof: Note first of all that the hypothesis implies $h^{2,0}(S) = 0$ (1.2). Consider now the Chow-Künneth decomposition of the diagonal Δ ([18]): $\Delta = \Delta_0 + \ldots + \Delta_4$, where the Δ_i's are algebraic cycles on $S \times S$, the classes of which are the projectors $H^{\cdot}(S) \longrightarrow H^i(S)$. The action of Δ_1 and Δ_0 on $CH_0(S)$ is trivial. The image of Δ_4 is isomorphic to $\mathbb{Z}$, and the image of Δ_3 is isomorphic to $Alb(S)$. Δ_2 is symmetric and its cohomology class is the projector on $H^2(S)$. Since $H^{2,0}(S) = \{0\}$, one has an equality of cohomology classes: $cl(\Delta_2) = \sum_{i,j} cl(Z_i) \otimes cl(Z'_j)$ where Z_i, Z'_j are divisors on S. One may assume that the cycle $\sum Z_i * Z'_j$ is a symmetric cycle of $S \times S$. Now consider on $S \times S \times S \times S$ the cycle $\Gamma = graph(Identity) + graph(\tau)$, where τ is the involution exchanging factors. By assumption, for $(s, s') \in S \times S, \Gamma((s, s'))$ is supported on $x_0 \times S \cup S \times x_0$ hence, following Bloch and Srinivas [6], we conclude that a multiple of Γ is rationally equivalent to a sum $\Gamma_1 + \Gamma_2$ where Γ_1 is supported on $D \times S \times S$ for some divisor D of $S \times S$ and Γ_2 is supported on $S \times S \times x_0 \times S \cup S \times S \times S \times x_0$.

Then, since $\Delta_2 - \sum Z_i * Z'_j$ is symmetric, we have ${}^t\Gamma(\Delta_2 - \sum Z_i * Z'_j) = 2(\Delta_2 - \sum Z_i * Z'_j)$, and from the decomposition above for $N\Gamma$ one concludes as in [6] that $2N(\Delta_2 - \sum Z_i * Z'_j)$ is algebraically equivalent to zero in $S \times S$.

Now one has the following lemma:

2.3 Lemma. *Let S be a variety, and Z a cycle of codimension $n = \dim S$ in $S \times S$ which is algebraically equivalent to zero. Then the morphism: $[Z] : CH_0(S) \longrightarrow CH_0(S)$ induced by Z is nilpotent.*

The lemma implies the theorem because $\Delta_2 - \sum Z_i * Z'_j$ induces by construction the identity map on Ker(alb). Since a multiple of it is nilpotent by 2.2 and 2.3, Ker(alb) must be of torsion; but it has no torsion by Roitman's theorem [5] so Ker(alb) $= 0$.

Proof of 2.3. Let $\circ$ denote the composition of correspondences in $S \times S$; we will prove in fact the following stronger statement:

2.3.1 *Z is nilpotent in $CH^n(S \times S)$ for the multiplication $\circ$.*

Proof. There exists a smooth curve C, a correspondence $\Gamma \subset C \times S \times S$ of codimension n, and a zero-cycle z of degree zero on C such that $Z = \Gamma(z) = pr_{23*}(\Gamma \cdot pr_1{}^*(z))$. Consider then the correspondence $\Gamma^{(k)}$ on $C^k \times S \times S$ defined as follows: Start with the product $\Gamma^{*k} \subset C^k \times (S \times S)^k$ and let Δ be the cycle of codimension $n(k-1)$ defined as $\{(x_1, \ldots, x_{2k}) \in (S \times S)^k / x_2 = x_3, \ldots, x_{2k-2} = x_{2k-1}\}$; then define $\Gamma^{(k)} = pr_{C^k \times S \times S*}(\Gamma^{*k} \cdot pr_{(S\times S)^k}{}^*(\Delta))$. In the projection $pr_{C^k \times S \times S}$ we are considering the first and last factors of $(S \times S)^k$. The meaning of $\Gamma^{(k)}$ is simply the following: Let $(c_1, \ldots, c_k) \in C^k$; then $\Gamma^{(k)}((c_1, \ldots, c_k)) = \Gamma(c_1) \circ \ldots \circ \Gamma(c_k)$.

Now let $z = \sum_{i=1}^m n_i c_i, c_i \in C, \sum n_i = 0$. Then:

$$z^{*k} = \sum_{f:\{1,\ldots,k\} \to \{1,\ldots,m\}} n_{f(1)} \ldots n_{f(k)} (c_{f(1)}, \ldots, c_{f(k)}),$$

so $\Gamma^{(k)}(z^{*k}) = \sum_{f:\{1,\ldots,k\} \to \{1,\ldots,m\}} n_{f(1)} \ldots n_{f(k)} \Gamma(c_{f(1)}) \circ \ldots \circ \Gamma(c_{f(k)}) = \Gamma(z)^{\circ k}$. Now this is finished because we know that $z^{*k} = 0$ in $CH_0(C^k)$ for $k \geq g(C) + 1$ (1.1, 2.1), hence $\Gamma(z)^{\circ k} = \Gamma^{(k)}(z^{*k}) = 0$ in $CH^n(S \times S)$ for $k \geq g(C) + 1$.

2.4 Remark. In fact, one can prove Theorem 2.2 without Lemma 2.3, just applying Bloch-Srinivas argument with more precision; However, we have proved in fact the following strengthening of 2.2:

2.4.1 *Assume the surface S satisfies: there exists a subset $\Sigma \subset S \times S$ of dimension at most two, such that symmetric zero-cycles of $S \times S$ are supported on Σ; Then $CH_0(S)_0 \cong Alb(S)$.*

2.5 Unfortunately, the proof of Theorem 2.2 does not extend to more general situations. There are the following generalizations of Theorem 2.2 that would be very interesting to know:

2.5.1 Conjecture. *Let S be a surface and suppose that for some $k, CH_0(S^k)^{\mathfrak{S}_k} = CH_0(S^{k+1})^{\mathfrak{S}_{k+1}}$. Then $CH_0(S)_0 \cong Alb(S)$.*

The next question concerns correspondences between any variety X and a surface Σ. Suppose we have a codimension two cycle $\Gamma \subset X \times \Sigma$; then we can consider Γ^{*k} in $X^k \times \Sigma^k$ and its effect on symmetric zero-cycles: $[\Gamma^{*k}] : CH_0(X^k)^{\mathfrak{S}_k} \to CH_0(\Sigma^k)^{\mathfrak{S}_k}$. Now the effect of Γ^{*k} on symmetric holomorphic $2k$-forms: $[\Gamma^{*k}]^{2k,0} : H^{2k,0}(\Sigma^k) \to H^{2k,0}(X^k)$ is just the map $S^k[\Gamma^{2,0}] : S^k H^{2,0}(\Sigma) \to S^k H^{2,0}(X)$ induced by $[\Gamma^{2,0}] : H^{2,0}(\Sigma) \to H^{2,0}(X)$. So if $[\Gamma^{*k}]^{2k,0} = 0$, $[\Gamma^{2,0}] = 0$ and Bloch's conjecture implies that $[\Gamma] : CH_0(X)_0 \to CH_0(\Sigma)_0$ factors through $Alb(X)$. Since $[\Gamma^{*k}]^{2k,0} = 0$ when $[\Gamma^{*k}] : CH_0(X^k)^{\mathfrak{S}_k} \to CH_0(\Sigma^k)^{\mathfrak{S}_k}$ vanishes on the subgroup $((CH_0(X)_0)^{*k})^{\mathfrak{S}_k}$, we have the following:

2.5.2 Conjecture. *Let $\Gamma \subset X \times \Sigma$ be a codimension two-cycle; suppose that for some k, $[\Gamma^{*k}] : CH_0(X^k)^{\mathfrak{S}_k} \to CH_0(\Sigma^k)^{\mathfrak{S}_k}$ vanishes on the subgroup $((CH_0(X)_0)^{*k})^{\mathfrak{S}_k}$. Then $[\Gamma] : CH_0(X)_0 \to CH_0(\Sigma)_0$ factors through $Alb(X)$.*

Of course 2.5.1, 2.5.2 are just particular cases of Bloch's conjecture, but the assumptions are much stronger so they could be more accessible. Concerning 2.5.2, Bloch-Srinivas argument gives the following proposition in the case $k = 2$. In 3.6 we construct threefolds with $h^{3,0} = 1$, and satisfying:

2.5.3 $CH_0(X^2)^{\mathfrak{S}_2} \cong CH_0(X)$, (so, as in 2.1 $((CH_0(X)_0)^{*2})^{\mathfrak{S}_2} = \{0\}$).

For them we have:

2.5.4 Proposition *Assume X satisfies $h^{1,0} = 0$ and the property 2.5.3; Then, for any correspondence $\Gamma \subset X \times \Sigma$, the composite ${}^t\Gamma \circ \Gamma : CH_0(X)_0 \to CH_1(X)_{alg}$ is equal to zero.*

We just sketch the proof of 2.5.4, since it is essentially an application of Bloch-Srinivas argument, as in the proof of 2.2; the point is that the assumptions imply that the codimension two cycle ${}^t\Gamma \circ \Gamma \subset X \times X$ is symmetric and homologous to zero modulo cycles coming from X via the projections. Now, also by assumption, symmetric zero-cycles on $X \times X$ are supported on the diagonal. Then, by [6], adapted as in the proof of 2.2, we conclude that a codimension two cycle Z symmetric and homologous to zero in $X \times X$ is rationally equivalent to zero iff its restriction to the diagonal is rationally equivalent to zero, and its Abel-Jacobi invariant vanishes. Finally, because $H^1(X) = \{0\}$, the invariant part under $\mathfrak{S}_2$ of $J^3(X \times X)$ identifies with $J^3(X)$ and the Abel-Jacobi map on codimension two symmetric cycles of $X \times X$ identifies with the restriction to the diagonal, followed by the Abel-Jacobi map of X. But then it is clear that ${}^t\Gamma \circ \Gamma$ is rationally equivalent to a cycle of the form $pr_1^*(Z) + pr_2^*(Z)$, which implies immediately 2.5.4.

2.6 2.5.2 could be a step for the study of the kernel of the Abel-Jacobi map for threefolds. We have the following:

2.6.1 Conjecture. *Let X be a threefold, and suppose that $h^{1,0}(X) = h^{2,0}(X) = 0$. Then the Abel-Jacobi map*

$$\Phi_X : CH_1(X)_{alg} \to J^3(X) = H^3(X, \mathbb{C})/(F^2H^3(X) + H^3(X, \mathbb{Z}))$$

is injective.

2.7 We will construct in the next section (3.6) threefolds satisfying the assumptions above, and such that the subgroup $((CH_0(X)_0)^{*k})^{\mathfrak{S}_k}$ of $CH_0(X^k)$ is zero for $k \geq h^{3,0}(X) + 1$. (This property should be satisfied by all threefolds with $h^{1,0} = h^{2,0} = 0$). These threefolds have $h^{3,0} \neq 0$, so they have a huge CH_0 group and are interesting examples to study 2.6.1. (Note that 2.6.1 is true if $CH_0(X)$ is representable [6]).

2.8 Consider such a threefold, and suppose we have a family of one-cycles in X parametrized by a surface Σ. So we have a correspondence $\Gamma \subset X \times \Sigma$ of codimension two; if 2.5.2 is true and X is as in 2.6, 2.7, the map $[\Gamma] : CH_0(X) \to CH_0(\Sigma)$ factors through the degree map. One concludes then as in Bloch-Srinivas [6], that a multiple $N\Gamma$ of Γ is supported on $D \times \Sigma \cup X \times \sigma_0$ for some divisor D of X and zero-cycle σ_0 of Σ, and then, that the map $N^t\Gamma : CH_0(\Sigma) \to CH_1(X)$ factors through a map: $CH_0(\Sigma) \to CH_1(D)$. Since D is a surface $CH_1(D)_{alg}$ is an abelian variety, so one concludes that $N^t\Gamma : CH_0(\Sigma)_0 \to CH_1(X)_{alg}$ factors through $Alb(\Sigma)$, hence also ${}^t\Gamma : CH_0(\Sigma)_0 \to CH_1(X)_{alg}$, since Ker($alb$) is divisible. So the next interesting question to solve would be the following:

2.9 Conjecture. *Let X be a threefold, and assume that for any codimension two correspondence $\Gamma \subset X \times Y$, the map ${}^t\Gamma : CH_0(Y)_0 \to CH_1(X)$ factors through $Alb(Y)$. Then*

$CH_1(X)_{alg}$ is representable, and $\Phi_X : CH_1(X)_{alg} \to J^3(X)$ is injective.

3. Skew-cycles on surfaces

We want to construct examples of surfaces with $h^{1,0} = 0$ and which satisfy the expected property:

(*) the map $CH_0(S)_0^{\otimes k} \to CH_0(S^k)$ given by $*$-product vanishes on $(CH_0(S)_0^{\otimes k})^{skew}$ for $k \geq h^{2,0}(S) + 1$ (see 1.3).

We use first 2.1 applied to curves, which satisfy the assumptions of 2.1 by 1.1; so we have:

3.0.1. *Let C be a curve of genus g; then for $z_1, \ldots, z_k$, zero-cycles of degree zero on C, $k \geq g+1$, one has: $\sum_{\sigma\in\mathfrak{S}_k} \sigma^*(z_1 * \ldots * z_k) = 0$ in $CH_0(C^k)$.*

This gives the first example as follows:

3.1 Proposition. *Let C_1 be an elliptic curve, C_2 a hyperelliptic curve of genus g_2, and denote by i_1, i_2 the hyperelliptic involutions of C_1, C_2. Then the surface $S = \widetilde{C_1 \times C_2/(i_1, i_2)}$ satisfies the property (*). Here $\widetilde{C_1 \times C_2}/(i_1, i_2)$ is the desingularization of $C_1 \times C_2$ obtained by blowing up the fixed points of (i_1, i_2).*

Proof: We have essentially to prove: let $z_1, \ldots, z_k$ be zero-cycles on $C_1 \times C_2$, of degree zero and invariant under (i_1, i_2). Then some multiple of $\sum_{\sigma\in\mathfrak{S}_k} \epsilon(\sigma)\sigma^*(z_1 * \ldots * z_k)$ vanishes for $k \geq g_2 + 1 = h^{2,0}(S) + 1$. This is sufficient by [5] which says that the Chow groups considered have no torsion. Now write $z_i = \sum_j n_i^j(c_i^j, d_i^j)$; then one verifies that $4z_i = 2(z_i + (i_1, i_2)z_i) = \sum_j n_i^j(c_i^j - i_1c_i^j) * (d_i^j - i_2d_i^j)$. So $4^k\, z_1 * \ldots * z_k = \sum_{(j_1,\ldots,j_k)} n_1^{j_1} \ldots n_k^{j_k}(c_1^{j_1} - i_1c_1^{j_1}) * \ldots * (c_k^{j_k} - i_1c_k^{j_k}) * (d_1^{j_1} - i_2d_1^{j_1}) * \ldots * (d_k^{j_k} - i_2d_k^{j_k})$. Now 3.0.1 applied to the elliptic curve C_1 shows that the $*$-product between zero-cycles of degree zero of C_1 is skew-symmetric. It follows that $4^k \sum_{\sigma\in\mathfrak{S}_k} \epsilon(\sigma)\sigma^*(z_1 * \ldots * z_k) = 4^k \sum_{(j_1,\ldots,j_k)} n_1^{j_1} \ldots n_k^{j_k}(c_1^{j_1} - i_1c_1^{j_1}) * \ldots * (c_k^{j_k} - i_1c_k^{j_k}) * \sum_{\sigma\in\mathfrak{S}_k} \sigma^*((d_1^{j_1} - i_2d_1^{j_1}) * \ldots * (d_k^{j_k} - i_2\, d_k^{j_k}))$. Finally the second factor in each term of the sum over $(j_1, \ldots, j_k)$ vanishes by 3.0.1 applied to C_2, because $k \geq g_2 + 1$. So 3.1 is proved.

Now consider surfaces with $h^{2,0} = 1$. In this case, property (*) means simply that the $*$-product between zero-cycles of degree zero of S is symmetric, which is a rather intriguing property.

Let us prove first of all that (*) holds for Kummer surfaces:

3.2 Proposition. *Let A be an abelian surface, and $S = \widetilde{A/\pm 1}$ be its Kummer surface. Then S satisfies property (*).*

Proof: We may assume that A is principally polarized. Consider the map $\phi : A\times A \to A\times A$ given by $\phi((x,y)) = (x+y, x-y)$. Since ϕ is an isogeny, the kernel of ϕ_* is made of torsion elements of $CH_0(A \times A)$ (See [3] or [7]), so it suffices to show: let $x, y \in A$, then $\phi_*((x + \{-x\} - 2\{0\}) * ((y + \{-y\} - 2\{0\}) - (y + \{-y\} - 2\{0\}) * ((x + \{-x\} - 2\{0\})) = 0$ in $CH_0(A \times A)$, that is:

3.2.1 $\phi_*((x,y) - (y,x) + (x,\{-y\}) - (\{-y\},x) + (\{-x\},y) - (y,\{-x\}) - 2(\{0\},y) + 2(y,\{0\}) - 2(\{0\},\{-y\}) + 2(\{-y\},\{0\}) - 2(x,\{0\}) + 2(\{0\},x) - 2(\{-x\},\{0\}) + 2(\{0\},\{-x\}) + (\{-x\},\{-y\}) - (\{-y\},\{-x\})) = 0$.

Now let $(u,v) \in A \times A$; then $\phi_*((u,v)-(v,u)) = \{u+v\}*(\{u-v\}-\{v-u\})$. But we know by [7] that for any $z \in CH_0(A)_0$ one has: $z-(-1)_*(z) = j_*(\tilde{z})$, for $j : C \hookrightarrow A$ the inclusion of a symmetric thêta divisor and $\tilde{z}$ the unique zero-cycle on C, such that $alb_C(\tilde{z}) = 2alb_A(z) \in A$. So $\phi_*((u,v)-(v,u)) = \{u+v\} * j_*(z_{u,v})$, where $alb(z_{u,v}) = 2(u-v) \in A$. It follows that the left hand side in 3.2.1 is equal to:

3.2.2 $\{x+y\} * j_*(z_{x,y}) + \{x-y\} * j_*(z_{x,-y}) + \{-x+y\} * j_*(z_{-x,y}) + \{-x-y\} * j_*(z_{-x,-y}) - 2y * j_*(z_{0,y}) - 2\{-y\} * j_*(z_{0,-y}) - 2x * j_*(z_{x,0}) - 2\{-x\} * j_*(z_{-x,0})$.

Now $z_{x,y} = -z_{-x,-y}$, $z_{x,-y} = -z_{-x,y}$, $z_{0,y} = -z_{0,-y}$ and $z_{x,0} = -z_{-x,0}$. So 3.2.2 is equal to:

3.2.3 $(\{x+y\} - \{-x-y\}) * j_*(z_{x,y}) + (\{x-y\} - \{-x+y\}) * j_*(z_{x,-y}) - 2(\{y\} - \{-y\}) * j_*(z_{0,y}) - 2(x - \{-x\}) * z_{x,0})$.

Applying again the formula $z-(-1)_*(z) = j_*(\tilde{z})$ one finds:

3.2.4 $j_*(z_{x+y}) * j_*(z_{x,y}) + j_*(z_{x-y}) * j_*(z_{x,-y}) - 2j_*(z_y) * j_*(z_{0,y}) - 2j_*(z_x) * j_*(z_{x,0})$, where $alb(z_u) = 2u$, $\forall u \in A$.

We have the following obvious relations between the cycles $z_.$, $z_{.,.}$ on C: $z_{x+y} = z_x + z_y$, $z_{x,y} = z_x - z_y$, $z_{x-y} = z_x - z_y$, $z_{x,-y} = z_x + z_y$, $z_{0,y} = -z_y$, $z_{x,0} = z_x$. So 3.2.4 is in fact equal to:

3.2.5 $(j,j)_*((z_x+z_y)*(z_x-z_y) + (z_x-z_y)*(z_x+z_y) + 2(z_y * z_y) - 2(z_x * z_x))$ which is clearly 0 as we wanted.

3.3 We will consider now the following much more general family of K3-surfaces (it is ten dimensional and the general member has Picard group of rank ten), and prove property (*) for the general member. We simply consider two degree three curves E_1, E_2 in $\mathbb{P}^2$, smooth and meeting transversally. Let $S \to \mathbb{P}^2$, be obtained by blowing up the double cover of $\mathbb{P}^2$ ramified along $E_1 \cup E_2$. Then we have:

3.4 Theorem. *Property (*) holds for S.*

Proof: Let F_1, F_2 be the equations for E_1, E_2; consider in $\mathbb{P}^5$ the sextic fourfold W with equation: $F_1(X_0,X_1,X_2)F_2(X_0,X_1,X_2) - F_1(Y_0,Y_1,Y_2)F_2(Y_0,Y_1,Y_2)$, where $X_i, Y_i, i = 0,1,2$ are homogeneous coordinates for $\mathbb{P}^5$. W has non degenerate quadratic singularities along the locus $\{F_i(X) = 0 = F_i(Y), i = 1,2\}$ and by blowing up this sublocus, we get a variety $\tilde{W}$ which is smooth and clearly invariant under the natural lifting $\tilde{i}$ of $i_{|W}$ where i is the involution on $\mathbb{P}^5$, defined by $i((x_0,\dots,x_2,y_0,\dots,y_2)) = (y_0,\dots,y_2,x_0,\dots,x_2)$ which leaves W invariant.

The first step is to prove:

3.4.1 Lemma. *There is a natural correspondence Γ between $S \times S$ and $\tilde{W}$, equivariant w.r.t. τ and $\tilde{i}$ (τ is the involution exchanging factors on $S \times S$), inducing an inclusion:* $[\Gamma] : \mathrm{Ker}(CH_0(S \times S)^- \xrightarrow{pr_{1*}} CH_0(S)) \hookrightarrow CH_0(\tilde{W})$.

Proof: Let $\Sigma \subset \mathbb{P}^3$ be defined by the equation: $U^6 = F_1(X)F_2(X)$. As in [26], there is a natural rational map $\phi : \Sigma \times \Sigma \to W$, defined by: $\phi((u,x),(u',x')) = (u'x_0,\dots,u'x_2,ux'_0,\dots,ux'_2)$. Finally using the natural map $\psi : \Sigma \to S$, given by $\psi((u,x)) = (u^3,x)$ (here (u,x) is a point in the total space of $\mathcal{O}(1)$ over $\mathbb{P}^2$, and (u^3,x) is a point in the total space of $\mathcal{O}(3)$ over $\mathbb{P}^2$), we get the correspondence

$$\begin{array}{ccc} \widetilde{\Sigma\times\Sigma} & \xrightarrow{\tilde{\phi}} & \tilde{W} \\ {\scriptstyle \psi\times\psi}\downarrow & & \\ S\times S & & \end{array}$$

where $\widetilde{\Sigma\times\Sigma}$ is a smooth blow-up of $\Sigma\times\Sigma$ on which $\tilde{\phi}$ is everywhere defined. To see that $\tilde{\phi}_*\circ(\psi\times\psi)^*$ is injective on $\mathrm{Ker}(CH_0(S\times S)^-\overset{pr_{1*}}{\to} CH_0(S))$, one notes simply that $\tilde{\phi}$ is essentially the quotient map for the diagonal action of $\mathbb{Z}/6\mathbb{Z}$ on $\widetilde{\Sigma\times\Sigma}$ (this is only a birational action but this is not a problem since we want to study zero-cycles which are birational invariants). Now on $(\psi\times\psi)^*(CH_0(S\times S))$ the diagonal action of $\mathbb{Z}/6\mathbb{Z}$ reduces to the diagonal action of $\mathbb{Z}/2\mathbb{Z}$ (that is of the involution j of S over $\mathbb{P}^2$). Since j acts by -1 on $CH_0(S)_0$, we see that the diagonal action of j is the identity on $CH_0(S)_0 * CH_0(S)_0$, hence we get:

$$CH_0(S)_0 * CH_0(S)_0 \overset{(\psi\times\psi)^*}{\hookrightarrow} CH_0(\Sigma\times\Sigma)_0^{\mathbb{Z}/6\mathbb{Z}} \overset{\tilde{\phi}_*}{\cong} CH_0(\tilde{W})_0.$$

Now, Theorem 3.4 will follow from:

3.4.2 Proposition. *$CH_0(\tilde{W})^-=0$, where $()^-$ denotes the -1-eigenspace for the action of $\tilde{i}$.*

Proof: The equation of W is $F_1(X)F_2(X)-F_1(Y)F_2(Y)=0$. So for each $\alpha\in\mathbb{C}$, W contains the threefold $X\subset\mathbb{P}^5$, with equations: $F_1(X)=\alpha F_2(Y), F_1(Y)=\alpha F_2(X)$. This is the complete intersection of two cubics in $\mathbb{P}^5$, in general smooth, and invariant under i. Since W is covered by such X's, it suffices to prove:

3.4.3 Proposition. *Let $X\subset\mathbb{P}^5$ be a threefold which is defined by $F^+=F^-=0$, where F^+ (resp. F^-) are cubic equations invariant (resp. antiinvariant) under i. Then $CH_0(X)^-=0$.*

We will not give the proof of 3.4.3: one has simply to note that X has trivial canonical bundle and that i^* acts by $+1$ on $H^{3,0}(X)$. The proof works then exactly as in [23], where a similar statement is proved for quintic threefolds in $\mathbb{P}^4$ invariant under an involution satisfying the same condition. So Theorem 3.4 is proved.

3.5 Note the following important consequences of Theorem 3.4, which are obtained using Bloch-Srinivas argument [6].

3.5.1 Corollary. *The generalized Hodge conjecture is true for the sub-Hodge structure of level two: $\bigwedge^2 H^2(S)\subset H^4(S\times S)$, for any S as in 3.3.*

More precisely, there exists a variety W and a family of one cycles

$$\begin{array}{ccc} \Gamma & \xrightarrow{q} & S\times S \\ {\scriptstyle p}\downarrow & & \\ C & & \end{array}$$

inducing an injective map: $\bigwedge^2 H^2(S)\overset{p_*q^*}{\to} H^2(W)$.

We have as well, again by [6], or directly from the corollary above:

3.5.2 Corollary. *The Hodge conjecture is true for Hodge classes in* $\bigwedge^2 H^2(S) \subset H^4(S \times S)$; *moreover for a skew two-cycle* $Z \in CH_2(S \times S)^-$ *one has:*

3.5.3 $Z \equiv_{rat} 0 \Leftrightarrow Z_{|s_0 \times S} \equiv_{rat} 0$, *and* Z *is homologous to zero.*

3.6 As mentioned in section 2, Theorem 3.4 or Proposition 3.1 allow us to construct threefolds with $h^{2,0} = h^{1,0} = 0, h^{3,0} \neq 0$ and $CH_0(X^k)^{\mathfrak{S}_k} = CH_0(X^{k+1})^{\mathfrak{S}_{k+1}}$ for $k \geq h^{3,0}(X)$. (As in 2.1, this is equivalent to the following: the $*$-product $CH_0(X)^{\otimes k} \to CH_0(X^k)$ vanishes on $(CH_0(X)_0^{\otimes k})^{\mathfrak{S}_k}$, for $k \geq h^{3,0}(X)+1$ or more simply to: for $z \in CH_0(X)_0, z^{*k} = 0$ in $CH_0(X^k)$ for $k \geq h^{3,0}(X) + 1$). The construction is exactly similar to that of 3.1. We start with the $K3$-surfaces S of 3.3, which have an involution i acting by -1 on $CH_0(S)_0$. By Theorem 3.4 one has (*):

For $z, z' \in CH_0(S)_0, z * z' = z' * z$ in $CH_0(S \times S)$.

Then, if C is a hyperelliptic curve of genus g, j the hyperelliptic involution of C, one constructs the threefold $X = \widetilde{C \times S}/(j,i)$ where $\tilde{}$ means desingularization by blowing-up the fixed curves of (j,i). Then X has $h^{3,0} = g, h^{1,0} = h^{2,0} = 0$ and we show:

3.6.1 Lemma. *Let* z *be a zero-cycle of degree zero on* X; *Then* $z^{*k} = 0$ *in* $CH_0(X^k)$ *for* $k \geq g + 1$.

Proof: Since these CH_0 groups have no torsion by Roitman's theorem [5], it suffices to show that for z a zero-cycle of degree zero on $C \times S$, one has: some multiple of $(z + (j,i)(z))^{*k}$ vanishes in $CH_0((C \times S)^k) = CH_0(C^k \times S^k)$ for $k \geq g+1$. Now let $z = \sum_{l=1}^m n_l(c_l, s_l), \sum n_l = 0$. Then from the fact that j acts by -1 on $CH_0(C)_0$, i acts by -1 on $CH_0(S)_0$, one deduces easily: $2(z + (j,i)_*(z)) \equiv \sum_{l=1}^m n_l(c_l - jc_l) * (s_l - is_l)$. So $2^k(z + (j,i)_*(z))^{*k} = \sum_{f:\{1,\dots,k\}\to\{1,\dots,m\}} n_{f(1)} \dots, n_{f(k)}(c_{f(1)} - jc_{f(1)}) * \dots * (c_{f(k)} - jc_{f(k)}) * ((s_{f(1)} - is_{f(1)}) * \dots * (s_{f(k)} - is_{f(k)})$.

Next one knows by 3.4 that the $*$-products $(s_{f(1)} - is_{f(1)}) * \dots * (s_{f(k)} - is_{f(k)})$ are invariant under any permutation of the factors: the common value of these products will be denoted by $*_{l\in I}(s_l - is_l)$ where I denotes the unordered set of indices (with multiplicities) corresponding to f. Then we have: $2^k(z + (j,i)_*(z))^{*k} = \sum_I \prod_{l\in I} n_l[\sum_{f:\{1,\dots,k\}\to I}(c_{f(1)} - jc_{f(1)}) * \dots * (c_{f(k)} - jc_{f(k)})] * *_{l\in I}(s_l - is_l)$, where the summation over f inside the bracket means that if $I = \{l_1, \dots, l_r\}$ with multiplicities $i_1, \dots, i_r$ one considers maps $f : \{1, \dots, k\} \to \{l_1, \dots, l_r\}$ such that l_j has exactly i_j preimages. Now each factor inside the bracket vanishes in $CH_0(C^k)$ for $k \geq g+1$ because it is the symmetrized $*$-product of $k \geq g+1$ 0-cycles of degree 0 of C (3.0.1), hence $2^k(z + (j,i)_*(z))^{*k} = 0$.

Now if we consider the surfaces S of 3.1, which satisfy the property (*): for $k \geq h^{2,0}+1 = g_2 + 1$, $\forall z_1, \dots, z_k \in CH_0(S)_0$, $\sum_{\sigma\in\mathfrak{S}_k} \epsilon(\sigma)\sigma^*(z_1 * \dots * z_k) = 0$, we can make the same construction with E an elliptic curve: E satisfies $\forall z_1, z_2 \in CH_0(E)_0, z_1 * z_2 = -z_2 * z_1$ in $CH_0(E \times E)$ (3.0.1). S has as before an involution i acting by -1 on $CH_0(S)_0$, and we define $X = \widetilde{E \times S}/(j,i)$ where j is a hyperelliptic involution on E. The same computation then shows:

3.6.2 Lemma *for* $k \geq h^{3,0}(X) + 1, (CH_0(X)_0^{*k})^{\mathfrak{S}_k} = 0$.

3.7 The $K3$-surfaces we have considered in 3.2 and 3.3 are special but, as was explained in the introduction, the property (*): $z * z' = z' * z$ in $CH_0(S \times S)$, $\forall z, z' \in CH_0(S)_0$

should hold for all algebraic $K3$-surfaces. In fact, one can even go further: one remarkable consequence of the property $h^{2,0} = 1$ is the following: for any k, the Hodge structure on $\bigwedge^k H^2(S) \subset H^{2k}(S^k)$ is of Hodge level two. Now note the following decomposition: $H^{\cdot}(S^k)^{skew} = \bigwedge^k H^2(S) \oplus \bigwedge^{k-1} H^2(S) \oplus \bigwedge^{k-1} H^2(S) \oplus \bigwedge^{k-2} H^2(S)$; In this decomposition, the degrees are respectively $2k, 2k-2, 2k+2, 2k$; According to Bloch-Beilinson's ideas, l-cycles on S^k should be governed by cohomology groups $H^{p,q}$ with $q \leq l$. From the fact that in the decomposition above, all terms are Hodge structures of level two, one concludes that the only piece of $H^{\cdot}(S^k)^{skew}$ having non zero (p,q)-component with $q \leq k-2$ is $\bigwedge^{k-1} H^2(S)$, which comes from $H^{\cdot}(S^{k-1})^{skew}$ via $pr^*_{1,\dots,k-1}$ followed by antisymmetrization. So one should have the following:

3.8 Conjecture. *Let S be an algebraic $K3$-surface. Then for any k, the projection $pr_{1,\dots,k-1} : S^k \to S^{k-1}$ induces an injection $pr_{1,\dots,k-1*} : CH_l(S^k)^{skew}_{\mathbb{Q}} \hookrightarrow CH_l(S^{k-1})^{skew}_{\mathbb{Q}}$, for $l \leq k-2$.*

Arguing as in Bloch-Srinivas [6] or better as in [20], this conjecture would have as a consequence that $\bigwedge^k H^2(S)$ is parametrized by $(k-1)$-cycles of S^k as is expected by the generalized Hodge conjecture (see 3.5.2). Now, assuming 3.8 is true for $k-1$, one checks easily that the map $\mu_{k-1,l} : CH_l(S^{k-1})^{skew}_{\mathbb{Q}} \otimes CH_0(S)_0 \to CH_l(S^k)^{skew}_{\mathbb{Q}}$, $\mu_{k-1,l}(Z \otimes z) = \sum_{\sigma \in \mathfrak{S}_k} \epsilon(\sigma)\sigma^*(Z * z)$ induces a surjective map: $\mathrm{Ker}(pr_{1,\dots,k-2*}) \otimes CH_0(S)_0 \to \mathrm{Ker}(pr_{1,\dots,k-1*})$ for $l \leq k-2$.

So one has essentially to prove the following generalization of the property (*):

3.9 Conjecture. *For any $k, l \leq k-1$, the map $\mu_{k,l}$ vanishes on* $\mathrm{Ker}(pr_{1,\dots,k-1*}) \otimes CH_0(S)_0$.

We want now to show 3.9 for $k = 2$ and for Kummer surfaces.

3.10 Theorem. *Let A be a principally polarized abelian surface and $S = \widetilde{A/\pm 1}$ be its Kummer surface; then the map $\mu_{2,1} : CH_1(S \times S)^-_{\mathbb{Q}} \otimes CH_0(S) \to CH_1(S \times S \times S)^{skew}_{\mathbb{Q}}$, $\mu_{2,1}(Z \otimes z) = \sum_{\sigma \in \mathfrak{S}_3} \epsilon(\sigma)\sigma^*(Z * z)$ vanishes on* $\mathrm{Ker}(pr_{1*}) \otimes CH_0(S)_0$.

Proof: The first thing to do is to describe $CH_1(S \times S)^-_{\mathbb{Q}}$. In fact we will work with cycles on $A \times A$ invariant under the action of -1 on each factor, and skew under τ, which is almost the same: (the remaining skew one-cycles on $S \times S$ are generated by $E * z - z * E$, $E \subset S$ a exceptional curve, $z \in CH_0(S)_0$; For them, Theorem 3.10 follows from 3.2). We will denote these subgroups of $CH.(A \times A)$ by $\widetilde{CH}.(A \times A)$. We have the following:

3.11 Lemma. *Write $A = JC$ and fix a symmetric embedding $C \hookrightarrow A$. For $(u,v) \in C \times C$ denote by $C_{u,v} \subset A \times A$ the curve $\{(x+u, x+v), x \in C\}$. Then $\mathrm{Ker}(\pi_* : \widetilde{CH}_1(A \times A)^-_{\mathbb{Q}} \to \widetilde{CH}_1(A)_{\mathbb{Q}})$ is generated by cycles $Z_{u,v} := C_{u,v} + (-1, Id)_*(C_{u,v}) + (Id, -1)_*(C_{u,v}) + (-1,-1)_*(C_{u,v}) - C_{v,u} - (-1, Id)_*(C_{v,u}) - (Id, -1)_*(C_{v,u}) - (-1,-1)_*(C_{v,u})$, and by cycles $C * z - z * C$, $z \in CH_0(A)_0$ invariant under -1.*

Proof: Let us consider the map $\sigma : A \times A \to A, \sigma((x,y)) = x + y$. The fiber $\sigma^{-1}(\alpha)$ is isomorphic to A, embedded in $A \times A$ by $j_\alpha(x) = (x, \alpha - x)$. The involution τ acts on the fiber $\sigma^{-1}(\alpha)$ by $\tau_\alpha(x) = \alpha - x$. As we already noted in 3.2, skew 0-cycles on A for τ_α are supported on $C_{\alpha/2} \subset A$, where $\alpha/2$ is any point of A such that $2\alpha/2 = \alpha$. It follows that if $Z \in CH_1(A \times A)^-_{\mathbb{Q}}$, $Z = Z_1 + Z_2$, where Z_2 is supported on fibers of σ, and Z_1 is supported on the divisor $D = \{(x+\mu, -x+\mu), x \in C, \mu \in A\}$. If now $Z \in \widetilde{CH_1}(A \times A)$, Z is in particular invariant under the multiplication by -1 on $A \times A$, which leaves D invariant. So we may assume that Z_1 is invariant under -1 acting on D. But D is covered by $C \times A$, via the map $(x, \mu) \rightsquigarrow (x+\mu, -x+\mu)$ and -1 on D lifts to the involution $(j, -1)$ on $C \times A$: now cycles on $C \times A$ split into invariants under (j, Id) and skew under (j, Id). Since $C/j \cong \mathbb{P}^1$, the first ones are generated by cycles $C \times \alpha, \alpha \in A$ and $x_0 \times C_\alpha, C_\alpha \in A$ a translate of $C, x_0 \in C$ one of the Weierstrass points of C. The cycles $x_0 \times C_\alpha$ are skew under the involution $(j, -1)$ modulo $x_0 \times C$.

Finally consider skew cycles under (j, Id)): if W is such a cycle, $W + (j, -1)_*(W) = W - (Id, -1)_*(W)$ is skew under $(Id, -1)_*$; So it must be the sum of a cycle supported on fibers of $C \times A \overset{pr_1}{\to} C$, and of a cycle supported on $C \times C \subset C \times A$, which itself decomposes into a sum $N\, diag(C)$+fibers. Summing up, $\widetilde{CH_1}(A \times A)^-_{\mathbb{Q}}$ is generated by the following cycles:

a) $\{(x+u, \alpha - x - u) \mid x \in C,\ \alpha \in A \text{ fixed } u \in A \text{ fixed}\}$, corresponding to cycles Z_2's above.

b) $\{(x+\alpha, x+\alpha) \mid x \in C, \alpha \in A \text{ fixed}\}$, corresponding to cycles $x_0 \times C_\alpha$.

c) $\{(x+\alpha, -x+\alpha) \mid x \in C, \alpha \in A \text{ fixed}\}$, corresponding to $C \times \alpha$.

d) $\{(y+x+\alpha, -y+x+\alpha) \mid y \in C \text{ fixed}, x \in C, \alpha \in A \text{ fixed}\}$, corresponding to $y \times C_\alpha$.

e) $\{(x, -x) \mid x \in C\}$, corresponding to $diag\,(C)$.

f) $\{(x+\alpha, -x+\alpha) \mid x \in C, \alpha \in A \text{ fixed}\}$, corresponding to fibers of $C \times C$.

Obviously b) does not contribute to $\widetilde{CH_1}(A \times A)^-$. As for c), note that $\{(x, -x), x \in C\} \equiv_{rat} -\{(x, x), x \in C\}$ modulo fibers of $C \times C \subset A \times A$. The same argument shows that all the other cycles are contained in the subgroup generated by cycles $(x+u, x+v)\ \ u, v \in A\ fixed, u \in C\}$, modulo cycles supported on fibers of $A \times A$. Note that cycles supported on fibers of $A \times A$ and invariant under $(-1, Id)_*$ and $(Id, -1)_*$ are exactly the cycles $C * z - z * C, z \in CH_0(A)$ invariant under $(-1)_*$. Taking now the projection of these cycles on the invariant subgroup $\widetilde{CH_1}(A \times A)^-_{\mathbb{Q}}$ gives the generators described in the statement of the lemma. (The fact that one may restrict to cycles $Z_{u,v}, u, v \in C$ is easy to prove).

3.12 Now, if one considers a cycle $W_z = C * z - z * C, z \in CH_0(A)_0, (-1)_*(z) = z$, and if one chooses $w \in CH_0(A)_0, (-1)_*(w) = w$, then $\sum_{\sigma \in \mathfrak{S}_3} \epsilon(\sigma)\sigma^*(W_z * w) = 0$ in $CH_1(A \times A \times A)$ by Proposition 3.2. So it remains only to show:

3.12.1 *For $Z_{u,v} \in \widetilde{CH_1}(A \times A)^-_{\mathbb{Q}}$ as in 3.11, and $w \in CH_0(A)_0, (-1)_*(w) = w$, one has $\sum_{\sigma \in \mathfrak{S}_3} \epsilon(\sigma)\sigma^*(Z_{u,v} * w) = 0$ in $CH_1(A \times A \times A)$,*

which is equivalent to:

3.12.2 $Z_{u,v} * w - \tau_{2,3_*}(Z_{u,v} * w) = -w * Z_{u,v}$.

Before we go to the proof of 3.12.2, we need the following lemma concerning the two families of one cycles generating $\widetilde{CH_1}(A \times A)^-_{\mathbb{Q}}$ provided by lemma 3.11. Consider the following two correspondences:

$$\begin{array}{ccc} C \times C \times C & \xrightarrow{q_1} & A \times A \\ {\scriptstyle p_1}\downarrow & & \\ C \times C & & \end{array}$$

where p_1 is the first projection, and $q_1((x, y, z)) = (x+z, y+z)$ and

$$\begin{array}{ccc} A \times C & \xrightarrow{q_2} & A \times A \\ {\scriptstyle p_2}\downarrow & & \\ A & & \end{array}$$

where p_2 is the first projection and q_2 is the identity on the first factors and the inclusion $C \hookrightarrow A$ chosen above on the second factors. We have the following:

3.13 Lemma.

i) $p_{2*}q_2{}^*(Z_{u,v}) = 0$ *in* $CH_0(A)$.

ii) $p_{2*}q_2{}^*(Z_\alpha) = -2\alpha,\ \forall \alpha \in CH_0(A)$.

iii) $p_{1*}q_1{}^*(Z_{u,v}) = -4((u,v)+(ju,jv)-(v,u)-(jv,ju))$, *where* j *is the hyperelliptic involution of* C.

Proof: i) Clearly $p_{2*}q_2{}^*(C_{u,v}) = t_{u-v}(C.C_v) = C_{u-v}\cdot C_u$, where $t_{(\cdot)}$ denotes translation by $(\cdot) \in A$, and $C_v := t_v(C)$. Similarly

$$p_{2*}q_2{}^*((-1,Id)(C_{u,v}) = (-1)_* p_{2*}q_2{}^*(C_{u,v}) = C_{v-u}\cdot C_{-u} \text{ and}$$
$$p_{2*}q_2{}^*((Id,-1)(C_{u,v})) = (-1)_* p_{2*}q_2{}^*((-1,-1)_*(C_{u,v})) = C_{u-v}\cdot C_u.$$

Hence

$$\begin{aligned} p_{2*}q_2{}^*(C_{u,v} &+ (-1,Id)_*(C_{u,v}) + (Id,-1)_*(C_{u,v}) + (-1,-1)_*(C_{u,v}) \\ &= 2(C_{u-v}\cdot C_u \ + \ C_{v-u}\cdot C_{-u}) \\ &= 2((C_u - C_v + C)\cdot C_u \ + \ (C_v - C_u + C)\cdot(-C_u + 2C)) \\ &= 4((C_u - C_v)\cdot(C_u - C) + C^2)). \end{aligned}$$

So

$$\begin{aligned} p_{2*}q_2{}^*(Z_{u,v}) &= 4(C_u - C_v)\cdot(C_u + C_v - 2C) \\ &= 4(C_u{}^2 - C_v{}^2 - C_u\cdot C + 2C_v\cdot C) \end{aligned}$$

and this is equal to zero since $C_u{}^2 = 2\{u\}, C_u\cdot C = \{u\}+\{0\}$.

ii) $C*\alpha$ does not meet $A\times C$ when $Sup\,\alpha$ does not meet C. So $p_{2*}q_2{}^*(Z_\alpha) = -p_{2*}q_2{}^*(\alpha * C)$. Since the divisor $A\times C$ of $A\times A$ is smooth, the result follows immediately from the excess intersection formula and from $C^2 = 2$.

iii) One has

$$p_{1*}q_1{}^*(C_{u,v} + (-1,-1)_*(C_{u,v}) - C_{v,u} - (-1,-1)_*(C_{v,u})) = 0,$$

which is proved as follows: Let $\tilde{Z}_{u,v}$ be the cycle above and let $\tilde{Z}^\epsilon_{u,v}$ be its translation by a torsion point $\epsilon \in A$, so $\tilde{Z}_{u,v} \equiv_{rat} \tilde{Z}^\epsilon_{u,v}$ up to torsion. Now $C_{\alpha,\beta}\cap C^\epsilon_{u,v} \neq 0 \Leftrightarrow \alpha + y = u + x + \epsilon$ and $\beta + y = v + x + \epsilon$ for some $x, y \in C$. This is also equivalent to: $\alpha - \beta = u - v$ and if this last condition is satisfied, there are exactly two couples $(x,y)\in C^2$ satisfying the above equalities, at least for generic (u,v). Now $\alpha - \beta = u - v \Leftrightarrow \alpha = u$ and $\beta = v$ or $\alpha = jv$ and $\beta = ju$; So $p_{1*}q_1{}^*(C^\epsilon_{u,v}) = 2((u,v)+(jv,ju))$. Since $(-1,-1)_*(C_{u,v}) = C_{ju,jv}$ it is then immediate that $p_{1*}q_1{}^*(\tilde{Z}^\epsilon_{u,v}) = 0$ in $CH_0(C\times C)$. So we have proved that

$$\begin{aligned} p_{1*}q_1{}^*(Z_{u,v}) =& p_{1*}q_1{}^*((-1,Id)_*(C_{u,v}) + (Id,-1)_*(C_{u,v}) \\ &- (-1,Id)_*(C_{v,u}) - (Id,-1)_*(C_{v,u})) \end{aligned}$$

with $(Id,-1)_*(C_{u,v}) = (-1,Id)_*(C_{ju,jv})$. Now $p_{1*}q_1{}^*((-1,Id)_*(C_{u,v})) = \sum_i(x_i,y_i)$ where the (x_i,y_i)'s are such that: $\exists\, w_i, z_i \in C$ with $(x_i+z_i, y_i+z_i) = (-u-w_i, v+w_i)$; the last equality is realized in the following cases:

a) $x_i = ju, z_i = jw_i, y_i = v, z_i = w_i$, with w_i a Weierstrass point of C.

b) $x_i = jw_i, z_i = ju, y_i = v, z_i = w_i$.

c) $x_i = ju, z_i = jw_i, y_i = w_i, z_i = v$.

d) $w_i = u, x_i = jz_i, y_i + jx_i = v + u$.

e) $w_i = jv, y_i = jz_i, x_i + z_i = -u - v$.

Since there are six Weierstrass points on C we get:

$$\begin{aligned} p_{1*}q_1^*((-1, Id)_*(C_{u,v})) =& 6(ju, v) + (u, v) + (ju, jv) \\ &+ (jv, u) + (ju, v) + (ju, v) + (jv, u), \end{aligned}$$

hence:

$$\begin{aligned} p_{1*}q_1^*(Z_{u,v}) =& 2((u, v) + (ju, jv) - (v, u) - (jv, ju)) \\ &+ 6((ju, v) - (jv, u) + (u, jv) - (v, ju)). \end{aligned}$$

Let $h = u + ju$, $\forall u \in C$ be the canonical divisor of C; Then $(ju, v) = h * v - (u, v)$, so:

$$p_{1*}q_1^*(Z_{u,v}) = -4((u, v) + (ju, jv) - (v, u) - (jv, ju))$$

as announced.

Now, consider a cycle $Z_{u,v} * w - \tau_{23*}(Z_{u,v} * w) \in CH_1(A \times A \times A)_{\mathbb{Q}}$, with $w \in CH_0(A)_{inv}$, i.e. w (-1)-invariant of degree zero, as in 3.12.1. This cycle is invariant under -1 acting on each factor of $A \times A \times A$, and is skew under τ_{23}. So by Proposition 3.2, one concludes that it lies in the image of $CH_0(A)_{inv} \otimes \widetilde{CH}_1(A \times A)^-_{\mathbb{Q}}$ in $CH_1(A \times A \times A)$. Then the proof of 3.12.2 will be finished by the following two lemmas: We consider now the two correspondences:

$$\begin{array}{ccc} A \times C \times C \times C & \xrightarrow{id \times q_1} & A \times A \times A \\ {\scriptstyle id \times p_1}\downarrow & & \\ A \times C \times C & & \end{array}$$

and

$$\begin{array}{ccc} A \times A \times C & \xrightarrow{id \times q_2} & A \times A \times A \\ {\scriptstyle id \times p_2}\downarrow & & \\ A \times A & & \end{array}$$

where the maps $p_i, q_i, i = 1, 2$ are defined in 3.12:

3.14 Lemma. *$(id \times p_2)_*(id \times q_2)^*(Z_{u,v} * w - \tau_{23*}(Z_{u,v} * w)) = 0$. (This is equal to $-(id \times p_2)_*(id \times q_2)^*(w * Z_{u,v})$ by 3.13, i)).*

3.15 Lemma. *$(id \times p_1)_*(id \times q_1)^*(Z_{u,v} * w - \tau_{23*}(Z_{u,v} * w)) = 4w * ((u, v) + (ju, jv) - (v, u) - (jv, ju))$ in $CH_0(A \times C \times C)$. (This is equal to $-(id \times p_1)_*(id \times q_1)^*(w * Z_{u,v})$ by 3.13, iii)).*

Indeed, Lemmas 3.11 and 3.13 show that the map $((id \times p_2)_*(id \times q_2)^*, (id \times p_1)_*(id \times q_1)^*)$ is injective on the image of $CH_0(A)_{inv} \otimes \widetilde{CH}_1(A \times A)^-_{\mathbb{Q}}$.

3.16 Proof of Lemma 3.14. Obviously $Z_{u,v} * w$ is annihilated by $(id \times p_2)_*(id \times q_2)^*$ since one may assume that the support of w does not meet C. Next one has clearly $(id \times p_2)_*(id \times q_2)^*(\tau_{23}(Z_{u,v})) = (p_{2*}q_2^*(Z_{u,v})) * w$ which is equal to zero by Lemma 3.13, i).

3.17 Proof of Lemma 3.15. Suppose first of all that w is a point of A. Then $(id \times p_1)_*(id \times q_1)^*(C_{u,v} \times w) = \{(a, \alpha, \beta) \mid \exists y, x \in C, (x + u, x + v, w) = (a, \alpha + y, \beta + y)\}$.

This is also the intersection of $A \times C \times C$ with $\{(x+u, x+v-y, w-y) \mid x, y \in C\}$ in $A \times A \times A$. Let $\sigma_1 : A \times C \times C \to A$ be defined by $\sigma_1((a, \alpha, \beta)) = \alpha - a - \beta$; it follows that:

3.17.1 $$(id \times p_1)_*(id \times q_1)^*(C_{u,v} \times w) = \pi_1{}^*(C_u) \cdot \pi_3^*(C_{w|C}) \cdot \sigma_1{}^*(\{v-u-w\}),$$

where the π's are the projections of $A \times C \times C$ on its factors.

So we get also:

3.17.2 $$\begin{aligned}(id \times p_1)_*(id \times q_1)^*((-1, Id)_*(C_{u,v} \times w)) \\ = (-1, Id, Id)_*(id \times p_1)_*(id \times q_1)^*(C_{u,v} \times w) \\ = \pi_1{}^*(C_{-u}) \cdot \pi_3^*(C_{w|C}) \cdot \sigma_2^*(\{v-u-w\}),\end{aligned}$$

where $\sigma_2((a, \alpha, \beta)) = \alpha + a - \beta$.

In the same way:

3.17.3 $$\begin{aligned}(id \times p_1)_*(id \times q_1)^*((Id, -1)_*(C_{u,v} \times w)) \\ = \pi_1{}^*(C_u) \cdot \pi_3^*(C_{w|C}) \cdot \sigma_2^*(\{-v+u-w\}) \quad \text{and} \\ (id \times p_1)_*(id \times q_1)^*((-1, -1)_*(C_{u,v} \times w)) \\ = \pi_1{}^*(C_{-u}) \cdot \pi_3^*(C_{w|C}) \cdot \sigma_1^*(\{-v+u-w\}).\end{aligned}$$

Write now $C_u = L_u + C$, with $L_u \in Pic^0(A)$. Then $L_{-u} = -L_u$, and we get:

3.17.4 $$\begin{aligned}(id \times p_1)_*(id \times q_1)^*(C_{u,v} \times w + (-1, Id)_*(C_{u,v} \times w) \\ + (Id, -1)_*(C_{u,v} \times w) + (-1, -1)_*(C_{u,v} \times w)) \\ = \pi_1{}^*(L_u) \cdot \pi_3^*(C_{w|C} \cdot (\sigma_1{}^*(\{v-u-w\}) - \sigma_2{}^*(\{v-u-w\}) \\ + \sigma_2{}^*(\{-v+u-w\}) - \sigma_1{}^*(\{-v+u-w\})) \\ + \pi_1{}^*(C) \cdot \pi_3{}^*(C_{w|C}) \cdot (\sigma_1{}^*(\{v-u-w\}) \\ + \sigma_2{}^*(\{v-u-w\}) + \sigma_2{}^*(\{-v+u-w\}) \\ + \sigma_1{}^*(\{-v+u-w\})).\end{aligned}$$

The last term $(\dots)$ in this sum is clearly symmetric in u and v, so will disappear in the computation of $(id \times p_1)_*(id \times q_1)^*(Z_{u,v} * w)$. Next note the following equality in $CH_0(A)$ (See 3.2.1):

$$\{v-u-w\} - \{-v+u-w\} = 2C_{-w} \cdot (C_v - C_u).$$

So 3.17.4 becomes

3.17.5 $2\pi_1{}^*(L_u) \cdot \pi_3{}^*(C_{w|C})(\sigma_1{}^*(C_{-w}) \cdot \sigma_1{}^*(C_v - C_u) - \sigma_2{}^*(C_{-w}) \cdot \sigma_2{}^*(C_v - C_u)) + (\dots)$.

In the expression above, one replaces C_{-w} by $C + L_{-w}$, and 3.17.5 gives:

3.17.6 $$\begin{aligned}2\pi_1{}^*(L_u) \cdot \pi_3{}^*(C_{w|C})(\sigma_1{}^*(L_{-w}) \cdot \sigma_1{}^*(C_v - C_u) \\ - \sigma_2{}^*(L_{-w}) \cdot \sigma_2{}^*(C_v - C_u)) + (\dots) \\ +2\pi_1{}^*(L_u) \cdot \pi_3{}^*(C_{w|C})(\sigma_1{}^*(C) \cdot \sigma_1{}^*(C_v - C_u) \\ - \sigma_2{}^*(C) \cdot \sigma_2{}^*(C_v - C_u)).\end{aligned}$$

The last term $((\dots))$ in this sum will vanish if we take for w a (-1)-invariant zero-cycle of degree zero on A instead of a point. Finally L_{-w} and $C_v - C_u$ are algebraically equivalent to zero, so we have:

$$\begin{aligned}
\sigma_1{}^*(L_{-w}) &= (\pi_2{}^* - \pi_1{}^* - \pi_3{}^*)(L_{-w}) \\
\sigma_1{}^*(C_v - C_u) &= (\pi_2{}^* - \pi_1{}^* - \pi_3{}^*)(C_v - C_u) \\
\sigma_2{}^*(L_{-w}) &= (\pi_2{}^* + \pi_1{}^* - \pi_3{}^*)(L_{-w}) \\
\sigma_2{}^*(C_v - C_u) &= (\pi_2{}^* + \pi_1{}^* - \pi_3{}^*)(C_v - C_u),
\end{aligned}$$

and 3.17.6 becomes:

3.17.7
$$\begin{aligned}
&-4\pi_1{}^*(L_u) \cdot \pi_3{}^*(C_{w|C}) \cdot (\pi_2{}^*(L_{-w}) \cdot \pi_1{}^*(C_v - C_u) \\
&\qquad + \pi_1{}^*(L_{-w}) \cdot \pi_2{}^*(C_v - C_u)) + (\dots) + ((\dots)).
\end{aligned}$$

Finally we want to compute $(id \times p_1)_*(id \times q_1)^*(Z_{u,v} * w - \tau_{23*}(Z_{u,v} * w))$, so we have to subtract $\{(u,v) \to (v,u)\}$ in 3.17.7, apply $Id \times \tau_{23}$ to $A \times C \times C$ and subtract. As mentioned above $(\dots)$ disappears and we get:

3.17.8
$$\begin{aligned}
&(id \times p_1)_*(id \times q_1)^*(Z_{u,v} * w - \tau_{23*}(Z_{u,v} * w)) \\
&= -4(\pi_1{}^*(L_u + L_v) \cdot \pi_3{}^*(C_{w|C}) \cdot (\pi_2{}^*(L_{-w}) \cdot \pi_1{}^*(C_v - C_u) \\
&\quad + \pi_1{}^*(L_{-w}) \cdot \pi_2{}^*(C_v - C_u)) \\
&\quad - \pi_1{}^*(L_u + L_v) \cdot \pi_2{}^*(C_{w|C}) \cdot (\pi_3{}^*(L_{-w}) \cdot \pi_1{}^*(C_v - C_u) \\
&\quad + \pi_1{}^*(L_{-w}) \cdot \pi_3{}^*(C_v - C_u))) + ((\dots)),
\end{aligned}$$

where $((\dots))$ will vanish if the point w is replaced by an element of $CH_0(A)_{inv}$ of degree zero.

One may also as in 3.17.5 replace $C_{w|C}$ by $L_{w|C} + C_{|C}$ and 3.17.8 becomes:

3.17.9
$$\begin{aligned}
&-4(\pi_1{}^*(L_u + L_v) \cdot \pi_3{}^*(L_{w|C}) \cdot (\pi_2{}^*(L_{-w}) \cdot \pi_1{}^*(C_v - C_u) \\
&\quad + \pi_1{}^*(L_{-w}) \cdot \pi_2{}^*(C_v - C_u)) \\
&\quad - \pi_1{}^*(L_u + L_v) \cdot \pi_2{}^*(L_{w|C}) \cdot (\pi_3{}^*(L_{-w}) \cdot \pi_1{}^*(C_v - C_u) \\
&\quad + \pi_1{}^*(L_{-w}) \cdot \pi_3{}^*(C_v - C_u))) + (((\dots))),
\end{aligned}$$

where $(((\dots)))$ vanishes when w is replaced by an element of $CH_0(A)_{inv}$ of degree zero.

Let now $\mu : C \times C \to A$ be the sum map: $\mu((c, c')) = c + c'$. The projections of C^4 to C will be denoted by $\pi'_0, \pi'_1, \pi_2, \pi_3$; In the sequel the line bundles $L_{|C}$ will still be denoted by L for $L \in Pic(A)$. Then for $L \in Pic^0(A), \mu^*(\pi_1{}^*(L)) = \pi'_0{}^*(L) + \pi'_1{}^*(L)$ so we get:

3.17.10
$$\begin{aligned}
&\mu^*((id \times p_1)_*(id \times q_1)^*(Z_{u,v} * w - \tau_{23*}(Z_{u,v} * w))) \\
&= -4((\pi'_0{}^*(L_u + L_v) \cdot \pi'_1{}^*(L_{-w}) \\
&\quad + \pi'_1{}^*(L_u + L_v) \cdot \pi'_0{}^*(L_{-w})) \cdot \pi_3{}^*(L_w) \cdot \pi_2{}^*(L_v - L_u) \\
&\quad - (\pi'_0{}^*(L_u + L_v) \cdot \pi'_1{}^*(L_{-w}) \\
&\quad + (\pi'_1{}^*(L_u + L_v) \cdot \pi'_0{}^*(L_{-w})) \cdot \pi_2{}^*(L_w) \cdot \pi_3{}^*(L_v - L_u)) \\
&\quad + (((\dots))).
\end{aligned}$$

Now $L_{-w} = -L_w$, and we can apply to C, a curve of genus two, the property 3.0.1, which gives:

$$\begin{aligned}
&\pi'_0{}^*(L_u + L_v) \cdot \pi'_1{}^*(L_{-w}) \cdot \pi_3{}^*(L_w) + \pi'_1{}^*(L_u + L_v) \cdot \pi'_0{}^*(L_{-w}) \cdot \pi_3{}^*(L_w) \\
&\qquad = -\pi'_0{}^*(L_{-w}) \cdot \pi'_1{}^*(L_w) \cdot \pi_3{}^*(L_u + L_v).
\end{aligned}$$

So 3.17.10 becomes:

3.17.11
$$4({\pi'_0}^*(L_{-w})\cdot{\pi'_1}^*(L_w)\cdot({\pi_3}^*(L_u+L_v)\cdot{\pi_2}^*(L_v-L_u) - ({\pi_2}^*(L_u+L_v)\cdot{\pi_3}^*(L_v-L_u))) + (((\dots))).$$

Now note that $\mu_*\mu^*=$ multiplication by 2 on $CH_0(A)$ and that: $\mu_*({\pi'_0}^*(L_{-w})\cdot{\pi'_1}^*(L_w)) = -(\{w\}+\{-w\}-2\{0\})$ for w a point of A; So we have:

3.17.12
$$\begin{aligned}&(id\times p_1)_*(id\times q_1)^*(Z_{u,v}*w-\tau_{23*}(Z_{u,v}*w))\\ &\quad=-\,4(\{w\}+\{-w\}-2\{0\})*({\pi_3}^*(L_u)\cdot{\pi_2}^*(L_v)\\ &\quad-{\pi_3}^*(L_v)\cdot{\pi_2}^*(L_u))+(((\dots))).\end{aligned}$$

Since the last term vanishes when w is replaced by an element $\tilde{w}$ of $CH_0(A)_{inv}$ of degree zero, we get:

3.17.13
$$\begin{aligned}&(id\times p_1)_*(id\times q_1)^*(Z_{u,v}*\tilde{w}-\tau_{23*}(Z_{u,v}*\tilde{w}))\\ &\quad=8\tilde{w}*({\pi_2}^*(L_u)\cdot{\pi_3}^*(L_v)-{\pi_2}^*(L_v)\cdot{\pi_3}^*(L_u)),\end{aligned}$$

which proves Lemma 3.15 since

$$2({\pi_2}^*(L_u)\cdot{\pi_3}^*(L_v)-({\pi_2}^*(L_v)\cdot{\pi_3}^*(L_u))=(u,v)+(ju,jv)-(v,u)-(jv,ju)$$

in $CH_0(C\times C)$.

4. Kuga-Satake varieties

To conclude, I want to explain one of the motivations for the study of skew cycles on S^k, for S a K3 surface; There is a beautiful construction, due to Kuga and Satake [14], which associates to a polarized K3 surface S (in fact one needs only a polarized Hodge structure of weight 2 with $h^{2,0}=1$) an abelian variety $KS(S)$ such that $H^2(S)^0$ is recovered as a direct factor Hodge structure of $H^1(KS(S))\otimes H^1(KS(S))$. Here ^{0} means the primitive part with respect to the polarization. In [8] Deligne proved, using the Mumford-Tate group of a Hodge structure, that there are many surfaces S (with $h^{2,0}\geq 2$) for which $H^2(S)^0$ cannot be realized as direct factor in the even dimensional cohomology of an abelian variety. Now the question is: How to realize geometrically the correspondence between $KS(S)\times KS(S)$ and S, that is to solve the Hodge conjecture for the Hodge class on $KS(S)\times KS(S)\times S$ of degree 4 corresponding to the inclusion $H^2(S)^0\hookrightarrow H^1(KS(S))\otimes H^1(KS(S))\hookrightarrow H^2(KS(S)\times KS(S))$?

In [15], Morrison has shown that for a Kummer surface $S=\widetilde{A/\pm 1}$, the Kuga-Satake variety is a sum of copies of A and $\check{A}$; In [21], Paranjape considers a 4-dimensional family of K3 surfaces S (so the general one is not a Kummer surface) and constructs geometrically the correspondence between $KS(S)^2$ and S. But the general case is quite mysterious.

4.1 This Kuga-Satake variety is essentially constructed as a complex torus $T^{0,1}/T_{\mathbb{Z}}$, where $T_{\mathbb{Z}}\otimes\mathbb{C}=T^{1,0}\oplus T^{0,1}, T^{1,0}=\overline{T^{0,1}}$. Now, starting from $H^2(S,\mathbb{Z})^0$ one defines

$$T_{\mathbb{Z}}=\bigwedge_{\mathbb{Z}}^{even}H^2(S,\mathbb{Z})^0=\bigoplus_k\bigwedge^{2k}H^2(S,\mathbb{Z})^0,$$

and the space $T^{1,0}$ is defined as follows: Let $\omega \in H^{2,0}$ be such that $<\omega,\overline{\omega}>=1$; then

$$T_{\mathbb{C}} = \bigwedge_{\mathbb{C}}^{even} H^2(S,\mathbb{C})^0 = \omega \wedge \bigwedge^{odd} H^{1,1} \oplus \overline{\omega} \wedge \bigwedge^{odd} H^{1,1} \oplus \bigwedge^{even} H^{1,1} \oplus i\omega \wedge \overline{\omega} \wedge \bigwedge^{even} H^{1,1}.$$

Then $T^{1,0} = \omega \wedge \bigwedge^{odd} H^{1,1} \oplus T'^{1,0}$, where $T'^{1,0}$ is the subspace of $\bigwedge^{even} H^{1,1} \oplus i\omega \wedge \overline{\omega} \wedge \bigwedge^{even} H^{1,1}$ defined as follows: $\bigwedge^{even} H^{1,1} \oplus i\omega \wedge \overline{\omega} \wedge \bigwedge^{even} H^{1,1}$ has a real structure, since $H^{1,1}$ has a real structure ($H^{1,1} = H^{1,1}_{\mathbb{R}} \otimes \mathbb{C}$), and $i\omega \wedge \overline{\omega}$ is real. Its real part is $\bigwedge^{even}_{\mathbb{R}} H^{1,1}_{\mathbb{R}} \oplus i\omega \wedge \overline{\omega} \wedge \bigwedge^{even}_{\mathbb{R}} H^{1,1}_{\mathbb{R}}$, which has a natural complex structure I given by $I((u, i\omega \wedge \overline{\omega} \wedge v)) = (-v, i\omega \wedge \overline{\omega} \wedge u)$. Then one takes for $T'^{1,0}$ the i-eigenspace of I in $(\bigwedge^{even}_{\mathbb{R}} H^{1,1}_{\mathbb{R}} \oplus i\omega \wedge \overline{\omega} \wedge \bigwedge^{even}_{\mathbb{R}} H^{1,1}_{\mathbb{R}}) \otimes \mathbb{C}$. Note finally that this construction does not depend on the choice of ω, because of the normalization $<\omega,\overline{\omega}>=1$, which determines ω up to multiplication by a complex number of modulus 1, hence determines $\omega \wedge \overline{\omega}$.

4.2 Example: Let us present one example (a four dimensional family of examples where the Kuga-Satake construction is easily constructed: Consider the family of cubic fourfolds in $\mathbb{P}^5$ with equation $U^3 + V^3 + F(X,Y,Z,T) = 0$, where F is of degree 3 and defines a smooth surface in $\mathbb{P}^3$. These cubics W_F contain many planes, so by [25], for each W_F one can find a K3 surface S_F (a double cover of $\mathbb{P}^2$ ramified along a sextic), and a correspondence $\Gamma_F \subset S_F \times W_F$, inducing an isomorphism: $CH_0(S_F)_0 \cong CH_1(W_F)_{alg}$. Note that, following [26], W_F is the quotient of the blow up of $E \times X_F$ along several copies of Σ_F by the diagonal action of $\mathbb{Z}/3\mathbb{Z}$, where E is the elliptic curve with equation $W^3 + V^3 + U^3 = 0$, the action of $\mathbb{Z}/3\mathbb{Z}$ being by multiplication of W by a third root of unity, X_F is the cubic threefold with equation $W'^3 + F(X,Y,Z,T) = 0$, $\mathbb{Z}/3\mathbb{Z}$ acting by multiplication by the same third root of unity on W', and $\Sigma_F \subset X_F$ is defined by $W' = 0$. It follows that the corresponding K3 surfaces S_F have the following property: The Hodge structure on the transcendental part $TH^2(S)$ of $H^2(S)$ identifies with the $\mathbb{Z}/3\mathbb{Z}$-invariant part of $H^1(E) \otimes H^1(A_F)$, where E is an elliptic curve with a $\mathbb{Z}/3\mathbb{Z}$-action, and A_F is a five dimensional abelian variety (the intermediate jacobian of the cubic threefold X_F) with a $\mathbb{Z}/3\mathbb{Z}$-action, satisfying: let g be a generator of $\mathbb{Z}/3\mathbb{Z}$, then for a certain $\zeta, \zeta^3 = 1, \zeta \neq 1, g^*_{|H^{1,0}(E)}$ is multiplication by ζ and $g^*_{|H^{1,0}(A_F)}$ has a one dimensional ζ^2-eigenspace and a four dimensional ζ-eigenspace. Now the transcendental lattice has even rank equal to 10, so one has to distinguish between $KS(S_F)^+$ and $KS(S_F)^-$ which are build as in 4.1 on the even and odd parts of the exterior algebra of $TH^2(S_F)$ respectively. In this case the Kuga-Satake correspondence is the map $TH^2(S_F) \to Hom(H_1(KS(S_F)^+, H_1(KS(S_F)^-)) \cong_{\mathbb{Q}} H^1(KS(S_F)^+) \otimes H^1(KS(S_F)^-)$ given by Clifford multiplication on the left: $TH^2(S_F) \to Hom(\bigwedge^{even} TH^2(S_F), \bigwedge^{odd} TH^2(S_F))$. Note that, although not canonically isogenous, $KS(S)^+$ and $KS(S)^-$ are isogenous in many ways, using Clifford multiplication on the right by elements of $\bigwedge^{odd} TH^2(S_F,\mathbb{Z})$. Now let us check:

4.3 Proposition: *$KS(S_F)^+$ contains A_F, and $KS(S_F)^-$ contains E.*

Notice that $KS(S_F)^+$ and $KS(S_F)^-$ have a big ring of endomorphisms, so in fact $KS(S_F)^+$ contains many copies of A_F and using isogenies $KS(S_F)^+ \cong_{\mathbb{Q}} KS(S_F)^-$, it contains also many copies of E. I do not know whether there are other abelian subvarieties of $KS(S_F)^+$; Also it would remain to verify that restricted to $A_F \times E \subset KS(S_F)^+ \times KS(S_F)^-$ the Kuga-Satake correspondence is actually the isomorphism $TH^2(S_F) \cong (H^1(A_F) \otimes H^1(E))_{inv}$, which is an algebraic correspondence (see 4.2).

Proof of Proposition 4.3: Let K denote the quadratic number field $\mathbb{Q}(\zeta)$. Working with rational cohomology, we have: $H^1(E)$ is a K-vector space of rank one, $H^1(A_F)$ is a K-vector space of rank 5, so $\bigwedge_K^5 H^1(A_F)$ is a K-vector space of rank 1. Let $\beta \in H^1(E)$ be a generator of $H^1(E)$ over K, and $\eta \in \bigwedge_K^5 H^1(A_F), \eta \neq 0$; then β gives an isomorphism $\psi_\beta : (H^1(A_F) \otimes H^1(E))_{inv} \cong H^1(A)$ obtained as the composite:

$$(H^1(A_F) \otimes H^1(E))_{inv} \cong Hom_K(H^1(E), H^1(A_F)) \xrightarrow{\phi_\beta} H^1(A_F),$$

where $\cong$ is given by the polarization of E, and ϕ_β is the evaluation $\psi \rightsquigarrow \psi(\beta)$. By the inclusion of $\mathbb{Q}$-vector spaces $\bigwedge_K^5 H^1(A_F) \hookrightarrow \bigwedge_{\mathbb{Q}}^5 H^1(A_F)$, we can consider η as an element of $\bigwedge_{\mathbb{Q}}^5 H^1(A_F)$, and then it gives two maps $int(\eta)$ (the interior product with $\eta, u \rightsquigarrow int_u(\eta)$ defined using the intersection form of $H^1(A)$) and $\wedge\eta$ (the exterior product with η) from $H^1(A)$ to $\bigwedge^{even} H^1(A)$. Now we show:

4.3.1 Lemma: *For an adequate choice of $\beta \in H^1(E), \eta_1, \eta_2 \in \bigwedge_K^5 H^1(A_F)$, the map $\gamma_{\beta,\eta_1,\eta_2} : H^1(A) \to \bigwedge^{even}((H^1(E) \otimes H^1(A_F))_{inv})$ given by the composite:*

$$H^1(A_F) \xrightarrow{int(\eta_1)+\wedge\eta_2} \bigwedge^{even} H^1(A) \xrightarrow{\psi_\beta} \bigwedge^{even} ((H^1(E) \otimes H^1(A_F))_{inv})$$

induces a morphism of abelian varieties: $A_F = \check{A}_F \to KS(S_F)^+$ (which is easily seen to be injective up to torsion).

To prove 4.3.1, we have to check that the map $\gamma_{\beta,\eta_1,\eta_2} \otimes \mathbb{C}$ sends $H^{1,0}(A_F)$ to $T^{1,0} \subset \bigwedge^{even}((H^1(E) \otimes H^1(A_F))_{inv}) \otimes \mathbb{C}$ (notations as in 4.1). Working over $\mathbb{C}$, we have the decomposition: $\beta = \beta^{1,0} + \overline{\beta^{1,0}}$ (in the sequel $\overline{\beta^{1,0}}$ will be denoted by $\beta^{0,1}$), with $g^*(\beta^{1,0}) = \zeta\,\beta^{1,0}$, and

$$H^1(A_F) \otimes \mathbb{C} = H^1(A_F)^\zeta \oplus H^1(A_F)^{\zeta^2} = H^{1,0}(A)^\zeta \oplus H^{0,1}(A_F)^\zeta \oplus H^{1,0}(A_F)^{\zeta^2} \oplus H^{0,1}(A_F)^{\zeta^2},$$

where $H^1(A_F)^\zeta$ is of rank 5, and $H^{1,0}(A_F)^\zeta$ is of rank 4. Let α_1 be a generator of $H^{1,0}(A)^{\zeta^2}$ (which is of rank one (4.2)): Over $\mathbb{C}$ we can write $\eta_1 = \eta'_1 + \overline{\eta'_1}$, where η'_1 generates $\bigwedge^5 H^1(A_F)^\zeta$. Then $\eta'_1 = \overline{\alpha_1} \wedge \eta''$, where η'' generates $\bigwedge^4 H^{1,0}(A_F)^\zeta$. Similarly $\eta'_2 = \overline{\alpha_2} \wedge \eta''$ and α_2 is obtained from α_1 by multiplication by a non zero element of $K \subset \mathbb{C}$.

The isomorphism $\psi_\beta : H^1(A_F) \otimes \mathbb{C} \cong (H^1(E)) \otimes H^1(A_F))_{inv} \otimes \mathbb{C}$ is given by:

$$H^1(A_F)^\zeta \oplus H^1(A_F)^{\zeta^2} \cong 1/c\,\beta^{0,1} \otimes H^1(A_F)^\zeta \oplus (-1/c)\beta^{1,0} \otimes H^1(A_F)^{\zeta^2},$$

where $c =< \beta^{1,0}, \beta^{0,1} >_E$.

On $H^{1,0}(A_F)^\zeta, int(\eta'_1) + \wedge\eta'_2$ vanishes, so for $u \in H^{1,0}(A_F)^\zeta$, we have: $\gamma_{\beta,\eta_1,\eta_2}(u) = \psi_\beta(int_u(\overline{\eta'_1}) + u \wedge (\overline{\eta'_2}))$. Now

$$int_u(\overline{\eta'_1}) = -\alpha_1 \wedge int_u(\overline{\eta''}) \in H^{1,0}(A_F)^{\zeta^2} \otimes \bigwedge^3 H^{0,1}(A_F)^{\zeta^2},$$

so

$$\psi_\beta(int_u(\overline{\eta'_1})) \in (-1/c)\beta^{1,0} \otimes H^{1,0}(A_F)^{\zeta^2} \otimes \bigwedge^3((-1/c)\beta^{1,0} \otimes H^{0,1}(A_F)^{\zeta^2})$$
$$\subset ((H^1(E) \otimes H^1(A_F))_{inv})^{2,0} \otimes \bigwedge^{odd}(((H^1(E) \otimes H^1(A_F))_{inv})^{1,1}) \subset T^{1,0} \quad \text{(see 4.1)}.$$

Similarly

$$u \wedge \overline{\eta'_2} = -\alpha_2 \wedge u \wedge \overline{\eta''} \in H^{1,0}(A)^{\zeta^2} \otimes H^{1,0}(A)^\zeta \otimes \bigwedge^4 H^{0,1}(A)^{\zeta^2},$$

so

$$\psi_\beta(u\wedge\overline{\eta'_2})\in(-1/c)\beta^{1,0}\otimes H^{1,0}(A)^{\zeta^2}\otimes 1/c\,\beta^{0,1}\otimes H^{1,0}(A)^{\zeta}\otimes\bigwedge^4((-1/c)\beta^{1,0}H^{0,1}(A)^{\zeta^2})$$
$$\subset((H^1(E)\otimes H^1(A_F))_{inv})^{2,0}\otimes\bigwedge^{odd}(((H^1(E)\otimes H^1(A_F))_{inv})^{1,1})\subset T^{1,0}.$$

Next assume $u\in H^{1,0}(A_F)^{\zeta^2}$; Then it is annihilated by $int(\overline{\eta'_1})+\wedge\overline{\eta'_2}$, so

$$\gamma_{\beta,\eta_1,\eta_2}(u)=\psi_\beta(int_u(\eta'_1)+u\wedge(\eta'_2))=\psi_\beta(<u,\overline{\alpha_1}>_{A_F}\eta''+u\wedge\overline{\alpha_2}\wedge\eta'')$$
$$=<u,\overline{\alpha_1}>_{A_F}\psi_\beta(\eta'')+(((-1/c)\beta^{1,0})\otimes u)\wedge((1/c\,\beta^{0,1})\otimes\overline{\alpha_2})\wedge\psi_\beta(\eta''),$$

where $\psi_\beta(\eta'')\in\wedge^4(1/c\,\beta^{0,1}\otimes H^{1,0}(A_F)^{\zeta})\subset\wedge^{even}(((H^1(E)\otimes H^1(A_F))_{inv})^{1,1})$. According to 4.1, to show that $\gamma_{\beta,\eta_1,\eta_2}(u)\in T^{1,0}$, we need the following equality:
4.3.2 $(-1/c^2)(\beta^{1,0}\otimes u)\wedge(\beta^{0,1}\otimes\overline{\alpha_2})=\omega\wedge\overline{\omega}<u,\overline{\alpha_1}>_{A_F}$ where $\omega\in(H^1(E)\otimes H^1(A_F)_{inv})^{2,0}$ satisfies $<\omega,\overline{\omega}>=3$. Here $<,>$ is the intersection form of $(H^1(E)\otimes H^1(A_F))_{inv}$; The reason for the coefficient 3 is that the intersection form on $TH^4(W_F)$ (4.2), which is the opposite of the one on $TH^2(S_F)$ is equal to $-1/3<,>$.
By $\mathbb{C}$-linearity w.r.t. u, and because $u\in H^{1,0}(A_F)^{\zeta^2}$ is proportional to α_1, which was chosen arbitrarily in $H^{1,0}(A_F)^{\zeta^2}$ we may assume $u=\alpha_1$ and $<\alpha_1,\overline{\alpha_1}>_{A_F}=-i$. So what we need is: $-i/c^2(\beta^{1,0}\otimes\alpha_1)\wedge(\beta^{0,1}\otimes\overline{\alpha_2})=\omega\wedge\overline{\omega}$, with $\alpha_2=\mu\alpha_1$, for some $\mu\in K$. For this it is sufficient that: $(-i/c^2)\overline{\mu}<\beta^{1,0},\beta^{0,1}>_E<\alpha_1,\overline{\alpha_1}>_{A_F}=3$, or that $\overline{\mu}=-3c$ with $c=<\beta^{1,0},\beta^{0,1}>$. But clearly $c\in K$, so $\mu=-3\overline{c}\in K$ and we have proved that for β,η_1,η_2 such that $\eta_2=\overline{\mu}\eta_1,\mu=-3\overline{<\beta^{1,0},\beta^{0,1}>}$ a multiple of $\gamma_{\beta,\eta_1,\eta_2}$ gives a map of abelian varieties $\check{A}_F\to KS(S_F)^+$.
4.3.3 It remains now to find copies of E in $KS(S_F)^+$; But this is easy: choose as before $\beta\in H^1(E)$ giving an isomorphism $\psi_\beta: H^1(A_F)\cong(H^1(E)\otimes H^1(A_F))_{inv}$. Now consider the composite map:

$$\bigwedge_K^5 H^1(A_F)\subset\bigwedge^5 H^1(A_F)\overset{\psi_\beta}{\cong}\bigwedge^5((H^1(E)\otimes H^1(A_F))_{inv})\subset\bigwedge^{odd}((H^1(E)\otimes H^1(A_F))_{inv}).$$

Now we have:

4.3.4 Lemma: *The weight one Hodge structure of $\wedge^{odd}((H^1(E)\otimes H^1(A_F))_{inv})$ induces a Hodge structure on the rank 2 $\mathbb{Q}$-vector space $\wedge^5_K H^1(A_F)$, and the corresponding elliptic curve is isogenous to E.*

Proof: One has $\wedge^5_K H^1(A_F)\otimes\mathbb{C}=<\{\eta,\overline{\eta}\}>$, where $\eta\in\wedge^5(H^1(A_F)^{\zeta^2}),\overline{\eta}\in\wedge^5(H^1(A_F)^{\zeta})$. Under the map ψ_β,η is sent to an element of $\wedge^5(\beta^{1,0}\otimes H^1(A_F)^{\zeta^2})$, and one has $H^1(A_F)^{\zeta^2}=H^{1,0}(A_F)^{\zeta^2}\oplus H^{0,1}(A_F)^{\zeta^2}$, the first piece being of rank 1, the second one being of rank 4. So

$$\bigwedge^5(\beta^{1,0}\otimes H^1(A_F)^{\zeta^2})\subset(H^1(E)\otimes H^1(A_F)_{inv})^{2,0}\otimes\bigwedge^{even}((H^1(E)\otimes H^1(A_F)_{inv})^{1,1})\subset T^{1,0}.$$

But this means exactly that $\wedge^5_K H^1(A_F)$ has an induced Hodge structure of level one, which is clearly invariant under the natural action of K on $\wedge^5_K H^1(A_F)$ so the corresponding elliptic curve is isogenous to E.

4.4 Now consider for any k the inclusion: $\wedge^{2k}H^2(S)^0\hookrightarrow\wedge^{even}H^2(S)^0=T_{\mathbb{Z}}\otimes\mathbb{C}$. This inclusion is defined over $\mathbb{Z}$, and one has:

$$\bigwedge^{2k}H^2(S)^0\cap T^{1,0}=\omega\wedge\bigwedge^{2k-1}H^{1,1}(S)=F^{2k+1}\bigwedge^{2k}H^2(S)^0.$$

So the Kuga-Satake variety contains naturally the partial torus

$$\bigwedge^{2k} H^2(S)^0/F^{2k+1}\bigwedge^{2k} H^2(S)^0 \oplus \bigwedge^{2k} H^2(S,\mathbb{Z})^0,$$

which is the essential part of the Deligne cohomology group $H_D^{4k+1}(S^k,\mathbb{Z}(2k+1))^{skew}$. Notice that, because rank $(TH^2(S)$ is odd for a general algebraic K3 surface, one may as well construct the Kuga-Satake variety using the odd part of the exterior algebra $\bigwedge^{odd} H^2(S)^0$, which shows that it contains as well the Deligne cohomology groups $H_D^{2k+1}(S^k,\mathbb{Z}(k+1))_0^{skew}$ for k odd $(()_0$ denotes here the part which is build on $\bigwedge^k H^2(S)^0 \hookrightarrow H^{2k}(S^k)^{skew})$. Each of these Deligne cohomology groups is a partial torus which is the target of the first regulator map ([10], or see 4.5). For $k=1$ it was proved by C. Oliva (to appear?), that the regulator map is non trivial, even modulo the obvious subgroup $NS(S)\otimes\mathbb{C}^*\subset H_D^3(S,\mathbb{Z}(2))$ and the torsion. This appears to be a special feature of surfaces with $h^{2,0}=1$ since for $h^{2,0}\geq 2$ one may find smooth projective surfaces for which the regulator map has torsion image modulo $NS(S)\otimes\mathbb{C}^*$(see [24]).

Note that the only continuous part in the image of the regulator map is $NS(S)\otimes\mathbb{C}^*$, so each of the Deligne cohomology groups $H_D^{2k+1}(S^k,\mathbb{Z}(k+1))_0^{skew}$ is a transcendental object. But the Kuga-Satake construction shows that they are part of an abelian variety, the points of which one would like to understand in terms of algebraic cycles on the various S^k's. It seems that Theorem 3.4 and its conjectural generalizations 3.8 would open the way toward this understanding because it shows that there are many equivalence relations between cycles in higher self-products of S, which are necessary to construct data on which a generalization of the regulator map could be defined, interpolating between the various regulator maps of the S^k's, and parametrizing the Kuga-Satake variety.

4.5 To be more precise the regulator map of

$$S^k : H^k(\mathcal{K}_{k+1}(S^k)) \to H_D^{2k+1}(S^k,\mathbb{Z}(k+1))$$

is defined on data $\sum_i(Z_i,\phi_i)$, $Z_i\subset S^k$ irreducible of codimension k, ϕ_i a rational function on Z_i, such that $\sum_i div(\phi_i)=0$ as a codimension $k+1$ cycle on S^k; the group $H^k(\mathcal{K}_{k+1}(S^k))$ is the group of such data modulo a certain subgroup. Let us consider the case $k=1$: the regulator map takes values in $H_D^3(S,\mathbb{Z}(2))$ and its image modulo $NS(S)\otimes\mathbb{C}^*$ is countable. In fact it seems possible that $H^1(\mathcal{K}_2(S))/Pic(S)\otimes\mathbb{C}^*$ itself is countable, or at least, there is the following evidence for that (this was the starting point for C. Oliva's work): if one wants to construct non trivial elements in $H^1(\mathcal{K}_2(S))$, it is natural to make the following construction: Let $|L|,|L'|$ be two linear systems on S; Consider the finite cover W of $|L|\times|L'|$ determined by an ordering of the points of intersection of $C\subset|L|, C'\subset|L'|$. Let N be the cardinal of this intersection, and let $n_1,\ldots,n_N$ be integers such that $\sum_i n_i=0$. Now let $Z=(C,C',p_1,\ldots,p_N)\in W$ and consider the conditions on Z : $\sum_i n_ip_i\equiv 0$ in $CH_0(C)$, and $\sum_i n_ip_i\equiv 0$ in $CH_0(C')$. If these conditions are satisfied one constructs an element α_Z of $H^1(\mathcal{K}_2(S))$ as follows: by assumption, there exists ϕ on C,ϕ' on C', such that: $\sum_i n_ip_i=div(\phi)$ as a zero-cycle of $C,\sum_i n_ip_i=div(\phi')$ as a zero-cycle on C'; Then $\alpha_Z=(C,\phi)+(C',\phi'^{-1})$ gives the desired element, well defined up to $Pic(S)\otimes\mathbb{C}^*$, reflecting the choice of ϕ and ϕ'. Now note that the expected dimension of this sublocus of W is zero, because $\dim(W)=\dim(|L|)+\dim(|L'|)$ and the expected codimension of the sublocus is $g(C)+g(C')$ which is also equal to $\dim(|L|)+\dim(|L'|)$ since we are on a K3 surface. So there are some reasons to expect that $H^1(\mathcal{K}_2(S))/Pic(S)\otimes\mathbb{C}^*$ is discrete. But if we work with the self-product S^2 we can construct positive dimensional families of data $\sum_i(C_i,\phi_i)$ such that $\sum_i div(\phi_i)=0$; For example, if we know the property (*): for

$(x, y) \in S^2, (x, y) - (y, x) \equiv_{rat} -x_0 * (x - y) + (x - y) * x_0$, there is a generically finite cover $G \xrightarrow{\pi} S \times S$ such that $\pi^{-1}((x, y))$ parametrizes curves $C_i \subset S \times S$ with a meromorphic function ϕ_i, such that $\sum_i div\phi_i = (x, y) - (y, x) + x_0 * (x - y) + (x - y) * x_0$. Then the surface $\pi^{-1}(\Delta)$ parametrizes data (c_i, ϕ_i) such that $\sum_i n_i div\phi_i = 0$.

References

[1] Y. André: On the Shafarevich and Tate conjecture for hyperkähler varieties (preprint).

[2] A. Beauville: Quelques remarques sur la transformée de Fourier dans l'anneau de Chow d'une variété abélienne, dans *Algebraic Geometry Tokyo/Kyoto 1982*, Lect. Notes in Math. **1016** (1983), 238–260, Berlin-Heidelberg-New York.

[3] A. Beauville: Sur l'anneau de Chow d'une variété abélienne, Math. Ann. **273** (1986), 647–651.

[4] S. Bloch: Algebraic cycles and higher K-theory, Adv. in Math. **61** (1986), 267–304.

[5] S. Bloch: Torsion algebraic cycles and a theorem of Roitman, Compositio Math. **39** (1979),107–127.

[6] S. Bloch and V. Srinivas: Remarks on correspondences and algebraic cycles, Amer. J. of Math. **105** (1983), 1235–1253.

[7] S. Bloch: Some elementary theorems about algebraic cycles on abelian varieties, Invent. Math. **37** (1976), 215–228.

[8] P. Deligne: La conjecture de Weil pour les surfaces K3, Invent. Math. **15** (1972), 206–226.

[9] H. Esnault and E. Viehweg: Deligne-Beilinson cohomology, in *Beilinson's conjectures on special values of L-functions* Ed. by Rapoport, Schappacher and Schneider, Perspectives in Math. Vol. **4** Academic Press (1988), 43–91.

[10] U. Jannsen: Deligne homology, Hodge D-conjecture and motives, same ref. as [9], 305–372.

[11] U. Jannsen: Motivic sheaves and filtration on Chow groups, in *Motives*, U. Jannsen, S. Kleiman, J.-P. Serre eds, Proc. Symp. Pure Math. **55** (1994), AMS Providence.

[12] S. Kimura and A. Vistoli: Chow rings of infinite symmetric products (preprint).

[13] C. Deninger and J. P. Murre: Motivic decomposition of abelian schemes and the Fourier transform, J. reine und angew. Math. **422** (1991), 201–219.

[14] M. Kuga and I. Satake: Abelian varieties attached to polarized K3 surfaces, Math. Ann. **169** (1967), 239–242.

[15] D. Morrison: The Kuga–Satake variety of an abelian surface, J. of Algebra **92** (1985), 454–476.

[16] D. Mumford: Rational equivalence of 0-cycles on surfaces, J. Math. Kyoto Univ. **9** (1968), 194–204.

[17] J. P. Murre: On a conjectural filtration on the Chow groups of an algebraic variety I, II, Indag. Math. N.S.**4**(2)(1993) 177–201.

[18] J. P. Murre: On the motive of an algebraic surface, J. reine und angew. Math. **409** (1990), 190–204.

[19] M. V. Nori: Algebraic cycles and Hodge theoretic connectivity, Invent. Math. **111**(2) (1993), 349–373.

[20] K. Paranjape: Cohomological and cycle theoretic connectivity, To appear in Annals of Math.

[21] K. Paranjape: Abelian varieties associated to certain K3 surfaces, Compositio Math. **68** (1988), 11–22.

[22] A. Roitman: Rational equivalence of zero–cycles, Math. USSR Sb. **18** (1972), 571–588.

[23] C. Voisin: Sur les zéro-cycles de certaines hypersurfaces munies d'un automorphisme, Annali della Scuola Norm. Sup. di Pisa, Vol. **29** (1993), 473–492.

[24] C. Voisin: Variations of Hodge structure and algebraic cycles, To appear in the proceedings of ICM, Zürich.

[25] C. Voisin: Théorème de Torelli pour les cubiques de $\mathbb{P}^5$, Inv. Math. **86** (1986), 577–601.

[26] T. Shioda: The Hodge conjecture for Fermat varieties, Math. Ann. **245** (1979), 175–184.

[27] S. Saito: Motives and filtrations on Chow groups, preprint Tokyo (1993).

MODULI OF PARABOLIC HIGGS SHEAVES

Kôji Yokogawa

Department of Mathematics, Graduate School of Science, Osaka University, Toyonaka, Osaka 560, Japan
e-mail: yokogawa@math.sci.osaka-u.ac.jp

INTRODUCTION

Recently many types of 'stable bundles' have been studied, for example, stable pairs, Higgs bundles, parabolic bundles and parabolic Higgs bundles and so on. In almost all cases, the moduli spaces for these 'stable bundles' were constructed using the geometric invariant theory or the gauge theory. To study the geometric properties of these moduli spaces, a recent technique, i.e. the 'flip' technique for the GIT quotients has been applied.

In some cases, stability of 'bundles' depends on parameters like weights in the case of parabolic bundles. Then the flip technique describes how these moduli spaces change with those parameters. In the case of Riemann surfaces, Boden and Hu [1] studied the flip phenomenon for the moduli of parabolic bundles. In the symposium, the author talked about a joint work with Boden (cf. [2]), i.e. we computed the Betti numbers of the moduli space of parabolic Higgs bundles $\mathcal{N}_\alpha$ for rank 2 cases, which turn out to be independent of the weights α. This behavior is in marked contrast to that exhibited by the moduli space $\mathcal{M}_\alpha$ of parabolic bundles, where the Betti numbers do depend on the weights α [1].

In this paper, first we shall see how various notions for usual vector bundles are generalized to those for parabolic bundles. Such a generalization enables us to treat parabolic bundles like usual vector bundles and would be useful in higher dimensional cases. Proofs of these general theory were given mainly in the paper [16]. Next, we shall introduce three moduli spaces; $\mathcal{M}_\alpha$ a moduli space of semi-stable parabolic sheaves, $\mathcal{N}_\alpha$ a moduli space of semi-stable parabolic Higgs sheaves and $\mathcal{P}_\alpha$ a moduli space of semi-stable parabolic pairs. All of them were constructed by the GIT [5], [15]. In the case of curves, further those moduli spaces are shown to be normal and irreducible. Finally, for the rank 2 case, the Betti numbers of the moduli space of parabolic Higgs sheaves will be given.

The author is very grateful to the Tanighchi Foundation for providing him with this opportunity.

1. GENERAL THEORY ON PARABOLIC SHEAVES

Parabolic bundles were originally defined by Seshadri for a Riemann surface with marked points as a vector bundle with weighted flag structure at each marked point (cf. [6]). To deal with parabolic bundles on higher dimensional varieties, we shall generalize various notions for usual coherent sheaves to notions for parabolic sheaves.

1.1. Parabolic Sheaf. Let D be an effective Cartier divisor on a scheme X. From now on, $\mathbb{R}$ is regarded as a category with $\mathrm{Ob}(\mathbb{R}) = \mathbb{R}$ (the set of all real numbers),

$$\mathrm{Mor}(\alpha, \beta) = \begin{cases} \{i^{\alpha,\beta}\} & \text{if } \alpha \geq \beta \\ \emptyset & \text{otherwise.} \end{cases}$$

An $\mathbb{R}$-*filtered* $\mathcal{O}_X$-*module* is a covariant functor E from $\mathbb{R}$ to the category of all $\mathcal{O}_X$-modules. Then $E(\alpha)$ is denoted by E_α and $E(i^{\alpha,\beta})$ is denoted by $i_E^{\alpha,\beta}$. We use a symbol E_* instead of the covariant functor E and sometimes E means the sheaf E_0. A natural transformation $f : E_* \to F_*$ is often denoted by f_* and $f(\alpha) : E_\alpha \to F_\alpha$ is denoted by f_α. Moreover, $E[\alpha]_*$ is defined by $E[\alpha]_\beta = E_{\alpha+\beta}$ and $i_{E[\alpha]}^{\beta,\gamma} = i_E^{\beta+\alpha,\gamma+\alpha}$. If $\alpha \geq \beta$, we have a natural transformation

$$i_E^{[\alpha,\beta]} : E[\alpha]_* \longrightarrow E[\beta]_*$$

defined by $i_E^{\alpha+\gamma,\beta+\gamma}$. Each $f_* : E_* \to F_*$ naturally induces $f[\alpha]_* : E[\alpha]_* \longrightarrow F[\alpha]_*$. For an $\mathcal{O}_X$-module F, a tensor product $E_* \otimes_X F$ is defined by setting $(E_* \otimes_X F)_\alpha = E_\alpha \otimes_X F$ and $i_{E_* \otimes F}^{\alpha,\beta} = i_E^{\alpha,\beta} \otimes id_F$.

Definition 1.1. A *parabolic* $\mathcal{O}_X$-*module* is an $\mathbb{R}$-filtered $\mathcal{O}_X$-module E_* together with an isomorphism of $\mathbb{R}$-filtered $\mathcal{O}_X$-modules

$$j_E : E_* \otimes_X \mathcal{O}_X(-D) \xrightarrow{\simeq} E[1]_*$$

which satisfies

$$\text{(1)} \qquad i_E^{[1,0]} \circ j_E = id_{E_*} \otimes i_D : E_* \otimes_X \mathcal{O}_X(-D) \to E_*$$

where $i_D : \mathcal{O}_X(-D) \to \mathcal{O}_X$ is a natural map defined by D.

For parabolic $\mathcal{O}_X$-modules E_* and F_*, a natural transformation $f_* : E_* \to F_*$ is called a *parabolic homomorphism* if the following diagram is commutative.

$$\text{(2)} \qquad \begin{array}{ccc} E_* \otimes_X \mathcal{O}_X(-D) & \xrightarrow{f_* \otimes id} & F_* \otimes_X \mathcal{O}_X(-D) \\ \simeq \downarrow j_E & & \simeq \downarrow j_F \\ E[1]_* & \xrightarrow{f[1]_*} & F[1]_* \end{array}$$

We use $\mathrm{Hom}_X(E_*, F_*)$ to denote the module of all parabolic homomorphisms. Also $\mathcal{H}om_X(E_*, F_*)$ denotes a sheaf with $\mathcal{H}om_X(E_*, F_*)(U) = \mathrm{Hom}_U(E_*|_U, F_*|_U)$ for each open subset U. Let us denote by $\mathcal{P}_{X/D}$ the category of parabolic $\mathcal{O}_X$-modules and their parabolic homomorphisms. Every $\mathcal{O}_X$-module E can be naturally regarded as a parabolic $\mathcal{O}_X$-module by setting

$$E_\alpha = E \otimes_X \mathcal{O}_X(-mD) \qquad \text{for } m - 1 < \alpha \leq m.$$

This structure is called a *special structure*.

Definition 1.2. A parabolic sheaf E_* is said to be *coherent* if the following conditions are satisfied.

(1) All E_α are coherent $\mathcal{O}_X$-modules.

(2) There is a sequence of real numbers $0 \leq \alpha_1 < \alpha_2 < \cdots < \alpha_l < 1$ such that $i_E^{\alpha_i,\alpha} : E_{\alpha_i} \to E_\alpha$ are isomorphisms for $\alpha_{i-1} < \alpha \leq \alpha_i$ ($\alpha_0 = 0$, $\alpha_{l+1} = 1$).

We call the set $\{\alpha_1, \alpha_2, \ldots, \alpha_l\}$ a *system of weights* of E_* if $i_E^{\alpha_{i+1},\alpha_i} : E_{\alpha_{i+1}} \to E_{\alpha_i}$ are not isomorphisms for $i = 1, 2, \ldots, l$. Moreover, when X is an integral scheme, a coherent parabolic sheaf E_* is said to be *torsion free* if all E_α are torsion free.

Remark 1.1. If E_* is torsion free, then all $i_E^{\alpha,\beta}$ are injective. Hence, we get a parabolic sheaf in the meaning of the original definition (cf. [5], [6])

$$E = E_0 \supsetneq E_{\alpha_1} \supsetneq \cdots \supsetneq E_{\alpha_l} \supsetneq E_1 \simeq E_0 \otimes_X \mathcal{O}_X(-D)$$

If E_* and F_* are torsion free, then the definition of parabolic homomorphisms is the same as that of the original one.

One reason why we introduce such an abstract definition of parabolic sheaves is the following.

Proposition 1.3 ([16] Prop. 1.1). *The category $\mathcal{P}_{X/D}$ is an abelian category with enough injective objects.*

Now, we can define various derived functors as in the case of usual $\mathcal{O}_X$-modules. For each E_*, $\mathrm{Ext}^i_X(E_*, -)$ (or, $\mathcal{E}xt^i_X(E_*, -)$) is the i-th right derived functor for $\mathrm{Hom}_X(E_*, -)$ (or, $\mathcal{H}om_X(E_*, -)$, respectively).

1.2. Parabolic Higgs Sheaf. Now let Ω be a locally free $\mathcal{O}_X$-module and let $\pi : Y = \mathrm{Spec}(S^*(\Omega^\vee)) \to X$ be a natural projection map.

Definition 1.4. A *parabolic Ω-pair* is a pair (E_*, φ) consisting of a parabolic $\mathcal{O}_X$-module E_* and a parabolic homomorphism $\varphi : E_* \to E_* \otimes_X \Omega$ with $\varphi \wedge \varphi = 0$. We say that (E_*, φ) is coherent (or, torsion free) if E_* is.

By the condition $\varphi \wedge \varphi = 0$, E_* has a structure of parabolic $\mathcal{O}_Y$-module with respect to $\pi^* D$. This correspondence gives us an equivalence between the category of parabolic Ω-pairs and $\mathcal{P}_{Y/\pi^*D}$. By Proposition 1.3, $\mathcal{P}_{Y/\pi^*D}$ has enough injective objects. For parabolic Ω-pairs (E_*, φ) and (E'_*, φ'), we denote by $\mathrm{Ext}^i_X((E_*, \varphi), (E'_*, \varphi'))$ the Ext-group $\mathrm{Ext}^i_Y(E_*, E'_*)$.

The modules $\mathrm{Ext}^i_X((E_*, \varphi), (E'_*, \varphi'))$ are described using Ext-groups for parabolic sheaves by the following spectral sequence.

Proposition 1.5 ([16], Prop. 4.2). *For parabolic Ω-pairs (E_*, φ) and (F_*, ψ), there is a spectral sequence*

$$E_1^{pq} = \mathrm{Ext}^q_X(E_*, F_* \otimes_X \wedge^p \Omega) \Rightarrow \mathrm{Ext}^{p+q}_X((E_*, \varphi), (F_*, \psi)) \tag{3}$$

To give a definition of parabolic Higgs sheaves, we need one more operation for parabolic sheaves. Let $\widehat{E}_\alpha$ be an inductive limit $\varinjlim_{\beta > \alpha} E_\beta$. For each $\alpha \geq \beta$, we have a natural homomorphism $i_{\widehat{E}}^{\alpha,\beta} : \widehat{E}_\alpha \to \widehat{E}_\beta$ induced by $i_E^{*,*}$ and have an isomorphism $j_{\widehat{E}}(\alpha) : \widehat{E}_\alpha \otimes_X \mathcal{O}_X(-D) \to \widehat{E}_{\alpha+1}$. Thus, we obtain a parabolic sheaf $\widehat{E}_*$. If E_* is a coherent parabolic sheaf corresponding to a filtration

$$E = E_0 \supsetneq E_{\alpha_1} \supsetneq \cdots \supsetneq E_{\alpha_l} \supsetneq E_1 \simeq E_0 \otimes_X \mathcal{O}_X(-D),$$

then $\widehat{E}_*$ is a parabolic sheaf with $\widehat{E}_\alpha = E_{\alpha_i}$ for all $\alpha_{i-1} \leq \alpha < \alpha_i$ ($i = 1, \ldots, l+1, \alpha_0 = \alpha_l - 1, \alpha_{l+1} = 1 + \alpha_1$).

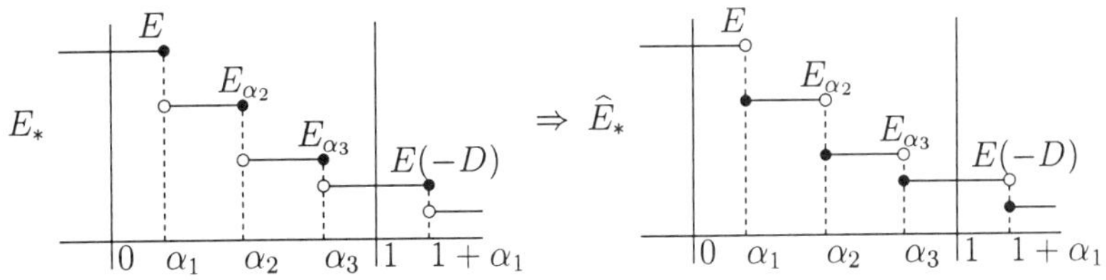

Notice that there is a canonical map from $\widehat{E}_*$ to E_*. Now suppose that X is a nonsingular variety over an algebraically closed field k and D is a simple normal crossing divisor so that $\Omega^1_X(\log D)$ the sheaf of one forms with logarithmic pole along D is locally free.

Definition 1.6. A *parabolic Higgs sheaf* is a pair (E_*, φ) consisting of a parabolic $\mathcal{O}_X$-module E_* and a parabolic homomorphism $\varphi : E_* \longrightarrow \widehat{E}_* \otimes_X \Omega^1_X(\log D)$ with $\varphi \wedge \varphi = 0$.

If E_* is torsion free, then the map $\widehat{E}_* \longrightarrow E_*$ is injective. Hence, the set of all parabolic Higgs sheaves can be naturally identified with a subset of all parabolic $\Omega^1_X(\log D)$-pairs.

1.3. Tensor products, Hom-sheaves and Serre duality. Now let us introduce parabolic tensor products and parabolic Hom-sheaves.

Definition 1.7. Let E_* and F_* be parabolic sheaves. For each α, set

$$\mathcal{H}om_X(E_*, F_*)_\alpha = \mathcal{H}om_X(E_*, F[\alpha]_*).$$

We can easily give natural homomorphisms $i^{[\alpha,\beta]}_{\mathcal{H}om_X(E_*,F_*)}$ and a natural isomorphism $j_{\mathcal{H}om_X(E_*,F_*)}$ so that $\mathcal{H}om_X(E_*, F_*)_*$ becomes a parabolic sheaf. It is called a *parabolic Hom-sheaf*.

Lemma 1.8 ([16] §3). *There exists a parabolic sheaf $(E_* \otimes_X F_*)_*$ which satisfies the following isomorphism of parabolic sheaves*

$$\mathcal{H}om_X((E_* \otimes_X F_*)_*, G_*)_* \simeq \mathcal{H}om_X(E_*, \mathcal{H}om_X(F_*, G_*)_*)_*.$$

The parabolic sheaf $(E_* \otimes_X F_*)_*$ is explicitly given in [16] as a quotient of type

$$(E_* \otimes_X F_*)_\alpha = (\bigoplus_{\alpha_1+\alpha_2=\alpha} (E_{\alpha_1} \otimes_X F_{\alpha_2}))/R_\alpha.$$

We call $(E_* \otimes_X F_*)_*$ a *parabolic tensor product* of E_* and F_*.

Proposition 1.9 ([16] Prop. 3.2, 3.3). *A functor $- \otimes_X E_*$ is a right exact functor. If E_* is coherent, then $- \otimes_X E_*$ is exact if and only if all E_α are locally free $\mathcal{O}_X$-modules and all $E_\alpha / i^{\beta,\alpha}_E(E_\beta)$ are locally free $\mathcal{O}_D$-modules for $\alpha \leq \beta < \alpha + 1$.*

If a coherent parabolic sheaf E_* satisfies the last equivalent conditions, it is said to be *locally free.* Locally free parabolic sheaves of rank one (call them parabolic invertible sheaves) are obtained locally by shifting invertible sheaves with special structure. Moreover, every locally free parabolic sheaf E_* is locally a direct sum of parabolic invertible sheaves. By virtue of the above proposition, tensor products of locally free parabolic sheaves are also locally free.

Example 1.1. Let E_* and F_* be parabolic sheaves. Assume that E_* is locally free. Set $U = X - D$. Let $i : U \rightarrow X$ be an inclusion map. Since $E_\alpha \otimes_X F_\beta \subset E_\alpha \otimes_X i_*i^*(F) = i_*i^*(E \otimes_X F)$, all $E_\alpha \otimes_X F_\beta$ are regarded as sub modules of $i_*i^*(E \otimes_X F)$. In this case,

$$(E_* \otimes_X F_*)_\alpha = \sum_{\alpha_1+\alpha_2=\alpha} (E_{\alpha_1} \otimes_X F_{\alpha_2})$$

as a sub-module of $i_*i^*(E \otimes_X F)$ and $i^{\alpha,\beta}_{E_*\otimes F_*}$ is the natural inclusion map.

Let us set $\vee(E_*)_\alpha = E^\vee_{-\alpha}$ (the dual sheaf of $E_{-\alpha}$), $i^{\alpha,\beta}_{\vee(E)} = (i^{-\beta,-\alpha}_E)^\vee$ for $\alpha \geq \beta$ and $j_{\vee(E)}(\alpha) = (j_E(-\alpha - 1) \otimes id_{\mathcal{O}_X(D)})^\vee$. Then $\vee(E_*)_*$ is also a parabolic sheaf.

Now let F be a parabolic sheaf with special structure. Then we have

$$\mathcal{H}om_X(E_*, F)_\alpha \simeq \mathcal{H}om_X(\widehat{E}[-1]_{-\alpha}, F) \simeq \mathcal{H}om_X(\widehat{E}_{-\alpha} \otimes_X \mathcal{O}_X(D), F).$$

In particular, we have a natural isomorphism

$$E_*^\vee \simeq \vee(\widehat{E}_* \otimes_X \mathcal{O}_X(D))_* \simeq \vee(\widehat{E}_*)_* \otimes_X \mathcal{O}_X(-D), \tag{4}$$

where $E_*^\vee = \mathcal{H}om_X(E_*, \mathcal{O}_X)_*$. We have a canonical homomorphism

$$E_* \longrightarrow E_*^{\vee\vee}.$$

If this is an isomorphism, E_* is said to be *reflexive*. A coherent parabolic sheaf E_* is reflexive if and only if all E_α are reflexive, since $(E_*^{\vee\vee})_\alpha = (E_\alpha)^{\vee\vee}$. In particular, locally free parabolic sheaves are reflexive.

Serre duality theorem can be generalized for parabolic sheaves as follows.

Proposition 1.10 ([16] Prop. 3.7). *Let X be a non-singular projective variety of dimension n over an algebraically closed field k. Let ω_X be a dualizing sheaf on X. Then for all locally free parabolic sheaves E_* and F_*, there are natural isomorphisms*

$$\theta^i : \operatorname{Ext}^i_X(E_*, F_* \otimes_X \omega_X(D)) \xrightarrow{\simeq} \operatorname{Ext}^{n-i}_X(F_*, \widehat{E}_*)^\vee \tag{5}$$

If, moreover, D is non-singular, then this formula holds for all coherent parabolic sheaves E_* and F_*.

Remark 1.2. Even for locally free parabolic sheaves E_*, in general,

$$\mathcal{H}om_X(E_*, E_*)^\vee \not\simeq \mathcal{H}om_X(E_*, E_*),$$

though we have $\mathcal{H}om_X(E_*, E_*)_*^\vee \simeq \mathcal{H}om_X(E_*, E_*)_*$.

2. Moduli Spaces

To construct moduli spaces of parabolic sheaves or parabolic Higgs sheaves, we need to generalize the notion of stability of vector bundles to parabolic cases. In curve cases, stability of parabolic bundles was measured by the parabolic degree. The following definition coincides with the definition of stability measured by parabolic degree when X is a curve. In fact, the second coefficient of the "parabolic Hilbert polynomial" $\int_0^1 \chi(E_\alpha(m))d\alpha$ is essentially the same with the parabolic degree.

Definition 2.1. Suppose that X is a nonsingular projective variety over an algebraically closed field k. Let $\mathcal{O}_X(1)$ be an ample line bundle on X. A coherent parabolic Ω-pair (E_*, φ) is said to be *stable* (or, *semi-stable*) if it is torsion free and for every coherent φ-invariant parabolic sub-sheaf F_* of E_* with $0 \neq F \neq E$,

$$\int_0^1 \frac{\chi(F_\alpha(m))}{\operatorname{rk} F} d\alpha < \int_0^1 \frac{\chi(E_\alpha(m))}{\operatorname{rk} E} d\alpha \quad (\text{or, } \leq, \text{ resp.})$$

for sufficiently large integers m. Here, "φ-invariant" means $\varphi(F_*) \subseteq F_* \otimes_X \Omega$.

Let $f : X \to S$ be a flat, quasi-projective morphism of noetherian schemes, let $\mathcal{O}_X(1)$ be an f-ample invertible sheaf and let D be a relative effective Cartier divisor on X/S. A coherent parabolic sheaf E_* is said to be *S-flat* if all E_α are S-flat. The support of E_* is *proper* over S if the support of each E_α is. For each S-scheme T, set

$$\operatorname{Par}_{X/D}(T) = \left\{ E_* \;\middle|\; \begin{array}{l} E_* \text{ is a } T\text{-flat, coherent} \\ \text{parabolic } \mathcal{O}_{X_T}\text{-module} \\ \text{with support proper over } T \end{array} \right\} \Big/ \sim$$

where $\sim$ is an equivalence relation such that $E_* \sim E'_*$ if and only if there is an invertible sheaf L on T and an isomorphism $E_* \simeq E'_* \otimes_T L$. Then $\mathrm{Par}_{X/D}$ defines a functor from (Sch/S) to $(Sets)$ in a natural way.

Let α_* be a set of real numbers $\{\alpha_1, \alpha_2, \dots, \alpha_l\}$ with $0 \leq \alpha_1 < \alpha_2 < \cdots < \alpha_l < 1$ and let H_* be a set of numerical polynomials $\{H, H_1, H_2, \dots, H_l\}$. Suppose that $f : X \to S$ is a smooth, projective and geometrically integral morphism. Let Ω be a locally free $\mathcal{O}_X$-module and let $\pi : Y = \mathrm{Spec}(S^*(\Omega^\vee)) \to X$ be a natural projection map. As in §1.2, we identify parabolic Ω-pairs on X with parabolic sheaves on Y. Let us denote by $\mathrm{Par}^{H_*,\alpha_*,s}_{Y/\pi^*D}$ (or, $\mathrm{Par}^{H_*,\alpha_*,ss}_{Y/\pi^*D}$) a sub-functor of Par_{Y/π^*D} such that $(E_*, \varphi) \in \mathrm{Par}_{Y/\pi^*D}(T)$ is a T-valued point of it if and only if for all geometric points t of T, the system of weights of $E_*|_{X_t}$ is α_*, $(E_*, \varphi)_{X_t}$ is stable (or, semi-stable, resp.), $\chi(E(m)_{X_s}) = H(m)$ and $\chi((E/i_E^{\alpha_{i+1},0}(E_{\alpha_{i+1}}))(m)_{X_s}) = H_i(m)$ for all i. Then the functors $\mathrm{Par}^{H_*,\alpha_*,s}_{Y/\pi^*D}$ and $\mathrm{Par}^{H_*,\alpha_*,ss}_{Y/\pi^*D}$ are open sub-functors of Par_{Y/π^*D} (cf. [15]).

Using the geometric invariant theory, moduli schemes of Seshadri equivalence classes of semi-stable parabolic Ω-pairs were constructed (cf. [5] and [15]). For simplicity, we assume that S is locally of finite type over an algebraically closed field k.

Theorem 2.2 (Theorem 4.6 of [15]). *Assume that all α_i are rational numbers. Then there exist an S-scheme $M^{H_*,\alpha_*,ss}_{\Omega/X/D}$ and a morphism $\Upsilon : \mathrm{Par}^{H_*,\alpha_*,ss}_{Y/\pi^*D} \to M^{H_*,\alpha_*,ss}_{\Omega/X/D}$ such that*

(i) *$M^{H_*,\alpha_*,ss}_{\Omega/X/D}$ is quasi-projective over S,*

(ii) *$M^{H_*,\alpha_*,ss}_{\Omega/X/D}$ contains an open sub-scheme $M^{H_*,\alpha_*,s}_{\Omega/X/D}$ which is a coarse moduli scheme for the functor $\mathrm{Par}^{H_*,\alpha_*,s}_{Y/\pi^*D}$ and*

(iii) *for each geometric point s of S, $\Upsilon(k(s))$ induces a bijection*

$$\mathrm{Par}^{H_*,\alpha_*,ss}_{Y/\pi^*D}(k(s))/\sim \xrightarrow{\simeq} M^{H_*,\alpha_*,ss}_{\Omega/X/D}(k(s)),$$

where $\sim$ means the Seshadri equivalence relation (cf. Definition 1.12 of [15]).

Thus $M^{H_*,\alpha_*,ss}_{\Omega/X/D}$ is a moduli scheme of Seshadri equivalence classes of semi-stable parabolic Ω-pairs on X with fixed weights α_* and Hilbert polynomials H_*. There exists a map, called a *Hitchin map*, from the moduli scheme $M^{H_*,\alpha_*,ss}_{\Omega/X/D}$ to an affine space. For simplicity, we assume that $S = \mathrm{Spec}(k)$. Then the Hitchin map

$$\Phi : M^{H_*,\alpha_*,ss}_{\Omega/X/D} \to \oplus_{i=0}^r H^0(X, S^i(\Omega))t^{r-i}$$

is defined by taking a characteristic polynomial of φ for each $(E_*, \varphi) \in M^{H_*,\alpha_*,ss}_{\Omega/X/D}(k)$ (r is the rank of E). When X is a curve and Hitchin defined the map for the moduli of Higgs bundles and proved that it is proper (cf. [3]). For the parabolic and higher dimensional case, the result is also true.

Proposition 2.3 ([15] §5). *The Hitchin map $\Phi : M^{H_*,\alpha_*,ss}_{\Omega/D/X} \to \oplus_{i=0}^r H^0(X, S^i(\Omega))t^{r-i}$ is a projective morphism.*

If $\Omega = 0$, then clearly the moduli scheme $M^{H_*,\alpha_*,ss}_{\Omega/D/X}$ is identified with the moduli scheme $M^{H_*,\alpha_*,ss}_{D/X}$ of semi-stable parabolic sheaves with the fixed invariants α_* and H_*. Since the characteristic polynomials are monic, the image of the Hitchin map is one point.

Corollary 2.4. *The moduli scheme $M^{H_*,\alpha_*,ss}_{D/X}$ is projective over S.*

For the usual sheaf case, the Zariski tangent space of moduli at E is described by $\mathrm{Ext}^1(E, E)$ and the obstruction class for smoothness at E is in $\mathrm{Ext}^2(E, E)$. Similar results hold for the parabolic case.

Theorem 2.5 ([16] Theorem 2.4). *Let (E_*, φ) be an Ω-pair corresponding to a k-valued point x of $M^{H_*,\alpha_*,s}_{\Omega/X/D}$. Then the Zariski tangent space of $M^{H_*,\alpha_*,s}_{\Omega/X/D}$ at x is naturally isomorphic to* $\mathrm{Ext}^1_X((E_*,\varphi),(E_*,\varphi))$. *If* $\mathrm{Ext}^2_{X_s}((E_*,\varphi),(E_*,\varphi)) = 0$ *for all semi-stable parabolic Ω-pairs (E_*,φ), then $M^{H_*,\alpha_*,ss}_{\Omega/X/D}$ is normal.*

Suppose, moreover, that X is a nonsingular variety over k and D is a simple normal crossing divisor. Let $\Omega = \Omega^1_X(\log D)$. In this case, a moduli functor for semi-stable parabolic Higgs sheaves is clearly a closed sub-functor of $\mathrm{Par}^{ss}_{Y/\pi^*D}$. Then a moduli scheme $N^{H_*,\alpha_*,ss}_{D/X}$ for Seshadri equivalence classes of semi-stable parabolic Higgs sheaves is constructed as a closed subscheme of $M^{H_*,\alpha_*,ss}_{\Omega/D/X}$.

Corollary 2.6. *The moduli scheme $N^{H_*,\alpha_*,ss}_{D/X}$ of semi-stable parabolic Higgs sheaves is projective over S.*

When X is a Riemann surface, the moduli space $N^{H_*,\alpha_*,ss}_{D/X}$ was constructed by Konno [4] using gauge theory (precisely speaking he constructed the locus of (E_*,φ) with fixed determinant and trace free Higgs field φ). He also proved that it is a hyper-Kähler manifold with twice dimension of the moduli scheme of semi-stable parabolic bundles with fixed determinant.

3. Moduli Spaces of Parabolic Higgs Bundles on Curves

In this section, X is a compact Riemann surface of genus g and $D = p_1 + \cdots + p_n$ is a reduced effective divisor. Let K be a canonical line bundle. In this case, $\Omega^1_X(\log D)$ is just $K(D)$.

Let E_* be a locally free, parabolic sheaf (call simply a parabolic bundle) of rank r on X. Then at each point $p \in D$, we get a weighted flag structure on the vector space E_p;

$$E_p = F_1(p) \supset \cdots \supset F_{s_p}(p) \supset 0$$
$$0 \le \alpha_1(p) < \cdots < \alpha_{s_p}(p) < 1.$$

For this E_*, a parabolic $K(D)$-pair (E_*,φ) becomes a parabolic Higgs bundle if and only if $\varphi(F_i(p)) \subset F_{i+1}(p) \otimes K(D)$ for all $p \in D$. Let $m_i(p) = \dim(F_i(p)) - \dim(F_{i+1}(p))$, the multiplicity of $\alpha_i(p)$.

Viewing α as a vector-valued function on D, we use it as an index to indicate the parabolic structure on E_*. For example, if $D = p + q$, $r = 3$, $m_1(p) = m_2(q) = 2$, then $\alpha(p) = (\alpha_1(p), \alpha_1(p), \alpha_2(p))$ and $\alpha(q) = (\alpha_1(q), \alpha_2(q), \alpha_2(q))$. Define the parabolic degree and slope of E_* by

$$\begin{aligned} \operatorname{pardeg} E_* &= \deg E + \sum_{p\in D}\sum_{i=1}^{s_p} m_i(p)\alpha_i(p), \\ \mu(E_*) &= \frac{\operatorname{pardeg} E_*}{\operatorname{rk} E}. \end{aligned}$$

Then Definition 2.1 of stability of parabolic bundles is equivalent to that measured by the slope $\mu(E_*)$.

3.1. First geometric properties of moduli spaces. Denote by $\mathcal{M}_\alpha$ ($\mathcal{N}_\alpha$ or $\mathcal{P}_\alpha$) the moduli space of the Seshadri equivalence classes of semi-stable parabolic bundles, (parabolic Higgs bundles or parabolic $K(D)$-pairs, resp.) with weighted flags of type α. All these are unions of some connected components of moduli schemes which were constructed in the last section. (There are some possibilities of α which belongs to the same Hilbert polynomials.)

The moduli space $\mathcal{M}_\alpha$ was first constructed by Mehta and Seshadri [6]. It is an irreducible normal projective variety of dimension

$$\dim \mathcal{M}_\alpha = (g-1)r^2 + 1 + \sum_{p\in D} f_p,$$

where $f_p = \frac{1}{2}(r^2 - \sum_{i=1}^{s_p}(m_i(p))^2)$, the dimension of the associated flag variety. (If $g = 0$, this holds only when $\mathcal{M}_\alpha \neq \emptyset$.) Also, we can deduce the normality and the dimension by calculating Ext-groups $\mathrm{Ext}^i_X(E_*, E_*)$. Since the dimension of X is one, we can easily prove the vanishing of $\mathrm{Ext}^2_X(E_*, E_*)$ as usual case.

Further, $\mathcal{P}_\alpha$ is an irreducible normal quasi-projective variety of dimension

$$\dim \mathcal{P}_\alpha = (2g-2+n)r^2 + 1$$

which contains $\mathcal{N}_\alpha$ as a closed subvariety of dimension

$$\dim \mathcal{N}_\alpha = 2(g-1)r^2 + 2 + 2\sum_{p\in D} f_p.$$

(cf. the proof of Theorem 5.2 of [16]). Notice that if $\mathcal{M}_\alpha$ is not empty, $\dim \mathcal{N}_\alpha = 2\dim \mathcal{M}_\alpha$. In fact, $\mathcal{N}_\alpha$ naturally contains the cotangent bundle of the moduli space of stable parabolic bundles $\mathcal{M}^s_\alpha$ (cf. Remark 5.1 of [16]).

In this case, vanishing of $\mathrm{Ext}^2_X((E_*, \varphi), (E_*, \varphi))$ is not clear because (E_*, φ) is considered as a parabolic sheaf on a two dimensional variety, i.e. a line bundle over X. But, by using Serre duality, $\mathrm{Ext}^2_X((E_*, \varphi), (E_*, \varphi))$ is isomorphic to the dual space of $\mathrm{Hom}_X((E_*, \varphi), (\widehat{E}_*, \widehat{\varphi}))$ and when (E_*, φ) is semi-stable, it is zero. The dimension of $\mathrm{Ext}^1_X((E_*, \varphi), (E_*, \varphi))$ is calculated by using the spectral sequence provided in Proposition 1.5.

For generic α, semi-stability is equivalent to stability. In this case, the moduli spaces $\mathcal{M}_\alpha, \mathcal{N}_\alpha$ and $\mathcal{P}_\alpha$ are nonsingular. The same is true for $\mathcal{M}^0_\alpha, \mathcal{N}^0_\alpha$ and $\mathcal{P}^0_\alpha$, the moduli spaces with fixed determinant and trace-free φ. The moduli space $\mathcal{N}^0_\alpha$ was first constructed by Konno [4] using gauge theory and was shown to be a hyperkähler manifold of complex dimension

$$\dim \mathcal{N}^0_\alpha = 2(g-1)(r^2-1) + 2\sum_{p\in D} f_p.$$

3.2. Variations of moduli spaces. In the paper [1], Boden and Hu studied how the topology of the moduli spaces $\mathcal{M}^0_\alpha$ varies with α. The space of weights α is divided into finite chambers by a finite number of hyperplanes called walls. Those walls are defined by the existence of strictly semi-stable parabolic bundles. Take two chambers C and C' which are separated by a wall H and choose $\alpha \in C$ and $\beta \in C'$. Suppose that $\gamma \in H$ is attached to both C and C'. Let Σ^γ be the locus of strictly semi-stable parabolic bundles in the moduli space $\mathcal{M}^0_\gamma$.

Theorem 3.1 (Boden and Hu [1]). *$\mathcal{M}^0_\alpha$ and $\mathcal{M}^0_\beta$ are related by a special birational transformation, that is, there are canonical projective maps*

$$\begin{array}{ccc} \mathcal{M}^0_\alpha & & \mathcal{M}^0_\beta \\ & \phi_\alpha \searrow \swarrow \phi_\beta & \\ & \mathcal{M}^0_\gamma & \end{array}$$

so that

(*a*) *along $\mathcal{M}^0_\gamma \setminus \Sigma_\gamma$, ϕ_α and ϕ_β are isomorphisms,*

(*b*) *along Σ_γ, ϕ_α and ϕ_β are $\mathbb{P}^{e_\alpha}$ and $\mathbb{P}^{e_\beta}$ fibrations, and*

(*c*) $e_\alpha + e_\beta + 1 = \operatorname{codim} \Sigma^\gamma$.

If γ corresponds to a parabolic bundle $E_* = L_* \oplus M_*$ with stable L_* and M_*, then one of e_α, e_β is $h^1(L_* \otimes M_*^\vee) - 1$ and the other is $h^1(M_* \otimes L_*^\vee) - 1$. Hence, in general, the topology of M_α^0 changes when α crosses the wall.

But in the case of parabolic $K(D)$-pairs or parabolic Higgs bundles, the topology of the moduli spaces does not change even if a similar 'flip' phenomenon occurs. This result was proved in a joint work with Boden [2] for the rank 2 case. Suppose $r = 2$ and flag structures at each $p \in D$ are full.

Theorem 3.2 ([2] §4). *The Poincaré polynomial of $\mathcal{N}_\alpha^0$ does not depend on the weights and is given by*

$$\begin{aligned} P_t(\mathcal{N}_\alpha^0) &= \frac{(1+t^3)^{2g}(1+t^2)^{n-1} + 2^{n-1}t^{2n+4g-3}(1+t)^{2g}[(2g-1)t-2g]}{(1-t^2)^2} \\ &\quad - \frac{2^{n-2}(n-1)t^{2n+4g-4}(1+t)^{2g}}{1-t^2} + 2^{n-1}(2^{2g}-1)t^{4g+2n-4}(1+t)^{2g-2}. \end{aligned}$$

Moreover, $\mathcal{N}_\alpha^0$ is connected and simply connected. If $g > 0$ or $g = 0$ and $n > 3$, then $\mathcal{N}_\alpha^0$ is noncompact.

Remark 3.1. A stronger result is proved by Nakajima's paper in this volume, i.e. $\mathcal{N}_\alpha^0$ are diffeomorphic to each other for generic weights (cf. [7]).

A key result for the proof of independence of the Betti numbers on weights is Serre duality. If $(E_*, \varphi) = (L_*, \phi) \oplus (M_*, \psi)$ corresponds to γ on a wall, then by considering similar results with Theorem 3.1, it will be enough to show that

$$\dim \operatorname{Ext}_X^1((L_*, \phi), (M_*, \psi)) = \dim \operatorname{Ext}_X^1((M_*, \psi), (L_*, \phi))$$

Using Serre duality and the spectral sequence in Proposition 1.5,

$$\dim \operatorname{Ext}_X^1((L_*, \phi), (M_*, \psi)) = \dim \operatorname{Ext}_X^1((M_*, \psi), (\widehat{L}_*, \widehat{\phi})).$$

Since the weights of L_* and those of M_* are different, we can show that

$$\dim \operatorname{Ext}_X^1((M_*, \psi), (L_*, \phi)) = \dim \operatorname{Ext}_X^1((M_*, \psi), (\widehat{L}_*, \widehat{\phi})).$$

Evaluating $P_t(\mathcal{N}_\alpha^0)$ at $t = -1$ shows the following.

Corollary 3.3 ([2] §4). *The Euler characteristic of $\mathcal{N}_\alpha^0$ is given by*

$$\chi(\mathcal{N}_\alpha^0) = \begin{cases} (n-1)(n-2)2^{n-4} & \text{if } g = 0, \\ 3 \cdot 2^n & \text{if } g = 1, \\ 0 & \text{if } g \geq 2 \end{cases}$$

References

[1] H. U. Boden and Y. Hu, *Variations of moduli of parabolic bundles*, Math. Ann. **301** (1995), 539–559.

[2] H. U. Boden and K. Yokogawa, *Moduli Spaces of Parabolic Higgs Bundles and Parabolic K(D) Pairs over Smooth Curves*, MPI-Preprint 94-107, (1994).

[3] N. J. Hitchin, *The self-duality equations on a Riemann surface*, Proc. London Math. Soc., (3) **55** (1987), 59–126.

[4] H. Konno, *Construction of the moduli space of stable parabolic Higgs bundles on a Riemann surface*, J. Math. Soc. Japan **45** (1993), 253–276.

[5] M. Maruyama and K. Yokogawa, *Moduli of parabolic stable sheaves*, Math. Ann. **293** (1992), 77–99.

[6] V. Mehta and C. Seshadri, *Moduli of vector bundles on curves with parabolic structures*, Math. Ann. **248** (1980), 205–239.

[7] H. Nakajima, *Hyper-Kähler structures on moduli spaces of parabolic Higgs bundles and filtered $\mathcal{D}$-modules*, in this volume.

[8] N. Nitsure, *Cohomology of the moduli of parabolic bundles*, Proc. Ind. Math. Soc. **95** (1986), 61-77.

[9] N. Nitsure, *Moduli space of semi-stable pairs on a curve*, Proc. London Math. Soc. **62** (1991), 275–300.

[10] P. Scheinost, *Metaplectic quantization of the moduli spaces of flat and parabolic bundles*, preprint (1992).

[11] C. S. Seshadri, *Moduli of vector bundles on curves with parabolic structures*, Bull. Amer. Math. Soc. **83** (1977), 124–126.

[12] C. T. Simpson, *Harmonic bundles on noncompact curves*, J. Am. Math. Soc. **3** (1990), 713–770.

[13] C. T. Simpson, *Higgs bundles and local systems*, Publ. Math. I.H.E.S. **75** (1992), 5–95.

[14] C. T. Simpson, *Moduli of representations of the fundamental group of a smooth projective variety I, II*, Preprint (1992).

[15] K. Yokogawa, *Compactification of moduli of parabolic sheaves and moduli of parabolic Higgs sheaves*, J. Math. Kyoto Univ. **33** (1993), 451–504.

[16] K. Yokogawa, *Infinitesimal deformation of parabolic Higgs sheaves*, Int. J. Math. **6** (1995), 125–148.

NUMBERS OF $\mathbb{F}_q$-RATIONAL POINTS OF THE MODULI OF STABLE SHEAVES ON ELLIPTIC SURFACES

Kōta Yoshioka

Department of Mathematics, Faculty of Science, Kyoto University, 606-01, Japan

0. INTRODUCTION

This note is a continuation of the talk at the Taniguchi symposium. Let X be a smooth projective surface defined over a finite field $\mathbb{F}_q$ and H an ample divisor on X. Let $M_H(c_1, c_2)$ be the moduli of stable sheaves of rank 2 with Chern classes $c_1 \in \mathrm{NS}(X)$ and $c_2 \in \mathbb{Z}$, where $\mathrm{NS}(X)$ is the Néron-Severi group of X. In case of $\kappa = -\infty$, we in principle computed the number of $\mathbb{F}_q$-rational points of $M_H(c_1, c_2)$ under the assumption that $(K_X, H) < 0$ and $M_H(c_1, c_2)$ is complete [Y1,2]. In order to treat these cases, we used the structure of $\mathbb{P}^1$-fibration. Let us recall this method. For simplicity, we assume that X is a $\mathbb{P}^1$-bundle over a curve C. Let η be the generic point of C and X_η the generic fibre. Let F be a vector bundle on X. Since X_η is a projective line, $F_{|X_\eta}$ is a direct sum of line bundles. If $F_{|X_\eta}$ is not semi-stable, then the Harder-Narasimhan filtration of $F_{|X_\eta}$ defines an exact sequence of F on X. In this case, we can reduce the computations to rank 1 case. If $F_{|X_\eta}$ is semi-stable, by using elementary transformations along divisors on fibres [L,M2], we can reduce those to considering vector bundles which are pull-backs of vector bundles on C.

We want to generalize this method to elliptic fibration $\pi : X \to C$. Vector bundles on elliptic curves were classified by Atiyah [A]. In rank 2 case, if the degree is even, then there are many semi-stable vector bundles with fixed determinant line bundle. If the degree is odd, then the semi-stable vector bundle with fixed determinant line bundle is unique. Taking the classification into account, we shall treat moduli spaces on X with (c_1, f) odd, where f is a general fibre. Let ξ be a π-ample divisor. Then $H_n := \xi + nf, n \gg 0$ is ample, the restriction $E_{|X_\eta}$ of any H_n-stable sheaf E, $n \gg c_2$ is stable and $M_{H_n}(c_1, c_2)$ does not depend on the choice of $n \gg c_2$. We denote this space by $M(c_1, c_2)$. Then every stable bundle in $M(c_1, c_2)$ is obtained by successive elementary transformations from a vector bundle which is stable for all fibres in the sense of Simpson [M3, 3.8]. Hence we can analyse the structure of $M(c_1, c_2)$ by using elementary transformations. This idea was already used by Friedman [F] to compute $SO(3)$-invariants of simply connected elliptic surfaces. For this purpose, he used elliptic surfaces $X \to \mathbb{P}^1$ such that singular fibres are irreducible curves with a node or multiples of elliptic curves. Moreover he showed that $M(c_1, c_2)$ is a smooth irreducible variety and it is birational to the Hilbert scheme $Hilb^t_{J^d(X)}$ parametrizing colength t ideal sheaves on $J^d(X)$, where $J^d(X)$ is the elliptic surface whose generic fibre is the set of line bundles of degree $d = ((c_1, f) + 1)/2$ and $2t = 4c_2 - (c_1^2) - 3\chi(\mathcal{O}_X)$.

In this paper, by using Friedman's results, we shall compute $\#M(c_1, c_2)(\mathbb{F}_q)$ under the same assumption on singular fibres as Friedman's. If singular fibres are not irreducible, then

Supported by JSPS Research Fellowships for Young Scientists.

the choice of center of elementary transformations seems to be complicated, which is the reason why we require the irreducibility of singular fibres. The author is also motivated by the work of Vafa and Witten on S-duality conjectures [V-W]. Roughly speaking, they conjectured that under suitable conditions, the generating function of "Euler-characteristics" of instanton moduli spaces $M_{v,k}$ with $v \in H^2(X, \mathbb{Z}/2\mathbb{Z})$ and instanton number k : $Z_v(\tau) := q^{-\chi(X)/12}\sum_k \chi(M_{v,k})q^k$, $q = \exp(2\pi i\tau)$ has modular properties. Moreover if $b_2^+ > 1$ and $M_H(c_1, c_2)$ is smooth and compact, then it is expected that $\chi(M_{c_1,c_2-(c_1^2)/4}) = \chi(M_H(c_1, c_2))$ (in case of $b_2^+ = 1$, moduli spaces depend heavily on H). Hence we may expect that the generating function of Poincaré polynomials $\sum_{c_2} P(M(c_1, c_2), z)t^{c_2}$ has a simple description and the generating function of Euler characteristics $\sum_{c_2} P(M(c_1, c_2), -1)t^{c_2}$ is written by using modular forms. By applying Weil conjectures [D], we shall verify these in case of regular elliptic surfaces. If X is K3 surfaces, then these are verified by Vafa and Witten [V-W] and Göttsche and Huybrechts [G-H], and moreover they computed the Hodge numbers. Although our method is not powerful enough to compute these invariants in case of K3 surfaces, it is interesting to compute Weil's zeta function of $M(c_1, c_2)$. We also give a remark on the asymptotic behaviors of Betti numbers, which is similar to those on ruled surfaces and $\mathbb{P}^2$.

One idea for computing the number of $\mathbb{F}_q$-rational points is to use the generating function $f(t) := \sum_{c_1,c_2} \#M(c_1, c_2)(\mathbb{F}_q)t^{c_2-(c_1^2)/4}$, where c_1 satisfy $c_1 \equiv \xi$, or $\xi + f$ (numerically equiv.) and $c_2 \in \mathbb{Z}$. Since $(\xi^2) \not\equiv ((\xi + f)^2) \bmod 4$ and $\#M(c_1, c_2)(\mathbb{F}_q) = \#M(c_1', c_2)(\mathbb{F}_q)$ for $c_1 \equiv c_1'$, our computations follow from $f(t)$. Another idea is to use the functional equation of Weil's zeta function. More precisely, since the contributions of elementary transformations are very complicated, we can not determine these contributions directly. Hence we shall partially determine these by direct computations and by using functional equations, we shall determine these completely.

I would like to thank Professor Maruyama for valuable suggestions.

1. Computations of rational points

Throughout this article, we shall maintain the following notation. Let $\pi : X \to C$ be an elliptic surface defined over a finite field $\mathbb{F}_q$ with q elements whose singular fibres are irreducible curves with a node or multiples of elliptic curves. We assume that $R^1\pi_*\mathcal{O}_X$ is torsion free. Let f be a smooth fibre of π and ξ a multisection of π with (ξ, f) odd. Since $(\xi^2) \not\equiv ((\xi + f)^2) \bmod 4$, we may denote $M(\xi + if, c_2)$ by $M(c_2 - ((\xi + if)^2)/4 - 3\chi(\mathcal{O}_X)/4)$ and the open subscheme parametrizing stable vector bundles by $M^0(c_2 - ((\xi + if)^2)/4 - 3\chi(\mathcal{O}_X)/4)$, $i = 0, 1$. For a torsion free sheaf E on X, we denote $c_2(E) - (c_1(E)^2)/4$ by $\Delta(E)$. For an algebraic scheme Y defined over $\mathbb{F}_q$, we denote Weil's zeta function $\exp(\sum_{r>0} \#Y(\mathbb{F}_{q^r})t^r/r)$ by $Z_q(Y, t)$.

Lemma 1.1. *Let V be the set of vector bundles F of rank 2 on X defined over $\mathbb{F}_q$ such that $F_{|f_{red}}$ is stable for all fibres f and $c_1(F) = \xi$ or $\xi + f$. Then $\#V = \#Pic_X^0(\mathbb{F}_q)$.*

Proof. We shall fix isomorphisms $Pic_X^{\xi+if} \cong Pic_X^0, i = 0, 1$. We shall consider the natural map $\delta : V \to \cup_{i=0}^1 Pic_X^{\xi+if}(\mathbb{F}_q) \to Pic_X^0(\mathbb{F}_q)/Pic_C^0(\mathbb{F}_q)$. Let L be an element of $Pic_X^0(\mathbb{F}_q)/Pic_C^0(\mathbb{F}_q)$. By [F, Lem. 3.5], there is a vector bundle F such that $F_{|f_{red}}$ is stable for all fibres f and $\delta(F)\otimes\mathcal{O}_X(D) = L$, where D is a vertical divisor. Let M be a line bundle with $M^{\otimes 2} \equiv \mathcal{O}_X(D) \bmod Pic_C^0(\mathbb{F}_q)$. Then $F \otimes M$ belongs to V. Thus δ is surjective. Let F, F' be elements of V such that $\det(F_{|f}) \cong \det(F'_{|f})$ for all fibres. In the same way as in [F, Cor. 1.5], we see that $F_{|f} \cong F'_{|f}$ and $H^0(f, (F^\vee \otimes F')_{|f}) \cong \mathbb{F}_q$ for all fibres. Hence $N := \pi_*(F^\vee \otimes F')$ is a line bundle on C and $F' \cong F \otimes \pi^*(N)$. Therefore $\#V = \#\mathrm{im}(\delta)\#Pic^0_C(\mathbb{F}_q) = \#Pic^0_X(\mathbb{F}_q)$. □

In the above lemma, $\dim Pic^0_X = \dim M(\Delta(F) - 3\chi(X)/4) = 4\Delta(F) - 3\chi(\mathcal{O}_X) + \dim Pic^0_X$, and hence $\Delta(F) = 3\chi(\mathcal{O}_X)/4$. Let T be the kernel of $\mathrm{NS}(X) \to \mathrm{Num}(X)$, where $\mathrm{Num}(X)$ is the set of numerical equivalence classes. For simplicity, we assume that $T = 0$. If $T \neq 0$, then we shall compute $\sum_{t \in T} \#M(c_1 + t, c_2)(\mathbb{F}_q) (= \#T\#M(c_1, c_2)(\mathbb{F}_q))$.

Definition 1.1. Let F be an element of V. $W_{F,P}(\Delta)_q$ denotes the set of rank 2 vector bundles E on X defined over $\mathbb{F}_q$ such that $\Delta(E) = \Delta + \Delta(F)$ and E is obtained by successive elementary transformations from F along stable sheaves of rank 1 with degree ≥ 1 on $\pi^{-1}(P)_{red}$.

Then $\#W_{F,P}(\Delta)_q$ does not depend on the choice of F. So we denote this number by $W_P(\Delta)_q$. Moreover it is easy to see

$$\sum_{\Delta} \#M^0(\Delta)(\mathbb{F}_q)t^\Delta = \prod_P \Big(\sum_{\Delta} W_P(\Delta)_q t^\Delta\Big)\#V, \tag{1.1}$$

where P runs over all closed points of C. By [Y1, Thm. 0.4], we get

$$\sum_{\Delta} \#M(\Delta)(\mathbb{F}_q)t^\Delta = \prod_P \Big(\sum_{\Delta} W_P(\Delta)_q t^\Delta\Big)\#V \prod_{a \geq 1} Z_q(X, q^{2a-2}t^a) Z_q(X, q^{2a-1}t^a). \tag{1.2}$$

Definition 1.2. Let C be an elliptic curve or an irreducible curve of genus 1 with a node defined over $\mathbb{F}_q$, and L a line bundle of degree 1 on C. Then $V(0)$ denotes the set of stable vector bundles E with $\det E = L$, $V(l)$ $l \geq 1$ does the set of vector bundles E of $\det E = L$ which is the direct sum of line bundles M_1, M_2 with $\deg M_1 = l$, and moreover $V'(l)$ $l \geq 1$ does the set of vector bundles E with $\det E = L$ which are defined by extensions $0 \to M_1 \to E \to M_2 \to 0$, where M_2 are non-locally free stable sheaf of rank 1 with $\deg M_2 = 1 - l$.

Let $\Phi_d(E)$ $(d \geq 1)$ be the set of surjections $\lambda : E \to N$ up to multiplication by constants such that N is a rank 1 torsion free sheaf with $\deg N = d$. By using [F, Lem. 1.3, 2.7], we get

Lemma 1.2. *Let $\Phi_d(E)_1$ be the subset of $\Phi_d(E)$ such that $\lambda : E \to N$ belongs to $\Phi_d(E)_1$ if and only if N is a line bundle, and $\Phi_d(E)_2$ the complement of $\Phi_d(E)_1$. If E belongs to $V(l)$, $l \geq 0$, then there are polynomials $f^l_d(u,v)_i$, $i = 1,2$ such that $\#\Phi_d(E)_i = f^l_d(q, \#C(\mathbb{F}_q))_i$ and if E belongs to $V'(l)$, $l \geq 1$, then there are polynomials $g^l_d(u,v)_i$, $i = 1,2$ such that $\#\Phi_d(E)_i = g^l_d(q, \#C(\mathbb{F}_q))_i$. If C is smooth, then we denote $f^l_d(u,v)_1$ by $f^l_d(u,v)$.*

Proof. We shall first treat the singular case. We shall prove this lemma by induction on d. For torsion free sheaves E and N, $\mathrm{Hom}(E,N)^{sur}$ denotes the subset of $\mathrm{Hom}(E,N)$ consisting of surjective homomorphisms. Assume that E is locally free of rank 2 and N is a torsion free sheaf of rank 1. Then we see that

$$\mathrm{Hom}(E,N) \setminus \{0\} = \coprod_{k \leq d} \coprod_{N_k \subset N} \mathrm{Hom}(E, N_k)^{sur}, \tag{1.3}$$

where N_k runs over the set of subsheaves of N with $\deg(N_k) = k$. We note that subsheaves $N_k \subset N$ are determined by giving torsion free sheaves N_k of rank 1 with $\deg(N_k) = k$ and non-zero homomorphisms $N_k \to N$ up to multiplication by constants. We first assume that N is locally free (i.e. we shall treat $\Phi_d(E)_1$). By [F, Lem. 2.2, 2.3], we see that $\dim \operatorname{Hom}(N_k, N) = d - k$ for $d > k$. Hence we can easily show that

$$\#(\operatorname{Hom}(E,N)\setminus\{0\}) = \#\operatorname{Hom}(E,N)^{sur} + \sum_{k<d} \#\Phi_k(E)(q^{d-k}-1). \tag{1.4}$$

If E belongs to $V(l) \cup V'(l)$ and $d > l$, then by [F, Lem. 2.7], we get $\dim \operatorname{Hom}(E,N) = 2d-1$. Since $\# Pic^0_C(\mathbb{F}_q) = \# C(\mathbb{F}_q) - 1$, we obtain that

$$\#\Phi_d(E)_1 = (\#C(\mathbb{F}_q)-1)\frac{q^{2d-1}-1}{q-1} - \left(\sum_{k<d} \#\Phi_k(E)(\#C(\mathbb{F}_q)-1)\frac{q^{d-k}-1}{q-1}\right). \tag{1.5}$$

By using [F, Lem. 2.7], we also obtain that $\#\Phi_d(E)_1 = q^{2d-1}$ for $E \in V(d)$ and $\#\Phi_d(E)_1 = 0$ for $E \in V'(d)$. We next assume that N is not locally free (i.e. we shall treat $\Phi_d(E)_2$). By [F, Lem. 2.2, 2.3], if N_k is not locally free, then $\dim \operatorname{Hom}(N_k, N) = d - k + 1$ for $d \geq l$. Hence, if E belongs to $V(l) \cup V'(l)$ and $d > l$, then we see that

$$\#\Phi_d(E)_2 = \frac{q^{2d-1}-1}{q-1} - \left(\sum_{k<d} \#\Phi_k(E)_1\frac{q^{d-k}-1}{q-1} + \sum_{k<d} \#\Phi_k(E)_2\frac{q^{d-k+1}-1}{q-1}\right). \tag{1.6}$$

If E belongs to $V(l) \cup V'(l)$ and $d = l$, then $\#\Phi_d(E)_2 = 0$. By these, we get our lemma.

In the smooth case, it is easy to see that if E belongs to $V(l)$ and $d > l$, then

$$\#\Phi_d(E) = \#C(\mathbb{F}_q)\frac{q^{2d-1}-1}{q-1} - \left(\sum_{k<d} \#\Phi_k(E)\#C(\mathbb{F}_q)\frac{q^{d-k}-1}{q-1}\right). \tag{1.7}$$

If E belongs to $V(l)$ and $d = l > 0$, then $\#\Phi_d(E) = q^{2d-1}$. □

By this lemma, we see that there is a polynomial $F_n(x)$ such that $W_P(rn/2)_q = F_n(\#\pi^{-1}(P)(k(P)))$ for any closed point P with smooth fibre $\pi^{-1}(P)$, where $r = [k(P) : \mathbb{F}_q]$.

Definition 1.3. For polynomials $p(x,y)$ and $p'(x,y)$, $p(x,y) = p'(x,y) + O(d)$ means that the degree of $p(x,y) - p'(x,y)$ is less than $d+1$.

We set $h(x,y) := (x-1)(y-1)$. By Lemma 1.2, we see that

$$\begin{cases} f^l_d(xy, h(x,y)) = h(x,y)(xy)^{2d-2} + O(4d-4), & l < d, \\ f^d_d(xy, h(x,y)) = (xy)^{2d-1} + O(4d-4). \end{cases} \tag{1.8}$$

By using these relations successively, we see that

$$\begin{aligned} F_n(h(x,y)) &= \sum_{\substack{0=a_0<a_1<a_2<\cdots<a_r \\ \sum_i n_i(2a_i-1)=n}} \prod_{i=1}^{r} \{ f^{a_{i-1}}_{a_i}(xy, h(x,y))(f^{a_i}_{a_i}(xy,h(x,y)))^{n_i-1}\} \\ &= \sum_{\substack{0=a_0<a_1<a_2<\cdots<a_r \\ \sum_i n_i(2a_i-1)=n}} \prod_{i=1}^{r} \{(xy-(x+y))(xy)^{(2a_i-1)n_i-1}\} + O(2n-2). \end{aligned}$$

We set

$$H(t) := \frac{(1-xt)(1-yt)}{(1-t)(1-xyt)},$$
$$\sum_{n\geq 0} W_n(x,y)t^{n/2} := \frac{\prod_{k\geq 1} H((xy)^{k-1}t^{k/2})}{\prod_{k\geq 1} H((xy)^{2k-2}t^k)H((xy)^{2k-1}t^k)}.$$

Then we get the following

$$\begin{aligned}\sum_{n\geq 0} F_n(h(x,y))t^{n/2} &= \prod_{k\geq 1} \frac{(1-x(xy)^{2k-2}t^{k-1/2})(1-y(xy)^{2k-2}t^{k-1/2})}{1-(xy)^{2k-1}t^{k-1/2}} + \sum_{n\geq 0} O(2n-2)t^{n/2} \\ &= \sum_{n\geq 0} W_n(x,y)t^{n/2} + \sum_{n\geq 0} O(2n-2)t^{n/2}.\end{aligned}$$

Proposition 1.3. *Let P be a $\mathbb{F}_q$-rational point of C over which π has a smooth fibre D. Let ω, ϖ be the solutions of $x^2-(1+q-\#D(\mathbb{F}_q))x+q=0$. Then*

$$\begin{aligned}\sum_{k\geq 0} W_P(k/2)_q t^{k/2} &= \sum_{k\geq 0} W_k(\omega,\varpi)t^{k/2} \\ &= \frac{\prod_{k\geq 1} Z_q(D,q^{k-1}t^{k/2})}{\prod_{k\geq 1} Z_q(D,q^{2k-2}t^k)Z_q(D,q^{2k-1}t^k)}.\end{aligned}$$

Proof. Since $W_P(k/2)$ depends only on D, we may assume that $X = D\times\mathbb{P}^1$. We note that $\#D(\mathbb{F}_{q^r}) = q^r+1-(\omega^r+\varpi^r)$. We shall prove that $W_{P_r}(k/2)_{q^r} = W_k(\omega^r,\varpi^r)$ by induction on k, where P_r is the $\mathbb{F}_{q^r}$-rational point of $X\otimes_{\mathbb{F}_q}\mathbb{F}_{q^r}$ defined by P. Assume that

$$\sum_{k\geq 0} W_{P_r}(k/2)_{q^r}t^{k/2} = \sum_{k\geq 0} F_k(h(\omega^r,\varpi^r))t^{k/2} \equiv \sum_{k\geq 0} W_k(\omega^r,\varpi^r)t^{k/2} \mod t^{n/2}$$

for $r\geq 1$. Let Q be a closed point of $\mathbb{P}^1$ with $k(Q)\cong\mathbb{F}_{q^r}$. Then $X\times_{\mathbb{P}^1}\operatorname{Spec}(k(Q))\cong D\otimes_{\mathbb{F}_q}\mathbb{F}_{q^r}$, and hence $W_Q(rk/2)_q = W_{P_r}(k/2)_{q^r}$. We set

$$p_r = \frac{\#(\mathbb{P}^1(\mathbb{F}_{q^r})\setminus\cup_{j<r}\mathbb{P}^1(\mathbb{F}_{q^j}))}{r}.$$

Then we see that

$$\begin{aligned}\prod_{r\geq 1}\left(\sum_{k\geq 0} W_k(\omega^r,\varpi^r)t^{rk/2}\right)^{p_r} &= \frac{\prod_{a\geq 1} Z_q(X,q^{a-1}t^{a/2})}{\prod_{a\geq 1} Z_q(X,q^{2a-2}t^a)Z_q(X,q^{2a-1}t^a)} \\ &= \frac{\sum_k \#Hilb_X^k(\mathbb{F}_q)t^{k/2}}{\prod_{a\geq 1} Z_q(X,q^{2a-2}t^a)Z_q(X,q^{2a-1}t^a)}\end{aligned}$$

(cf. [G1]). Let σ be a constant section and we shall consider the following closed subscheme of $M(k/2)$.

$$M^\sigma(k/2) = \{E|E\in M(k/2), \det E = \sigma, \sigma+f\}.$$

Then we see that

$$\begin{aligned}
&\sum_{k\geq 0}\#M^{\sigma}(k/2)(\mathbb{F}_q)t^{k/2}-\sum_{k\geq 0}\#Hilb_X^k(\mathbb{F}_q)t^{k/2}\\
&=\prod_{a\geq 1}Z_q(X,q^{2a-2}t^a)Z_q(X,q^{2a-1}t^a)\left[\prod_{r\geq 1}\prod_{\substack{P\\ \#k(P)=q^r}}\left(\sum_{k\geq 0}W_P(rk/2)_q t^{rk/2}\right)\right.\\
&\qquad\qquad\left.-\prod_{r\geq 1}\left(\prod_{a\geq 1}W_k(\omega^r,\varpi^r)t^{rk/2}\right)^{p_r}\right]\\
&\equiv(q+1)(F_n(h(\omega,\varpi))-W_n(\omega,\varpi))t^{n/2}\mod t^{(n+1)/2}.
\end{aligned}$$

Thus we obtain that $\#M^{\sigma}(k/2)(\mathbb{F}_q)-\#Hilb_X^k(\mathbb{F}_q)=(q+1)(F_n(h(\omega,\varpi))-W_n(\omega,\varpi))$. Replacing q by q^r, we get

$$\#M^{\sigma}(n/2)(\mathbb{F}_{q^r})-\#Hilb_X^n(\mathbb{F}_{q^r})=(q^r+1)(F_n(h(\omega^r,\varpi^r))-W_n(\omega^r,\varpi^r)).$$

By Lemma 1.2, $(xy+1)(F_n(h(x,y))-W_n(x,y))$ is a polynomial of degree less than $2n+1$. Since $\dim M^{\sigma}(n/2)=\dim Hilb_X^n=2n$, the functional equation of zeta functions implies that there are algebraic integers ω_i of $|\omega_i|=q^n$ and satisfy the following.

$$(q^r+1)(F_n(h(\omega^r,\varpi^r))-W_n(\omega^r,\varpi^r))=\sum_i p_i\omega_i^r,\, p_i\in\mathbb{Z}.$$

Let $F_n(h(x,y))-W_n(x,y)=\sum_{i=0}^{2n-2}f_i(x,y)$ be the decomposition such that $f_i(x,y)$ is a homogeneous polynomial of degree i. Then we obtain the following relations:

$$\begin{cases} f_0(\omega^r,\varpi^r)=0\\ qf_{j-1}(\omega^r,\varpi^r)+f_j(\omega^r,\varpi^r)=0,\ 1\leq j\leq 2n-2\\ qf_{2n-2}(\omega^r,\varpi^r)=\sum_i p_i\omega_i^r.\end{cases}\tag{1.9}$$

Hence we get $f_i(\omega^r,\varpi^r)=0$ for all i. Thus $F_n(h(\omega^r,\varpi^r))-W_n(\omega^r,\varpi^r)=0$, which implies that our claim holds for $k=n$. □

Corollary 1.4. *Let S be the closed subset of critical points of π on C and C^0 the complement of S. We set $X^0=X\times_C C^0$. Then,*

$$\prod_{P\subset C^0}\left(\sum_{k\geq 0}W_P(k/2)_q t^{k/2}\right)=\frac{\prod_{k\geq 1}Z_q(X^0,q^{k-1}t^{k/2})}{\prod_{k\geq 1}Z_q(X^0,q^{2k-2}t^k)Z_q(X^0,q^{2k-1}t^k)}.$$

Proof. Let P_r be a closed point of C^0 with $k(P_r)\cong\mathbb{F}_{q^r}$. By the definition of $W_{P_r}(k/2)_q$, we get that $\sum_{k\geq 0}W_{P_r}(k/2)_q t^{k/2}=\sum_{k\geq 0}W_{P_r}(rk/2)_q t^{rk/2}$. By Proposition 1.3, we get

(1.10)

$$\begin{aligned}
\sum_{k\geq 0}W_{P_r}(rk/2)_q t^{rk/2}&=\frac{\prod_{k\geq 1}Z_{q^r}(\pi^{-1}(P_r),q^{r(k-1)}t^{rk/2})}{\prod_{k\geq 1}Z_{q^r}(\pi^{-1}(P_r),q^{r(2k-2)}t^{rk})Z_{q^r}(\pi^{-1}(P_r),q^{r(2k-1)}t^{rk})}\\
&=\prod_{s\geq 1}\prod_{Q_{rs}\in\pi^{-1}(P_r)}\frac{\prod_{k\geq 1}Z_{q^{rs}}(Q_{rs},q^{rs(k-1)}t^{rsk/2})}{\prod_{k\geq 1}Z_{q^{rs}}(Q_{rs},q^{rs(2k-2)}t^{rsk})Z_{q^{rs}}(Q_{rs},q^{rs(2k-1)}t^{rsk})},
\end{aligned}$$

where Q_{rs} are closed points of $\pi^{-1}(P_r)$ with $k(Q_{rs}) \cong \mathbb{F}_{q^{rs}}$. Hence we see that

(1.11)
$$\begin{aligned}
&\prod_{r\geq 1}\prod_{P_r\in C^0}\left(\sum_{k\geq 0} W_{P_r}(rk/2)_q t^{rk/2}\right)\\
&=\prod_{r,s\geq 1}\prod_{P_r\in C^0}\prod_{Q_{rs}\in\pi^{-1}(P_r)}\frac{\prod_{k\geq 1} Z_{q^{rs}}(Q_{rs}, q^{rs(k-1)}t^{rsk/2})}{\prod_{k\geq 1} Z_{q^{rs}}(Q_{rs}, q^{rs(2k-2)}t^{rsk})Z_{q^{rs}}(Q_{rs}, q^{rs(2k-1)}t^{rsk})}\\
&=\prod_{n\geq 1}\prod_{Q_n\in X^0}\frac{\prod_{k\geq 1} Z_{q^n}(Q_n, q^{n(k-1)}t^{nk/2})}{\prod_{k\geq 1} Z_{q^n}(Q_n, q^{n(2k-2)}t^{nk})Z_{q^n}(Q_n, q^{n(2k-1)}t^{nk})}\\
&=\frac{\prod_{k\geq 1} Z_q(X^0, q^{k-1}t^{k/2})}{\prod_{k\geq 1} Z_q(X^0, q^{2k-2}t^k)Z_q(X^0, q^{2k-1}t^k)},
\end{aligned}$$

where Q_n are closed points of X^0 with $k(Q_n) \cong \mathbb{F}_{q^n}$. Therefore we obtain our corollary. □

Let P be a $\mathbb{F}_q$-rational point of C such that $\pi^{-1}(P)$ is not smooth. Then in the same way as in the smooth case, there is a polynomial $F'_n(x)$ such that $W_P(n/2)_q = F'_n(\pi^{-1}(P)(\mathbb{F}_q)) = F'_n(q)$. Moreover, we see that

$$\sum_{k\geq 0} F'_k(x)t^{k/2} = \prod_{a\geq 1}\frac{(1-x^{2a-1}t^a)(1-x^{2a}t^a)}{(1-x^a t^{a/2})} + \sum_{k\geq 0} O(k-1)t^{k/2}. \tag{1.12}$$

Then in the same way as in the proof of Proposition 1.3, we obtain the following proposition.

Proposition 1.5.

$$\sum_{k\geq 0} F'_k(x)t^{k/2} = \prod_{a\geq 1}\frac{(1-x^{2a-1}t^a)(1-x^{2a}t^a)}{(1-x^a t^{a/2})}. \tag{1.13}$$

Proof. Let $X \to C$ be an elliptic K3 surface whose singular fibres are irreducible curve with a node. By using Corollary 1.4 and the functional equation, we can easily prove this proposition. □

By Corollary 1.4 and Proposition 1.5, we get the following theorem.

Theorem 1.6. *Let $\pi : X \to C$ be an elliptic surface defined over $\mathbb{F}_q$ whose singular fibres are irreducible curves with a node or multiples of elliptic curves. We assume that $R^1\pi_*\mathcal{O}_X$ is torsion free. Then for a multisection ξ of π with (ξ, f) odd,*

$$\begin{aligned}
\sum_{k\geq 0} \#M(k/2)(\mathbb{F}_q)t^{k/2} &= \#Pic^0_X(\mathbb{F}_q)\prod_{a\geq 1} Z_q(X, q^{a-1}t^{a/2})\\
&= \#Pic^0_X(\mathbb{F}_q)\left(\sum_{k\geq 0}\#Hilb^k_X(\mathbb{F}_q)t^{k/2}\right).
\end{aligned}$$

Corollary 1.7. *We assume, in addition to the conditions in Theorem 1.6, that X is Abelian, K3 or hyperelliptic surface. Then for a general polarization H, we obtain that*

$$\#M_H(c_1, c_2)(\mathbb{F}_q) = \#Pic^0_X(\mathbb{F}_q)\#Hilb_X^{2c_2-(c_1^2)/2-3\chi(\mathcal{O}_X)/2}(\mathbb{F}_q)$$

for c_1 with (c_1, f) odd.

Proof. By [G2, Thm. 3.4], the number of $\mathbb{F}_q$-rational points does not depend on a general H (cf. [Y3, Rem. 3.1, 3.2]). □

Let $\pi : X \to \mathbb{P}^1$ be an elliptic K3 surface defined over $\mathbb{F}_q$ which has a section σ and whose singular fibres are irreducible curves with a node. Let $L := \sigma + 2f$ be a nef and big divisor on X and $M_L(c_1, c_2)(\mathbb{F}_q)$ be the set of μ-semi-stable sheaves of rank 2 with Chern classes c_1, c_2 defined over $\mathbb{F}_q$. Considering elementary transformations along divisors on σ, we obtain

Corollary 1.8. *For a general polarization H, we get $\#M_H(f, c_2)(\mathbb{F}_q) = \#Hilb_X^{2c_2-3}(\mathbb{F}_q)$.*

2. Betti numbers and Euler characteristics of moduli spaces

Let $X \to C$ be an elliptic surface defined over $\mathbb{C}$ whose singular fibres are irreducible curve with a node or multiples of elliptic curves.

Theorem 2.1. *The generating function of Poincaré polynomials satisfies that*

$$\sum_{k\geq 0} P(M(k/2), z)t^{k/2} = (1+z)^{b_1} \prod_{a\geq 1} \frac{(1+z^{2a-1}t^{a/2})^{b_1}(1+z^{2a+1}t^{a/2})^{b_3}}{(1-z^{2a-2}t^{a/2})^{b_0}(1-z^{2a}t^{a/2})^{b_2}(1-z^{2a+2}t^{a/2})^{b_4}},$$

where b_i is the i-th Betti number of X. If X is a regular elliptic surface, then the generating function of Euler characteristics satisfies the following:

$$q^{-\chi(X)/12} \sum_{\substack{c_1=\xi, \xi+f \\ c_2\in\mathbb{Z}}} \chi(M(c_1, c_2))q^{c_2-(c_1^2)/4} = \frac{1}{\eta(q^{1/2})^{\chi(X)}},$$

where $q = \exp(2\pi i\tau)$ and $\eta(q) = q^{1/24}\prod_{a\geq 1}(1-q^a)$.

Proof. Let Q be an open subscheme of a quot scheme $Quot_{\mathcal{O}_X^N/X}$ such that $M(\Delta) = Q/SL(N)$, (for more details, see [M1, or Y2, Sect. 4]). Applying [Y2, Prop. 4.3] (due to Kirwan [K1]) to this situation, we obtain the first equality. By Noether's formula, we have $\chi(\mathcal{O}_X) = \chi(X)/12$. Then a simple calculation shows the second equality. □

Corollary 2.2. *If X is regular, then for $c_1 = \xi$, or $\xi + f$,*

$$Z_{c_1}(\tau) := q^{-\chi(X)/12} \sum_{c_2} \chi(M(c_1, c_2))q^{c_2-(c_1^2)/4}$$
$$= \frac{1}{2}\left(\frac{1}{\eta(q^{1/2})^{\chi(X)}} + \frac{i^{(c_1^2)}}{\eta(-q^{1/2})^{\chi(X)}}\right).$$

Remark 2.1. In a ruled surface case, we can prove that

$$\#M(c_1, c_2)(\mathbb{F}_q) = Pic_X^0(\mathbb{F}_q)\#M(L, c_2)(\mathbb{F}_q),$$

where L is a line bundle with $c_1(L) = c_1$ and $M(L, c_2)$ is the moduli space with determinant L. Unfortunately, this does not hold for elliptic surfaces in general. In fact, let $X = C \times C$ be a product of elliptic curves and $p_i : X \to C$, $i = 1, 2$ projections. We shall consider the elliptic fibration $p_1 : X \to C$. Let M be a line bundle of degree 1 on C and E a stable vector bundle of $\det E = M$. Then $F = p_2^*E$ belongs to $M(p_2^*M, 0)$. Assume that 2-torsion points of Pic_C^0 are defined over $\mathbb{F}_q$. Let L be a 2-torsion point of Pic_C^0. Then $F \otimes p_1^*L$ also belongs to $M(p_2^*M, 0)(\mathbb{F}_q)$. Therefore $\#M(p_2^*M, 0)(\mathbb{F}_q) = 4$, which implies that the above equality does not hold for this case.

Remark 2.2. By using [K2, Cor. 6.4], we can show the following. Let B be a compact subset of the ample cone. Then there is an integer $N(B)$ such that for a polarization which belongs to B and for c_1 with (c_1, f) odd,

$$P(M_H(c_1, c_2), z) \equiv (1 - z^2) \prod_{a \geq 1} \frac{(1 + z^{2a-1})^{b_1+b_3}}{(1 - z^{2a})^{b_0+b_2+b_4}} \mod z^{c_2 - N(B)}.$$

If X is a surface with $\kappa(X) = -\infty$, then the same results also hold for any c_1.

REFERENCES

[A] Atiyah, M. F., *Vector bundles over an elliptic curve,* Proc. Lond. Math. Soc. (3) VII **27** (1957), pp. 414–452.

[D] Deligne, P., *La conjecture de Weil. I,* Inst. Hautes Etudes Sci. Publ. Math. **43** (1974), pp. 273–307.

[F] Friedman, R., *Vector bundles and SO(3)-invariants for elliptic surfaces III,* preprint (1993).

[G1] Göttsche, L., *The Betti numbers of the Hilbert scheme of points on a smooth projective surface,* Math. Ann. **286** (1990), pp. 193–207.

[G2] Göttsche, L., *Change of polarization and Hodge numbers of moduli spaces of torsion free sheaves on surfaces,* MPI 94-9 (1994).

[G-H] Göttsche, L., Huybrechts, D., *Hodge numbers of moduli spaces of stable bundles on K3 surfaces,* preprint (1994).

[K1] Kirwan, F., *Cohomology of quotients in symplectic and algebraic geometry,* Princeton Math. Notes 31, Princeton N.J. (1984).

[K2] Kirwan, F., *On spaces of maps from Riemann surfaces to Grassmannians and applications to the cohomology of moduli of vector bundles,* Ark. Math. **24** (1986), pp. 221–275.

[L] Langton, S. G., *Valuative criteria for families of vector bundles on an algebraic varieties,* Ann. of Math. **101** (1975), pp. 88–110.

[M1] Maruyama, M., *Moduli of stable sheaves II,* J. Math. Kyoto Univ. **18** (1978), pp. 557–614.

[M2] Maruyama, M., *Singularities of the curve of jumping lines of a vector bundle of rank 2 on* $\mathbb{P}^2$, Alg. Geom., Proc. of Japan-France Conf., Tokyo and Kyoto 1982, Lect. Notes in Math. 1016, Springer-Verlag 1983.

[M3] Maruyama, M., *Moduli of algebraic vector bundles,* in preparation.

[V-W] Vafa, C., Witten, E., *A strong coupling test of S-duality,* hep-th/9408074.

[Y1] Yoshioka, K., *The Betti numbers of the moduli space of stable sheaves of rank 2 on* $\mathbb{P}^2$, J. reine angew. Math. **453** (1994), 193–220.

[Y2] Yoshioka, K., *The Betti numbers of the moduli space of stable sheaves of rank 2 on a ruled surface,* Math. Ann. to appear.

[Y3] Yoshioka, K., *Chamber structure of polarizations and the moduli of stable sheaves on a ruled surface,* preprint (1994).